Student Solutions Manual

College Algebra with Applications for Business and the Life Sciences

SECOND EDITION

Ron Larson
The Pennsylvania State University
The Behrend College

Anne V. Hodgkins
Phoenix College

BROOKS/COLE
CENGAGE Learning

Australia • Brazil • Japan • Korea • Mexico • Singapore • Spain • United Kingdom • United States

BROOKS/COLE
CENGAGE Learning

For product information and technology assistance, contact us at **Cengage Learning Customer & Sales Support, 1-800-354-9706**

For permission to use material from this text or product, submit all requests online at **www.cengage.com/permissions**
Further permissions questions can be emailed to **permissionrequest@cengage.com**

ISBN-13: 978-1-133-10901-3
ISBN-10: 1-133-10901-2

Brooks/Cole
20 Channel Center Street
Boston, MA 02210
USA

Cengage Learning is a leading provider of customized learning solutions with office locations around the globe, including Singapore, the United Kingdom, Australia, Mexico, Brazil, and Japan. Locate your local office at: **www.cengage.com/global**

Cengage Learning products are represented in Canada by Nelson Education, Ltd.

To learn more about Brooks/Cole, visit **www.cengage.com/brookscole**

Purchase any of our products at your local college store or at our preferred online store **www.cengagebrain.com**

Printed in the United States of America
1 2 3 4 5 6 7 15 14 13 12 11

Preface

This *Student Solutions Manual* is designed as a supplement to *College Algebra with Applications for Business and the Life Sciences*, Second Edition, by Ron Larson and Anne Hodgkins. All references to chapters, theorems, and exercises relate to the main text. Solutions to every odd-numbered exercise in the text are given with all essential algebraic steps included. Although this supplement is not a substitute for good study habits, it can be valuable when incorporated into a well planned course of study.

We have made every effort to see that the solutions are correct. However, we would appreciate hearing about any errors or other suggestions for improvement. Good luck with your study of algebra.

Ron Larson
Anne Hodgkins

CONTENTS

CHAPTER 0
Fundamental Concepts of Algebra

CHAPTER 0
Fundamental Concepts of Algebra

Section 0.1 Real Numbers: Order and Absolute Value

1. $\left\{-9, -\frac{7}{2}, 5, \frac{2}{3}, \sqrt{2}, 0.1\right\}$

(a) Natural numbers: $\{5\}$

(b) Integers: $\{-9, 5\}$

(c) Rational numbers: $\left\{-9, -\frac{7}{2}, 5, \frac{2}{3}, 0.1\right\}$

(d) Irrational numbers: $\left\{\sqrt{2}\right\}$

3. $\left\{12, -13, 1, \sqrt{4}, \sqrt{6}, \frac{3}{2}\right\}$

Note: $\sqrt{4} = 2$

(a) Natural numbers: $\left\{12, 1, \sqrt{4}\right\}$

(b) Integers: $\left\{12, -13, 1, \sqrt{4}\right\}$

(c) Rational numbers: $\left\{12, -13, 1, \sqrt{4}, \frac{3}{2}\right\}$

(d) Irrational numbers: $\left\{\sqrt{6}\right\}$

5. $\left\{\frac{8}{2}, -\frac{8}{3}, \sqrt{10}, -4, 9, 14.2\right\}$

Note: $\frac{8}{2} = 4$

(a) Natural numbers: $\left\{\frac{8}{2}, 9\right\}$

(b) Integers: $\left\{\frac{8}{2}, -4, 9\right\}$

(c) Rational numbers: $\left\{\frac{8}{2}, -\frac{8}{3}, -4, 9, 14.2\right\}$

(d) Irrational numbers: $\left\{\sqrt{10}\right\}$

7. $\frac{2}{3} = 0.\overline{6}$

9. $\frac{14}{111} = 0.\overline{126}$

11. $-1 < 2.5$

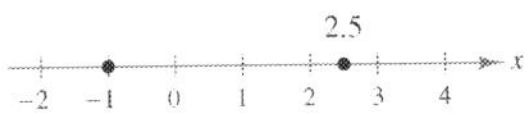

13. $-6 < 7$

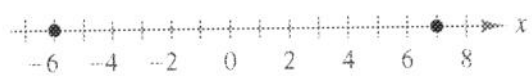

15. $1 > -3.5$

17. $\frac{5}{6} > \frac{2}{3}$

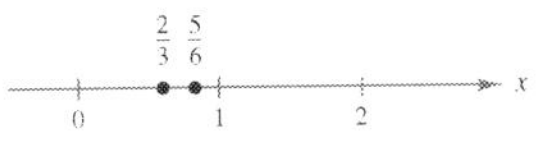

19. $\frac{204}{60} = 3.4,\ \frac{31}{9} \approx 3.44,\ 3.45,\ 2\sqrt{3} \approx 3.46,\ \frac{7}{2} = 3.5$

21. $\frac{127}{90} \approx 1.41111,\ \frac{584}{413} \approx 1.41404,$

$\frac{7071}{5000} = 1.4142,\ \sqrt{2} \approx 1.41421,$

$\frac{47}{33} \approx 1.42424$

23. $x \le 4$

25. $0 < x \le 3$

27. The inequality $x < 0$ denotes all real numbers that are less than zero.

29. The inequality $x \le 5$ denotes all real numbers that are less than or equal to 5.

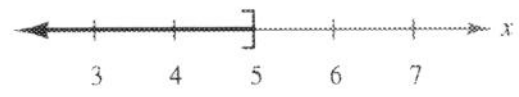

31. The inequality $x > 3.5$ denotes all real numbers that are greater than 3.5.

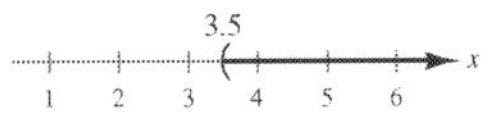

33. The inequality $-2 < x < 2$ denotes all real numbers between -2 and 2, not including -2 or 2.

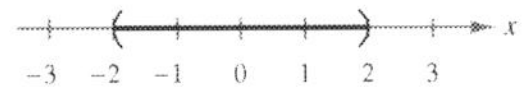

35. The inequality $-1 \le x < 0$ denotes all real numbers less than zero and greater than or equal to -1.

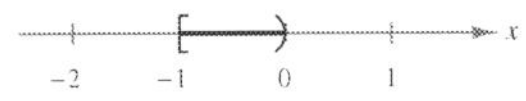

37. $x > 0$

39. $5 < y \le 12$

41. $A \ge 21$

43. $-0.5\% \le r \le 3.8\%$

45. $|-10| = 10$

47. $-\frac{3}{2} - \left|-\frac{3}{2}\right| = -\frac{3}{2} - \frac{3}{2} = -3$

49. $-0.8|-2| = -0.8(2) = -1.6$

51. $\frac{-5}{|-5|} = \frac{-5}{5} = -1$

53. $|3 - \pi| = -(3 - \pi) = \pi - 3$

55. $|-7| = |7|$ because $7 = 7$.

57. $|-3| > -|-3|$ because $3 > -3$.

59. $-|-2| = -|2|$ because $-2 = -2$.

61. $d(-1, 3) = |3 - (-1)| = |3 + 1| = 4$

63. $d(-4, -3) = |-3 - (-4)| = |-3 + 4| = 1$

65. $d(-7, 0) = |0 - (-7)| = |7| = 7$

67. $d(126, 75) = |75 - 126| = 51$

69. $d(2.22, -1.45) = |2.22 - (-1.45)| = 3.67$

71. $|z - 6| > 1$

73. $|x - (-10)| \geq 6 \Rightarrow |x + 10| \geq 6$

75. $|y - 0| \geq 6 \Rightarrow |y| \geq 6$

77. $|x - m| > 5$

79. $|27 - 108| = 81$ miles

(a) ii (b) i

81. $|63 - (-26)| = 89°\text{F}$

83.

Budgeted Expense, b	*Actual Expense, a*	$\lvert a - b \rvert$	$0.05b$
\$30,000	\$29,123.45	\$876.55	\$1500

The actual expense difference is greater than \$500 but is less than 5% of the budgeted amount, so it does not pass the test.

85.

Budgeted Expense, b	*Actual Expense, a*	$\lvert a - b \rvert$	$0.05b$
\$12,000	\$11,735.68	\$264.32	\$600

The actual expense passes both tests.

87.

Expected Diameter, b	*Actual Diameter, a*	$\lvert a - b \rvert$	$0.0012b$
14 in.	13.998 in.	0.002 in.	0.0168 in.

The tire passes inspector's guidelines.

89.

Expected Diameter, b	*Actual Diameter, a*	$\lvert a - b \rvert$	$0.0012b$
16 in.	15.973 in.	0.027 in.	0.0192 in.

The tire fails both guidelines and is rejected.

91.

Expected Diameter, b	*Actual Diameter, a*	$\lvert a - b \rvert$	$0.0012b$
13 in.	13.022 in.	0.022 in.	$0.01\overline{56}$ in.

The tire fails both guidelines and is rejected.

93. The expression $|u + v|$ can never be greater than $|u| + |v|$. $|u + v|$ will equal $|u| + |v|$ if u and v are both negative or both positive.

$|u + v| < |u| + |v|$ if u and v differ in sign. For example, if $u = -5$ and $v = 2$ then

$|u + v| = |-5 + 2| = |-3| = 3 < |-5| + |2| = 7.$

95. $a \leq 0$; If a is negative, then $|a|$ results in a positive number, $-a$. If $a = 0$, then $|0| = -0 = 0$. If a is positive, then $|a| = -a$ can never be true.

Section 0.2 The Basic Rules of Algebra

Skills Warm Up

1. $-4 < -2$

2. $0 > -3$

3. $-\frac{11}{9} > -\frac{13}{9}$

4. $-\frac{12}{33} < \frac{11}{33}$

5. $d(4, 6) = |4 - 6| = 2$

6. $d(-2, 2) = |-2 - 2| = 4$

7. $d(0, -5) = |0 - (-5)| = 5$

8. $d(-1, 3) = |-1 - 3| = 4$

1. $7x + 4$

Terms: $7x, 4$

3. $x^2 - 4x + 8$

Terms: $x^2, -4x, 8$

5. $2x^2 - 9x + 13$

Terms: $2x^2, -9x, 13$

7. $(2 - 7) + 3 = -5 + 3 = -2$

9. $(4 - 7)(-2) = (-3)(-2) = 6$

11. $4x - 6$

(a) $4(-1) - 6 = -4 - 6 = -10$

(b) $4(0) - 6 = 0 - 6 = -6$

13. $x^2 - 3x + 4$

(a) $(-2)^2 - 3(-2) + 4 = 4 + 6 + 4 = 14$

(b) $2^2 - 3(2) + 4 = 4 - 6 + 4 = 2$

15. $\dfrac{x}{x - 2}$

(a) $\dfrac{-2}{-2 - 2} = \dfrac{-2}{-4} = \dfrac{1}{2}$

(b) $\dfrac{2}{2 - 2} = \dfrac{2}{0}$ is undefined.

17. $x + 3y + z = 3 + 3(-2) + 4 = 3 - 6 + 4 = 1$

19. $x^2 - 5y + 4z = 3^2 - 5(-2) + 4(4)$

$= 9 + 10 + 16$

$= 35$

21. $\dfrac{x - y}{5z} = \dfrac{3 - (-2)}{5(4)} = \dfrac{5}{20} = \dfrac{1}{4}$

23. $3 + 4 = 4 + 3$ Commutative (addition)

25. $-15 + 15 = 0$ Inverse (addition)

27. $2(x + 3) = 2x + 6$ Distributive Property

29. $2\left(\frac{1}{2}\right) = 1$ Inverse (multiplication)

31. $h + 0 = h$ Identity (addition)

33. $57 \cdot 1 = 57$ Identity (multiplication)

35. $6 + (7 + 8) = (6 + 7) + 8$ Associative (addition)

37. $x(3y) = (x \cdot 3)y$ Associative (multiplication)

$= (3x)y$ Commutative (multiplication)

39. $48 = 16 \cdot 3 = 4 \cdot 4 \cdot 3 = 2 \cdot 2 \cdot 2 \cdot 2 \cdot 3$

41. $240 = 16 \cdot 15 = 4 \cdot 4 \cdot 3 \cdot 5 = 2 \cdot 2 \cdot 2 \cdot 2 \cdot 3 \cdot 5$

43. $\frac{0}{7} + 3x = 0 + 3x = 3x$

45. $\dfrac{x}{4} + \dfrac{3x}{5} = \dfrac{x}{4} \cdot \dfrac{5}{5} + \dfrac{3x}{5} \cdot \dfrac{4}{4} = \dfrac{5x + 12x}{20} = \dfrac{17x}{20}$

47. $\dfrac{8}{x} \div \dfrac{2}{3} = \dfrac{8}{x} \cdot \dfrac{3}{2} = \dfrac{4 \cdot \cancel{2} \cdot 3}{x \cdot \cancel{2}} = \dfrac{12}{x}$

49. $\dfrac{2 - 79}{-11} = \dfrac{-77}{-11} = \dfrac{7(\cancel{-11})}{\cancel{-11}} = 7$

51. $\dfrac{5}{8} + \dfrac{1}{4} - \dfrac{5}{6} = \dfrac{5}{8} \cdot \dfrac{3}{3} + \dfrac{1}{4} \cdot \dfrac{6}{6} - \dfrac{5}{6} \cdot \dfrac{4}{4}$

$= \dfrac{15 + 6 - 20}{24} = \dfrac{1}{24}$

53. $\dfrac{2}{5} \cdot \dfrac{7}{8} = \dfrac{\cancel{2} \cdot 7}{5 \cdot \cancel{2} \cdot 4} = \dfrac{7}{20}$

55. $\left(\dfrac{2}{3} \div 8\right) + \left(3 \cdot \dfrac{3}{12}\right) - \left(\dfrac{2}{3} \cdot \dfrac{1}{8}\right) + \dfrac{9}{12}$

$= \dfrac{1}{12} + \dfrac{9}{12} = \dfrac{10}{12}$

$= \dfrac{\cancel{2} \cdot 5}{\cancel{2} \cdot 6} = \dfrac{5}{6}$

57. $3\left(-\frac{5}{12} + \frac{3}{8}\right) \approx -0.13$

59. $\dfrac{11.46 - 5.37}{3.91} \approx 1.56$

61. $-108 \div 3^3 = -4$

63. $5(18 - 2^3) \div 10 = 5$

65. $0.35(68) = 23.8$

67. $1.25(37) = 46.25$

69. (a) $1 - (0.207 + 0.166 + 0.145 + 0.05 + 0.225) = 0.207 = 20.7\%$

In 2008, Social Security was 20.7% of total expenses.

(b) National Defense: $0.207(2982.9) \approx \$617$ billion

Social Security: $0.207(2982.9) \approx \$617$ billion

Health and Medicare: $0.225(2982.9) \approx \$671$ billion

Education and Veterans' Benefits: $0.05(2982.9) \approx \$149$ billion

Income Security: $0.145(2982.9) \approx \$433$ billion

Other: $0.166(2982.9) \approx \$495$ billion

71. $0.492(3445) \approx 1695$ people

73. Food: $\dfrac{17.56}{45.53} \approx 0.386 = 38.6\%$

Vet care: $\dfrac{12.04}{45.53} \approx 0.264 = 26.4\%$

Supplies/OTC medicine: $\dfrac{10.41}{45.53} \approx 0.229 = 22.9\%$

Live animal purchases: $\dfrac{2.16}{45.53} \approx 0.047 = 4.7\%$

Grooming and boarding: $\dfrac{3.36}{45.53} \approx 0.074 = 7.4\%$

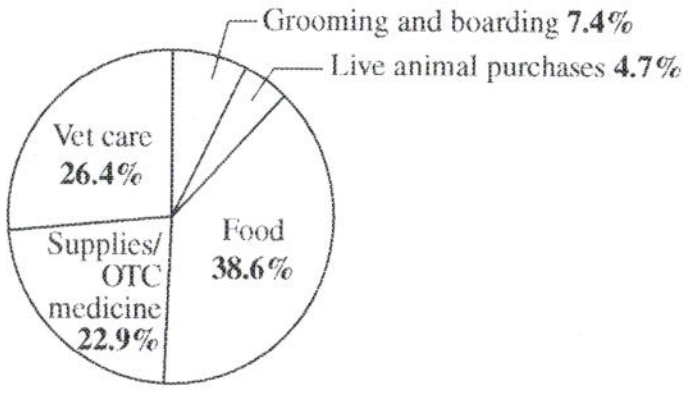

75. Answers will vary.

Section 0.3 Integer Exponents

Skills Warm Up

1. $\left(\frac{2}{3}\right)\left(\frac{3}{2}\right) = \frac{6}{6} = 1$

2. $\left(\frac{1}{4}\right)(5)(4) = \left(\frac{1}{4}\right)(4)(5) = 1(5) = 5$

3. $3\left(-\frac{2}{7}\right) + 11\left(-\frac{2}{7}\right) = -\frac{6}{7} + -\frac{22}{7} = -\frac{28}{7} = -4$

4. $11\left(-\frac{1}{4}\right) + \frac{5}{4} = -\frac{11}{4} + \frac{5}{4} = -\frac{6}{4} = -\frac{3}{2}$

5. $\frac{1}{2} \div (-2) = \frac{1}{2}\left(-\frac{1}{2}\right) = -\frac{1}{4}$

6. $-\dfrac{1}{3} \div \left(-\dfrac{1}{3}\right) = -\dfrac{1}{3} \cdot -\dfrac{3}{1} = \dfrac{-3}{-3} = 1$

7. $$\begin{aligned}\frac{1}{7} + \frac{1}{3} - \frac{1}{21} &= \frac{1}{7}\cdot\frac{3}{3} + \frac{1}{3}\cdot\frac{7}{7} - \frac{1}{21}\\ &= \frac{3+7-1}{21}\\ &= \frac{9}{21} = \frac{3}{7}\end{aligned}$$

8. $$\begin{aligned}\frac{1}{3} + \frac{1}{2} - \frac{5}{6} &= \frac{1}{3}\cdot\frac{2}{2} + \frac{1}{2}\cdot\frac{3}{3} - \frac{5}{6}\\ &= \frac{2+3-5}{6}\\ &= \frac{0}{6}\\ &= 0\end{aligned}$$

9. $$\begin{aligned}\frac{1}{12} - \left(\frac{1}{3} + \frac{1}{8}\right) &= \frac{1}{12} - \left(\frac{1}{3}\cdot\frac{8}{8} + \frac{1}{8}\cdot\frac{3}{3}\right)\\ &= \frac{1}{12} - \left(\frac{8+3}{24}\right)\\ &= \frac{1}{12} - \frac{11}{24}\\ &= \frac{1}{12}\cdot\frac{2}{2} - \frac{11}{24}\\ &= \frac{2-11}{24} = -\frac{9}{24} = -\frac{3}{8}\end{aligned}$$

10. $\left(\frac{1}{2} - \frac{1}{3}\right) \div \frac{1}{6} = \left(\frac{1}{2} - \frac{1}{3}\right)\cdot 6 = 3 - 2 = 1$

1. $2^2 \cdot 2^4 = 2^{2+4} = 2^6 = 64$

3. $\dfrac{2^6}{2^3} = 2^{6-3} = 2^3 = 8$

5. $\left(3^3\right)^2 = 3^{3\cdot 2} = 3^6 = 729$

7. $-3^4 = -(3)^4 = -81$

9. $\left(2^3 \cdot 3^2\right)^2 = \left(2^3\right)^2 \cdot \left(3^2\right)^2 = 2^6 \cdot 3^4 = 64 \cdot 81 = 5184$

11. $\left(-\dfrac{3}{5}\right)^3\left(\dfrac{5}{3}\right)^2 = (-1)^3 \cdot \dfrac{3^3}{5^3} \cdot \dfrac{5^2}{3^2}$

$= -1 \cdot \dfrac{3^3}{3^2} \cdot \dfrac{5^2}{5^3}$

$= -1 \cdot 3 \cdot \dfrac{1}{5}$

$= -\dfrac{3}{5}$

13. $3^0 = 1$

15. $(-5z)^3 = (-5)^3 z^3 = -125z^3$

17. $\left(8x^4\right)\left(2x^3\right) = 8(2)\left(x^{4+3}\right) = 16x^7$

19. $10\left(x^2\right)^2 = 10x^{2\cdot 2} = 10x^4$

21. $(-z)^3\left(3z^4\right) = (-1)^3\left(z^3\right)\left(3z^4\right) = -1(3)\left(z^{3+4}\right) = -3z^7$

23. $\dfrac{25y^8}{15y^4} = \dfrac{5}{3}y^{8-4} = \dfrac{5}{3}y^4$

25. $\left(\dfrac{4}{y}\right)^3\left(\dfrac{3}{y}\right)^4 = \left(\dfrac{4^3}{y^3}\right)\left(\dfrac{3^4}{y^4}\right) = \dfrac{4^3 \cdot 3^4}{y^{3+4}} = \dfrac{64 \cdot 81}{y^7} = \dfrac{5184}{y^7}$

27. $\dfrac{7x^2}{x^3} = 7x^{2-3} = 7x^{-1} = \dfrac{7}{x}$

29. $\dfrac{x^2 \cdot x^n}{x^3 \cdot x^n} = x^{2-3} \cdot x^{n-n} = x^{-1} \cdot x^0 = \dfrac{1}{x} \cdot 1 = \dfrac{1}{x}$

31. $3^n \cdot 3^{2n} = 3^{n+2n} = 3^{3n}$

33. $\dfrac{15(x+3)^3}{9(x+3)^2} = \dfrac{5}{3}(x+3)^{3-2} = \dfrac{5}{3}(x+3)$

35. $\left(2x^5\right)^0 = 2^0 x^0 = 1 \cdot 1 = 1, x \neq 0$

37. $8 \cdot (-2)^{-2} \cdot 4^{-1} = 8 \cdot \dfrac{1}{(-2)^2} \cdot \dfrac{1}{4} = 8 \cdot \dfrac{1}{4} \cdot \dfrac{1}{4} = \dfrac{8}{16} = \dfrac{1}{2}$

39. $\left(\dfrac{1}{2}\right)^{-3} = \dfrac{1^{-3}}{2^{-3}} = \dfrac{2^3}{1^3} = 8$

41. $5^{-1} - 2^{-1} = \dfrac{1}{5} - \dfrac{1}{2} = \dfrac{1}{5} \cdot \dfrac{2}{2} - \dfrac{1}{2} \cdot \dfrac{5}{5} = \dfrac{2-5}{10} = -\dfrac{3}{10}$

43. $4x^{-2} = 4 \cdot \dfrac{1}{x^2} = \dfrac{4}{x^2}$

45. $\dfrac{9a^2b^{-2}}{3a^{-1}b} = \dfrac{3a^2 \cdot a}{b \cdot b^2} = \dfrac{3a^{2+1}}{b^{1+2}} = \dfrac{3a^3}{b^3}$

47. $\left(\dfrac{x}{10}\right)^{-1} = \dfrac{10}{x}$

49. $\left(\dfrac{x^{-3}y^4}{5}\right)^{-3} = \left(\dfrac{5}{x^{-3}y^4}\right)^3 = \left(\dfrac{5x^3}{y^4}\right)^3 = \dfrac{125x^9}{y^{12}}$

51. When $x = -6, \dfrac{x^4}{2x^2} = \dfrac{(-6)^4}{2(-6)^2} = \dfrac{(-6)^2}{2} = \dfrac{36}{2} = 18.$

53. When $x = 4, 7x^{-2} = 7(4)^{-2} = \frac{7}{16}.$

55. $10^3 = 1000$

$10^2 = 100$

$10^1 = 10$

$10^0 = 1$

$10^{-1} = \dfrac{1}{10} = 0.1$

$10^{-2} = \dfrac{1}{10^2} = \dfrac{1}{100} = 0.01$

As the power of the integer exponent decreases by a factor of 10.

57. $57{,}300{,}000 = 5.73 \times 10^7$ square miles

59. $9{,}460{,}000{,}000{,}000 = 9.46 \times 10^{12}$ kilometers

61. $0.0000001 = 1 \times 10^{-7}$ meter

63. $3.5 \times 10^8 = 350{,}000{,}000$ air sacs

65. $1.602 \times 10^{-19} = 0.0000000000000000001602$ coulomb

67. (a) $\left(1.2 \times 10^7\right)\left(5 \times 10^{-3}\right) = 6 \times 10^4$

(b) $\left(\dfrac{6.0 \times 10^8}{3.0 \times 10^{-3}}\right) = 2.0 \times 10^{11}$

69. (a) $48{,}000{,}000{,}000(250{,}000{,}000)$

$= \left(4.8 \times 10^{10}\right)\left(2.5 \times 10^8\right) = 1.2 \times 10^{19}$

(b) $\dfrac{0.000000012}{0.0000064} = \dfrac{1.2 \times 10^{-8}}{6.4 \times 10^{-6}} = 1.875 \times 10^{-3}$

71. $11^5 + 49 = 161{,}100$

73. $2^{-3} + 5^{-1} = 0.325$

75. $\dfrac{4^4 + 1}{4^4 - 1} = 1.007843137$

77. (a) $0.000345(8{,}900{,}000{,}000) \approx 3.071 \times 10^6$

(b) $\dfrac{67{,}000{,}000 + 93{,}000{,}000}{0.0052} \approx 3.077 \times 10^{10}$

79. (a) $(9.3 \times 10^6)^3(6.1 \times 10^{-4}) \approx 4.907 \times 10^{17}$

(b) $\dfrac{(2.414 \times 10^4)^6}{(1.68 \times 10^5)^5} \approx 1.479 \times 10^0$ or 1.479

81. (a) Daily: $A = 10{,}000\left(1 + \dfrac{0.0375}{365}\right)^{365(12)}$

$\approx \$15{,}682.76$

(b) Weekly: $A = 10{,}000\left(1 + \dfrac{0.0375}{52}\right)^{52(12)}$

$\approx \$15{,}680.58$

(c) Monthly: $A = 10{,}000\left(1 + \dfrac{0.0375}{12}\right)^{12(12)}$

$\approx \$15{,}672.12$

(d) Quarterly: $A = 10{,}000\left(1 + \dfrac{0.0375}{4}\right)^{4(12)}$

$\approx \$15{,}650.28$

The balance is larger when the interest is compounded more frequently.

83. $A = P\left(1 + \dfrac{r}{n}\right)^{nt}$

$30{,}778 = 22{,}896(1 + r)^6$

If $r = 0.05$, then $22{,}896(1 + 0.05)^6 \approx 30{,}683$.

Because $30{,}683 < 30{,}778$, try a greater value of r.

If $r = 0.051$, then $22{,}896(1 + 0.051)^6 \approx 30{,}859$.

Because $30{,}859 > 30{,}778$, try a smaller value of r.

If $r = 0.0505$, then $22{,}896(1 + 0.0505)^6 \approx 30{,}771$.

Because 30,771 is closer to 30,778 than 30,683 or 30,859, the average yearly inflation rate is about 0.0505, or 5.05%.

85. $\dfrac{2.68 \times 10^9 \text{ mi}}{1.86 \times 10^5 \text{ mi/sec}} = 1.4409 \times 10^4$ seconds

$= 14{,}409$ seconds ≈ 4 hours

Section 0.4 Radicals and Rational Exponents

Skills Warm Up

1. $\left(\dfrac{1}{3}\right)\left(\dfrac{2}{3}\right)^2 = \dfrac{1}{3}\left(\dfrac{2^2}{3^2}\right) = \dfrac{1}{3}\left(\dfrac{4}{9}\right) = \dfrac{4}{27}$

2. $3(-4)^2 = 3(16) = 48$

3. $(-2x)^3 = (-2)^3x^3 = -8x^3$

4. $(-2x^3)(-3x^4) = -2(-3)(x^{3+4}) = 6x^7$

5. $(7x^5)(4x) = 7(4)(x^{5+1}) = 28x^6$

6. $(5x^4)(25x^2)^{-1} = 5x^4\left(\dfrac{1}{25x^2}\right) = \dfrac{5x^4}{25x^2}$

$= \dfrac{1}{5}x^{4-2} = \dfrac{1}{5}x^2, x \neq 0$

7. $\dfrac{12z^6}{4z^2} = 3z^{6-2} = 3z^4, z \neq 0$

8. $\left(\dfrac{2x}{5}\right)^2\left(\dfrac{2x}{5}\right)^{-4} = \left(\dfrac{2x}{5}\right)^{2+(-4)} = \left(\dfrac{2x}{5}\right)^{-2} = \left(\dfrac{5}{2x}\right)^2$

$= \dfrac{5^2}{2^2x^2} = \dfrac{25}{4x^2}$

9. $\left(\dfrac{3y^2}{x}\right)^0 = 1$

10. $\left[(x + 2)^2(x + 2)^3\right]^2 = \left[(x + 2)^{2+3}\right]^2 = \left[(x + 2)^5\right]^2$

$= (x + 2)^{5 \cdot 2} = (x + 2)^{10}$

1. $\sqrt{4} = 2$
The principal square root is 2 and the other real root is –2.

3. $\sqrt[3]{-8} = -2$
The principal cube root is –2 and there is no other real root.

5. $\sqrt[3]{27} = 3$
The principal cube root is 3 and there is no other real root.

7. $\sqrt[4]{-1}$ is not a real number because there is no real number that can be raised to the fourth power to produce –1.

9. $\sqrt[6]{64} = 2$
The principal 6th root is 2 and the other real root is –2.

11. $\sqrt{9} = \sqrt{3^2} = 3$

13. $-\sqrt[3]{-27} = -\sqrt[3]{(-3)^3} = -(-3) = 3$

15. $\dfrac{14}{\sqrt{49}} = \dfrac{14}{\sqrt{7^2}} = \dfrac{14}{7} = 2$

17. $\sqrt[4]{-16}$ is not a real number because there is no real number that can be raised to the 4th power to produce –16.

19. $\sqrt[3]{\dfrac{-27}{125}} = \dfrac{\sqrt[3]{-27}}{\sqrt[3]{125}} = \dfrac{-3}{5} = -\dfrac{3}{5}$

21. $\sqrt[3]{-54} = \sqrt[3]{(-27)\cdot 2} = \sqrt[3]{(-3)^3\cdot 2} = -3\sqrt[3]{2}$

23. $\left(\sqrt[3]{-125}\right)^3 = \left[(-125)^{1/3}\right]^3 = -125$

25. $\sqrt[3]{16x^5} = \sqrt[3]{8x^3\cdot 2x^2} = \sqrt[3]{(2x)^3\cdot 2x^2} = 2x\sqrt[3]{2x^2}$

27. $\sqrt{\left(4x^3\right)^2} = 4x^3$

29. $\sqrt{\dfrac{z^2}{25}} = \dfrac{\sqrt{z^2}}{\sqrt{25}} = \dfrac{z}{5}$

31. $\sqrt{2xy^4z^2} = y^2|z|\sqrt{2x}$
When $x = 2, y = 3, z = 5$:
$y^2|z|\sqrt{2x} = 3^2\cdot|5|\sqrt{2(2)} = 45\sqrt{4} = 90$

33. $\sqrt[4]{16x^{-4}y^8z^4} = 2y^2\left|\dfrac{z}{x}\right|$
When $x = 2, y = 3, z = 5$:
$2y^2\left|\dfrac{z}{x}\right| = 2(3^2)\left|\dfrac{5}{2}\right| = 18\left(\dfrac{5}{2}\right) = 45$

35. $\sqrt[3]{z}\cdot\sqrt[3]{z^2} = \sqrt[3]{z^{1+2}} = \sqrt[3]{z^3} = z = 5$

37. $\dfrac{1}{\sqrt{5}} = \dfrac{1}{\sqrt{5}}\cdot\dfrac{\sqrt{5}}{\sqrt{5}} = \dfrac{\sqrt{5}}{5}$

39. $\dfrac{8}{\sqrt[3]{2}} = \dfrac{8}{\sqrt[3]{2}}\cdot\dfrac{\sqrt[3]{4}}{\sqrt[3]{4}} = \dfrac{8\sqrt[3]{4}}{2} = 4\sqrt[3]{4}$

41.
$$\frac{2x}{5-\sqrt{3}} = \frac{2x}{5-\sqrt{3}}\cdot\frac{5+\sqrt{3}}{5+\sqrt{3}} = \frac{2x\left(5+\sqrt{3}\right)}{25-3} = \frac{x\left(5+\sqrt{3}\right)}{11}$$

43.
$$\frac{3}{\sqrt{5}+\sqrt{6}} = \frac{3}{\sqrt{5}+\sqrt{6}}\cdot\frac{\sqrt{5}-\sqrt{6}}{\sqrt{5}-\sqrt{6}} = \frac{3\left(\sqrt{5}-\sqrt{6}\right)}{5-6} = -3\left(\sqrt{5}-\sqrt{6}\right) = 3\left(\sqrt{6}-\sqrt{5}\right)$$

45. Radical form: $\sqrt{9} = 3$
Rational exponent form: $9^{1/2} = 3$

47. Radical form: $\sqrt[5]{32} = 2$
Rational exponent form: $32^{1/5} = 2$

49. Radical form: $\sqrt[3]{-216} = -6$
Rational exponent form: $(-216)^{1/3} = -6$

51. Radical form: $\left(\sqrt[4]{81}\right)^3 = 27$
Rational exponent form: $\left(81^{1/4}\right)^3 = 81^{3/4} = 27$

53. Radical form: $\sqrt[3]{125^2} = 25$
Rational exponent form: $125^{2/3} = 25$

55. $16^{1/2} = \sqrt{16} = 4$

57. $\left(\dfrac{16}{81}\right)^{-3/4} = \left(\dfrac{81}{16}\right)^{3/4} = \left(\sqrt[4]{\dfrac{81}{16}}\right)^3 = \left(\dfrac{3}{2}\right)^3 = \dfrac{27}{8}$

59. $\left(-\dfrac{1}{64}\right)^{-1/3} = (-64)^{1/3} = \sqrt[3]{-64} = -4$

61. $5^{1/2}\cdot 5^{3/2} = 5^{1/2+3/2} = 5^{4/2} = 5^2 = 25$

63. $\dfrac{2^{3/2}}{2} = 2^{3/2-1} = 2^{1/2}$

65. $\dfrac{x^2}{x^{1/2}\cdot x^{3/2}} = \dfrac{x^2}{x^{1/2+3/2}} = \dfrac{x^2}{x^2} = 1;\ x \neq 0$

67. $\left(x^6x^3\right)^{1/3} = \left(x^{6+3}\right)^{1/3} = \left(x^9\right)^{1/3} = x^{9\cdot(1/3)} = x^3$

69. $\left(16x^8y^4\right)^{3/4} = \left(2^4x^8y^4\right)^{3/4}$

$= 2^{4\cdot(3/4)}\cdot x^{8\cdot(3/4)}\cdot y^{4\cdot(3/4)}$

$= 2^3x^6y^3 = 8x^6y^3$

71. $\sqrt[4]{y^2} = y^{2/4} = y^{1/2} = \sqrt{y}$

73. $\sqrt{\sqrt{32}} = \left(32^{1/2}\right)^{1/2} = 32^{1/4} = \sqrt[4]{32}$

$= \sqrt[4]{2\cdot 16} = 2\sqrt[4]{2}$

75. $\sqrt[8]{(x+1)^6} = (x+1)^{6/8} = (x+1)^{3/4} = \sqrt[4]{(x+1)^3}$

77. $5\sqrt{x} - 3\sqrt{x} = (5-3)\sqrt{x} = 2\sqrt{x}$

79. $5\sqrt{50} + 3\sqrt{8} = 5\sqrt{25\cdot 2} + 3\sqrt{4\cdot 2}$

$= 25\sqrt{2} + 6\sqrt{2}$

$= 31\sqrt{2}$

81. $2\sqrt{4y} - 2\sqrt{9y} = 4\sqrt{y} - 6\sqrt{y} = -2\sqrt{y}$

83. $\sqrt[3]{45} \approx 3.557$

85. $6.7^{2/5} \approx 2.140$

87. $(0.38)^{-0.6} \approx 1.787$

89. $\dfrac{3-\sqrt{5}}{2} \approx 0.382$

91. Sample answer: [(] 4 [−] [√] 7 [)] [)] [÷] 3

93. $\sqrt{5} + \sqrt{3} \approx 3.968;\ \sqrt{5+3} \approx 2.828$

Because $3.968 > 2.828$, $\sqrt{5} + \sqrt{3} > \sqrt{5+3}$.

95. $\sqrt{3}\cdot\sqrt[4]{3} \approx 2.280$, $\sqrt[8]{3} \approx 1.147$

Because $2.280 > 1.147$, $\sqrt{3}\cdot\sqrt[4]{3} > \sqrt[8]{3}$.

97. $x = \sqrt{1100} = 10\sqrt{11}$ feet

The dimensions of the square classroom floor are $10\sqrt{11}$ feet by $10\sqrt{11}$ feet.

99. $r = 1 - \left(\dfrac{S}{C}\right)^{1/n} = 1 - \left(\dfrac{2500}{25{,}000}\right)^{1/8}$

$= 1 - \left(\dfrac{1}{10}\right)^{1/8} \approx 0.2501 = 25.01\%$

101. $T = 2\pi\sqrt{\dfrac{L}{32}} = 2\pi\sqrt{\dfrac{4}{32}} = 2\pi\sqrt{\dfrac{1}{8}} \approx 2.221$ seconds

103. $\sqrt{\dfrac{2\left(6.67\times 10^{-11}\right)\left(7.36\times 10^{22}\right)}{1.74\times 10^6}}$

≈ 2375.43 meters per second

Because 2000 meters per second is less than 2375.43 meters per second, the rocket will not escape the gravitational field of the moon.

105. $S = \sqrt{30Df}$

$= \sqrt{30(60)(0.90)}$

$= \sqrt{1620}$

≈ 40.2 miles per hour

107. $W = 35.75 + 0.6215T - 35.75v^{0.16} + 0.4275Tv^{0.16}$

$= 35.75 + 0.6215(30) - 35.75\left(20^{0.16}\right)$

$+ 0.4275(30)\left(20^{0.16}\right)$

$\approx 17.4°\text{F}$

109. $F_2 = F_1\left(2^{1/12}\right)$

$= 440\left(2^{1/12}\right)$

$F_3 = F_2\left(2^{1/12}\right)$

$= 440\left(2^{1/12}\right)\left(2^{1/12}\right)$

$= 440\left(2^{1/6}\right)$

≈ 494 vibrations per second

111. Musical note E one octave above the E above middle C has a higher frequency (choice a).

Sample answer: The frequency of note E one octave above the E above middle C is denoted by F_8 and the frequency of note D one octave above the D above middle C is denoted by F_6. $F_7 = F_6\cdot 2^{1/12}$ and $F_8 = F_6\cdot 2^{1/12}$, so $F_8 = F_6\cdot 2^{1/6}$. Because $2^{1/6} \approx 1.12 > 1$, you know that $F_8 > F_6$.

113. $\left(\dfrac{2}{\sqrt{5}}\right)^2 = \dfrac{4}{5}$

This is *not* the same as rationalizing the denominator because $\dfrac{2}{\sqrt{5}}\cdot\dfrac{\sqrt{5}}{\sqrt{5}} = \dfrac{2\sqrt{5}}{5} \neq \dfrac{4}{5}$.

115. No; for negative values of x, $2x < 0$, and $\sqrt{4x^2}$ can never be negative. So, $\sqrt{4x^2} = 2x$ for $x \geq 0$.

Chapter 0 Quiz Yourself

1. $-|-7| < |-7|$ because $-7 < 7$.

2. $-(-3) = |-3|$ because $3 = 3$.

3. $\sqrt{1}$ when simplified to 1 is a natural number.

   -6 and $\sqrt{1}$ are integers.

   $-6, \sqrt{1}, 3/4$, and $-1/2$ are rational numbers.

   $\sqrt{11}$ is an irrational number.

4. $x \geq 0$

5. $d(-6, 5) = |5 - (-6)| = 11$

6. $3x^2 - 7x + 2$

   Terms: $3x^2, -7x, 2$

7. $-4 - 0 \cdot 7 = -4 - 0 = -4$

8. $\frac{31 - 5}{-2} = \frac{26}{-2} = -13$

9. $\frac{2}{3} \cdot \frac{5}{4} \cdot \frac{3}{7} = \frac{\cancel{2} \cdot 5 \cdot \cancel{3}}{\cancel{3} \cdot \cancel{2} \cdot 2 \cdot 7} = \frac{5}{14}$

10. $\frac{11}{15} \div \frac{3}{5} = \frac{11}{15} \cdot \frac{5}{3} = \frac{11 \cdot \cancel{5}}{\cancel{5} \cdot 3 \cdot 3} = \frac{11}{9}$

11. $(-x)^3(2x^{-4}) = -x^3 \cdot \frac{2}{x^4} = \frac{-2x^3}{x^4} = -\frac{2}{x}, x \neq 0$

12. $\frac{5y^7}{15y^{-3}} = \frac{y^7 \cdot y^3}{3} = \frac{y^{7+3}}{3} = \frac{y^{10}}{3}$

13. $0.000014 = 1.4 \times 10^{-5}$ liters

14. $A = P\left(1 + \frac{r}{n}\right)^{nt}$;

    $A = 5000\left(1 + \frac{0.022}{4}\right)^{4 \cdot 5} = 5000(1.0055)^{20} \approx \$5579.71$

15. $\frac{-\sqrt[4]{81}}{3} = \frac{-\sqrt[4]{3^4}}{3} = \frac{-3}{3} = -1$

16. $\left(\sqrt[3]{-64}\right)^3 = \left[(-64)^{1/3}\right]^3 = -64$

17. $3^{1/2} \cdot 3^{3/2} = 3^{1/2+3/2} = 3^2 = 9$

18. $\sqrt[3]{81} - 4\sqrt[3]{3} = \sqrt[3]{27 \cdot 3} - 4\sqrt[3]{3}$

    $= 3\sqrt[3]{3} - 4\sqrt[3]{3}$

    $= -\sqrt[3]{3}$

19. $\sqrt[10]{12^5} = 12^{5/10} = 12^{1/2} = \sqrt{12} = \sqrt{4 \cdot 3} = 2\sqrt{3}$

20. $x = \sqrt[3]{10{,}648} = 22$ centimeters

    The dimensions of the cube are 22 cm × 22 cm × 22 cm.

Section 0.5 Polynomials and Special Products

Skills Warm Up

1. $(7x^2)(6x) = 42x^{2+1} = 42x^3$

2. $(5x^3)(-9x^2) = 5(-9)x^{3+2} = -45x^5$

3. $(10z^3)(-2z^{-1}) = -20z^{3+(-1)} = -20z^2, z \neq 0$

4. $(-3x^4)(5x^{-5}) = -3(5)x^{4-5} = -15x^{-1} = -\frac{15}{x}$

5. $(-3x^2)^3 = (-3)^3(x^2)^3 = -27x^{2 \cdot 3} = -27x^6$

6. $-3(x^2)^3 = -3(x^{2 \cdot 3}) = -3x^6$

7. $\frac{27z^5}{12z^2} = \frac{9}{4}z^{5-2} = \frac{9}{4}z^3, z \neq 0$

8. $\frac{32x^4}{20x} = \frac{8}{5}x^{4-1} = \frac{8}{5}x^3, x \neq 0$

9. $\left(\frac{2x}{3}\right)^{-2} = \left(\frac{3}{2x}\right)^2 = \frac{3^2}{(2x)^2} = \frac{9}{4x^2}$

10. $\left(\frac{x^{-1}}{y^2}\right)^{-1} = \frac{x^{-1(-1)}}{y^{2(-1)}} = \frac{x^1}{y^{-2}} = xy^2$

1. Standard form: $4x^2 + 3x + 2$

   Degree: 2

   Leading coefficient: 4

3. Standard form: $x^7 - 8$

   Degree: 5

   Leading coefficient: 1

5. Standard form: $-2x^5 + 6x^4 - x + 1$

Degree: 5

Leading coefficient: -2

7. $2x - 3x^3 + 8$ *is* a polynomial.

Standard form: $-3x^3 + 2x + 8$

Degree: 3

9. $\dfrac{3x + 4}{x} = 3 + \dfrac{4}{x}$ is *not* a polynomial.

11. $w^2 - w^4 + 2w^3$ *is* a polynomial.

Standard form: $-w^4 + 2w^3 + w^2$

Degree: 4

13. $4x + 5$

(a) $4(-2) + 5 = -8 + 5 = -3$

(b) $4(-1) + 5 = -4 + 5 = 1$

(c) $4(0) + 5 = 0 + 5 = 5$

(d) $4(3) + 5 = 12 + 5 = 17$

15. $-2x^2 + 3x + 4$

(a) $-2(-2)^2 + 3(-2) + 4 = -8 - 6 + 4 = -10$

(b) $-2(-1)^2 + 3(-1) + 4 = -2 - 3 + 4 = -1$

(c) $-2(0)^2 + 3(0) + 4 = 0 + 0 + 4 = 4$

(d) $-2(1)^2 + 3(1) + 4 = -2 + 3 + 4 = 5$

17. $(6x + 5) - (8x + 15) = 6x + 5 - 8x - 15$

$= (6x - 8x) + (5 - 15)$

$= -2x - 10$

19. $-(x^3 + 5) + (3x^3 - 4x) = -x^3 - 5 + 3x^3 - 4x$

$= (-x^3 + 3x^3) - 4x - 5$

$= 2x^3 - 4x - 5$

21. $(15x^2 - 6) - (-8x^3 - 14x^2 - 17)$

$= 15x^2 - 6 + 8x^3 + 14x^2 + 17$

$= 8x^3 + (15x^2 + 14x^2) + (-6 + 17)$

$= 8x^3 + 29x^2 + 11$

23. $3x(x^2 - 2x + 1) = 3x(x^2) + 3x(-2x) + 3x(1)$

$= 3x^3 - 6x^2 + 3x$

25. $-3x(-x)(3x - 7) = 3x^2(3x - 7)$

$= 3x^2(3x) - 3x^2(7)$

$= 9x^3 - 21x^2$

27. $(x + 3)(x + 4) = x^2 + 4x + 3x + 12$

$= x^2 + 7x + 12$

29. $(3x - 5)(2x + 1) = 6x^2 + 3x - 10x - 5$

$= 6x^2 - 7x - 5$

31. $(x^2 + 2x)(x^2 + x - 3)$

$= x^2(x^2 + x - 3) + 2x(x^2 + x - 3)$

$= x^4 + x^3 - 3x^2 + 2x^3 + 2x^2 - 6x$

$= x^4 + 3x^3 - x^2 - 6x$

33. $(-x^2 + x - 5)(3x^2 + 4x + 1) = -x^2(3x^2 + 4x + 1) + x(3x^2 + 4x + 1) - 5(3x^2 + 4x + 1)$

$= -3x^4 - 4x^3 - x^2 + 3x^3 + 4x^2 + x - 15x^2 - 20x - 5$

$= -3x^4 - x^3 - 12x^2 - 19x - 5$

35. $(x + 5)(x - 5) = x^2 - 5x + 5x - 25 = x^2 - 25$

37. $(x + 6)^2 = x^2 + 2(6)(x) + 36 = x^2 + 12x + 36$

39. $(2x - 5y)^2 = 4x^2 - 2(5y)(2x) + 25y^2$

$= 4x^2 - 20xy + 25y^2$

41. $[(x - 3) + y]^2 = (x - 3)^2 + 2y(x - 3) + y^2$

$= x^2 - 6x + 9 + 2xy - 6y + y^2$

$= x^2 + 2xy + y^2 - 6x - 6y + 9$

43. $(x + 1)^3 = x^3 + 3x^2(1) + 3x(1)^2 + 1^3$

$= x^3 + 3x^2 + 3x + 1$

45. $(2x - y)^3 = (2x)^3 - 3(2x)^2y + 3(2x)y^2 - y^3$

$= 8x^3 - 12x^2y + 6xy^2 - y^3$

47. $(3y^2 - 1)(3y^2 + 1) = 9y^4 + 3y^2 - 3y^2 - 1$

$= 9y^4 - 1$

49. $(m - 3 + n)(m - 3 - n) = (m - 3)^2 - n^2$

$= m^2 - n^2 - 6m + 9$

51. $(x^2 - x + 1)(x^2 + x + 1)$

$= [(x^2 + 1) - x][(x^2 + 1) + x]$

$= (x^2 + 1)^2 - x^2$

$= x^4 + 2x^2 + 1 - x^2$

$= x^4 + x^2 + 1$

53. $(3x - 7)(8x + 1) = 24x^2 + 3x - 56x - 7$

$= 24x^2 - 53x - 7$

55. $(4x^3 + 6) + (6x^2 - 8x - 11) = 4x^3 + 6 + 6x^2 - 8x - 11$

$= 4x^3 + 6x^2 - 8x - 5$

57. The student omitted the middle term when squaring the binomial.

$(x - 3)^2 = x^2 - 2(x)(3) + 9 = x^2 - 6x + 9$

59. $1000(1 + r)^3 = 1000(r + 1)^3$

$= 1000(r^3 + 3r^2 + 3r + 1)$

$= 1000r^3 + 3000r^2 + 3000r + 1000$

61. When $r = 4\%$:

Balance $= 1200(1.04)^4 + 1700(1.04)^3 + 1500(1.04)^2 + 2200(1.04)^1 + 3000(1.04)^0$

$= 10{,}226.50$; Yes

When $r = 3\%$:

Balance $= 1200(1.03)^4 + 1700(1.03)^3 + 1500(1.03)^2 + 2200(1.03)^1 + 3000(1.03)^0$

$= 10{,}065.60$; Yes

When $r = 2\%$:

Balance $= 1200(1.02)^4 + 1700(1.02)^3 + 1500(1.02)^2 + 2200(1.02)^1 + 3000(1.02)^0$

$= 9907.57$; No

63. When $x = 6$:

$-164.14(6)^2 + 6777.8(6) + 43{,}161 = 77{,}918.76$

When $x = 7$:

$-164.14(7)^2 + 6777.8(7) + 43{,}161 = 82{,}562.74$

In 2006, the total amount of federal student aid disbursed was about \$77,918.76 million. In 2007, the total amount of federal student aid disbursed was about \$82,562.74 million.

65. $V = l \cdot w \cdot h$

$= (57 - 2x)(39 - 2x)x$

$= 2223x - 192x^2 + 4x^3$

$= 4x^3 - 192x^2 + 2223x$

When $x = 4$:

$V = 4(4)^3 - 192(4)^2 + 2223(4) = 6076 \text{ in.}^3$

When $x = 6$:

$V = 4(6)^3 - 192(6)^2 + 2223(6) = 7290 \text{ in.}^3$

When $x = 10$:

$V = 4(10)^3 - 192(10)^2 + 2223(10) = 7030 \text{ in.}^3$

The maximum volume occurs when $x = 6$ inches.

67. Answers will vary.

Section 0.6 Factoring

Skills Warm Up

1. $3x(5x - 2) = 15x^2 - 6x$

2. $-2y(y + 1) = -2y^2 - 2y$

3. $(2x + 3)^2 = (2x + 3)(2x + 3)$

$= 4x^2 + 6x + 6x + 9$

$= 4x^2 + 12x + 9$

4. $(3x - 8)^2 = (3x - 8)(3x - 8)$

$= 9x^2 - 24x - 24x + 64$

$= 9x^2 - 48x + 64$

5. $(2x - 3)(x + 8) = 2x^2 + 16x - 3x - 24$

$= 2x^2 + 13x - 24$

6. $(4 - 5z)(1 + z) = 4 + 4z - 5z - 5z^2$

$= -5z^2 - z + 4$

7. $(2y + 1)(2y - 1) = 4y^2 - 2y + 2y - 1$

$= 4y^2 - 1$

8. $(x + a)(x - a) = x^2 - a^2$

9. $(x + 4)^3 = (x + 4)(x + 4)(x + 4)$

$= (x^2 + 8x + 16)(x + 4)$

$= x^3 + 8x^2 + 16x + 4x^2 + 32x + 64$

$= x^3 + 12x^2 + 48x + 64$

10. $(2x - 3)^3 = (2x - 3)(2x - 3)(2x - 3)$

$= (4x^2 - 12x + 9)(2x - 3)$

$= 8x^3 - 24x^2 + 18x - 12x^2 + 36x - 27$

$= 8x^3 - 36x^2 + 54x - 27$

1. $3x + 6 = 3(x + 2)$

3. $3x^3 - 6x = 3x(x^2 - 2)$

5. $3x(x - 5) + 8(x - 5) = (x - 5)(3x + 8)$

7. $x^2 - 36 = x^2 - 6^2 = (x + 6)(x - 6)$

9. $16x^2 - 9y^2 = (4x)^2 - (3y)^2 = (4x - 3y)(4x + 3y)$

11. $(x - 1)^2 - 4 = [(x - 1) + 2][(x - 1) - 2]$
$= (x + 1)(x - 3)$

13. $81x^4 - 1 = (9x^2)^2 - 1$
$= (9x^2 - 1)(9x^2 + 1)$
$= (3x - 1)(3x + 1)(9x^2 + 1)$

15. $3x^2 - 3 = 3(x^2 - 1) = 3(x^2 - 1^2) = 3(x + 1)(x - 1)$

17. $8x^2 - 50y^2 = 2(4x^2 - 25y^2)$
$= 2[(2x)^2 - (5y)^2]$
$= 2(2x + 5y)(2x - 5y)$

19. $25(x - 1) - x^2(x - 1) = (x - 1)(25 - x^2)$
$= (x - 1)(5^2 - x^2)$
$= (x - 1)(5 + x)(5 - x)$

21. $x^2 - 4x + 4 = x^2 - 2(2)x + 2^2 = (x - 2)^2$

23. $4y^2 + 12y + 9 = 4y^2 + 2(2)(3)y + 3^2 = (2y + 3)^2$

25. $y^2 - \frac{2}{3}y + \frac{1}{9} = \left(y - \frac{1}{3}\right)^2$

27. $x^3 - 8 = x^3 - 2^3 = (x - 2)(x^2 + 2x + 4)$

29. $y^3 + 125 = y^3 + 5^3 = (y + 5)(y^2 - 5y + 25)$

31. $x^3 - \frac{8}{27} = x^3 - \left(\frac{2}{3}\right)^3 = \left(x - \frac{2}{3}\right)\left(x^2 + \frac{2}{3}x + \frac{4}{9}\right)$

33. $x^2 + x - 2 = (x + 2)(x - 1)$

35. $w^2 - 5w + 6 = (w - 2)(w - 3)$

37. $y^2 + y - 20 = (y + 5)(y - 4)$

39. $x^2 - 30x + 200 = (x - 20)(x - 10)$

41. $3x^2 - 5x + 2 = (3x - 2)(x - 1)$

43. $9x^2 - 3x - 2 = (3x + 1)(3x - 2)$

45. $6x^2 + 37x + 6 = (6x + 1)(x + 6)$

47. $x^3 - x^2 + 2x - 2 = x^2(x - 1) + 2(x - 1)$
$= (x - 1)(x^2 + 2)$

49. $2x^3 - x^2 - 6x + 3 = x^2(2x - 1) - 3(2x - 1)$
$= (2x - 1)(x^2 - 3)$

51. $6 + 2y - 3y^3 - y^4 = 2(3 + y) - y^3(3 + y)$
$= (2 - y^3)(3 + y)$
or
$= (2 - y^3)(y + 3)$

53. $2x^2 + 3x - 2 = 2x^2 + 4x - x - 2$
$= 2x(x + 2) - 1(x + 2)$
$= (x + 2)(2x - 1)$

55. $6x^2 - 7x - 10 = 6x^2 - 12x + 5x - 10$
$= 6x(x - 2) + 5(x - 2)$
$= (x - 2)(6x + 5)$

57. $4x^2 - 8x = 4x(x - 2)$

59. $y^3 - \frac{1}{9}y = y\left(y^2 - \frac{1}{9}\right) = y\left(y - \frac{1}{3}\right)\left(y + \frac{1}{3}\right)$

61. $3x^2 - 48 = 3(x - 4)(x + 4)$

63. $x^2 - 2x + 1 = (x - 1)^2$

65. $1 - 4x + 4x^2 = (1 - 2x)^2$

67. $2y^3 - 7y^2 - 15y = y(2y + 3)(y - 5)$

69. $5x^2 - 4x + 3$ will not factor.

71. $3x^3 + x^2 + 15x + 5 = x^2(3x + 1) + 5(3x + 1)$
$= (3x + 1)(x^2 + 5)$

73. $x^4 - 4x^3 + x^2 - 4x = x(x^3 - 4x^2 + x - 4)$
$= x[x^2(x - 4) + (x - 4)]$
$= x(x - 4)(x^2 + 1)$

75. $25 - (x + 5)^2 = [5 + (x + 5)][5 - (x + 5)]$
$= -x(x + 10)$

77. $(x^2+1)^2 - 4x^2 = \left[(x^2+1)+2x\right]\left[(x^2+1)-2x\right]$
$= (x^2+2x+1)(x^2-2x+1)$
$= (x+1)^2(x-1)^2$

79. $2t^3 - 16 = 2(t^3-8) = 2(t-2)(t^2+2t+4)$

81. $x^2+3x+2 = (x+2)(x+1)$

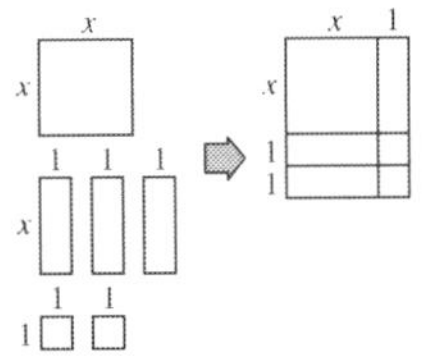

83. $2x^2+7x+3 = (2x+1)(x+3)$

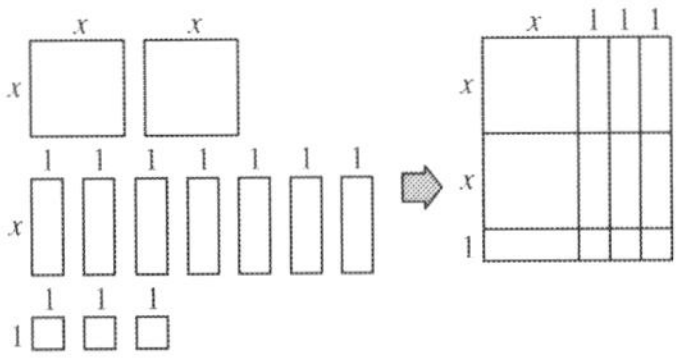

85. $2x^2 - x - 3 = (x+1)(2x-3)$

The length is $(2x-3)$ feet.

87. (a) $x^2+11x+24 = (x+8)(x+3)$

This polynomial was factored by grouping.

(b) $3x^2+7x-20 = (3x-5)(x+4)$

This polynomial was factored by trial and error.

89. $c = 1 \cdot 7 = 7$
$c = 2 \cdot 6 = 12$
$c = 3 \cdot 5 = 15$
$c = 4 \cdot 4 = 16$

The middle term, $8x$, is found by adding each factor c times x.

Because $8 = 1+7 = 2+6 = 3+5 = 4+4$, the values of c could be 7, 12, 15, or 16.

91. Answers will vary. The polynomial $2x^2-7x-15$ can be factored by trial and error or by grouping. Using trial and error, you know

$2x^2-7x-15 = (2x+3)(x-5)$ because

$-2x^2 \neq 2x(x)$ and $-5(2x)+3(x) = -7x$.

Using the grouping method, you rewrite

$2x^2-7x-15 = 2x^2-10x+3x-15$

and then

$2x^2-10x+3x-15 = 2x(x-5)+3(x-5)$
$= (2x+3)(x-5)$

or

$2x^2+3x-10x-15 = x(2x+3)-5(2x+3)$
$= (2x+3)(x-5).$

Regardless of which method is used, the results are mathematically equivalent.

93. The volume of Box 1 is $a^2(a-b)$.

The volume of Box 2 is $ab(a-b)$.

The volume of Box 3 is $b^2(a-b)$.

The total volume = volume of Box 1 + volume of Box 2 + volume of Box 3
$= a^2(a-b)+ab(a-b)+b^2(a-b)$
$= a^3-a^2b+a^2b-ab^2+ab^2-b^3$
$= a^3-b^3$

Section 0.7 Fractional Expressions

Skills Warm Up

1. $5x^2-15x^3 = 5x^2(1-3x)$

2. $16x^2-9 = (4x+3)(4x-3)$

3. $9x^2-6x+1 = (3x-1)^2$

4. $9+12y+4y^2 = (2y+3)^2$

5. $z^2+4z+3 = (z+3)(z+1)$

6. $2x^2-30x+100 = 2(x^2-15x+50)$
$= 2(x-5)(x-10)$

Skills Warm Up —continued—

7. $3 + 8x - 3x^2 = (3 - x)(1 + 3x)$

8. $3x^2 - 46x + 15 = (x - 15)(3x - 1)$

9. $s^3 + s^2 - 4s - 4 = s^2(s + 1) - 4(s + 1)$
$= (s^2 - 4)(s + 1)$
$= (s + 2)(s - 2)(s + 1)$

10. $2x^3 - 128 = 2(x^3 - 64) = 2(x - 4)(x^2 + 4x + 16)$

1. (a) $\dfrac{-\frac{1}{2} + 2}{4\left(-\frac{1}{2}\right) + 2} = \dfrac{\frac{3}{2}}{-2 + 2} = \dfrac{\frac{3}{2}}{0}$ No

(b) $\dfrac{2 + 2}{2(2) + 2} = \dfrac{4}{4 + 2} = \dfrac{4}{6} = \dfrac{2}{3}$ Yes

3. (a) $\sqrt{2(-2) + 4} = \sqrt{-4 + 4} = \sqrt{0}$ Yes

(b) $\sqrt{2(2) + 4} = \sqrt{4 + 4} = \sqrt{8}$ Yes

5. The domain of the polynomial $3x^2 - 4x + 7$ is the set of all real numbers.

7. The domain of $\dfrac{1}{3 - x}$ is the set of all real numbers except $x = 3$.

9. The domain of $\dfrac{x - 1}{x(x - 4)}$ is the set of all real numbers except $x = 0$ and $x = 4$.

11. The domain of $\sqrt{x + 1}$ is all real numbers greater than or equal to -1.

13. $\dfrac{5}{2x} = \dfrac{5(3x)}{(2x)(3x)} = \dfrac{5(3x)}{6x^2}, x \neq 0$

15. $\dfrac{x + 1}{x} = \dfrac{(x + 1)(x - 2)}{x(x - 2)}, x \neq 0, 2$

17. $\dfrac{3x}{x - 3} = \dfrac{3x(x + 2)}{(x - 3)(x + 2)} = \dfrac{3x(x + 2)}{x^2 - x - 6}, x \neq -2, 3$

19. $\dfrac{15x^2}{10x} = \dfrac{5x(3x)}{5x(2)} = \dfrac{3x}{2}, x \neq 0$

21. $\dfrac{2x}{4x + 4} = \dfrac{2x}{4(x + 1)} = \dfrac{x}{2(x + 1)}, x \neq -1$

23. $\dfrac{x - 5}{10 - 2x} = \dfrac{x - 5}{-2(x - 5)} = -\dfrac{1}{2}, x \neq 5$

25. $\dfrac{x^2 - 25}{5 - x} = \dfrac{(x - 5)(x + 5)}{-1(x - 5)} = -(x + 5), x \neq 5$

27. $\dfrac{x^3 + 5x^2 + 6x}{x^2 - 4} = \dfrac{x(x + 2)(x + 3)}{(x + 2)(x - 2)}$
$= \dfrac{x(x + 3)}{x - 2}, x \neq \pm 2$

29. $\dfrac{y^2 - 7y + 12}{y^2 + 3y - 18} = \dfrac{(y - 3)(y - 4)}{(y + 6)(y - 3)}$
$= \dfrac{y - 4}{y + 6}, y \neq -6, 3$

31. $\dfrac{2 - x + 2x^2 - x^3}{x - 2} = \dfrac{-1(x - 2)(x^2 + 1)}{x - 2}$
$= -(x^2 + 1), x \neq 2$

33. $\dfrac{z^3 - 27}{z^2 + 3z + 9} = \dfrac{(z - 3)(z^2 + 3z + 9)}{z^2 + 3z + 9} = z - 3$

35. $\dfrac{5}{x - 1} \cdot \dfrac{x - 1}{25(x - 2)} = \dfrac{1}{5(x - 2)}, x \neq 1, 2$

37. $\dfrac{(x - 9)(x + 7)}{x + 1} \cdot \dfrac{x}{9 - x} = \dfrac{x(x - 9)(x + 7)}{(x + 1)(-1)(x - 9)}$
$= -\dfrac{x(x + 7)}{x + 1}, x \neq -1, 9$

39. $\dfrac{r}{r - 1} \cdot \dfrac{r^2 - 1}{r^2} = \dfrac{r(r + 1)(r - 1)}{r^2(r - 1)} = \dfrac{r + 1}{r}, r \neq 0, 1$

41. $\dfrac{t^2 - t - 6}{t^2 + 6t + 9} \cdot \dfrac{t + 3}{t^2 - 4} = \dfrac{(t - 3)(t + 2)(t + 3)}{(t + 3)^2(t + 2)(t - 2)}$
$= \dfrac{t - 3}{(t + 3)(t - 2)}, t \neq \pm 2, -3$

43. $\dfrac{x^2 + x - 2}{x^3 + x^2} \cdot \dfrac{x}{x^2 + 3x + 2}$
$= \dfrac{(x + 2)(x - 1)}{x^2(x + 1)} \cdot \dfrac{x}{(x + 1)(x + 2)}$
$= \dfrac{x - 1}{x(x + 1)^2}, x \neq -2, -1, 0$

45. $\dfrac{3(x+y)}{4} \div \dfrac{x+y}{2} = \dfrac{3(x+y)}{4} \cdot \dfrac{2}{x+y} = \dfrac{3}{2}, x \neq -y$

47. $\dfrac{\left[\dfrac{x^2}{(x+1)^2}\right]}{\left[\dfrac{x}{(x+1)^3}\right]} = \dfrac{x^2}{(x+1)^2} \cdot \dfrac{(x+1)^3}{x}$

$= x(x+1), x \neq -1, 0$

49. Factors of $x + 4$: $x + 4$

Factors of $x^2 - 2x - 24$: $(x+4)(x-6)$

LCD $= (x+4)(x-6)$

51. Factors of x^2: x, x

Factors of $x - 1$: $x - 1$

Factors of $x^2 - x$: $x, x - 1$

LCD $= x^2(x-1)$

53. $\dfrac{4x}{x-2} + \dfrac{x}{x-2} = \dfrac{4x+x}{x-2} = \dfrac{5x}{x-2}, x \neq 2$

55. $\dfrac{3x}{x-4} - \dfrac{x}{4-x} = \dfrac{3x}{x-4} - \dfrac{x}{-(x-4)}$

$= \dfrac{3x}{x-4} + \dfrac{x}{x-4}$

$= \dfrac{4x}{x-4}$

57. $4 + \dfrac{3}{x-5} = 4 \cdot \dfrac{x-5}{x-5} + \dfrac{3}{x-5}$

$= \dfrac{4x - 20 + 3}{x-5} = \dfrac{4x-17}{x-5}$

59. $\dfrac{2}{x^2-4} - \dfrac{1}{x^2-3x+2}$

$= \dfrac{2}{(x+2)(x-2)} - \dfrac{1}{(x-1)(x-2)}$

$= \dfrac{2(x-1)-(x+2)}{(x+2)(x-2)(x-1)}$

$= \dfrac{x-4}{(x+2)(x-2)(x-1)}, x \neq 1, \pm 2$

61. $-\dfrac{1}{x} + \dfrac{2}{x^2+1} + \dfrac{1}{x^3+x}$

$= \dfrac{-(x^2+1)}{x(x^2+1)} + \dfrac{2x}{x(x^2+1)} + \dfrac{1}{x(x^2+1)}$

$= \dfrac{-x^2 - 1 + 2x + 1}{x(x^2+1)} = \dfrac{-x^2+2x}{x(x^2+1)}$

$= \dfrac{-x(x-2)}{x(x^2+1)} = \dfrac{x-2}{x^2+1} = \dfrac{2-x}{x^2+1}, x \neq 0$

63. $\dfrac{\left(\dfrac{x}{2} - 1\right)}{(x-2)} = \dfrac{\left(\dfrac{x}{2} - \dfrac{2}{2}\right)}{\left(\dfrac{x-2}{1}\right)} = \dfrac{x-2}{2} \cdot \dfrac{1}{x-2} = \dfrac{1}{2}, x \neq 2$

65. $\dfrac{\left(\dfrac{1}{x} - \dfrac{1}{x+1}\right)}{\left(\dfrac{1}{x+1}\right)} = \dfrac{\dfrac{(x+1)-x}{x(x+1)}}{\dfrac{1}{x+1}} = \dfrac{1}{x(x+1)} \cdot \dfrac{x+1}{1}$

$= \dfrac{1}{x}, x \neq -1, 0$

67. $\dfrac{\left(\sqrt{x} - \dfrac{1}{2\sqrt{x}}\right)}{\sqrt{x}} = \dfrac{\left(\sqrt{x} - \dfrac{1}{2\sqrt{x}}\right)}{\sqrt{x}} \cdot \dfrac{2\sqrt{x}}{2\sqrt{x}}$

$= \dfrac{2x-1}{2x}, x > 0$

69. (a) $r = \dfrac{\dfrac{24(48(415) - 18{,}000)}{48}}{\left(18{,}000 + \dfrac{48(415)}{12}\right)} \approx 0.0488 = 4.88\%$

(b) $r = \dfrac{\left[\dfrac{24(NM-P)}{N}\right]}{\left(P + \dfrac{NM}{12}\right)} = \dfrac{\left[\dfrac{24(NM-P)}{N}\right]}{\left(\dfrac{12P+NM}{12}\right)} = \dfrac{24(NM-P)}{N} \cdot \dfrac{12}{12P+NM}$

$r = \dfrac{288(NM-P)}{N(12P+NM)}$

$r = \dfrac{288[48(415) - 18{,}000]}{48[12(18{,}000) + 48(415)]} \approx 0.0488 = 4.88\%$

71.

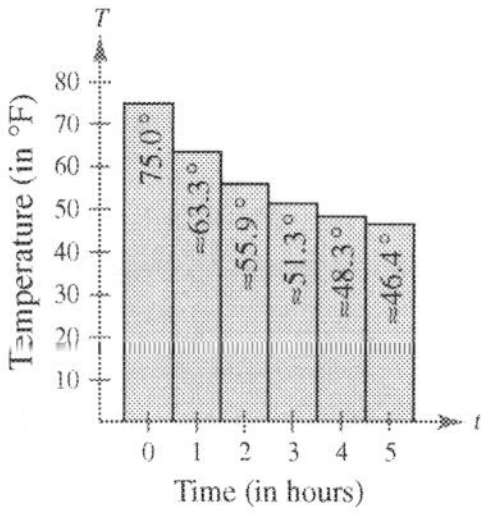

It is unlikely that the temperature of the food will drop to 40°F within 7 hours because the differences in heights of consecutive bars in the bar graph are getting smaller and smaller as t increases.

73. True;

$$\frac{x^{2n} - 1^{2n}}{x^n - 1^n} = \frac{\cancel{(x^n - 1^n)}(x^n + 1^n)}{\cancel{x^n - 1^n}} = x^n + 1^n$$

75. When simplifying a rational expression, domain restrictions must be given for divided-out factors that caused the original expression to be undefined. For example,

$$\frac{x^2 - 7x + 10}{x - 2} = \frac{\cancel{(x - 2)}(x - 5)}{\cancel{x - 2}} = x - 5, x \neq 2.$$

Review Exercises for Chapter 0

1. $\left\{9, -7, -\frac{8}{9}, \frac{5}{2}, \sqrt{6}, 0.4\right\}$

(a) Natural numbers: $\{9\}$

(b) Integers: $\{9, -7\}$

(c) Rational numbers: $\left\{9, -7, -\frac{8}{9}, \frac{5}{2}, 0.4\right\}$

(d) Irrational numbers: $\left\{\sqrt{6}\right\}$

3. $-4 < -3$

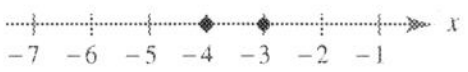

5. The inequality $x \leq -6$ denotes all real numbers less than or equal to -6.

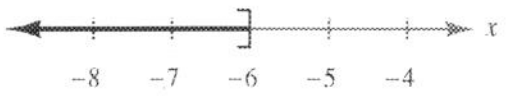

7. $x \geq 0$

9. $2 < x \leq 5$

11. $-|-14| = -14$

13. $d(-14, -18) = |-14 - (-18)| = |4| = 4$

15. $d(2, -8) = |-8 - 2| = |-10| = 10$

17. $|x - 7| \geq 4$

19. $5x^3 - 2x + 7$

Terms: $5x^3, -2x, 7$

21. $-4x^2 - 6x$

(a) $-4(-1)^2 - 6(-1) = -4 + 6 = 2$

(b) $-4(0)^2 - 6(0) = 0$

23. $5(x^2 + x) = 5x^2 + 5x$ Distributive Property

25. $3x + 7 = 7 + 3x$ Commutative (addition)

27. $-3(0) - 2(4 - 5) = 0 - 2(-1) = 2$

29.
$$\begin{aligned}\frac{1}{2} + \frac{1}{3} - \frac{1}{6} &= \frac{1}{2}\cdot\frac{3}{3} + \frac{1}{3}\cdot\frac{2}{2} - \frac{1}{6}\\ &= \frac{3}{6} + \frac{2}{6} - \frac{1}{6}\\ &= \frac{3 + 2 - 1}{6}\\ &= \frac{4}{6}\\ &= \frac{2}{3}\end{aligned}$$

31. $4\left(\frac{1}{6} - \frac{1}{7}\right) \approx 0.10$

33. $\frac{(4x)^2}{2x} = \frac{16x^2}{2x} = 8x, x \neq 0$

35. $\frac{10x^2}{2x^6} = \frac{5}{x^4}$

37. $308,700,000 = 3.087 \times 10^8$

39. $7.28 \times 10^6 = 7,280,000$

41. $1800(1 + 0.08)^{24} \approx 11,414.125$

43.

Year	Balance
5	\$10,014.37
10	\$12,535.94
15	\$15,692.44
20	\$19,643.73
25	\$24,589.94

$$A = 8000\left(1 + \frac{0.045}{12}\right)^{12(5)} = 10{,}014.37$$
$$A = 8000\left(1 + \frac{0.045}{12}\right)^{12(10)} = 12{,}535.94$$
$$A = 8000\left(1 + \frac{0.045}{12}\right)^{12(15)} = 15{,}692.44$$
$$A = 8000\left(1 + \frac{0.045}{12}\right)^{12(20)} = 19{,}643.73$$
$$A = 8000\left(1 + \frac{0.045}{12}\right)^{12(25)} = 24{,}589.94$$

45. Radical Form: $\sqrt{16} = 4$

Rational Exponent Form: $16^{1/2} = 4$

47. $\sqrt{169} = \sqrt{13^2} = 13$

49. $\sqrt{4x^4} = \sqrt{\left(2x^2\right)^2} = 2x^2$

51. $\sqrt{6}\sqrt{2} = \sqrt{12} = \sqrt{4 \cdot 3} = 2\sqrt{3}$

53. $\dfrac{1}{2 - \sqrt{3}} = \dfrac{1}{2 - \sqrt{3}} \cdot \dfrac{2 + \sqrt{3}}{2 + \sqrt{3}}$

$= \dfrac{2 + \sqrt{3}}{4 - 3}$

$= \dfrac{2 + \sqrt{3}}{1}$

$= 2 + \sqrt{3}$

55. $2\sqrt{x} - 5\sqrt{x} = -3\sqrt{x}$

57. $(64)^{-2/3} = \dfrac{1}{64^{2/3}} = \dfrac{1}{16}$

59. $\sqrt[4]{5^2} = 5^{2/4} = 5^{1/2} = \sqrt{5}$

61. $\sqrt{127} \approx 11.269$

63. $(x - 3) - (2x - 8) = x - 3 - 2x + 8 = -x + 5$

65. $x(x - 2) - 2(3x + 7) = x^2 - 2x - 6x - 14$

$= x^2 - 8x - 14$

67. $(x + 1)(x - 2) = x^2 - 2x + x - 2 = x^2 - x - 2$

69. $(x + 4)(x^2 - 4x + 16)$

$= x^3 + 4x^2 - 4x^2 - 16x + 16x + 64$

$= x^3 + 64$

71. $(x + 4)^2 = (x + 4)(x + 4)$

$= x^2 + 4x + 4x + 16$

$= x^2 + 8x + 16$

73. When $x = 10$:

$14{,}945.3x + 105{,}238 = 14{,}945.3(10) + 105{,}238 = 254{,}691$

In 2010, there were about 254,691 mobile telephone cell sites in the United States.

75. $4x^2 - 36 = 4\left(x^2 - 9\right) = 4(x - 3)(x + 3)$

77. $8x^3 - 125 = (2x)^3 - 5^3 = (2x - 5)\left(4x^2 + 10x + 25\right)$

79. $x^2 - 4x - 5 = (x - 5)(x + 1)$

81. $x^3 - 4x^2 - 2x + 8 = x^2(x - 4) - 2(x - 4)$

$= \left(x^2 - 2\right)(x - 4)$

83. The domain of $\dfrac{2x + 1}{x - 4}$ is all real numbers except $x = 4$.

85. The domain of $2\sqrt{x + 1}$ is all real numbers greater than or equal to -1.

87. $\dfrac{4}{3x} = \dfrac{4(3x)}{9x^2},\ x \neq 0$

89. $\dfrac{x^2 - 4}{2x + 4} = \dfrac{(x - 2)(x + 2)}{2(x + 2)} = \dfrac{x - 2}{2},\ x \neq -2$

91. $\dfrac{x^3 - 9x}{x^3 - 4x^2 + 3x} = \dfrac{x(x^2 - 9)}{x(x^2 - 4x + 3)}$

$= \dfrac{x(x - 3)(x + 3)}{x(x - 3)(x - 1)}$

$= \dfrac{x + 3}{x - 1},\ x \neq 0, 3$

93. $\dfrac{2x - 1}{x + 1} \cdot \dfrac{x^2 - 1}{2x^2 - 7x + 3} = \dfrac{2x - 1}{x + 1} \cdot \dfrac{(x + 1)(x - 1)}{(2x - 1)(x - 3)}$

$= \dfrac{x - 1}{x - 3},\ x \neq \dfrac{1}{2},\ x \neq -1$

95. $\dfrac{x}{x-1} + \dfrac{2x}{x-2} = \dfrac{x}{x-1} \cdot \dfrac{x-2}{x-2} + \dfrac{2x}{x-2} \cdot \dfrac{x-1}{x-1}$

$= \dfrac{x^2 - 2x}{(x-1)(x-2)} + \dfrac{2x^2 - 2x}{(x-1)(x-2)}$

$= \dfrac{x^2 - 2x + 2x^2 - 2x}{(x-1)(x-2)}$

$= \dfrac{3x^2 - 4x}{(x-1)(x-2)}$

97. $\dfrac{2}{x-1} + \dfrac{4}{x+1} + \dfrac{8}{x^2-1}$

$= \dfrac{2(x+1)}{(x-1)(x+1)} + \dfrac{4(x-1)}{(x-1)(x+1)} + \dfrac{8}{(x-1)(x+1)}$

$= \dfrac{2x + 2 + 4x - 4 + 8}{(x-1)(x+1)}$

$= \dfrac{6x + 6}{(x-1)(x+1)}$

$= \dfrac{6(x+1)}{(x-1)(x+1)}$

$= \dfrac{6}{x-1}, x \neq -1$

99. $\dfrac{\left(\dfrac{x^2-1}{x}\right)}{\left[\dfrac{(x-1)^2}{x}\right]} = \dfrac{x^2-1}{x} \div \dfrac{(x-1)^2}{x}$

$= \dfrac{x^2 - 1}{x} \cdot \dfrac{x}{(x-1)^2}$

$= \dfrac{(x-1)(x+1)}{x} \cdot \dfrac{x}{(x-1)(x-1)}$

$= \dfrac{x+1}{x-1}, x \neq 0$

Chapter 0 Test Yourself

1. $-3(-3)^2 - 5(-3) = -3(9) + 15 = -27 + 15 = -12$

2.

Year	Balance
5	\$4763.77
10	\$5673.38
15	\$6756.67
20	\$8046.81
25	\$9583.29

$A = 4000\left(1 + \dfrac{0.035}{12}\right)^{12(5)} = 4763.77$

$A = 4000\left(1 + \dfrac{0.035}{12}\right)^{12(10)} = 5673.38$

$A = 4000\left(1 + \dfrac{0.035}{12}\right)^{12(15)} = 6756.67$

$A = 4000\left(1 + \dfrac{0.035}{12}\right)^{12(20)} = 8046.81$

$A = 4000\left(1 + \dfrac{0.035}{12}\right)^{12(25)} = 9583.29$

3. $8(-2x^2)^3 = 8 \cdot (-1)^3 \cdot 2^3 x^6 = -64x^6$

4. $3\sqrt{x} - 7\sqrt{x} = -4\sqrt{x}$

5. $5^{1/4} \cdot 5^{7/4} = 5^{8/4} = 5^2 = 25$

6. $\sqrt{48} - \sqrt{80} = 4\sqrt{3} - 4\sqrt{5}$

7. $\sqrt{12x^3} = \sqrt{4 \cdot 3x^3} = 2x\sqrt{3x}, x \geq 0$

8. $\dfrac{2}{5 - \sqrt{7}} = \dfrac{2}{5 - \sqrt{7}} \cdot \dfrac{5 + \sqrt{7}}{5 + \sqrt{7}}$

$= \dfrac{2(5 + \sqrt{7})}{25 - 7}$

$= \dfrac{2(5 + \sqrt{7})}{18}$

$= \dfrac{5 + \sqrt{7}}{9}$

9. $(3x + 7)^2 = (3x + 7)(3x + 7) = 9x^2 + 42x + 49$

10. $3x(x + 5) - 2x(4x - 7) = 3x^2 + 15x - 8x^2 + 14x$
$= -5x^2 + 29x$

11. $5x^2 - 80 = 5(x - 4)(x + 4)$

12. $4x^2 + 12x + 9 = (2x + 3)(2x + 3) = (2x + 3)^2$

13. $x^3 + 6x^2 - 3x + 18 = x^2(x - 6) - 3(x - 6)$
$= (x^2 - 3)(x - 6)$

14. $x^3 + 2x^2 - 4x - 8 = x^2(x + 2) - 4(x + 2)$
$= (x^2 - 4)(x + 2)$
$= (x - 2)(x + 2)(x + 2)$

15. $\dfrac{x^2 - 16}{3x + 12} = \dfrac{(x - 4)(x + 4)}{3(x + 4)} = \dfrac{x - 4}{3}, x \neq -4$

16. $\dfrac{3x - 5}{x + 3} \cdot \dfrac{x^2 + 7x + 12}{9x^2 - 25} = \dfrac{3x - 5}{x + 3} \cdot \dfrac{(x + 3)(x + 4)}{(3x - 5)(3x + 5)}$
$= \dfrac{x + 4}{3x + 5}, x \neq -3, x \neq \dfrac{5}{3}$

17. $\dfrac{x}{x - 3} + \dfrac{3x}{x - 4} = \dfrac{x}{x - 3} \cdot \dfrac{x - 4}{x - 4} + \dfrac{3x}{x - 4} \cdot \dfrac{x - 3}{x - 3}$
$= \dfrac{x(x - 4) + 3x(x - 3)}{(x - 3)(x - 4)}$
$= \dfrac{x^2 - 4x + 3x^2 - 9x}{(x - 3)(x - 4)}$
$= \dfrac{4x^2 - 13x}{(x - 3)(x - 4)}$

18. $\dfrac{3}{x + 5} - \dfrac{4}{x - 2} = \dfrac{3}{x + 5} \cdot \dfrac{x - 2}{x - 2} - \dfrac{4}{x - 2} \cdot \dfrac{x + 5}{x + 5}$
$= \dfrac{3(x - 2) - 4(x + 5)}{(x + 5)(x - 2)}$
$= \dfrac{-x - 26}{(x + 5)(x - 2)}$

19. The domain of $4\sqrt{x - 10}$ is all real numbers greater than or equal to 10.

20. The domain of $\dfrac{3}{x + 9}$ is all real numbers except $x = -9$.

21. $\dfrac{\left(\dfrac{2x - 9}{x - 1}\right)}{\left(\dfrac{3}{x - 1} + \dfrac{1 - x}{x + 2}\right)}$
$= \dfrac{\left(\dfrac{2x - 9}{x - 1}\right)}{\left(\dfrac{3}{x - 1} + \dfrac{1 - x}{x + 2}\right)} \cdot \dfrac{(x - 1)(x + 2)}{(x - 1)(x + 2)}$
$= \dfrac{(2x - 9)(x + 2)}{3(x + 2) + (1 - x)(x - 1)}$
$= \dfrac{2x^2 - 5x - 18}{3x + 6 - (x^2 - 2x + 1)}$
$= \dfrac{2x^2 - 5x - 18}{3x + 6 - x^2 + 2x - 1}$
$= \dfrac{2x^2 - 5x - 18}{-x^2 + 5x + 5}$ or $-\dfrac{2x^2 - 5x - 18}{x^2 - 5x - 5}$
or $-\dfrac{(x + 2)(2x - 9)}{x^2 - 5x + 5}, x \neq -2, 1$

22. When $x = 0$:

$0.223x + 5.35 = 0.223(0) + 5.35 = 5.35$

When $x = 9$:

$0.223x + 5.35 = 0.223(9) + 5.35 = 7.36$

The average price of a movie ticket in the United States increased from $5.35 in 2000 to $7.36 in 2009.

CHAPTER 1
Equations and Inequalities

CHAPTER 1
Equations and Inequalities

Section 1.1 Linear Equations

Skills Warm Up

1. $(2x - 4) - (5x + 6) = 2x - 4 - 5x - 6$
$= -3x - 10$

2. $(3x - 5) + (2x - 7) = 5x - 12$

3. $2(x + 1) - (x + 2) = 2x + 2 - x - 2 = x$

4. $-3(2x - 4) + 7(x + 2) = -6x + 12 + 7x + 14$
$= x + 26$

5. $\frac{x}{2} + \frac{x}{7} = \frac{7x + 2x}{14} = \frac{9x}{14}$

6. $x - \frac{x}{4} = \frac{4x - x}{4} = \frac{3x}{4}$

7. $\frac{1}{x + 1} - \frac{1}{x} = \frac{x - (x + 1)}{x(x + 1)} = \frac{x - x - 1}{x(x + 1)} = -\frac{1}{x(x + 1)}$

8. $\frac{5}{x} + \frac{3}{x} = \frac{5 + 3}{x} = \frac{8}{x}$

9. $\frac{4}{x} + \frac{3}{x - 2} = \frac{4(x - 2) + 3x}{x(x - 2)} = \frac{4x - 8 + 3x}{x(x - 2)}$
$= \frac{7x - 8}{x(x - 2)}$

10. $\frac{1}{x + 1} - \frac{1}{x - 1} = \frac{x - 1 - (x + 1)}{(x + 1)(x - 1)} = \frac{x - 1 - x - 1}{(x + 1)(x - 1)}$
$= -\frac{2}{x^2 - 1}$

1. The equation $2(x - 1) = 2x - 2$ is an *identity* because (by the Distributive Property) it is true for every real value of x.

3. The equation $2(x - 1) = 3x + 4$ is *conditional* since there are real number values of x for which the equation is not true.

5. The equation $2(x + 1) = 2x + 1$ is *conditional* since there are *no* real number values of x for which the equation is true.

7. $5x - 3 = 3x + 5$

(a) $5(0) - 3 \stackrel{?}{=} 3(0) + 5$
$-3 \neq 5$
$x = 0$ *is not* a solution.

(b) $5(-5) - 3 \stackrel{?}{=} 3(-5) + 5$
$-28 \neq -10$
$x = -5$ *is not* a solution.

(c) $5(4) - 3 \stackrel{?}{=} 3(4) + 5$
$17 = 17$
$x = 4$ *is* a solution.

(d) $5(10) - 3 \stackrel{?}{=} 3(10) + 5$
$47 \neq 35$
$x = 10$ *is not* a solution.

9. $3x^2 + 2x - 5 = 2x^2 - 2$

(a) $3(-3)^2 + 2(-3) - 5 \stackrel{?}{=} 2(-3)^2 - 2$
$16 = 16$
$x = -3$ *is* a solution.

(b) $3(1)^2 + 2(1) - 5 \stackrel{?}{=} 2(1)^2 - 2$
$0 = 0$
$x = 1$ *is* a solution.

(c) $3(4)^2 + 2(4) - 5 \stackrel{?}{=} 2(4)^2 - 2$
$51 \neq 30$
$x = 4$ *is not* a solution.

(d) $3(-5)^2 + 2(-5) - 5 \stackrel{?}{=} 2(-5)^2 - 2$
$60 \neq 48$
$x = -5$ *is not* a solution.

11. $\dfrac{5}{2x} - \dfrac{4}{x} = 3$

(a) $\dfrac{5}{2(-1/2)} - \dfrac{4}{(-1/2)} \stackrel{?}{=} 3$

$3 = 3$

$x = -\dfrac{1}{2}$ *is* a solution.

(b) $\dfrac{5}{2(4)} - \dfrac{4}{4} \stackrel{?}{=} 3$

$-\dfrac{3}{8} \neq 3$

$x = 4$ *is not* a solution.

(c) $\dfrac{5}{2(0)} - \dfrac{4}{0}$ is undefined.

$x = 0$ *is not* a solution.

(d) $\dfrac{5}{2(1/4)} - \dfrac{4}{1/4} \stackrel{?}{=} 3$

$-6 \neq 3$

$x = \dfrac{1}{4}$ *is not* a solution.

13. $(x + 5)(x - 3) = 20$

(a) $(3 + 5)(3 - 3) \stackrel{?}{=} 20$

$0 \neq 20$

$x = 3$ *is not* a solution.

(b) $(-2 + 5)(-2 - 3) \stackrel{?}{=} 20$

$-15 \neq 20$

$x = -2$ *is not* a solution.

(c) $(0 + 5)(0 - 3) \stackrel{?}{=} 20$

$-15 \neq 20$

$x = 0$ *is not* a solution.

(d) $(-7 + 5)(-7 - 3) \stackrel{?}{=} 20$

$20 = 20$

$x = -7$ *is* a solution.

15. $\sqrt{2x - 3} = 3$

(a) $\sqrt{2(6) - 3} \stackrel{?}{=} 3$

$\sqrt{9} \stackrel{?}{=} 3$

$3 = 3$

$x = 6$ *is* a solution.

(b) $\sqrt{2(-3) - 3} \stackrel{?}{=} 3$

$\sqrt{-9} \stackrel{?}{=} 3$

$\sqrt{-9}$ is undefined.

$x = -3$ *is not* a solution.

(c) $\sqrt{2\left(-\frac{1}{3}\right) - 3}$ is undefined.

$x = -\frac{1}{3}$ *is not* a solution.

(d) $\sqrt{2(-2) - 3}$ is undefined.

$x = -2$ *is not* a solution.

17.
$$\begin{aligned} x + 10 &= 15 \\ x + 10 - 10 &= 15 - 10 \\ x &= 5 \end{aligned}$$

19.
$$\begin{aligned} 7 - 2x &= 15 \\ 7 - 2x - 7 &= 15 - 7 \\ -2x &= 8 \\ x &= -4 \end{aligned}$$

21.
$$\begin{aligned} 8x - 5 &= 3x + 10 \\ 8x - 3x - 5 + 5 &= 3x - 3x + 10 + 5 \\ 5x &= 15 \\ x &= 3 \end{aligned}$$

23.
$$\begin{aligned} 5x - x &= 4x - 4 \\ 4x &= 4x - 4 \\ 4x - 4x &= 4x - 4x - 4 \\ 0 &\neq -4 \end{aligned}$$

No solution

25.
$$\begin{aligned} 4x - 8 &= -4(2 - x) \\ 4x - 8 &= -8 + 4x \\ 0 &= 0 \end{aligned}$$

Infinitely many solutions

27.
$$\begin{aligned} 2(x + 5) - 7 &= 3(x - 2) \\ 2x + 10 - 7 &= 3x - 6 \\ 2x + 3 &= 3x - 6 \\ -x &= -9 \\ x &= 9 \end{aligned}$$

29.
$$\begin{aligned} 6[x - (2x + 3)] &= 8 - 5x \\ 6[-x - 3] &= 8 - 5x \\ -6x - 18 &= 8 - 5x \\ -x &= 26 \\ x &= -26 \end{aligned}$$

31.
$$\begin{aligned} x - 3(2x + 3) &= 8 - 5x \\ x - 6x - 9 &= 8 - 5x \\ -5x - 9 &= 8 - 5x \\ 0 &\neq 17 \end{aligned}$$

No solution

33.
$$\begin{aligned} \frac{5x}{4} + \frac{1}{2} &= x - \frac{1}{2} \\ 4\left(\frac{5x}{4}\right) + 4\left(\frac{1}{2}\right) &= 4(x) - 4\left(\frac{1}{2}\right) \\ 5x + 2 &= 4x - 2 \\ x &= -4 \end{aligned}$$

35.
$$\begin{aligned} \tfrac{3}{2}(z + 5) &= \tfrac{1}{4}(z + 24) \\ 4\left[\tfrac{3}{2}(z + 5)\right] &= 4\left[\tfrac{1}{4}(z + 24)\right] \\ 6(z + 5) &= z + 24 \\ 6z + 30 &= z + 24 \\ 5z &= -6 \\ z &= -\tfrac{6}{5} \end{aligned}$$

37.
$$\begin{aligned} \frac{100 - 4u}{3} &= \frac{5u + 6}{4} + 6 \\ 12\left(\frac{100 - 4u}{3}\right) &= 12\left(\frac{5u + 6}{4}\right) + 12(6) \\ 4(100 - 4u) &= 3(5u + 6) + 72 \\ 400 - 16u &= 15u + 18 + 72 \\ -31u &= -310 \\ u &= 10 \end{aligned}$$

39.
$$\begin{aligned} \frac{5x - 4}{5x + 4} &= \frac{2}{3} \\ 3(5x - 4) &= 2(5x + 4) \\ 15x - 12 &= 10x + 8 \\ 5x &= 20 \\ x &= 4 \end{aligned}$$

41.
$$\begin{aligned} 10 - \frac{13}{x} &= 4 + \frac{5}{x} \\ x(10) - x\left(\frac{13}{x}\right) &= x(4) + x\left(\frac{5}{x}\right) \\ 10x - 13 &= 4x + 5 \\ 6x &= 18 \\ x &= 3 \end{aligned}$$

43.
$$\begin{aligned} \frac{1}{x - 3} + \frac{1}{x + 3} &= \frac{10}{x^2 - 9} \\ \frac{1}{x - 3}(x - 3)(x + 3) + \frac{1}{x + 3}(x - 3)(x + 3) &= \frac{10}{x^2 - 9}(x - 3)(x + 3) \\ x + 3 + x - 3 &= 10 \\ 2x &= 10 \\ x &= 5 \end{aligned}$$

45.
$$\begin{aligned} \frac{6}{(x - 3)(x - 1)} &= \frac{3}{x - 3} + \frac{4}{x - 1} \\ 6 &= 3(x - 1) + 4(x - 3) \\ 6 &= 3x - 3 + 4x - 12 \\ 6 &= 7x - 15 \\ -7x &= -21 \\ x &= 3 \end{aligned}$$

A check reveals that $x = 3$ is an extraneous solution, so there is no solution.

47.
$$\begin{aligned} \frac{x - 3}{2x + 1} &= \frac{x - 5}{2x + 4} \\ (2x + 1)(x - 5) &= (x - 3)(2x + 4) \\ 2x^2 - 9x - 5 &= 2x^2 - 2x - 12 \\ -9x - 5 &= -2x - 12 \\ -7x - 5 &= -12 \\ -7x &= -7 \\ x &= 1 \end{aligned}$$

49.
$$\begin{aligned} 0.25x + 0.75(10 - x) &= 3 \\ 4(0.25x) + 4(0.75)(10 - x) &= 4(3) \\ x + 3(10 - x) &= 12 \\ x + 30 - 3x &= 12 \\ 2x &= -18 \\ x &= 9 \end{aligned}$$

51.
$$\begin{aligned} (x + 2)^2 + 5 &= (x + 3)^2 \\ x^2 + 4x + 4 + 5 &= x^2 + 6x + 9 \\ 4x + 9 &= 6x + 9 \\ -2x &= 0 \\ x &= 0 \end{aligned}$$

53.
$$\begin{aligned} (x + 2)^2 - x^2 &= 4(x + 1) \\ x^2 + 4x + 4 - x^2 &= 4x + 4 \\ 4 &= 4 \end{aligned}$$

The equation is an identity; every real number is a solution.

55.
$$\begin{aligned} (2x + 1)^2 &= 4(x^2 + x + 1) \\ 4x^2 + 4x + 1 &= 4x^2 + 4x + 4 \\ 1 &\neq 4 \end{aligned}$$

No solution

57. When you check $x = 2$ in the original equation, you get division by zero, which is undefined. So, $x = 2$ is an extraneous solution and the equation has no solution.

59. Extraneous solutions may arise when a fractional expression is multiplied by factors involving the variable. For example, in Exercise 47, we multiplied by $x(x - 3)$. If x or $x - 3$ are actually zero, an extraneous solution is introduced. In Exercise 47, an extraneous solution was $x = 3$ since multiplying the equation by $(x - 3)$ actually multiplied the equation by zero.

61. Equivalent equations are equations that have the same solutions as the original equation. Usually, as part of the process of solving an equation, we rewrite the equation (using the properties and rules of algebra) into an equivalent form. For example, to solve Exercise 49, we would rewrite $(x + 2)^2 + 5 = (x + 3)^2$ as $x^2 + 4x + 4 + 5 = x^2 + 6x + 9$. Next we might combine terms on the left so we would have equivalent equation $x^2 + 4x + 9 = x^2 + 6x + 9$. We would continue the solution process to the equation $2x = 0$ and finally $x = 0$.

63.
$$\begin{aligned} 0.275x + 0.725(500 - x) &= 300 \\ 0.275x + 362.5 - 0.725x &= 300 \\ -0.45x &= -62.5 \\ x &= \frac{62.5}{0.45} \\ &\approx 138.889 \end{aligned}$$

65.
$$\begin{aligned} \frac{x}{0.6321} + \frac{x}{0.0692} &= 1000 \\ 0.0692x + 0.6321x &= 1000(0.6321)(0.0692) \\ 0.7013x &= 43.74132 \\ x &= \frac{43.74132}{0.7013} \\ &\approx 62.372 \end{aligned}$$

67.
$$\begin{aligned} \frac{2}{7.398} - \frac{4.405}{x} &= \frac{1}{x} \\ 2x - (4.405)(7.398) &= 7.398 \\ 2x &= (4.405)(7.398) + 7.398 \\ 2x &= (5.405)(7.398) \\ x &= \frac{(5.405)(7.398)}{2} \\ x &\approx 19.993 \end{aligned}$$

69. Use the *table* feature in ASK mode or evaluate using the scientific calculator part of a graphing utility. (Answers will vary.)

71. (a) $\dfrac{1 + 0.73205}{1 - 0.73205} \approx 6.46$

(b) $\dfrac{1 + 0.73205}{1 - 0.73205} \approx \dfrac{1.73}{0.27} \approx 6.41$

(c) The second method introduced an additional round-off error. A result should be rounded only after the final calculation.

73. (a) $\dfrac{333 + \dfrac{1.98}{0.74}}{4 + \dfrac{6.25}{3.15}} \approx 56.09$

(b) $\dfrac{333 + \dfrac{1.98}{0.74}}{4 + \dfrac{6.25}{3.15}} \approx \dfrac{335.68}{5.98} \approx 56.13$

(c) The second method introduced an additional round-off error. A result should be rounded only after the final calculation.

75.
$$\begin{aligned} y &= 1114.4t + 21{,}952 \\ 32{,}000 &= 1114.4t + 21{,}952 \\ 10{,}048 &= 1114.4t \\ \frac{10{,}048}{1114.4} &= t \\ 9.0 &\approx t \end{aligned}$$

The per capita personal income in West Virginia was about \$32,000 in 2009.

77.
$$\begin{aligned} y &= 18 \\ 0.432x - 10.44 &= 18 \\ 0.432x &= 28.44 \\ x &= 65.8\overline{3} \end{aligned}$$

The female was about 65.8 inches tall.

79. (a)
$$\begin{aligned} y &= 322.5t + 4788 \\ 7000 &= 322.5t + 4788 \\ 2212 &= 322.5t \\ \frac{2212}{322.5} &= t \\ t &\approx 6.9 \end{aligned}$$

The total number of CHIP enrollees exceeded 7 million in 2007.

(b)
$$\begin{aligned} 322.5t + 4788 &= 8{,}300 \\ 322.5t &= 3512 \\ t &\approx 10.89 \approx 11 \end{aligned}$$

The total number of CHIP enrollees will exceed 8.3 million in 2011.

Section 1.2 Mathematical Modeling

Skills Warm Up

1.
$$\begin{aligned} 3x - 42 &= 0 \\ 3x &= 42 \\ x &= 14 \end{aligned}$$

2.
$$\begin{aligned} 64 - 16x &= 0 \\ -16x &= -64 \\ x &= 4 \end{aligned}$$

3.
$$\begin{aligned} 2 - 3x &= 14 + x \\ 2 - 4x &= 14 \\ -4x &= 12 \\ x &= -3 \end{aligned}$$

4.
$$\begin{aligned} 7 + 5x &= 7x - 1 \\ 7 - 2x &= -1 \\ -2x &= -8 \\ x &= 4 \end{aligned}$$

5.
$$\begin{aligned} 5[1 + 2(x + 3)] &= 6 - 3(x - 1) \\ 5[1 + 2x + 6] &= 6 - 3x + 3 \\ 5(2x + 7) &= -3x + 9 \\ 10x + 35 &= -3x + 9 \\ 13x + 35 &= 9 \\ 13x &= -26 \\ x &= -2 \end{aligned}$$

6.
$$\begin{aligned} 2 - 5(x - 1) &= 2[x + 10(x - 1)] \\ 2 - 5x + 5 &= 2[x + 10x - 10] \\ -5x + 7 &= 2(11x - 10) \\ -5x + 7 &= 22x - 20 \\ -27x + 7 &= -20 \\ -27x &= -27 \\ x &= 1 \end{aligned}$$

7.
$$\begin{aligned} \frac{x}{3} + \frac{x}{2} &= \frac{1}{3} \\ 6\left(\frac{y}{3}\right) + 6\left(\frac{y}{2}\right) &= 6\left(\frac{1}{3}\right) \\ 2x + 3x &= 2 \\ 5x &= 2 \\ x &= \frac{2}{5} \end{aligned}$$

8.
$$\begin{aligned} \frac{2}{x} + \frac{2}{5} &= 1 \\ \frac{2}{x} &= \frac{3}{5} \\ 3x &= 10 \\ x &= \frac{10}{3} \end{aligned}$$

Skills Warm Up —*continued*—

9. $$\begin{aligned} 1 - \frac{2}{z} &= \frac{z}{z+3} \\ \frac{z-2}{2} &= \frac{z}{z+3} \\ (z-2)(z+3) &= z^2 \\ z^2 + z - 6 &= z^2 \\ z - 6 &= 0 \\ z &= 6 \end{aligned}$$

10. $$\begin{aligned} \frac{x}{x+1} - \frac{1}{2} &= \frac{4}{3} \\ \frac{x}{x+1} &= \frac{11}{6} \\ 6x &= 11(x+1) \\ 6x &= 11x + 11 \\ -5x &= 1 \\ x &= -\frac{11}{5} \end{aligned}$$

1. (a) (First number) + (Second number)

(b) First number = n, Second number = $n + 1$

(c) $n + (n + 1) = 2n + 1$

3. (a) (Rate) × (Time)

(b) Rate = 50 mph, Time = t

(c) $50t$

5. (a) 2(Width) + 2(Length)

(b) Width = x, Length = 2(Width) = $2x$

(c) $2x + 2(2x) = 6x$

7. $5 + x = 8$

9. $\frac{r}{2} = 9$

11. $60{,}000 + 0.08(60{,}000) = S$

13. (a) Model: Sum = (First number) + (Second number)

(b) Labels: Sum = 525, First number = n,
Second number = $n + 1$

(c) Equation: $$\begin{aligned} 525 &= n + (n+1) \\ 524 &= 2n \\ n &= 262 \end{aligned}$$

(d) Answer: First number = n = 262,
Second number = $n + 1$ = 263

15. (a) Model: Difference = (One number) − (Another number)

(b) Labels: Difference = 148, One number = $5x$,
Another number = x

(c) Equation: $$\begin{aligned} 148 &= 5x - x \\ 148 &= 4x \\ x &= 37 \end{aligned}$$

(d) Answer: One number = $5x$ = 185,
Another number = x = 37

17. (a) Model: Raise = (Percent) · (Old wage)

(b) Labels: Old wage = \$440, Raise = \$506 − \$440 = \$66, Percent = r

(c) Equation: $$\begin{aligned} 66 &= 440r \\ \frac{66}{440} &= r \\ 0.15 &= r \end{aligned}$$

(d) Answer: Raise = r = 0.15 or 15%

19. (a) Model: Change in price $=$ (Percent) $\cdot$ (Original price)

(b) Labels: Change in price $= \$30$, Percent $= r$, Original price $= \$119 + \$30 = \$149$

(c) Equation:
$$30 = 149r$$
$$\frac{30}{149} = r$$
$$0.2013 \approx r$$

(d) Answer: Percent change is about 20.13%.

21. (a) Model: (Total of paychecks) $=$ (Coworker's paycheck) $+$ (Your paycheck)

(b) Labels: Total of paychecks $= \$848$, Coworker's paycheck $= x$,
Your paycheck $=$ (Coworker's paycheck) $+ 12\% \times$ (Coworker's paycheck) $= x + 0.12x = 1.12x$

(c) Equation:
$$848 = x + 1.12x$$
$$848 = 2.12x$$
$$400 = x$$

(d) Answer: Coworker's paycheck $= x = \$400$, Your paycheck $= 1.12x = \$448$

23. Model: (Chamber of Secrets) $=$ (Percent change)(Sorcerer's Stone) $+$ (Sorcerer's Stone)

Labels: Sorcerer's Stone $= 317{,}557{,}891$
Percent change $= p$
Chamber of Secrets $= 261{,}987{,}880$

Equation:
$$261{,}987{,}880 = p(317{,}557{,}891) + 317{,}557{,}891$$
$$-55{,}570{,}011 = 317{,}557{,}891p$$
$$-0.175 \approx p$$

Answer: The percent decrease was about 17.5%.

25. Model: (Goblet of Fire) $=$ (Percent change)(Prisoner of Azkaban) $+$ (Prisoner of Azkaban)

Labels: Prisoner of Azkaban $= 249{,}538{,}952$
Percent change $= p$
Goblet of Fire $= 290{,}013{,}036$

Equation:
$$290{,}013{,}036 = p(249{,}538{,}952) + 249{,}538{,}952$$
$$40{,}474{,}084 = 249{,}538{,}952p$$
$$0.162 \approx p$$

Answer: The percent increase was about 16.2%.

27. Model: (Half-Blood Prince) $=$ (Percent change)(Order of the Phoenix) $+$ (Order of the Phoenix)

Labels: Half-Blood Prince $= 301{,}959{,}197$
Percent change $= p$
Order of the Phoenix $= 292{,}004{,}738$

Equation:
$$301{,}959{,}197 = p(292{,}004{,}738) + 292{,}004{,}738$$
$$9{,}954{,}459 = 292{,}004{,}738p$$
$$0.034 \approx p$$

Answer: The percent increase was about 3.4%.

29. Model: (2011 size) = (Percent change)(Past size) + (Past size)

Labels: 2011 size = 16, Percent change = p, Past size = 7

Equation: $16 = p(7) + 7$

$$\frac{9}{7} = p$$

$$p \approx 1.286 = 128.6\%$$

Answer: The percent increase in size was about 128.6%.

31. Model: (2011 size) = (Percent change)(Past size) + (Past size)

Labels: 2011 size = 5.4 oz

Percent change = p

Past size = 3.5 oz

Equation:

$$5.4 = p(3.5) + 3.5$$

$$1.9 = 3.5p$$

$$0.543 \approx p$$

Answer: The percent increase was about 54.3%.

33. Model: (Smaller lunch) = (Percent change)(Larger lunch) + (Larger lunch)

Labels: Smaller lunch = 680

Percent change = p

Larger lunch = 1560

Equation:

$$680 = p(1560) + 1560$$

$$-880 = 1560p$$

$$-0.564 \approx p$$

Answer: The percent decrease is about 56.4%.

35. Model: (Salary) = (Percent increase)(Salary previous year) + (Salary previous year)

Labels: S_2 = Salary second year

S_3 = Salary third year

S_4 = Salary fourth year

(a) $S_2 = 0.08(35{,}000) + 35{,}000 = 37{,}800$

Your salary for the second year is $37,800.

(b) $S_3 = 0.078(37{,}800) + 37{,}800 = 40{,}748.40$

Your salary for the third year is $40,748.40.

(c) $S_4 = 0.094(40{,}748.40) + 40{,}748.40 \approx 44{,}578.75$

Your salary for the fourth year is $44,578.75.

37. (a) Model: Total = Cars + Light trucks + Other trucks

Labels: Cars = 51.5%

Light trucks = x%

Other trucks = 1.9%

Equation: $100 = 51.5 + x + 1.9$

$100 = x + 53.4$

$46.6 = x$

Answer: The percent of new motor vehicles that were light trucks is 46.6%.

(b) Model: Number of vehicles = (Percent)(10,601,000)

Labels: % of cars = 51.5

% of light trucks = 46.6

% of other trucks = 1.9

x = number of type of vehicle

For new cars:

$x = 0.515(10{,}601{,}000) = 5{,}459{,}515$

So, there were 5,459,515 new cars sold or leased in 2009.

For light trucks:

$x = 0.466(10{,}601{,}000) = 4{,}940{,}066$

So, there were 4,940,066 new light trucks sold or leased in 2009.

For other trucks:

$x = 0.019(10{,}601{,}000) = 201{,}419$

So, there were 201,419 new other trucks sold or leased in 2009.

(c) Model: (Number of light trucks) = (Percent)(Total number of trucks)

Labels: Number of light trucks = 4,940,066

Percent = x

Total number of trucks = 4,940,066 + 201,419 = 5,141,485

Equation: $4{,}940{,}066 = x(5{,}141{,}485)$

$$\frac{4{,}940{,}066}{5{,}141{,}485} = x$$

$$0.961 \approx x$$

The percent of new trucks sold or leased in 2009 that were light trucks is 96.1%.

39. Model: Perimeter = 2(Width) + 2(Length)

Labels: Perimeter = 3 feet,

Width (in feet) = 0.62(length) = $0.62x$,

Length (in feet) = x

Equation: $3 = 2(0.62x) + 2x$

$3 = 3.24x$

$$x = \frac{3}{3.24} \approx 0.93$$

Answer: Width = $0.62x \approx 0.57$ feet, Length = $x \approx 0.93$ feet

41. Model: Interest = (Principal)(Rate)(Time)

Labels: Interest = \$1000

Principal = \$2500

Rate = 0.045

Time = x

Equation:
$$1000 = (2500)(0.045)(x)$$
$$1000 = 112.5x$$
$$x = \frac{1000}{112.5} \approx 8.9$$

It will take about 8.9 years.

43. Model: $\text{Average} = \dfrac{(\text{Test }\#1) + (\text{Test }\#2) + (\text{Test }\#3) + (\text{Test }\#4)}{4}$

Labels: Average = 90, Test #1 = 87, Test #2 = 92, Test #3 = 84, Test #4 = x

Equation:
$$90 = \frac{87 + 92 + 84 + x}{4}$$
$$360 = 263 + x$$
$$x = 97$$

Answer: You must score 97 or more on test 4 to have an average of at least 90%.

45. Model: (Sale Price) = (List price) – (Discount percent)(List price)

Labels: x = list price, 1200 = sale price, 15% = discount

Equation:
$$1200 = x - (0.15)x$$
$$1200 = 0.85x$$
$$x = \frac{1200}{0.85} = \$1411.76$$

Answer: The list price was \$1411.76.

47. Model: (Sale price) = (List price) – (Discount percent)(List price)

Labels: Sale price = 21.60, List price = (Wholesale price)(0.60) + Wholesale price, Discount percent = 0.25, Wholesale price = w

Equation:
$$21.60 = w(0.60) + w - 0.25[w(0.60) + w]$$
$$21.60 = 1.60w - 0.25(1.60w)$$
$$21.60 = 1.2w$$
$$\frac{21.60}{1.2} = w$$
$$18 = w$$

Answer: The wholesale price of the power drill is \$18.

49. (a) (Reduced salary) = (Original weekly salary) – (Discount percent)(Original weekly salary)
$$x = 425 - 0.15(425)$$
$$x = \$361.25$$

The reduced salary is \$361.25.

(b) (Original salary) = (Reduced salary) + (Percent raise)(Reduced salary)
$$425 = 361.25 + x(361.25)$$
$$x \approx 0.1765$$

The raise should be about 17.65%. A higher percentage raise is necessary because you are taking a percentage of a smaller number.

51. $(\text{Time}) = \dfrac{(\text{Distance})}{(\text{Rate})}$

$(\text{Rate}) = \dfrac{30 \text{ miles}}{28 \text{ minutes}} = \dfrac{30 \text{ miles}}{7/15 \text{ hour}}$

$t = \dfrac{(150)}{[30/(7/15)]} = 2\frac{1}{3} \text{ hours}$

The entire trip will take about $2\frac{1}{3}$ hours.

53. $\dfrac{\text{Height of tree}}{\text{Tree's shadow}} = \dfrac{\text{Height of lamp post}}{\text{Lamp post's shadow}}$

$\dfrac{x}{25} = \dfrac{5}{2}$

$x = 25\left(\dfrac{5}{2}\right) = 62.5 \text{ feet tall}$

The tree is 62.5 feet tall.

55. $\text{Time} = \dfrac{\text{Distance}}{\text{Rate}}$

$t = \dfrac{3.84 \times 10^8 \text{ meters}}{3.0 \times 10^8 \text{ meters per second}}$

$t = 1.28 \text{ seconds}$

It will take 1.28 seconds for a radio wave to travel from Houston to the surface of the moon.

57. Model: $(\text{Interest from } 6.5\%) + (\text{Interest from } 7.5\%) = (\text{Total interest})$

Labels: Amount invested at 6.5% $= x$, Amount invested at 7.5% $= 15{,}000 - x$,

Interest from 6.5% $= 0.065x$, Interest from 7.5% $= 0.075(15{,}000 - x)$, Total interest $= 1020$

Equation: $0.065x + 0.075(15{,}000 - x) = 1020$

$0.065x + 1125 - 0.075x = 1020$

$-0.01x = -105$

$x = 10{,}500$

Answer: The amount invested at 6.5% is \$10,500 and the amount invested at 7.5% is $15{,}000 - x = 15{,}000 - 10{,}500 = \$4500$.

59. Model: (Amount earned by stock A) + (Amount earned by stock B) = (Total amount earned)

Labels: Amount invested in stock A = x, Amount invested in stock B = $5000 - x$, Amount earned by stock A = $0.098x$,

Amount earned by stock B = $0.062(5000 - x)$, Total amount earned = 389.20

Equation: $0.098x + 0.062(5000 - x) = 389.20$

$0.098x + 310 - 0.062x = 389.20$

$0.036x = 79.20$

$x = 2200$

Answer: The amount invested in stock A is \$2200 and the amount invested in stock B is $5000 - x = 5000 - 2200 = \$2800$.

61. $\text{Cost} = (\text{Fixed costs}) + (\text{Variable cost})(\text{Number of units})$

$\$90{,}000 = \$15{,}000 + \$8.75x$

$x = \dfrac{75{,}000}{8.75} \approx 8571$

The company can manufacture about 8571 units.

63. $A = \dfrac{1}{2}bh$

$2A = bh$

$\dfrac{2A}{b} = h$

65. $P = 2l + 2w$

$P - 2w = 2l$

$\dfrac{P - 2w}{2} = l$

67. $s = L - RL$

$s = L(1 - R)$

$\dfrac{s}{1 - R} = L$

69.
$$A = P\left(1 + \frac{r}{n}\right)^{nt}$$
$$A\left(1 + \frac{r}{n}\right)^{-nt} = P$$

71. Volume $= \pi r^2 h$

Diameter $= 2r = 4 \Rightarrow r = 2$

$$603.2 = \pi(2)^2 h$$
$$h = \frac{603.2}{4\pi} \approx 48 \text{ feet long}$$

73. (Final concentration)(Amount) = (Sol. 1 concentration)(Amount) + (Sol. 2 concentration)(Amount)

$$(75\%)(55\text{gal}) = (40\%)(55 - x) + (100\%)x$$
$$41.25 = 0.60x + 22$$
$$x \approx 32.1 \text{ gal}$$

75. Williams' average $= \dfrac{25{,}000 + 28{,}800 + 21{,}000}{3} = \dfrac{74{,}800}{3} \approx \$24{,}933$

Gonzalez's average $= \dfrac{26{,}200 + 27{,}800 + 29{,}500}{3} = \dfrac{83{,}500}{3} \approx \$27{,}833$

Walters' average $= \dfrac{26{,}600 + 23{,}400 + 26{,}900}{3} = \dfrac{76{,}900}{3} \approx \$25{,}633$

Gilbert's average $= \dfrac{27{,}100 + 22{,}200 + 29{,}000}{3} = \dfrac{78{,}300}{3} = \$26{,}100$

Hart's average $= \dfrac{23{,}100 + 27{,}400 + 22{,}800}{3} = \dfrac{73{,}300}{3} \approx \$24{,}433$

April's average $= \dfrac{25{,}000 + 26{,}200 + 26{,}600 + 27{,}100 + 23{,}100}{5} = \dfrac{128{,}000}{5} = \$25{,}600$

May's average $= \dfrac{28{,}800 + 27{,}800 + 23{,}400 + 22{,}200 + 27{,}400}{5} = \dfrac{129{,}600}{5} = \$25{,}920$

June's average $= \dfrac{21{,}000 + 29{,}500 + 26{,}900 + 29{,}000 + 22{,}800}{5} = \dfrac{129{,}200}{5} = \$25{,}840$

77. The time the diver takes to descend from 25 feet to 150 feet and the depth of 150 feet are red herrings.

Section 1.3 Quadratic Equations

Skills Warm Up

1. $\sqrt{\dfrac{7}{50}} = \dfrac{\sqrt{7}}{\sqrt{50}} \cdot \dfrac{\sqrt{50}}{\sqrt{50}} = \dfrac{\sqrt{7} \cdot \sqrt{50}}{50}$

$= \dfrac{\sqrt{7} \cdot \sqrt{25 \cdot 2}}{50} = \dfrac{5\sqrt{7 \cdot 2}}{50} = \dfrac{\sqrt{14}}{10}$

2. $\sqrt{32} = \sqrt{16 \cdot 2} = 4\sqrt{2}$

3. $\sqrt{7^2 + 3 \cdot 7^2} = \sqrt{49 + 147} = \sqrt{196} = 14$

4. $\sqrt{\dfrac{1}{4} + \dfrac{3}{8}} = \sqrt{\dfrac{5}{8}} = \dfrac{\sqrt{5}}{\sqrt{8}} \cdot \dfrac{\sqrt{8}}{\sqrt{8}} = \dfrac{\sqrt{5} \cdot \sqrt{8}}{8}$

$= \dfrac{\sqrt{5} \cdot \sqrt{4 \cdot 2}}{8} = \dfrac{2\sqrt{5 \cdot 2}}{8} = \dfrac{\sqrt{10}}{4}$

5. $3x^2 + 5x = x(3x + 5)$

6. $4x^2 - 25 = (2x + 5)(2x - 5)$

7. $16 - (x - 11)^2 = [4 + (x - 11)][4 - (x - 11)]$

$= (x - 7)(-x + 15)$

$= -(x - 7)(x - 15)$

8. $x^2 + 7x - 18 = (x + 9)(x - 2)$

9. $10x^2 + 13x - 3 = (5x - 1)(2x + 3)$

10. $6x^2 - 73x + 12 = (6x - 1)(x - 12)$

1. $2x^2 = 3 - 5x$

$2x^2 + 5x - 3 = 0$

3. $x^2 = 19x$

$x^2 - 19x = 0$

5. $(x - 3)^2 = 5$

$(x - 3)(x - 3) = 5$

$x^2 - 6x + 9 = 5$

$x^2 - 6x + 4 = 0$

7. $x(x + 2) = 3x^2 + 1$

$x^2 + 2x = 3x^2 + 1$

$-2x^2 + 2x - 1 = 0$

$2x^2 - 2x + 1 = 0$

9. $\dfrac{3x^2 - 10}{5} = 12x$

$3x^2 - 10 = 60x$

$3x^2 - 60x - 10 = 0$

11. $x^2 - 2x - 8 = 0$

$(x - 4)(x + 2) = 0$

$x - 4 = 0$ or $x + 2 = 0$

$x = 4$ or $x = -2$

13. $6x^2 + 3x = 0$

$3x(2x + 1) = 0$

$3x = 0$ or $2x + 1 = 0$

$x = 0$ or $x = -\frac{1}{2}$

15. $x^2 + 10x + 25 = 0$

$(x + 5)^2 = 0$

$x + 5 = 0$

$x = -5$

17. $3 + 5x - 2x^2 = 0$

$(3 - x)(1 + 2x) = 0$

$3 - x = 0$ or $1 + 2x = 0$

$x = 3$ or $x = -\frac{1}{2}$

19. $x^2 + 4x = 12$

$x^2 + 4x - 12 = 0$

$(x + 6)(x - 2) = 0$

$x + 6 = 0$ or $x - 2 = 0$

$x = -6$ or $x = 2$

21. $-x^2 - 7x = 10$

$0 = x^2 + 7x + 10$

$0 = (x + 5)(x + 2)$

$x + 5 = 0$ or $x + 2 = 0$

$x = -5$ or $x = -2$

23. $x^2 = 16$

$x = \pm\sqrt{16}$

$x = \pm 4$

25. $x^2 = 7$

$x = \pm\sqrt{7}$

$\approx \pm 2.65$

27. $3x^2 = 36$

$x^2 = 12$

$x = \pm 2\sqrt{3}$

$\approx \pm 3.46$

29. $(x - 12)^2 = 144$

$x - 12 = \pm\sqrt{144}$

$x - 12 = \pm 12$

$x = 12 \pm 12$

$x = 24$ or

$x = 0$

31. $(x + 2)^2 = 12$

$x + 2 = \pm 2\sqrt{3}$

$x = -2 \pm 2\sqrt{3}$

$x \approx 1.46$ or

$x \approx -5.46$

33. $12x^2 = 300$

$x^2 = 25$

$x = \pm\sqrt{25}$

$x = \pm 5$

35. $5x^2 = 190$

$x^2 = 38$

$x = \pm\sqrt{38}$

$x \approx \pm 6.16$

37. $3x^2 + 2(x^2 - 4) = 15$

$3x^2 + 2x^2 - 8 = 15$

$5x^2 = 23$

$x^2 = \frac{23}{5}$

$x = \pm\sqrt{\frac{23}{5}}$

$x = \pm\frac{\sqrt{115}}{5}$

$x \approx \pm 2.14$

39. $6x^2 - 3(x^2 + 1) = 23$

$6x^2 - 3x^2 - 3 = 23$

$3x^2 = 26$

$x^2 = \frac{26}{3}$

$x = \pm\sqrt{\frac{26}{3}}$

$x = \pm\frac{\sqrt{78}}{3}$

$x \approx \pm 2.94$

41. $x^2 = 64$

$x = \pm 8$

43. $x^2 - 2x + 1 = 0$

$(x - 1)^2 = 0$

$x - 1 = 0$

$x = 1$

45. $16x^2 - 9 = 0$

$(4x + 3)(4x - 3) = 0$

$4x + 3 = 0$ or $4x - 3 = 0$

$x = -\frac{3}{4}$ or $x = \frac{3}{4}$

47. $4x^2 - 12x + 9 = 0$

$(2x - 3)^2 = 0$

$2x - 3 = 0$

$x = \frac{3}{2}$

49. $(x + 4)^2 = 49$

$x + 4 = \pm 7$

$x + 4 = 7$ or $x + 4 = -7$

$x = 3$ or $x = -11$

51. $4x = 4x^2 - 3$

$4x^2 - 4x - 3 = 0$

$(2x + 1)(2x - 3) = 0$

$2x + 1 = 0$ or $2x - 3 = 0$

$x = -\frac{1}{2}$ or $x = \frac{3}{2}$

53. $50 + 5x = 3x^2$

$3x^2 - 5x - 50 = 0$

$(3x + 10)(x - 5) = 0$

$3x + 10 = 0$ or $x - 5 = 0$

$x = -\frac{10}{3}$ or $x = 5$

55. $12x = x^2 + 27$

$0 = x^2 - 12x + 27$

$0 = (x - 9)(x - 3)$

$x - 9 = 0$ or $x - 3 = 0$

$x = 9$ or $x = 3$

57. $50x^2 - 60x + 10 = 0$

$10(5x^2 - 6x + 1) = 0$

$10(5x - 1)(x - 1) = 0$

$5x - 1 = 0$ or $x - 1 = 0$

$x = \frac{1}{5}$ or $x = 1$

59. $(x + 3)^2 - 4 = 0$

$(x + 3)^2 = 4$

$x + 3 = \pm 2$

$x = -3 \pm 2$

$x = -3 + 2$ or $x = -3 - 2$

$x = -1$ or $x = -5$

61. $(x + 1)^2 = x^2$

$(x + 1)^2 - x^2 = 0$

$x^2 + 2x + 1 - x^2 = 0$

$2x + 1 = 0$

$x = -\frac{1}{2}$

63. Answers will vary. Sample answer:

Algebra argument:

$(x + 2)^2 = (x + 2)(x + 2)$ Definition of an exponent

$= x^2 + 2x + 2x + 4$ FOIL

$= x^2 + 4x + 4$ Simplify.

So, $(x + 2)^2 \neq x^2 + 4$ for any real x.

Graphing utility argument:

Let $y_1 = (x + 2)^2$ and $y_2 = x^2 + 4$. Use the *table* feature with a value of x other than zero.

The table will show y_1 is not the same as y_2.

OR Use a scientific calculator to show that if $x = 5$ then $(5 + 2)^2 = 49$ and $5^2 + 4 = 29$ so $(x + 2)^2$ is not the same as $x^2 + 4$.

65. Area = (Length)(Width)

$1632 = (w + 14)(w)$

$0 = w^2 + 14w - 1632$

$0 = (w + 48)(w - 34)$

$w + 48 = 0$ or $w - 34 = 0$

$w = -48$ or $w = 34$

Extraneous

Solution: Length $= w + 14 = 48$ feet,

Width $= w = 34$ feet

The building has a length of 48 feet and a width of 34 feet.

67. Area $= \frac{1}{2}$ (Base)(Height)

$4 = \frac{1}{2}(b)(b)$

$8 = b^2$

$\pm\sqrt{8} = b$

$b = 2\sqrt{2}\left(-2\sqrt{2} \text{ is extraneous.}\right)$

Solution: Base = Height

$= 2\sqrt{2} \approx 2.83$ feet

The sign has a base length of about 2.83 feet and a height of about 2.83 feet.

69. Area = (Length)(Width)

$1200 = (2x + 30)(2x + 20)$

$1200 = 4x^2 + 100x + 600$

$0 = 4x^2 + 100x - 600$

$0 = 4(x^2 + 25x - 150)$

$0 = 4(x + 30)(x - 5)$

$x + 30 = 0$ or $x - 5 = 0$

$x = -30$ or $x = 5$

Extraneous

Solution: Width = x = 5 feet

The width of the path is 5 feet.

71. $s = -16t^2 + v_0t + s_0$

$v_0 = 0$ and $s_0 = 200$

$s = 0$ when the rock hits the ocean.

$0 = -16t^2 + 200$

$16t^2 = 200$

$t^2 = \frac{25}{2}$

$t = \pm\frac{5}{\sqrt{2}}$

$t = \frac{5}{\sqrt{2}} \approx 3.54 \left(-\frac{5}{\sqrt{2}} \text{ is extraneous.}\right)$

The rock hits the ocean in about 3.54 seconds.

73. $s = -16t^2 + v_0t + s_0$

$v_0 = 0$ and $s_0 = 10$ meters ≈ 32.808 feet

$s = 0$ when the diver hits the water.

$0 = -16t^2 + 32.808$

$16t^2 = 32.808$

$t^2 = 2.0505$

$t = \pm\sqrt{2.0505}\left(-\sqrt{2.0505} \text{ is extraneous.}\right)$

$t \approx 1.43$

The diver will be in the air about 1.43 seconds.

75. $s = -16t^2 + v_0t + s_0$

$s = 3000$, $v_0 = 0$, and $s_0 = 13{,}000$

$3000 = -16t^2 + 13{,}000$

$16t^2 = 10{,}000$

$t^2 = 625$

$t = \pm\sqrt{625} = \pm 25$ (-25 is extraneous.)

$t = 25$

The ball reaches a height of 3000 feet $67 - 25 = 42$ seconds faster than the skydiver.

77. $a^2 + b^2 = c^2$

$x^2 + x^2 = 6^2$

$2x^2 = 36$

$x^2 = 18$

$x \approx 4.24$

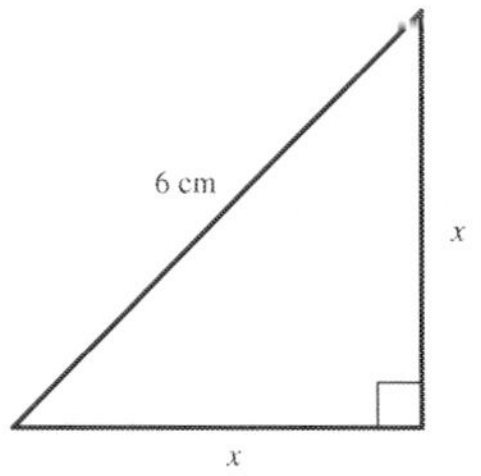

The sides of the isosceles right triangle are about 4.24 centimeters in length.

79. $a^2 + b^2 = c^2$

$a^2 + 583^2 = 703^2$

$a^2 + 339{,}889 = 494{,}209$

$a^2 = 154{,}320$

$a \approx 393$

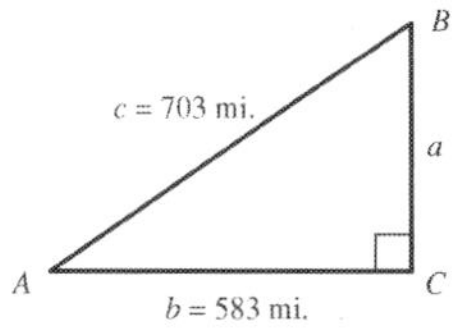

The flying distance from Atlanta to Buffalo by way of Chicago = $a + b = 976$ miles.

81. Since the angle is 45°, the triangle is isosceles.

$x^2 + x^2 = 700^2$

$2x^2 = 490{,}000$

$x^2 = 245{,}000$

$x \approx 494.97$

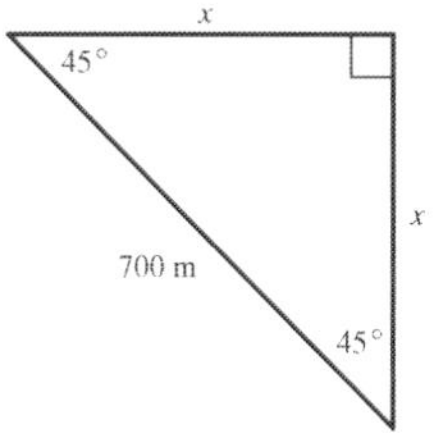

The whale shark is about 494.97 meters deep.

83. $R = xp = x(36 - 0.0003x)$

$1{,}080{,}000 = 36x - 0.0003x^2$

$0.0003x^2 - 36x + 1{,}080{,}000 = 0$

$x^2 - 120{,}000x + 3{,}600{,}000{,}000 = 0$

$(x - 60{,}000)^2 = 0$

$x = 60{,}000$

To produce a revenue of \$1,080,000, a total of 60,000 units must be sold.

85. $35.65t^2 + 7205 = 10{,}000$

$35.65t^2 = 2795$

$t^2 = \dfrac{2795}{35.65}$

$t = \pm\sqrt{\dfrac{2795}{35.65}} \approx \pm 8.85$

In the context of the exercise, only $t \approx 8.85$ makes sense. So, the average monthly cost will be about \$10,000 in 2019.

87. (a) $P = 694.59t^2 + 6179, \ 0 \le t \le 9$

($t = 0$ corresponds to 1800, $t = 1$ corresponds to 1810, etc.)

$250{,}000 = 694.59t^2 + 6179$

$243{,}821 = 694.59t^2$

$351.03 \approx t^2$

$18.74 \approx t$

The resident population would have reached 250,000,000 in 1987.

$\left(t = 18.74 \text{ represents } 1800 + 18.74(10) = 1987.4.\right)$

(b) Because the lengths of the bars that represent the actual data are close to the lengths of the bars that represent the data values given by the model, the model is a good representation of the resident population through 1890.

(c) For 2010, $t = 21$.

$P = 694.59(21)^2 + 6179 = 312{,}493.19$

The model predicts a population of 312,493,190 people, which is close to the actual 2010 population of 310,000,000 people.

(d) For 2050, $t = 25$.

$P = 694.59(25)^2 + 6179 = 440{,}297.75$

The model supports this prediction because 440,297,750 is close to the U.S. Census Bureau's estimate of 439,010,253.

89. $25{,}000 = 135.47t^2 + 13{,}702$

$11{,}298 = 135.47t^2$

$\dfrac{11{,}298}{135.47} = t^2$

$t = \pm\sqrt{\dfrac{11{,}298}{135.47}} \approx \pm 9.13$

Choose the positive value. The average monthly cost will be about \$25,000 in 2019.

91. $T = 0.31t^2 + 32.9, \ 7 \le t \le 15$

($t = 7$ corresponds to 7 A.M.)

$85 = 0.31t^2 + 32.9$

$52.1 = 0.31t^2$

$168.06 \approx t^2$

$12.96 \approx t$

The temperature was 85°F at about 1 P.M.

When $t = 19$ (7 P.M.):

$T = 0.31(19)^2 + 32.9 = 144.81°\text{F}$

Because the temperature is extremely high, the answer is unreasonable.

So, the model should not be used to predict the temperature at 7 P.M.

Section 1.4 The Quadratic Formula

Skills Warm Up

1. $\sqrt{25-4(3)(-12)} = \sqrt{25-(-144)} = \sqrt{169} = 13$

2. $\sqrt{49-4(2)(3)} = \sqrt{49-24} = \sqrt{25} = 5$

3. $\sqrt{12^2-4(3)(4)} = \sqrt{144-48} = \sqrt{96} = 4\sqrt{6}$

4. $\sqrt{15^2+4(9)(12)} = \sqrt{225+432} = \sqrt{657} = 3\sqrt{73}$

5. $x^2 - x - 2 = 0$
$(x-2)(x+1) = 0$
$x - 2 = 0$ or $x + 1 = 0$
$x = 2$ or $x = -1$

6. $2x^2 + 3x - 9 = 0$
$(2x-3)(x+3) = 0$
$2x - 3 = 0$ or $x + 3 = 0$
$x = \frac{3}{2}$ or $x = -3$

7. $x^2 - 4x = 5$
$x^2 - 4x - 5 = 0$
$(x-5)(x+1) = 0$
$x - 5 = 0$ or $x + 1 = 0$
$x = 5$ or $x = -1$

8. $2x^2 + 13x = 7$
$2x^2 + 13x - 7 = 0$
$(2x-1)(x+7) = 0$
$2x - 1 = 0$ or $x + 7 = 0$
$x = \frac{1}{2}$ or $x = -7$

9. $x(x-7) = -10$
$x^2 - 7x + 10 = 0$
$(x-5)(x-2) = 0$
$x - 5 = 0$ or $x - 2 = 0$
$x = 5$ or $x = 2$

10. $x(x-3) = 4$
$x^2 - 3x = 4$
$x^2 - 3x - 4 = 0$
$(x+1)(x-4) = 0$
$x + 1 = 0$ or $x - 4 = 0$
$x = -1$ or $x = 4$

1. $x^2 - 4x + 4 = 0$
$b^2 - 4ac = (-4)^2 - 4(1)(4) = 0$
One real solution

3. $3x^2 + 4x + 1 = 0$
$b^2 - 4ac = (4)^2 - 4(3)(1) = 4 > 0$
Two real solutions

5. $2x^2 - 5x = -5$
$2x^2 - 5x + 5 = 0$
$b^2 - 4ac = (-5)^2 - 4(2)(5) = -15 < 0$
No real solutions

7. $\frac{1}{5}x^2 + \frac{6}{5}x - 8 = 0$
$b^2 - 4ac = \left(\frac{6}{5}\right)^2 - 4\left(\frac{1}{5}\right)(-8) = \frac{196}{25} > 0$
Two real solutions

9. $2x^2 + x - 1 = 0$

$$x = \frac{-b \pm \sqrt{b^2-4ac}}{2a} = \frac{-1 \pm \sqrt{1^2 - 4(2)(-1)}}{2(2)} = \frac{-1 \pm 3}{4} = \frac{1}{2}, -1$$

11. $16x^2 + 8x - 3 = 0$

$$x = \frac{-b \pm \sqrt{b^2-4ac}}{2a} = \frac{-8 \pm \sqrt{8^2 - 4(16)(-3)}}{2(16)} = \frac{-8 \pm 16}{32} = \frac{1}{4}, -\frac{3}{4}$$

13. $6x = 4 - x^2$

$x^2 + 6x - 4 = 0$

$$x = \frac{-b \pm \sqrt{b^2 - 4ac}}{2a} = \frac{-6 \pm \sqrt{6^2 - 4(1)(-4)}}{2(1)} = \frac{-6 \pm 2\sqrt{13}}{2} = -3 \pm \sqrt{13}$$

15. $x^2 - 10x + 22 = 0$

$$x = \frac{-b \pm \sqrt{b^2 - 4ac}}{2a} = \frac{-(-10) \pm \sqrt{(-10)^2 - 4(1)(22)}}{2(1)} = \frac{10 \pm 2\sqrt{3}}{2} = 5 \pm \sqrt{3}$$

17. $x^2 + 8x - 4 = 0$

$$x = \frac{-b \pm \sqrt{b^2 - 4ac}}{2a} = \frac{-8 \pm \sqrt{8^2 - 4(1)(-4)}}{2(1)} = \frac{-8 \pm 4\sqrt{5}}{2} = -4 \pm 2\sqrt{5}$$

19. $12x - 9x^2 = -3$

$-9x^2 + 12x + 3 = 0$

$$x = \frac{-b \pm \sqrt{b^2 - 4ac}}{2a} = \frac{-12 \pm \sqrt{12^2 - 4(-9)(3)}}{2(-9)} = \frac{-12 \pm 6\sqrt{7}}{-18} = \frac{2}{3} \pm \frac{\sqrt{7}}{3}$$

21. $36x^2 + 24x = 7$

$36x^2 + 24x - 7 = 0$

$$x = \frac{-b \pm \sqrt{b^2 - 4ac}}{2a} = \frac{-24 \pm \sqrt{24^2 - 4(36)(-7)}}{2(36)} = \frac{-24 \pm 12\sqrt{11}}{72} = -\frac{1}{3} \pm \frac{\sqrt{11}}{6}$$

23. $4x^2 + 4x = 7$

$4x^2 + 4x - 7 = 0$

$$x = \frac{-b \pm \sqrt{b^2 - 4ac}}{2a} = \frac{-4 \pm \sqrt{4^2 - 4(4)(-7)}}{2(4)} = \frac{-4 \pm 8\sqrt{2}}{8} = -\frac{1}{2} \pm \sqrt{2}$$

25. $28x - 49x^2 = 4$

$-49x^2 + 28x - 4 = 0$

$$x = \frac{-b \pm \sqrt{b^2 - 4ac}}{2a} = \frac{-28 \pm \sqrt{28^2 - 4(-49)(-4)}}{2(-49)} = \frac{-28 \pm 0}{-98} = \frac{2}{7}$$

27. $8t = 5 + 2t^2$

$-2t^2 + 8t - 5 = 0$

$$t = \frac{-b \pm \sqrt{b^2 - 4ac}}{2a} = \frac{-8 \pm \sqrt{8^2 - 4(-2)(-5)}}{2(-2)} = \frac{-8 \pm 2\sqrt{6}}{-4} = 2 \pm \frac{\sqrt{6}}{2}$$

29. $(y - 5)^2 = 2y$

$y^2 - 12y + 25 = 0$

$$y = \frac{-b \pm \sqrt{b^2 - 4ac}}{2a} = \frac{12 \pm \sqrt{(-12)^2 - 4(1)(25)}}{2(1)} = \frac{12 \pm 2\sqrt{11}}{2} = 6 \pm \sqrt{11}$$

31. $5.1x^2 - 1.7x - 3.2 = 0$

$$x = \frac{1.7 \pm \sqrt{(-1.7)^2 - 4(5.1)(-3.2)}}{2(5.1)}$$

$x \approx 0.976, -0.643$

33. $12.67x^2 + 31.55x + 8.09 = 0$

$$x = \frac{-31.55 \pm \sqrt{31.55^2 - 4(12.67)(8.09)}}{2(12.67)}$$

$x \approx -0.290, -2.200$

35. $-0.003x^2 + 0.025x - 0.98 = 0$

$$x = \frac{-0.025 \pm \sqrt{0.025^2 - 4(-0.003)(-0.98)}}{2(-0.003)}$$

No real solution

37. $2x^2 + 7 = 2x^2 - x - 4$

$7 = -x - 4$

$x = -11$

39. $4x^2 - 15 = 25$

$4x^2 = 40$

$x^2 = 10$

$x = \pm\sqrt{10}$

41. $x^2 - 2x - 1 = 0$

$$x = \frac{-(-2) \pm \sqrt{(-2)^2 - 4(1)(-1)}}{2(1)}$$

$$= \frac{2 \pm \sqrt{4 + 4}}{2}$$

$$= \frac{2 \pm 2\sqrt{2}}{2}$$

$$= 1 \pm \sqrt{2}$$

43. $(x - 1)^2 = 9$

$x - 1 = \pm 3$

$x = 1 \pm 3$

$x = 1 + 3 = 4$

OR

$x = 1 - 3 = -2$

45. $3x^2 + 5x - 11 = 4(x - 2)$

$3x^2 + 5x - 11 = 4x - 8$

$3x^2 + x - 3 = 0$

$$x = \frac{-1 \pm \sqrt{1^2 - 4(3)(-3)}}{2(3)}$$

$$= \frac{-1 \pm \sqrt{37}}{6} = -\frac{1}{6} \pm \frac{\sqrt{37}}{6}$$

47. Let x = one integer, $100 - x$ = the other integer.

$x(100 - x) = 2500$

$0 = x^2 - 100x + 2500$

$0 = (x - 50)^2$

$x = 50$

$100 - x = 50$

Real-life problems will vary.

49. Let x = an integer, $x + 1$ = next integer.

$x^2 + (x + 1)^2 = 113$

$2x^2 + 2x - 112 = 0$

$x^2 + x - 56 = 0$

$(x - 7)(x + 8) = 0$

$x = 7$ OR $x = -8$

$x + 1 = 8$ $\quad x = -7$

Real-life problems will vary.

51. $C = 14{,}000 = 0.125x^2 + 20x + 5000$

$0.125x^2 + 20x - 9000 = 0$

$$x = \frac{-20 \pm \sqrt{20^2 - 4(0.125)(-9000)}}{2(0.125)}$$

$$= \frac{-20 \pm 70}{0.25}$$

Choosing the positive value of x, we have $x = 200$ units.

53. $C = 1680 = 800 + 0.04x + 0.002x^2$

$0.002x^2 + 0.04x - 880 = 0$

$$x = \frac{-0.04 \pm \sqrt{(0.04)^2 - 4(0.002)(-880)}}{2(0.002)}$$

$$= \frac{-0.04 \pm \sqrt{7.0416}}{0.004}$$

Choosing the positive value for x, we have

$$x = \frac{-0.04 + \sqrt{7.0416}}{0.004} \approx 653 \text{ units.}$$

55. Number of rows $= x$

Number of seats/row $= \dfrac{108}{x}$

$\left(\dfrac{108}{x} + 6\right)(x - 3) = 108$

$(108 + 6x)(x - 3) = 108x$

$6x^2 - 18x - 324 = 0$

$6(x^2 - 3x - 54) = 0$

$x^2 - 3x - 54 = 0$

$(x - 9)(x + 6) = 0$

$x - 9 = 0$ or $x + 6 = 0$

$x = 9$ $\quad x = -6$

There are now $x - 3 = 9 - 3 = 6$ rows. The number of chairs per row is $\dfrac{108}{6} = 18$.

57. Volume $= 2x^2 = 200$

$$x^2 = 100$$

$$x = 10$$

The original piece of material was $x + 4 = 14$ inches by 14 inches.

59. $s = -16t^2 + v_0t + s_0$

$v_0 = -5$ and $s_0 = 866$

$s = 0$ when the coin hits the ground.

$$s = -16t^2 - 5t + 866 = 0$$

$$t = \frac{-b \pm \sqrt{b^2 - 4ac}}{2a}$$

$$= \frac{-(-5) \pm \sqrt{(-5)^2 - 4(-16)(866)}}{2(-16)}$$

$$= \frac{5 \pm \sqrt{55{,}449}}{-32}$$

Choose the positive value.

$t \approx 7.2$

The coin hits the ground in about 7.2 seconds.

61. Moon: $s = -2.7t^2 + v_0t + s_0$

Let $v_0 = 40$, $s_0 = 5$ and set $s = 0$.

$$0 = -2.7t^2 + 40t + 5$$

$$t = \frac{-40 \pm \sqrt{40^2 - 4(-2.7)(5)}}{2(-2.7)}$$

$$t = \frac{-40 - \sqrt{1654}}{-5.4}$$

$t \approx 14.9$ seconds

Earth: $s = -16t^2 + v_0t + s_0$

Let $v_0 = 40$, $s_0 = 5$ and set $s = 0$.

$$0 = -16t^2 + 40t + 5$$

$$t = \frac{-40 \pm \sqrt{40^2 - 4(-16)(5)}}{2(-16)}$$

$$t = \frac{-40 - \sqrt{1920}}{-32}$$

$t \approx 2.6$ seconds

On the moon, it would take about 14.9 seconds to hit the surface while on Earth it would take only about 2.6 seconds.

63. Moon: $s = -2.7t^2 + v_0t + s_0$

Let $v_0 = 27$, $s_0 = 6$ and set $s = 0$.

$$0 = -2.7t^2 + 27t + 6$$

$$t = \frac{-27 \pm \sqrt{27^2 - 4(-2.7)(6)}}{2(-2.7)}$$

$$t = \frac{-27 - \sqrt{793.8}}{-5.4}$$

$t \approx 10.2$ seconds

Earth: $s = -16t^2 + v_0t + s_0$

Let $v_0 = 27$, $s_0 = 6$ and set $s = 0$.

$$0 = -16t^2 + 27t + 6$$

$$t = \frac{-27 \pm \sqrt{27^2 - 4(-16)(6)}}{2(-16)}$$

$$t = \frac{-27 - \sqrt{1113}}{-32}$$

$t \approx 1.9$ seconds

The rock will take longer to reach the ground on the moon.

65. The distance from Oklahoma City to Austin and then to New Orleans is $1348 - 560 = 788$ miles.

Let x = distance from Austin to New Orleans then $788 - x$ = distance from Oklahoma City to Austin.

By the Pythagorean Theorem:

$$x^2 + (788 - x)^2 = 560^2$$

$$x^2 + x^2 - 1576x + 620{,}944 = 313{,}600$$

$$2x^2 - 1576x + 307{,}344 = 0$$

$$2(x^2 - 788x + 153{,}672) = 0$$

$$x = \frac{-(-788) \pm \sqrt{(-788)^2 - 4(1)(153{,}672)}}{2(1)}$$

$$x = \frac{788 \pm \sqrt{6256}}{2}$$

$x \approx 433.55$ or $x \approx 354.45$

The distance from Oklahoma City to Austin is about 354.45 mi and the distance from Austin to New Orleans is about 433.55 mi.

67. $L = -0.270t^2 + 3.59t + 83.1$, $2 \le t \le 7$

($t = 2$ corresponds to 2:00 P.M.)

$$93 = -0.270t^2 + 3.59t + 83.1$$

$$0 = -0.270t^2 + 3.59t - 9.9$$

$$t = \frac{-3.59 \pm \sqrt{3.59^2 - 4(-0.270)(-9.9)}}{2(-0.270)}$$

$t \approx 4$ or $t \approx 9$

Because $t = 9$ is not in the domain, choose $t \approx 4$. The patient's blood oxygen level was 93% at about 4:00 P.M.

69. $d_E = (3 \text{ hours})(r + 50 \text{ mph})$

$d_S = (3 \text{ hours})(r \text{ mph})$

$$d_E^{\,2} + d_S^{\,2} = 2440^2$$

$$9(r + 50)^2 + 9r^2 = 2440^2$$

$$18r^2 + 900r - 5{,}931{,}100 = 0$$

$$r = \frac{-900 \pm \sqrt{900^2 - 4(18)(-5{,}931{,}100)}}{2(18)}$$

$$= \frac{-900 \pm 60\sqrt{118{,}847}}{36}$$

Thus, the eastbound plane is moving at $r + 50 \approx 600$ mph and the southbound at $r \approx 550$ mph.

71. (a) From the graph, there were about 60% of homes with VCRs in 2011 $(t \approx 11)$.

(b) From the graph, there were about 80% of homes with VCRs in 2008 $(t \approx 8)$.

(c) $-0.496t^2 + 3.01t + 87.2 = 0$

$$t = \frac{-3.01 \pm \sqrt{(3.01)^2 - 4(-0.496)(87.2)}}{2(-0.496)}$$

$t \approx 16.636$ or $t \approx -10.568$

Choose the positive value. The last year that VCRs will exist in U.S. households is 2016. The answer seems reasonable based on the trends in the data.

73. $C = 0.45x^2 - 1.65x + 50.75,\ 10 \le x \le 25$

(a) $150 = 0.45x^2 - 1.65x + 50.75$

$0 = 0.45x^2 - 1.65x - 99.25$

$$x = \frac{1.65 \pm \sqrt{(-1.65)^2 - 4(0.45)(-99.25)}}{2(0.45)}$$

≈ 16.80

The air temperature is about 16.80°C.

(b) When $x = 10$:

$$C = 0.45(10)^2 - 1.65(10) + 50.75 = 79.25$$

When $x = 20$: $C = 0.45(20)^2 - 1.65(20) + 50.75$

$= 197.75$

$$= \frac{197.75}{79.25} \approx 2.5$$

Oxygen consumption increased by a factor of 2.5.

75.

$$\text{Revenue} = xp = x(50 - 0.0005x)$$

$$250{,}000 = 50x - 0.0005x^2$$

$$0.0005x^2 - 50x + 250{,}000 = 0$$

$$x = \frac{-(-50) \pm \sqrt{(-50)^2 - 4(0.0005)(250{,}000)}}{2(0.0005)}$$

$$x = \frac{50 + \sqrt{2000}}{0.001}$$

$x \approx 94{,}721$ units

or

$$x = \frac{50 - \sqrt{2000}}{0.001}$$

$x \approx 5279$ units

To produce a revenue of \$250,000, approximately 5279 or approximately 94,721 units must be sold.

77. No, the two quadratic equations cannot have the same solutions. A discriminant of 5 means there are two real solutions. A discriminant of -10 means there are no real solutions.

79. Answers will vary.

Chapter 1 Quiz Yourself

1. $$\begin{aligned} 3(x-2)-4(2x+5) &= 4 \\ 3x-6-8x-20 &= 4 \\ -5x &= 30 \\ x &= -6 \end{aligned}$$

2. $$\begin{aligned} \frac{3x+3}{5x-2} &= \frac{3}{4} \\ 4(3x+3) &= 3(5x-2) \\ 12x+12 &= 15x-6 \\ -3x &= -18 \\ x &= 6 \end{aligned}$$

3. $$\begin{aligned} \frac{2}{x(x-1)}+\frac{1}{x} &= \frac{1}{x-4} \\ x(x-1)(x-4)\left[\frac{2}{x(x-1)}+\frac{1}{x}\right] &= x(x-1)(x-4)\left(\frac{1}{x-4}\right) \\ 2(x-4)+(x-1)(x-4) &= x(x-1) \\ 2x-8+x^2-5x+4 &= x^2-x \\ x^2-3x-4 &= x^2-x \\ -2x &= 4 \\ x &= -2 \end{aligned}$$

4. $$\begin{aligned} (x+3)^2-x^2 &= 6(x+2) \\ x^2+6x+9-x^2 &= 6x+12 \\ 6x+9 &= 6x+12 \\ 9 &\neq 12 \end{aligned}$$

Contradiction: No solution

5. One method would be to use the scientific calculator portion of a graphing utility to check for a true statement. Another method would be to use a table feature.

6. $$\begin{aligned} \frac{x}{2.004}-\frac{x}{5.128} &= 100 \\ (2.004)(5.128)\left[\frac{x}{2.004}-\frac{x}{5.128}\right] &= (2.004)(5.128)(100) \\ 5.128x-2.004x &= 1027.6512 \\ 3.124x &= 1027.6512 \\ x &= \frac{1027.6512}{3.124} \\ x &\approx 328.954 \end{aligned}$$

7. $$\begin{aligned} 0.378x+0.757(500-x) &= 215 \\ 0.378x+378.5-0.757x &= 215 \\ -0.379x &= -163.5 \\ x &= \frac{163.5}{0.379} \\ x &\approx 431.398 \end{aligned}$$

8. (Total cost) = (Fixed costs) + (Unit cost)(Number of units)

$$\begin{aligned} 200{,}000 &= 30{,}000+8.50x \\ 170{,}000 &= 8.50x \\ 20{,}000 &= x \end{aligned}$$

The company can manufacture 20,000 units.

9.

$$\begin{aligned} R &= xp \\ 300{,}000 &= x(75 - 0.0002x) \\ 300{,}000 &= 75x - 0.0002x^2 \\ 0.0002x^2 - 75x + 300{,}000 &= 0 \end{aligned}$$

$$x = \frac{75 \pm \sqrt{(-75)^2 - 4(0.0002)(300{,}000)}}{2(0.0002)} = \frac{75 \pm \sqrt{5385}}{0.0004}$$

$$x \approx 4044 \text{ or } x \approx 370{,}956$$

To produce a revenue of \$300,000 about 4044 units or about 370,956 units must be sold.

10.

$$\begin{aligned} 3x^2 + 13x &= 10 \\ 3x^2 + 13x - 10 &= 0 \\ (3x - 2)(x + 5) &= 0 \end{aligned}$$

$$3x - 2 = 0 \quad \text{or} \quad x + 5 = 0$$

$$x = \tfrac{2}{3} \quad \text{or} \quad x = -5$$

11. $3x^2 = 15$

$x^2 = 5$

$x = \pm\sqrt{5} \approx \pm 2.24$

12. $(x + 3)^2 = 17$

$x + 3 = \pm\sqrt{17}$

$x = -3 \pm \sqrt{17}$

$x \approx 1.12$ or $x \approx -7.12$

13.

$$\begin{aligned} 2x + x^2 &= 6 \\ x^2 + 2x - 6 &= 0 \end{aligned}$$

$$x = \frac{-2 \pm \sqrt{2^2 - 4(1)(-6)}}{2(1)} = \frac{-2 \pm \sqrt{4 + 24}}{2} = \frac{-2 \pm \sqrt{28}}{2} = \frac{-2 \pm 2\sqrt{7}}{2} = -1 \pm \sqrt{7} \approx -3.65, 1.65$$

14. $3x^2 + 7x - 2 = 0$

$$x = \frac{-7 \pm \sqrt{(7)^2 - 4(3)(-2)}}{2(3)} = \frac{-7 \pm \sqrt{73}}{6} \approx 0.26, -2.59$$

15. $3x^2 - 4.50x - 0.32 = 0$

$$x = \frac{-(-4.50) \pm \sqrt{(-4.50)^2 - 4(3)(-0.32)}}{2(3)} = \frac{4.50 \pm \sqrt{24.09}}{6}$$

$x \approx 1.568$ or $x \approx -0.068$

16. $2x^2 - 4x + 9 = 0$

$b^2 - 4ac = (-4)^2 - 4(2)(9) = -56 < 0$

No real solutions

17. $9x^2 - 3x + \frac{1}{4} = 0$

$b^2 - 4ac = (-3)^2 - 4(9)\left(\frac{1}{4}\right) = 0$

There is one real solution.

18. Answers will vary.

One method would be to use the definition of an exponent, multiply and simplify:

$(x + 3)^2 = (x + 3)(x + 3) = x^2 + 3x + 3x + 9 = x^2 + 6x + 9$

Another method would be to use the table feature of a graphing utility and let

$y_1 = (x + 3)^2$ and $y_2 = x^2 + 6x + 9$.

As real values of x are selected y_1 will always equal y_2.

A third method would be to use the scientific calculator portion of a graphing utility and demonstrate that $(x + 3)^2 = x^2 + 6x + 9$ with real values of x.

19. $s = -16t^2 + v_0 t + s_0$

$v_0 = 0$ and $s_0 = 324$

$s = 0$ when the rock hits the ground.

$$s = -16t^2 + 324 = 0$$
$$-16t^2 + 324 = 0$$
$$-16t^2 = -324$$
$$t^2 = \frac{81}{4}$$
$$t = \pm\sqrt{\frac{81}{4}} = \pm\frac{9}{2}$$

Choose the positive value. The rock will hit the ground in $\frac{9}{2} = 4.5$ seconds.

20.

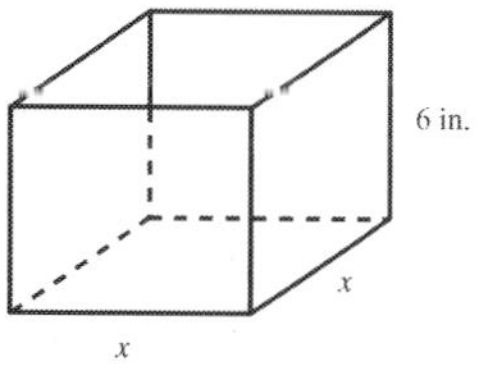

$$\text{Volume} = \text{Length} \times \text{Width} \times \text{Height}$$
$$486 = x(x)6$$
$$486 = 6x^2$$
$$x^2 = 81$$
$$x = 9$$

The box is 9 inches × 9 inches × 6 inches.

Section 1.5 Other Types of Equations

Skills Warm Up

1. $x^2 - 22x + 121 = 0$

$$(x - 11)(x - 11) = 0$$
$$(x - 11)^2 = 0$$
$$x - 11 = 0$$
$$x = 11$$

2. $x(x - 20) + 3(x - 20) = 0$

$$(x - 20)(x + 3) = 0$$
$$x - 20 = 0 \quad \text{or} \quad x + 3 = 0$$
$$x = 20 \quad \text{or} \quad x = -3$$

3. $(x + 20)^2 = 625$

$$x + 20 = \pm 25$$
$$x = -20 + 25 \quad \text{or} \quad x = -20 - 25$$
$$x = 5 \quad \text{or} \quad x = -45$$

4. $5x^2 + x = 0$

$$x(5x + 1) = 0$$
$$x = 0 \quad \text{or} \quad 5x + 1 = 0$$
$$x = -\frac{1}{5}$$

5. $3x^2 + 4x - 4 = 0$

$$(3x - 2)(x + 2) = 0$$
$$3x - 2 = 0 \quad \text{or} \quad x + 2 = 0$$
$$x = \frac{2}{3} \quad \text{or} \quad x = -2$$

6. $12x^2 + 8x - 55 = 0$

$$(6x - 11)(2x + 5) = 0$$
$$6x - 11 = 0 \quad \text{or} \quad 2x + 5 = 0$$
$$x = \frac{11}{6} \quad \text{or} \quad x = -\frac{5}{2}$$

7. $x^2 + 4x - 5 = 0$

$$(x + 5)(x - 1) = 0$$
$$x + 5 = 0 \quad \text{or} \quad x - 1 = 0$$
$$x = -5 \quad \text{or} \quad x = 1$$

8. $4x^2 + 4x - 15 = 0$

$$(2x + 5)(2x - 3) = 0$$
$$2x + 5 = 0 \quad \text{or} \quad 2x - 3 = 0$$
$$x = -\frac{5}{2} \quad \text{or} \quad x = \frac{3}{2}$$

9. $x^2 - 3x + 1 = 0$

$$x = \frac{-b \pm \sqrt{b^2 - 4ac}}{2a}$$
$$x = \frac{3 \pm \sqrt{(-3)^2 - 4(1)(1)}}{2(1)} = \frac{3 \pm \sqrt{5}}{2} = \frac{3}{2} \pm \frac{\sqrt{5}}{2}$$

10. $x^2 - 4x + 2 = 0$

$$x = \frac{-b \pm \sqrt{b^2 - 4ac}}{2a}$$
$$x = \frac{4 \pm \sqrt{(-4)^2 - 4(1)(2)}}{2(1)}$$
$$= \frac{4 \pm \sqrt{8}}{2} = \frac{4 \pm 2\sqrt{2}}{2} = 2 \pm \sqrt{2}$$

1. $x^3 - 2x^2 - 3x = 0$

$x(x^2 - 2x - 3) = 0$

$x(x - 3)(x + 1) = 0$

$x = 0$

$x - 3 = 0 \Rightarrow x = 3$

$x + 1 = 0 \Rightarrow x = -1$

3. $4x^4 - 36x^2 = 0$

$4x^2(x^2 - 9) = 0$

$4x^2(x - 3)(x + 3) = 0$

$4x^2 = 0 \Rightarrow x = 0$

$x - 3 = 0 \Rightarrow x = 3$

$x + 3 = 0 \Rightarrow x = -3$

5. $x^4 - 81 = 0$

$(x^2 + 9)(x + 3)(x - 3) = 0$

$x^2 + 9 = 0 \Rightarrow$ No real solution

$x + 3 = 0 \Rightarrow x = -3$

$x - 3 = 0 \Rightarrow x = 3$

7. $5x^3 + 30x^2 + 45x = 0$

$5x(x^2 + 6x + 9) = 0$

$5x(x + 3)^2 = 0$

$5x = 0 \Rightarrow x = 0$

$x + 3 = 0 \Rightarrow x = -3$

9. $x^3 - 7x^2 - 4x + 28 = 0$

$x^2(x - 7) - 4(x - 7) = 0$

$(x^2 - 4)(x - 7) = 0$

$(x - 2)(x + 2)(x - 7) = 0$

$x - 2 = 0 \Rightarrow x = 2$

$x + 2 = 0 \Rightarrow x = -2$

$x - 7 = 0 \Rightarrow x = 7$

11. $x^4 - x^3 + x - 1 = 0$

$x^3(x - 1) + (x - 1) = 0$

$(x - 1)(x^3 + 1) = 0$

$(x - 1)(x + 1)(x^2 - x + 1) = 0$

$x - 1 = 0 \Rightarrow x = 1$

$x + 1 = 0 \Rightarrow x = -1$

$x^2 - x + 1 = 0 \Rightarrow$ No real solution

13. $x^4 - 12x^2 + 11 = 0$

$(x^2 - 11)(x^2 - 1) = 0$

$x^2 - 11 = 0 \Rightarrow x = \pm\sqrt{11} \approx \pm 3.317$

$x^2 - 1 = 0 \Rightarrow x = \pm 1$

15. $x^4 + 5x^2 - 36 = 0$

$(x^2 + 9)(x^2 - 4) = 0$

$(x^2 + 9)(x + 2)(x - 2) = 0$

$x + 2 = 0 \Rightarrow x = -2$

$x - 2 = 0 \Rightarrow x = 2$

$x^2 + 9 = 0 \Rightarrow$ No real solution

17. $4x^4 - 65x^2 + 16 = 0$

$(4x^2 - 1)(x^2 - 16) = 0$

$(2x + 1)(2x - 1)(x + 4)(x - 4) = 0$

$2x + 1 = 0 \Rightarrow x = -\frac{1}{2}$

$2x - 1 = 0 \Rightarrow x = \frac{1}{2}$

$x + 4 = 0 \Rightarrow x = -4$

$x - 4 = 0 \Rightarrow x = 4$

19. $x^6 + 7x^3 - 8 = 0$

$(x^3 + 8)(x^3 - 1) = 0$

$(x + 2)(x^2 - 2x + 4)(x - 1)(x^2 + x + 1) = 0$

$x + 2 = 0 \Rightarrow x = -2$

$x^2 - 2x + 4 = 0 \Rightarrow$ No real solution

$x - 1 = 0 \Rightarrow x = 1$

$x^2 + x + 1 = 0 \Rightarrow$ No real solution

21. $\sqrt{2x} - 10 = 0$

$\sqrt{2x} = 10$

$2x = 100$

$x = 50$

23. $\sqrt{x - 10} - 4 = 0$

$\sqrt{x - 10} = 4$

$x - 10 = 16$

$x = 26$

25. $\sqrt[3]{2x + 5} + 3 = 0$

$\sqrt[3]{2x + 5} = -3$

$2x + 5 = -27$

$2x = -32$

$x = -16$

27. $2x + 9\sqrt{x} - 5 = 0$

$$9\sqrt{x} = 5 - 2x$$
$$81x = 25 - 20x + 4x^2$$
$$0 = 25 - 101x + 4x^2$$
$$0 = (25 - x)(1 - 4x)$$
$$1 - 4x = 0 \Rightarrow x = \tfrac{1}{4}$$

$x = 25$ is not a solution.

29. $x = \sqrt{11x - 30}$

$$x^2 = 11x - 30$$
$$x^2 - 11x + 30 = 0$$
$$(x - 5)(x - 6) = 0$$
$$x - 5 = 0 \Rightarrow x = 5$$
$$x - 6 = 0 \Rightarrow x = 6$$

31. $-\sqrt{26 - 11x} + 4 = x$

$$4 - x = \sqrt{26 - 11x}$$
$$16 - 8x + x^2 = 26 - 11x$$
$$x^2 + 3x - 10 = 0$$
$$(x + 5)(x - 2) = 0$$
$$x + 5 = 0 \Rightarrow x = -5$$
$$x - 2 = 0 \Rightarrow x = 2$$

33. $\sqrt{x + 1} - 3x = 1$

$$\sqrt{x + 1} = 3x + 1$$
$$x + 1 = 9x^2 + 6x + 1$$
$$0 = 9x^2 + 5x$$
$$0 = x(9x + 5)$$
$$x = 0$$

$x = -\frac{5}{9}$ is not a solution.

35. $(x - 5)^{2/3} = 16$

$$x - 5 = \pm 16^{3/2}$$
$$x - 5 = \pm 64$$
$$x = 69, -59$$

37. $(x + 3)^{3/2} = 8$

$$x + 3 = 8^{2/3}$$
$$x + 3 = 4$$
$$x = 1$$

39. $(x^2 - 5)^{2/3} = 16$

$$x^2 - 5 = \pm 16^{3/2}$$
$$x^2 - 5 = \pm 64$$
$$x^2 = 69 \Rightarrow x = \pm\sqrt{69}$$
$$x^2 = -59 \Rightarrow \text{No real solution}$$

41. $\frac{1}{x} = \frac{4}{x - 1} + 1$

$$x - 1 = 4x + x(x - 1),\ x \neq 0, 1$$
$$x - 1 = x^2 + 3x$$
$$x^2 + 2x + 1 = 0$$
$$(x + 1)^2 = 0$$
$$x + 1 = 0 \Rightarrow x = -1$$

43. $\frac{4}{x} - \frac{5}{3} = \frac{x}{6}$

$$24 - 10x = x^2,\ x \neq 0$$
$$-x^2 - 10x + 24 = 0$$
$$x = \frac{-b \pm \sqrt{b^2 - 4ac}}{2a}$$
$$= \frac{-(-10) \pm \sqrt{(-10)^2 - 4(-1)(24)}}{2(-1)}$$
$$= \frac{10 \pm 14}{-2} = -12 \text{ or } 2$$

45. $\frac{4}{x + 1} - \frac{3}{x + 2} = 1$

$$4(x + 2) - 3(x + 1) = (x + 1)(x + 2),\ x \neq -2, -1$$
$$4x + 8 - 3x - 3 = x^2 + 3x + 2$$
$$x^2 + 2x - 3 = 0$$
$$(x - 1)(x + 3) = 0$$
$$x - 1 = 0 \Rightarrow x = 1$$
$$x + 3 = 0 \Rightarrow x = -3$$

47. $\frac{x + 2}{x - 2} - \frac{x}{3} = 0$

$$3(x + 2) - x(x - 2) = 0,\ x \neq 2$$
$$3x + 6 - x^2 + 2x = 0$$
$$-x^2 + 5x + 6 = 0$$
$$x^2 - 5x - 6 = 0$$
$$(x - 6)(x + 1) = 0$$
$$x - 6 = 0 \Rightarrow x = 6$$
$$x + 1 = 0 \Rightarrow x = -1$$

49. $|x + 1| = 2$

$$x + 1 = 2 \Rightarrow x = 1$$
$$-(x + 1) = 2 \Rightarrow x = -3$$

51. $|2x - 1| = 5$

$$2x - 1 = 5 \Rightarrow x = 3$$
$$-(2x - 1) = 5 \Rightarrow x = -2$$

53. $|x| = x^2 + x - 3$

$x = x^2 + x - 3$

$x^2 - 3 = 0$

$x = \pm\sqrt{3}$

OR

$-x = x^2 + x - 3$

$x^2 + 2x - 3 = 0$

$(x - 1)(x + 3) = 0$

$x - 1 = 0 \Rightarrow x = 1$

$x + 3 = 0 \Rightarrow x = -3$

Only $x = \sqrt{3}$ and $x = -3$ are solutions.

55. $|x - 10| = x^2 - 10x$

$x - 10 = x^2 - 10x$

$x^2 - 11x + 10 = 0$

$(x - 1)(x - 10) = 0$

$x - 1 = 0 \Rightarrow x = 1$

$x - 10 = 0 \Rightarrow x = 10$

OR

$-(x - 10) = x^2 - 10x$

$-x + 10 = x^2 - 10x$

$x^2 - 9x - 10 = 0$

$(x + 1)(x - 10) = 0$

$x + 1 = 0 \Rightarrow x = -1$

$x - 10 = 0 \Rightarrow x = 10$

Only $x = -1$ and $x = 10$ are solutions.

63. Number of students $= x$

Cost per student $= f$

$$fx = 1700 \Rightarrow f = \frac{1700}{x}$$

$$(f - 7.50)(x + 6) = 1700$$

$$\left(\frac{1700}{x} - 7.50\right)(x + 6) = 1700$$

$$(3400 - 15x)(x + 6) = 3400x$$

$$-15x^2 - 90x + 20{,}400 = 0$$

$$x = \frac{90 \pm \sqrt{(-90)^2 - 4(-15)(20{,}400)}}{2(-15)} = \frac{90 \pm 1110}{-30}$$

The original number was $x = 34$.

57. The quadratic equation was not written in standard form. As a result, the substitutions in the quadratic formula are incorrect. The solution should be:

$3x^2 - 7x - 4 = 0$

$$x = \frac{-(-7) \pm \sqrt{(-7)^2 - 4(3)(-4)}}{2(3)}$$

$x = \dfrac{7 + \sqrt{97}}{6}$ is the only solution.

$x = \dfrac{7 - \sqrt{97}}{6}$ is not a solution.

59. $3.2x^4 - 1.5x^2 - 2.1 = 0$

$$x^2 = \frac{1.5 \pm \sqrt{(-1.5)^2 - 4(3.2)(-2.1)}}{2(3.2)}$$

$$x = \pm\sqrt{\frac{1.5 + \sqrt{29.13}}{6.4}} \approx \pm 1.038$$

61. $1.8x - 6\sqrt{x} - 5.6 = 0$

$$\sqrt{x} = \frac{6 \pm \sqrt{(-6)^2 - 4(1.8)(-5.6)}}{2(1.8)}$$

$$x = \left[\frac{6 + \sqrt{76.32}}{3.6}\right]^2 \approx 16.756$$

65.
$$A = P\left(1 + \frac{r}{n}\right)^{nt}$$
$$3422.91 = 3000\left(1 + \frac{r}{12}\right)^{12(6)}$$
$$1.14097 = \left(1 + \frac{r}{12}\right)^{72}$$
$$(1.14097)^{1/72} = 1 + \frac{r}{12}$$
$$1.001833328 \approx 1 + \frac{r}{12}$$
$$0.001833328 \approx \frac{r}{12}$$
$$0.022 \approx r$$

The interest rate is about 2.2%.

67.
$$A = P\left(1 + \frac{r}{n}\right)^{nt}$$
$$320 = 300\left(1 + \frac{r}{12}\right)^{12(1/2)}$$
$$\frac{16}{15} = \left(1 + \frac{r}{12}\right)^{6}$$
$$\left(\frac{16}{15}\right)^{1/6} = 1 + \frac{r}{12}$$
$$\left[\left(\frac{16}{15}\right)^{1/6} - 1\right](12) = r$$
$$r \approx 0.1298 = 12.98\%$$

You are paying an annual interest rate of about 12.98%.

69.
$$C = \sqrt{0.25x + 1}$$
$$3.5 = \sqrt{0.25x + 1}$$
$$12.25 = 0.25x + 1$$
$$11.25 = 0.25x$$
$$45 = x$$
$$45(1000) = 45{,}000$$

There were 45,000 passengers that flew in the month.

71.
$$\sqrt{0.828x^2 - 135.71x + 5672.1} = 25$$
$$0.828x^2 - 135.71x + 5672.1 = 625$$
$$0.828x^2 - 135.71x + 5047.1 = 0$$
$$x = \frac{-(-135.71) \pm \sqrt{(-135.71)^2 - 4(0.828)(5047.1)}}{2(0.828)}$$
$$x \approx 106.857,\ 57.043$$

Only $x \approx 57.043$ lies in the domain of the function, so the person is about 57 years old.

73.
$$p = 40 - \sqrt{0.01x + 1},\ 0 \le x \le 159{,}900$$
$$13.95 = 40 - \sqrt{0.01x + 1}$$
$$\sqrt{0.01x + 1} = 26.05$$
$$0.01x + 1 = 678.6025$$
$$0.01x = 677.6025$$
$$x = 67{,}760.25 \approx 67{,}760 \text{ units}$$

When the price is set at \$13.95, the demand for the product is 67,760 units.

This model is only valid for $0 \le x \le 159{,}900$ because it does not make sense in the context of the problem to have a demand less than 0 or a price less than 0.

75.
$$|x - 16| = 0.4$$
$$x - 16 = 0.4 \Rightarrow x = 16.4$$
$$-(x - 16) = 0.4 \Rightarrow x = 15.6$$

The least acceptable weight is 15.6 ounces and the greatest acceptable weight is 16.4 ounces.

77.
$$\frac{1}{8} + \frac{1}{t} = \frac{1}{5}$$
$$40t \cdot \frac{1}{8} + 40t \cdot \frac{1}{t} = 40t \cdot \frac{1}{5}$$
$$5t + 40 = 8t$$
$$-3t = -40$$
$$t = \frac{40}{3} = 13\frac{1}{3}$$

It will take $13\frac{1}{3}$ minutes (13 minutes and 20 seconds) for the tub of the washing machine to fill with only hot water.

79. A polynomial of degree n has at most n real solutions.

Section 1.6 Linear Inequalities

Skills Warm Up

1.

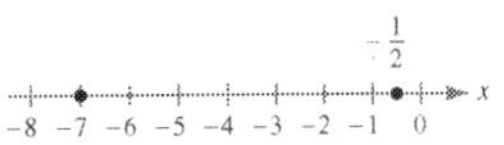

Because −7 is to the left of $-\frac{1}{2}$, $-\frac{1}{2}$ is greater than −7.

2.

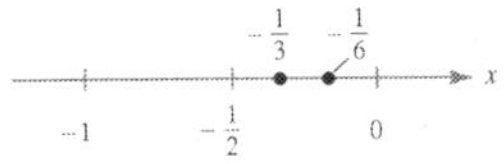

Because $-\frac{1}{3}$ is to the left of $-\frac{1}{6}$, $-\frac{1}{6}$ is greater than $-\frac{1}{3}$.

3.

Because $-\pi(\pi \approx 3.14)$ is to the left of −3, −3 is greater than $-\pi$.

4.

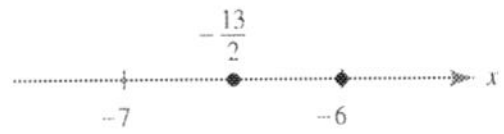

Because $-\frac{13}{2}$ is to the left of − 6, − 6 is greater than $-\frac{13}{2}$.

5. The statement "x is nonnegative" can be represented by $x \geq 0$.

6. The statement "z is strictly between − 3 and 10" can be represented by $-3 < z < 10$.

7. The statement "P is no more than 2" can be represented by $P \leq 2$.

8. The statement "W is at least 200" can be represented by $W \geq 200$.

9. $|x - 10|$

When $x = 12$: $|12 - 10| = |2| = 2$

When $x = 3$: $|3 - 10| = |-7| = 7$

10. $|2x - 3|$

When $x = \frac{3}{2}$: $|2(\frac{3}{2}) - 3| = |3 - 3| = |0| = 0$

When $x = 1$: $|2(1) - 3| = |2 - 3| = |-1| = 1$

---

1. $[-1, 5]$ corresponds to $-1 \leq x \leq 5$.

The interval is bounded.

3. $(11, \infty)$ corresponds to $x > 11$.

The interval is unbounded.

5. $(-\infty, -2)$ corresponds to $x < -2$.

The interval is unbounded.

7. $x < 4$ indicates all points to the left of $x = 4$; graph (c).

9. $-2 < x \leq 5$ indicates all points from, but not including $x = -2$ up to and including $x = 5$; graph (f).

11. $|x| < 4$ indicates all points less than 4 units from $x = 0$ in both directions; graph (g).

13. $|x - 5| > 2$ indicates all points more than 2 units from $x = 5$; graph (b).

15. $5x - 12 > 0$

(a) $5(3) - 12 = 3 > 0 \Rightarrow x = 3$ is a solution of the inequality.

(b) $5(-3) - 12 = -27 < 0 \Rightarrow x = -3$ is not a solution of the inequality.

(c) $5(\frac{5}{2}) - 12 = \frac{1}{2} > 0 \Rightarrow x = \frac{5}{2}$ is a solution of the inequality.

(d) $5(\frac{3}{2}) - 12 = -\frac{9}{2} < 0 \Rightarrow x = \frac{3}{2}$ is not a solution of the inequality.

17. $0 < \dfrac{x-2}{4} < 2$

(a) $0 < \dfrac{4-2}{4} < 2 \Rightarrow 0 < \dfrac{1}{2} < 2 \Rightarrow x = 4$ is a solution of the inequality.

(b) $0 < \dfrac{10-2}{4} < 2 \Rightarrow 0 < 2 \not< 2 \Rightarrow x = 10$ is not a solution of the inequality.

(c) $0 < \dfrac{0-2}{4} < 2 \Rightarrow 0 \not< -\dfrac{1}{2} < 2 \Rightarrow x = 0$ is not a solution of the inequality.

(d) $0 < \dfrac{7/2-2}{4} < 2 \Rightarrow 0 < \dfrac{3}{8} < 2 \Rightarrow x = \dfrac{7}{2}$ is a solution of the inequality.

19. $|x + 2| \le 10$

(a) $|-15 + 2| = |-13| = 13 \Rightarrow 13 \not\le 10 \Rightarrow x = -15$ is not a solution of the inequality.

(b) $|-4 + 2| = |-2| = 2 \Rightarrow 2 \le 10 \Rightarrow x = -4$ is a solution of the inequality.

(c) $|1 + 2| = |3| = 3 \Rightarrow 3 \le 10 \Rightarrow x = 1$ is a solution of the inequality.

(d) $|8 + 2| = |10| = 10 \Rightarrow 10 \le 10 \Rightarrow x = 10$ is a solution of the inequality.

21. $|2x - 3| < 15$

(a) $|2(-6) - 3| \stackrel{?}{<} 15$

$15 \not< 15 \Rightarrow x = -6$ is not a solution of the inequality.

(b) $|2(0) - 3| \stackrel{?}{<} 15$

$3 < 15 \Rightarrow x = 0$ is a solution of the inequality.

(c) $|2(12) - 3| \stackrel{?}{<} 15$

$21 \not< 15 \Rightarrow x = 12$ is not a solution of the inequality.

(d) $|2(7) - 3| \stackrel{?}{<} 15$

$11 < 15 \Rightarrow x = 7$ is a solution of the inequality.

23. If $2x > 6$, then $x > 3$.

25. If $2x \le -8$, then $x \le -4$.

27. $2 - 4x > -10$

$-4x > -12$

$x < 3$

If $2 - 4x > -10$, then $x < 3$.

29. If $-\frac{2}{3}x \ge -6$, then $x \le 9$.

31. $\frac{3}{2}x \ge 9$

$\frac{2}{3}\left(\frac{3}{2}x\right) \ge \frac{2}{3}(9)$

$x \ge 6$

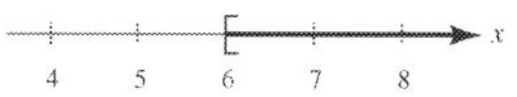

33. $-10x < 40$

$-\frac{1}{10}(-10x) > -\frac{1}{10}(40)$

$x > -4$

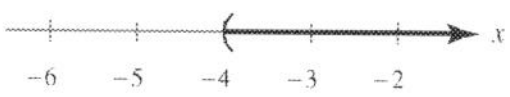

35. $\frac{3}{5}x - 7 < 8$

$\frac{3}{5}x < 15$

$x < 25$

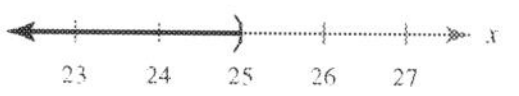

37. $2x + 7 < 3 + 4x$

$-2x + 7 < 3$

$-2x < -4$

$x > 2$

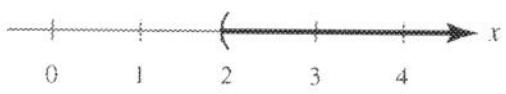

39. $2x - 1 \geq 5x + 1$

$-3x - 1 \geq 1$

$-3x \geq 2$

$x \leq -\frac{2}{3}$

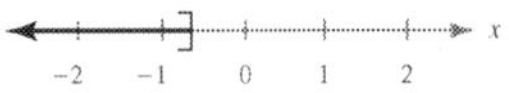

41. $3(x + 2) + 7 < 2x - 5$

$3x + 6 + 7 < 2x - 5$

$3x + 13 < 2x - 5$

$x < -18$

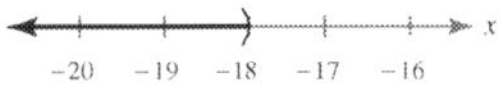

43. $-3(x - 1) + 7 < 2x + 8$

$-3x + 3 + 7 < 2x + 8$

$-3x + 10 < 2x + 8$

$-5x < -2$

$x > \frac{2}{5}$

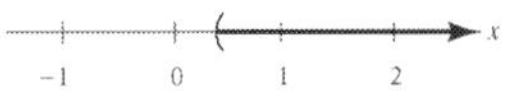

45. $3 \leq 2x - 1 < 7$

$4 \leq 2x < 8$

$2 \leq x < 4$

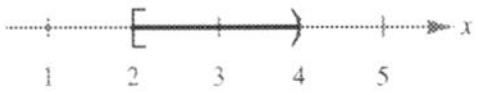

47. $-11 \leq 1 + 4(x + 1) < 13$

$-11 \leq 1 + 4x + 4 < 13$

$-11 \leq 4x + 5 < 13$

$-16 \leq 4x < 8$

$-4 \leq x < 2$

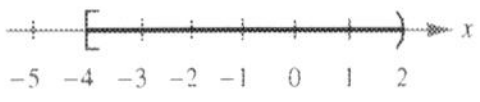

49. $-4 < \dfrac{x - 3}{3} < 4$

$-12 < x - 3 < 12$

$-9 < x < 15$

51. $\frac{3}{4} > x + 1 > \frac{1}{4}$

$-\frac{1}{4} > x > -\frac{3}{4}$

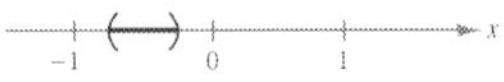

53. $|x| < 6$

$-6 < x < 6$

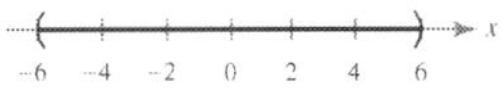

55. $\left|\dfrac{x}{2}\right| > 3$

$\dfrac{x}{2} < -3$ or $\dfrac{x}{2} > 3$

$x < -6$ $\quad$ $x > 6$

57. $|x + 3| < 5$

$-5 < x + 3 < 5$

$-8 < x < 2$

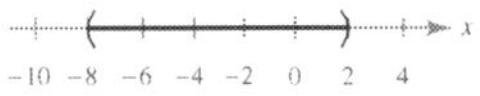

59. $|x - 5| < 0$

No solution

61. $|2x - 5| > 6$

$2x - 5 < -6$ or $2x - 5 > 6$

$2x < -1$ $\quad$ $2x > 11$

$x < -\frac{1}{2}$ $\quad$ $x > \frac{11}{2}$

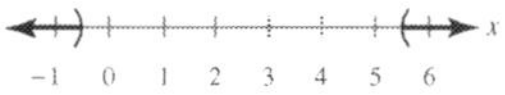

63. $2|x + 10| \geq 9$

$|x + 10| \geq \frac{9}{2}$

$x + 10 \leq -\frac{9}{2}$ or $x + 10 \geq \frac{9}{2}$

$x \leq -\frac{29}{2}$ $\quad$ $x \geq -\frac{11}{2}$

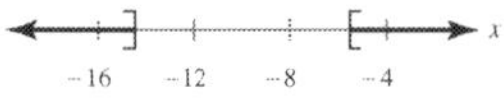

65. $\left|\dfrac{x - 3}{2}\right| \geq 5$

$\dfrac{x - 3}{2} \leq -5$ or $\dfrac{x - 3}{2} \geq 5$

$x - 3 \leq -10$ $\quad$ $x - 3 \geq 10$

$x \leq -7$ $\quad$ $x \geq 13$

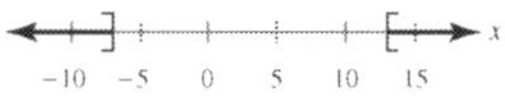

67. $|9 - x| - 2 < -1$

$|9 - x| < 1$

$-1 < 9 - x < 1$

$-10 < -x < -8$

$10 > x > 8$

or

$8 < x < 10$

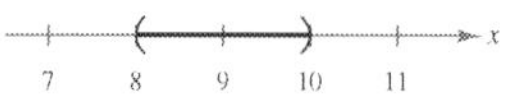

69. $2|2x - 5| - 4 > 6$

$2|2x - 5| > 10$

$|2x - 5| > 5$

$2x - 5 < -5$ or $2x - 5 > 5$

$2x < 0$ $\quad$ $2x > 10$

$x < 0$ $\quad$ $x > 5$

71. All real numbers no more than 2 units from 0 yields $|x - 0| \leq 2$ or $|x| \leq 2$.

73. All real numbers no less than 3 units from 9 yields $|x - 9| \geq 3$.

75. $|x - 12| \le 10$

77. $|x - (-3)| > 5$

$|x + 3| > 5$

79. $0.52m + 2500 < 12{,}000$

$0.52m < 9500$

$m < 18{,}269.23$

The annual operating cost will be less than \$12,000 if the miles are less than 18,269.23 miles.

81. $14.516t + 24.65 > 140$

$14.516t > 115.35$

$t > \dfrac{115.35}{14.516}$

$t > 7.9$

The average price of brand name prescription drugs exceeded \$140 in 2008.

83. Rental fee for Company B > Rental fee for Company A

$199 + 0.32x > 279$

$0.32x > 80$

$x > 250$

You must drive more than 250 miles in a week for the rental fee for Company B to be greater than that for Company A.

85. (a) $0.1215t - 0.351 < 1.5$

$0.1215t < 1.851$

$t < \dfrac{1.851}{0.1215}$

$t < 15.23$

According to the model the average National Football Leagues player's salary was less than \$1.5 million from 1998 to 2005. Yes, the results agree with the bar graph which shows that the average salary was less than \$1.5 million from 1998 to 2005.

(b) $0.1215t - 0.351 > 2.2$

$0.1215t > 2.551$

$t > \dfrac{2.551}{0.1215}$

$t > 20.996$

The average salary will exceed \$2.2 million in 2011. This result seems reasonable based on the increasing salaries shown in the bar graph.

87. $180 - 1.5x \ge 130$

$50 \ge 1.5x$

$33\tfrac{1}{3} \ge x$

The person must be in the program $33\tfrac{1}{3}$ or less weeks.

89. $4x + 72 \le 132$

$4x \le 60$

$x \le 15$

The square base must be less than or equal to 15 inches on a side.

91. $R > C$

$25.95x > 13.95x + 125{,}000$

$12x > 125{,}000$

$x \ge 10{,}417$

The product will return a profit for $x \ge 10{,}417$ units.

93. $I = Prt,\ I = 1680 - 1500 = 180$

$1500(r)(3) > 180$

$4500r > 180$

$r > 0.04$

The interest rate must be greater than 4%.

95. $|x - 10.4| \le 0.0625$

$-0.0625 \le x - 10.4 \le 0.0625$

$10.3375 \le x \le 10.4625$

Area $= x^2$

Interval: $[106.86 \text{ in.}^2, 109.46 \text{ in.}^2]$

97. $|h - 68.5| \le 2.7$

$-2.7 \le h - 68.5 \le 2.7$

$65.8 \le h \le 71.2$

Interval: $[65.8, 71.2]$

99. $|h - 50| \le 30$

$-30 \le h - 50 \le 30$

$20 \le h \le 80$

The minimum relative humidity is 20% and the maximum relative humidity is 80%.

101. $\dfrac{1}{2}\text{ oz} \cdot \dfrac{1 \text{ lb}}{16 \text{ oz}} = \dfrac{1}{32}\text{ lb}$

$|x - 5.72| \le \dfrac{1}{32}$

$-\dfrac{1}{32} \le x - 5.72 \le \dfrac{1}{32}$

$5.68875 \le x \le 5.75125$

You could have been undercharged

$5.72(8.99) - 5.68875(8.99) = \$0.28$

or you could have been overcharged

$5.75125(8.99) - 5.72(8.99) = \$0.28.$

Section 1.7 Other Types of Inequalities

Skills Warm Up

1. $-\frac{y}{3} > 2$

$y < -6$

2. $-6z < 27$

$z > -\frac{9}{2}$

3. $-3 \le 2x + 3 < 5$

$-6 \le 2x < 2$

$-3 \le x < 1$

4. $-3x + 5 \ge 20$

$-3x \ge 15$

$x \le -5$

5. $10 > 4 - 3(x + 1)$

$6 > -3(x + 1)$

$-2 < x + 1$

$-3 < x$ or $x > -3$

6. $3 < 1 + 2(x - 4) < 7$

$2 < 2(x - 4) < 6$

$1 < x - 4 < 3$

$5 < x < 7$

7. $2|x| \le 7$

$|x| \le \frac{7}{2}$

$-\frac{7}{2} \le x \le \frac{7}{2}$

8. $|x - 3| > 1$

$x - 3 < -1$ or $x - 3 > 1$

$x < 2$ or $x > 4$

9. $|x + 4| > 2$

$x + 4 < -2$ or $x + 4 > 2$

$x < -6$ or $x > -2$

10. $|2 - x| \le 4$

$-4 \le 2 - x \le 4$

$-6 \le -x \le 2$

$6 \ge x \ge -2$

$-2 \le x \le 6$

---

1. $x^2 - 25 < 0$

$x^2 - 25 = 0$

$x^2 = 25$

$x = \pm 5$

Critical numbers: $x = \pm 5$

Test intervals: $(-\infty, -5), (-5, 5), (5, \infty)$

3. $2x^2 + 7x + 16 \ge 20$

$2x^2 + 7x + 16 = 20$

$2x^2 + 7x - 4 = 0$

$(2x - 1)(x + 4) = 0$

$2x - 1 = 0 \Rightarrow x = \frac{1}{2}$

$x + 4 = 0 \Rightarrow x = -4$

Critical numbers: $x = -4, x = \frac{1}{2}$

Test intervals: $(-\infty, -4), \left(-4, \frac{1}{2}\right), \left(\frac{1}{2}, \infty\right)$

5. $\frac{x - 3}{x - 1} < 2$

$\frac{x - 3}{x - 1} = 2, x \ne 1$

$x - 3 = 2(x - 1)$

$x - 3 = 2x - 2$

$-1 = x$

Critical numbers: $x = -1, x = 1$

Test intervals: $(-\infty, -1), (-1, 1), (1, \infty)$

7. $x^2 \le 9$

$x^2 - 9 \le 0$

$(x + 3)(x - 3) \le 0$

Critical numbers: $x = \pm 3$

Test intervals: $(-\infty, -3), (-3, 3), (3, \infty)$

Test: Is $(x + 3)(x - 3) \le 0$?

Solution set: $[-3, 3]$

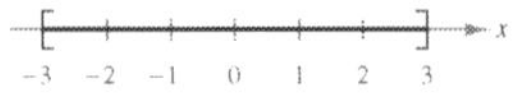

9. $x^2 > 4$

$x^2 - 4 > 0$

$(x + 2)(x - 2) > 0$

Critical numbers: $x = \pm 2$

Test intervals: $(-\infty, -2), (-2, 2), (2, \infty)$

Test: Is $x^2 - 4 > 0$?

Solution set: $(-\infty, -2) \cup (2, \infty)$

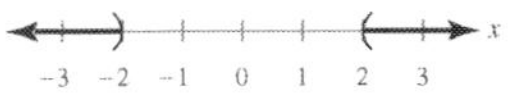

11. $(x + 2)^2 < 25$

$x^2 + 4x + 4 < 25$

$x^2 + 4x - 21 < 0$

$(x + 7)(x - 3) < 0$

Critical numbers: $x = -7, x = 3$

Test intervals: $(-\infty, -7), (-7, 3), (3, \infty)$

Test: Is $(x + 7)(x - 3) < 0$?

Solution set: $(-7, 3)$

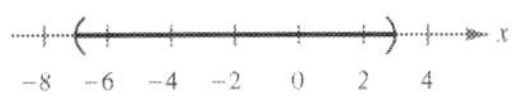

13. $x^2 + 4x + 4 \ge 9$

$x^2 + 4x - 5 \ge 0$

$(x + 5)(x - 1) \ge 0$

Critical numbers: $x = -5, x = 1$

Test intervals: $(-\infty, -5), (-5, 1), (1, \infty)$

Test: Is $(x + 5)(x - 1) \ge 0$?

Solution set: $(-\infty, -5] \cup [1, \infty)$

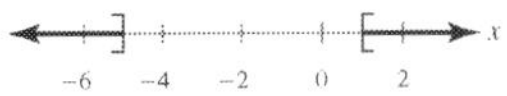

15. $x^2 + x < 6$

$x^2 + x - 6 < 0$

$(x + 3)(x - 2) < 0$

Critical numbers: $x = -3, x = 2$

Test intervals: $(-\infty, -3), (-3, 2), (2, \infty)$

Test: Is $(x + 3)(x - 2) < 0$?

Solution set: $(-3, 2)$

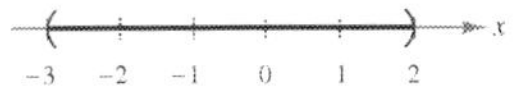

17. $3(x - 1)(x + 1) > 0$

Critical numbers: $x = \pm 1$

Test intervals: $(-\infty, -1), (-1, 1), (1, \infty)$

Test: Is $3(x - 1)(x + 1) > 0$?

Solution set: $(-\infty, -1) \cup (1, \infty)$

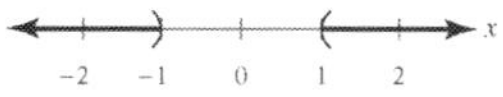

19. $x^2 + 2x - 3 < 0$

$(x + 3)(x - 1) < 0$

Critical numbers: $x = -3, x = 1$

Test intervals: $(-\infty, -3), (-3, 1), (1, \infty)$

Test: Is $x^2 + 2x - 3 < 0$?

Solution set: $(-3, 1)$

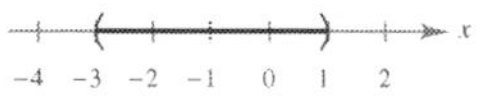

21. $x^2 - 8x + 16 > 0$

$(x - 4)(x - 4) > 0$

Critical number: $x = 4$

Test intervals: $(-\infty, 4), (4, \infty)$

Test: Is $(x - 4)(x - 4) > 0$?

Solution set: $(-\infty, 4) \cup (4, \infty)$

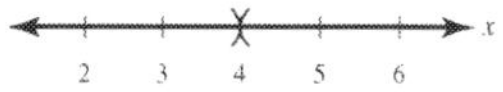

23. $4x^2 - 4x + 1 \le 0$

$(2x - 1)(2x - 1) \le 0$

Critical numbers: $x = \frac{1}{2}$

Test intervals: $\left(-\infty, \frac{1}{2}\right)\left(\frac{1}{2}, \infty\right)$

Test: Is $(2x - 1)(2x - 1) \le 0$?

Solution set: $x = \frac{1}{2}$

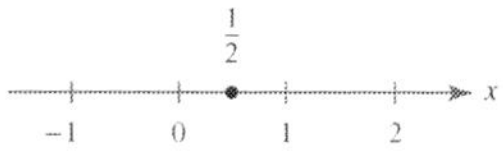

25. $x^2 + 4 > 0$

Critical numbers: none

Test interval: $(-\infty, \infty)$

Test: Is $x^2 + 4 > 0$?

Solution set: $(-\infty, \infty)$

27. $x^2 < -1$

$x^2 + 1 < 0$

Critical numbers: none

Test interval: $(-\infty, \infty)$

Test: Is $x^2 + 1 < 0$?

Solution set: No solution

29. $4x^3 - 6x^2 < 0$

$2x^2(2x - 3) < 0$

Critical numbers: $x = 0, x = \frac{3}{2}$

Test intervals: $(-\infty, 0), \left(0, \frac{3}{2}\right), \left(\frac{3}{2}, \infty\right)$

Test: Is $2x^2(2x - 3) < 0$?

Solution set: $(-\infty, 0) \cup \left(0, \frac{3}{2}\right)$

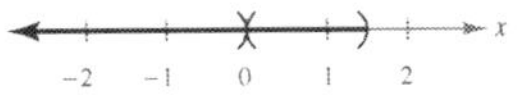

31. $x^3 - 4x \geq 0$

$x(x + 2)(x - 2) \geq 0$

Critical numbers: $x = 0, x = \pm 2$

Test intervals: $(-\infty, -2), (-2, 0), (0, 2), (2, \infty)$

Test: Is $x(x + 2)(x - 2) \geq 0$?

Solution set: $[-2, 0] \cup [2, \infty)$

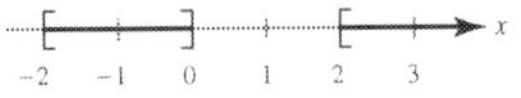

33. $x^3 - 2x^2 - x + 2 \geq 0$

$(x - 1)(x + 1)(x - 2) \geq 0$

Critical numbers: $x = 2, \pm 1$

Test intervals: $(-\infty, -1), (-1, 1), (1, 2), (2, \infty)$

Test: Is $(x - 1)(x + 1)(x - 2) \geq 0$?

Solution: $[-1, 1], [2, \infty)$

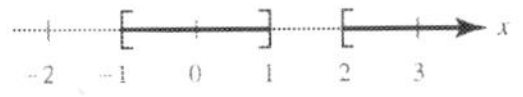

35. $\frac{1}{x} > x$

$\frac{1}{x} - x > 0$

$\frac{1 - x^2}{x} > 0$

Critical numbers: $x = 0, x = \pm 1$

Test intervals: $(-\infty, -1), (-1, 0), (0, 1), (1, \infty)$

Test: Is $\frac{1 - x^2}{x} > 0$?

Solution set: $(-\infty, -1) \cup (0, 1)$

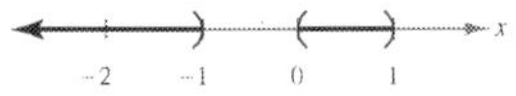

37. $\frac{x + 6}{x + 1} < 2$

$\frac{x + 6}{x + 1} - 2 < 0$

$\frac{x + 6 - 2(x + 1)}{x + 1} < 0$

$\frac{4 - x}{x + 1} < 0$

Critical numbers: $x = -1, x = 4$

Test intervals: $(-\infty, -1), (-1, 4), (4, \infty)$

Test: Is $\frac{4 - x}{x + 1} < 0$?

Solution set: $(-\infty, -1) \cup (4, \infty)$

39. $\frac{3x - 5}{x - 5} > 4$

$\frac{3x - 5}{x - 5} - 4 > 0$

$\frac{3x - 5 - 4(x - 5)}{x - 5} > 0$

$\frac{15 - x}{x - 5} > 0$

Critical numbers: $x = 5, x = 15$

Test intervals: $(-\infty, 5), (5, 15), (15, \infty)$

Test: Is $\frac{15 - x}{x - 5} > 0$?

Solution set: $(5, 15)$

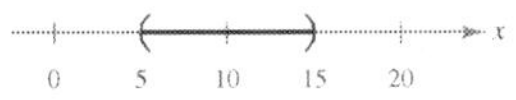

41. $$\frac{4}{x+5} > \frac{1}{2x+3}$$
$$\frac{4}{x+5} - \frac{1}{2x+3} > 0$$
$$\frac{4(2x+3)-(x+5)}{(x+5)(2x+3)} > 0$$
$$\frac{7x+7}{(x+5)(2x+3)} > 0$$

Critical numbers: $x = -1, x = -5, x = -\frac{3}{2}$

Test intervals: $(-\infty, -5), \left(-5, -\frac{3}{2}\right)$
$\left(-\frac{3}{2}, -1\right), (-1, \infty)$

Test: Is $\dfrac{7(x+1)}{(x+5)(2x+3)} > 0$?

Solution set: $\left(-5, -\frac{3}{2}\right) \cup (-1, \infty)$

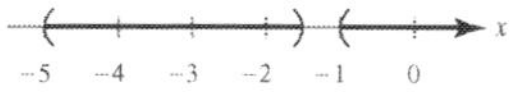

43. $$\frac{1}{x-3} \le \frac{9}{4x+3}$$
$$\frac{1}{x-3} - \frac{9}{4x+3} \le 0$$
$$\frac{4x+3-9(x-3)}{(x-3)(4x+3)} \le 0$$
$$\frac{30-5x}{(x-3)(4x+3)} \le 0$$

Critical numbers: $x = 3, x = -\frac{3}{4}, x = 6$

Test intervals: $\left(-\infty, -\frac{3}{4}\right), \left(-\frac{3}{4}, 3\right), (3, 6), (6, \infty)$

Test: Is $\dfrac{5(6-x)}{(x-3)(4x+3)} \le 0$?

Solution set: $\left(-\frac{3}{4}, 3\right) \cup [6, \infty)$

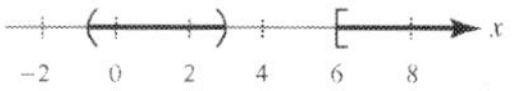

45. $x^2 - 9 \ge 0$
$(x+3)(x-3) \ge 0$

Critical numbers: $x = \pm 3$

Test intervals: $(-\infty, -3), (-3, 3), (3, \infty)$

Test: Is $x^2 - 9 \ge 0$?

Domain: $(-\infty, -3] \cup [3, \infty)$

47. $6 + x^2 \ge 0$

No critical numbers because
$b^2 - 4ac = 0^2 - 4(1)(6) < 0.$

Test interval: $(-\infty, \infty)$

Test: Is $6 + x^2 \ge 0$?

Domain: $(-\infty, \infty)$ or all real numbers

49. $81 - 4x^2 \ge 0$
$(9+2x)(9-2x) \ge 0$

Critical numbers: $x = \pm\frac{9}{2}$

Test intervals: $\left(-\infty, -\frac{9}{2}\right), \left(-\frac{9}{2}, \frac{9}{2}\right), \left(\frac{9}{2}, \infty\right)$

Test: Is $(9+2x)(9-2x) \ge 0$?

Domain: $\left[-\frac{9}{2}, \frac{9}{2}\right]$

51. $x^2 - 7x + 10 \ge 0$
$(x-5)(x-2) \ge 0$

Critical numbers: $x = 5, x = 2$

Test intervals: $(-\infty, 2), (2, 5), (5, \infty)$

Test: Is $(x-5)(x-2) \ge 0$?

Domain: $(-\infty, 2] \cup [5, \infty)$

53. $x^2 - 3x + 3 \ge 0$

No critical numbers since
$b^2 - 4ac = (-3)^2 - 4(1)(3) < 0$

Test interval: $(-\infty, \infty)$

Test: Is $x^2 - 3x + 3 \ge 0$?

Domain: $(-\infty, \infty)$ or all real numbers

55. $6x^3 - 10x^2 > 0$
$2x^2(3x-5) > 0$

Critical numbers: $x = 0, x = \frac{5}{3}$

Test intervals: $(-\infty, 0), \left(0, \frac{5}{3}\right), \left(\frac{5}{3}, \infty\right)$

Test: Is $2x^2(3x-5) > 0$?

Solution set: $\left(\frac{5}{3}, \infty\right)$

57. $x^3 - 9x \le 0$

$x(x^2 - 9) \le 0$

$x(x + 3)(x - 3) \le 0$

Critical numbers: $x = 0,\ x = \pm 3$

Test intervals: $(-\infty, -3), (-3, 0), (0, 3), (3, \infty)$

Test: Is $x(x + 3)(x - 3) \le 0$?

Solution set: $(-\infty, -3] \cup [0, 3]$

59. $(x - 1)^2(x + 2)^3 \ge 0$

Critical numbers: $x = 1,\ x = -2$

Test intervals: $(-\infty, -2), (-2, 1), (1, \infty)$

Test: Is $(x - 1)^2(x + 3)^3 \ge 0$?

Solution set: $[-2, \infty)$

61. $0.4x^2 + 5.26 < 10.2$

$0.4x^2 - 4.94 < 0$

$0.4(x^2 - 12.35) < 0$

Critical numbers: $x \approx \pm 3.51$

Test intervals: $(-\infty, -3.51), (-3.51, 3.51), (3.51, \infty)$

Solution set: $(-3.51, 3.51)$

63. $-0.5x^2 + 12.5x + 1.6 > 0$

The zeros are $x = \dfrac{-12.5 \pm \sqrt{(12.5)^2 - 4(-0.5)(1.6)}}{2(-0.5)}$.

Critical numbers: $x \approx -0.13$

$x \approx 25.13$

Test intervals: $(-\infty, -0.13), (-0.13, 25.13), (25.13, \infty)$

Solution set: $(-0.13, 25.13)$

65. $\dfrac{1}{2.3x - 5.2} > 3.4$

$\dfrac{1}{2.3x - 5.2} - 3.4 > 0$

$\dfrac{-7.82x + 18.68}{2.3x - 5.2} > 0$

Critical numbers: $x \approx 2.39,\ x \approx 2.26$

Test intervals: $(-\infty, -2.26), (2.26, 2.39), (2.39, \infty)$

Solution set: $(2.26, 2.39)$

67. $2l + 2w = 100 \Rightarrow w = 50 - l$

$lw \ge 500$

$l(50 - l) \ge 500$

$-l^2 + 50l - 500 \ge 0$

$\left[l - \left(25 - 5\sqrt{5}\right)\right]\left[l - \left(25 + 5\sqrt{5}\right)\right] \le 0$

13.8 meters $\approx 25 - 5\sqrt{5} \le l \le 25 + 5\sqrt{5} \approx 36.2$ meters. (Use the Quadratic Formula to find the critical numbers.)

69. (a) Profit = Revenue − Cost $\ge 1{,}650{,}000$

$x(50 - 0.0002x) - (12x + 150{,}000) \ge 1{,}650{,}000$

$-0.0002x^2 + 38x - 1{,}800{,}000 \ge 0$

$0.0002x^2 - 38x + 1{,}800{,}000 \le 0$

Critical numbers:

$x = \dfrac{38 \pm \sqrt{38^2 - 4(0.0002)(1{,}800{,}000)}}{2(0.0002)}$

$x = 100{,}000$ or $x = 90{,}000$

90,000 units $\le x \le$ 100,000 units

(b) $p = 50 - 0.0002x$

When $x = 90{,}000$, $p = \$32$.

When $x = 100{,}000$, $p = \$30$.

$\$30 \le p \le \$32$

(c) Revenue = Cost

$x(50 - 0.0002x) = 12x + 150{,}000$

$50x - 0.0002x^2 = 12x + 150{,}000$

$0 = 0.0002x^2 - 38x + 150{,}000$

$x = \dfrac{38 \pm \sqrt{(-38)^2 - 4(0.0002)(150{,}000)}}{2(0.0002)}$

$x \approx 4033$ or $x \approx 185{,}967$

Choose the larger value of x.

After revenue starts to decrease, the revenue is approximately equal to the cost when 185,967 units are sold.

Producing more than 185,967 units will result in the cost being greater than revenue. So, the company would incur a loss instead of making a profit.

71. $A = P(1 + r)^3 > 1150$

$1000(1 + r)^3 > 1150$

$(1 + r)^3 > 1.15$

$1 + r > \sqrt[3]{1.15}$

$r > \sqrt[3]{1.15} - 1 \approx 0.048$

The interest rate must be greater than 4.8%.

73. (a) When $t = 4$:

$P = 0.06(4)^2 + 75.9(4) + 6089 = 6393.56$

The population in 2004 was 6393.56 million. This value matches the value shown in the bar graph.

(b) $0.06t^2 + 75.9t + 6089 > 7000$

$0.06t^2 + 75.9t - 911 > 0$

The zeros are

$$t = \frac{-75.9 \pm \sqrt{75.9^2 - 4(0.06)(-911)}}{2(0.06)}.$$

Critical numbers: $t \approx 11.89, t \approx -1276.89$

The world population will exceed 7 billion people in 2012. The result seems reasonable.

75. $7.20t^2 - 9.7t + 1828 > 6000$

$7.20t^2 - 9.7t - 4172 > 0$

The zeros are

$$t = \frac{-(-9.7) \pm \sqrt{(-9.7)^2 - 4(7.20)(-4172)}}{2(7.20)}.$$

Critical numbers: $t \approx 24.755, t \approx -23.407$

In 2014/2015, the average yearly dormitory charge will be greater than \$6,000.

77. $\frac{1}{R} = \frac{1}{R_1} + \frac{1}{2}$

$2R_1 = 2R + RR_1$

$2R_1 = R(2 + R_1)$

$\frac{2R_1}{2 + R_1} = R$

Since $R \geq 1$, we have

$\frac{2R_1}{2 + R_1} \geq 1$

$\frac{2R_1}{2 + R_1} - 1 \geq 0$

$\frac{R_1 - 2}{2 + R_1} \geq 0.$

Since $R_1 > 0$, the only critical number is $R_1 = 2$. The inequality is satisfied when $R_1 \geq 2$ ohms.

Review Exercises for Chapter 1

1. $5(x - 3) = 2x + 9$

$5x - 15 = 2x + 9$

$3x = 24$

$x = 8$

Conditional

3. $4(x + 2) = 4x + 8$

$4x + 8 = 4x + 8$

$0 = 0$

Identity

5. $5(x - 1) = 3x + 1$

(a) $x = 0$

$5(0 - 1) \stackrel{?}{=} 3(0) + 1$

$-5 \neq 1$

$x = 0$ *is not* a solution.

(b) $x = -1$

$5(-1 - 1) \stackrel{?}{=} 3(-1) + 1$

$-10 \neq -2$

$x = -1$ *is not* a solution.

(c) $x = 3$

$5(3 - 1) \stackrel{?}{=} 3(3) + 1$

$10 = 10$

$x = 3$ *is* a solution.

(d) $x = 2$

$5(2 - 1) \stackrel{?}{=} 3(2) + 1$

$5 \neq 7$

$x = 2$ *is not* a solution.

7. (a) $6 + \dfrac{3}{(5) - 4} \stackrel{?}{=} 5$

$9 \neq 5$

$x = 5$ *is not* a solution.

(b) $6 + \dfrac{3}{(0) - 4} \stackrel{?}{=} 5$

$5.25 \neq 5$

$x = 0$ *is not* a solution.

(c) $6 + \dfrac{3}{(-2) - 4} \stackrel{?}{=} 5$

$5.5 \neq 5$

$x = -2$ *is not* a solution.

(d) $6 + \dfrac{3}{(7) - 4} \stackrel{?}{=} 5$

$7 \neq 5$

$x = 7$ *is not* a solution.

9. $x + 7 = 20$

$x = 13$

11. $4(x + 3) - 3 = 2(4 - 3x) - 4$

$4x + 12 - 3 = 8 - 6x - 4$

$4x + 9 = -6x + 4$

$10x = -5$

$x = -\frac{1}{2}$

13. $5x + x = 6x + 3$

$6x = 6x + 3$

$0 \neq 3$

There is no solution.

15. $8(x - 2) = 8x - 16$

$8x - 16 = 8x - 16$

$0 = 0$

This is an identity and the solution is all real numbers.

17. $\dfrac{3x - 2}{5x - 1} = \dfrac{3}{4}$

$4(3x - 2) = 3(5x - 1)$

$12x - 8 = 15x - 3$

$-3x = 5$

$x = -\dfrac{5}{3}$

19. $\dfrac{x}{x + 3} - \dfrac{4}{x + 3} + 2 = 0$

$x - 4 + 2(x + 3) = 0;\ x \neq -3$

$3x + 2 = 0$

$x = -\dfrac{2}{3}$

21. $0.375x - 0.75(300 - x) = 200$

$0.375x - 225 + 0.75x = 200$

$1.125x = 425$

$x \approx 377.778$

23. $\dfrac{x}{0.055} + \dfrac{x}{0.085} = 1$

$0.085x + 0.055x = 0.004675$

$0.14x = 0.004675$

$x \approx 0.033$

25. $-15.5t + 1859 = 1720$

$-15.5t = -139$

$t = \dfrac{-139}{-15.5} \approx 8.97$

The number of Safeway Stores decreased to 1720 in 2009.

27. (a) (First integer) + (Second integer) + (Third integer) = Sum

(b) First integer $= x$

Second integer $= x + 2$

Third integer $= x + 4$

Sum $= 42$

(c) $x + (x + 2) + (x + 4) = 42$

$3x + 6 = 42$

(d) $3x + 6 = 42$

$3x = 36$

$x = 12$

The smallest even integer is 12.

29. (a) New salary $=$ (Percent)(Old salary) $+$ (Old salary)

(b) Old salary $= 28{,}900$ (dollars)

Percent $= 0.07$ (percent in decimal form)

New salary $= S$ (in dollars)

(c) $S = 0.07(28{,}900) + 28{,}900$

(d) $S = 0.07(28{,}900) + 28{,}900$

$S = 30{,}923$

Your new annual salary is \$30,923.

31. Sale price $=$ List price $-$ 15% of List price

$$139 = L - 0.15L$$
$$139 = 0.85L$$
$$L \approx 163.53$$

The outdoor grill was originally priced \$163.53.

33. Distance difference $=$ (Rate of faster car) $\times$ (Time) $-$ (Rate of slower car) $\times$ (Time)

$$10 = 50t - 45t$$
$$10 = 5t$$
$$t = 2 \text{ hours}$$

35. Model: Avatar $=$ (Percent)(Titanic) $+$ Titanic

Labels: Avatar $= 760{,}507{,}625$

Percent $= p$

Titanic $= 600{,}788{,}188$

$$760{,}507{,}625 = p(600{,}788{,}188) + 600{,}788{,}188$$
$$159{,}719{,}437 = 600{,}788{,}188p$$
$$0.2658 = p$$

The percent increase was about 26.58%.

37. 10% of $(10 - x)$ + 100% of x = 30% of 10

$$0.10(10 - x) + 1.00x = 0.30(10)$$
$$1 - 0.10x + 1.00x = 3$$
$$0.90x = 2$$
$$x \approx 2.2 \text{ quarts}$$

About 2.2 $\left(\text{or } 2\frac{2}{9}\right)$ quarts will have to be replaced with pure antifreeze.

39. Perimeter $= 2(\text{length}) + 2(\text{width})$

$$177 = 2(2w) + 2w$$
$$177 = 6w$$
$$29.5 = w$$
$$\text{width} = w = 29.5 \text{ feet}$$
$$\text{length} = 2w = 59 \text{ feet}$$

41. $I = Prt$

$I = 500(0.04)(1) = 20$

After 1 year, you will earn \$20 in interest.

43.
$$6x^2 = 5x + 4$$
$$6x^2 - 5x - 4 = 0$$
$$(3x - 4)(2x + 1) = 0$$
$$3x - 4 = 0 \quad \text{or} \quad 2x + 1 = 0$$
$$x = \tfrac{4}{3} \quad \text{or} \quad x = -\tfrac{1}{2}$$

45.
$$x^2 - 11x + 24 = 0$$
$$(x - 3)(x - 8) = 0$$
$$x - 3 = 0 \quad \text{or} \quad x - 8 = 0$$
$$x = 3 \quad \text{or} \quad x = 8$$

47.
$$x^2 = 11$$
$$x = \pm\sqrt{11} \approx \pm 3.32$$

49.
$$(x + 4)^2 = 18$$
$$x + 4 = \pm 3\sqrt{2}$$
$$x = -4 \pm 3\sqrt{2}$$
$$x \approx 0.24 \quad \text{or} \quad x \approx -8.24$$

51.
$$\text{Area} = (\text{length})(\text{height})$$
$$405 = (h + 12)h$$
$$0 = h^2 + 12h - 405$$
$$0 = (h + 27)(h - 15)$$
$$h + 27 = 0$$
$$h = -27$$
$$h - 15 = 0$$
$$h = 15$$
$$\text{Length} = h + 12 = 27 \text{ feet}$$
$$\text{Height} = h = 15 \text{ feet}$$

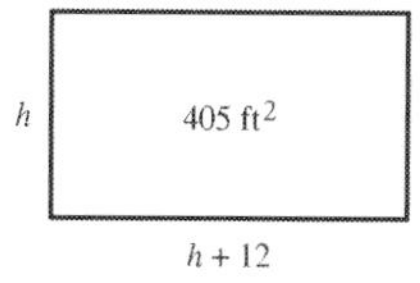

53. $x(60 - 0.0001x) = 8{,}000{,}000$

$$0 = 0.0001x^2 - 60x + 8{,}000{,}000$$

$$x = \frac{60 \pm \sqrt{(-60)^2 - 4(0.0001)(8{,}000{,}000)}}{2(0.0001)}$$

$x = 200{,}000$ units or $x = 400{,}000$ units

To produce a revenue of \$8,000,000, the company should produce 200,000 units or 400,000 units.

55. $x^2 + 11x + 24 = 0$

$b^2 - 4ac = (11)^2 - 4(1)(24) = 25 > 0$

Two real solutions

57. $3x^2 - 4x + 2 = 0$

$b^2 - 4ac = (-4)^2 - 4(3)(2) = -8$

No real solutions

59. $x^2 - 12x + 30 = 0$

$$x = \frac{-(-12) \pm \sqrt{(-12)^2 - 4(1)(30)}}{2(1)} = \frac{12 \pm \sqrt{24}}{2} = \frac{12 \pm 2\sqrt{6}}{2} = 6 \pm \sqrt{6}$$

61.

$$(y + 7)^2 = -5y$$

$$y^2 + 14y + 49 = -5y$$

$$y^2 + 19y + 49 = 0$$

$$y = \frac{-19 \pm \sqrt{19^2 - 4(1)(49)}}{2(1)} = \frac{-19 \pm \sqrt{165}}{2}$$

63. $x^2 + 6x - 3 = 0$

$$x = \frac{-6 \pm \sqrt{6^2 - 4(1)(-3)}}{2(1)} = \frac{-6 \pm \sqrt{48}}{2} = \frac{-6 \pm 4\sqrt{3}}{2} = -3 \pm 2\sqrt{3}$$

65. $3.6x^2 - 5.7x - 1.9 = 0$

$$x = \frac{-(-5.7) \pm \sqrt{(-5.7)^2 - 4(3.6)(-1.9)}}{2(3.6)}$$

$x \approx 1.866$ or $x \approx -0.283$

67. $34x^2 - 296x + 47 = 0$

$$x = \frac{-(-296) \pm \sqrt{(-296)^2 - 4(34)(47)}}{2(34)} = \frac{296 \pm \sqrt{81{,}224}}{68}$$

$x \approx 8.544$ or $x \approx 0.162$

69. On the moon

$$-2.7t^2 + 200 = 0$$

$$t^2 = \frac{200}{2.7}$$

$$t = \sqrt{\frac{200}{2.7}} \approx 8.61 \text{ seconds}$$

On the Earth

$$-16t^2 + 200 = 0$$

$$t^2 = \sqrt{\frac{25}{2}} \approx 3.54 \text{ seconds}$$

When the rock is dropped off a 200 foot cliff on the moon it takes about 8.61 seconds for it to hit the lunar surface. On Earth it would only take about 3.54 seconds.

71. $3x^3 - 9x^2 - 12x = 0$

$$3x(x^2 - 3x - 4) = 0$$

$$3x(x - 4)(x + 1) = 0$$

$x = 0$, $x = 4$ or $x = -1$

73.

$$x^4 - 5x^2 + 4 = 0$$

$$(x^2 - 1)(x^2 - 4) = 0$$

$$(x + 1)(x - 1)(x + 2)(x - 2) = 0$$

$x = \pm 1$, $x = \pm 2$

75. $2\sqrt{x} - 5 = 0$

$$\sqrt{x} = \tfrac{5}{2}$$

$$x = \tfrac{25}{4}$$

77. $2\sqrt{x-3} - 4 = 3x$

$\left(2\sqrt{x-3}\right)^2 = (3x+4)^2$

$4x - 12 = 9x^2 + 24x + 16$

$9x^2 + 20x + 28 = 0$

There are no real zeros since

$b^2 - 4ac = 20^2 - 4(9)(28) < 0.$

No solution

79. $\left(x^2 - 5\right)^{2/3} = 9$

$x^2 - 5 = 9^{3/2}$

$x^2 - 5 = \pm 27$

$x^2 = 32$ or $x^2 = -22 \Rightarrow$ No real solution

$x = \pm\sqrt{32} = \pm 4\sqrt{2}$

81. $|5x + 4| = 11$

$5x + 4 = 11$ or $5x + 4 = -11$

$5x = 7$ or $5x = -15$

$x = \frac{7}{5}$ or $x = -3$

83. $\dfrac{5}{x+1} + \dfrac{3}{x+3} = 1$

$5(x+3) + 3(x+1) = (x+1)(x+3),\ x \neq -1,\ x \neq -3$

$5x + 15 + 3x + 3 = x^2 + 4x + 3$

$0 = x^2 - 4x - 15$

$x = \dfrac{-(-4) \pm \sqrt{(-4)^2 - 4(1)(-15)}}{2(1)}$

$= \dfrac{4 \pm \sqrt{76}}{2} = \dfrac{4 \pm 2\sqrt{19}}{2}$

$= 2 \pm \sqrt{19}$

85. Let r = rent per student per week if three students rent the condominium.

Weekly rent: $3r = 4(r - 75)$

$3r = 4r - 300$

$300 = r$

The weekly rent is $3(300) = \$900$.

87. $p = 45 - \sqrt{0.002x + 1}$

$19.95 = 45 - \sqrt{0.002x + 1}$

$-25.05 = -\sqrt{0.002x + 1}$

$627.5025 = 0.002x + 1$

$626.5025 = 0.002x$

$313{,}251 \approx x$

When the price is set at \$19.95, the demand is 313,251 units.

89. $[-4, 7]$

$-4 \le x \le 7$

The interval is bounded.

91. $(-\infty, 1000]$

$x \le 1000$

The interval is unbounded.

93. $3(x - 1) < 2x + 8$

$3x - 3 < 2x + 8$

$x < 11$

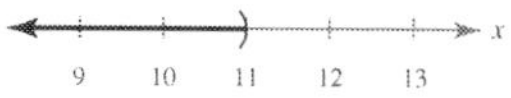

95. $-3 < \dfrac{2x+1}{4} < 3$

$-12 < 2x + 1 < 12$

$-13 < 2x < 11$

$-\dfrac{13}{2} < x < \dfrac{11}{2}$

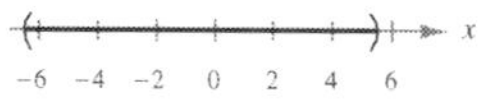

97. $|x + 10| + 3 < 5$

$|x + 10| < 2$

$-2 < x + 10 < 2$

$-12 < x < -8$

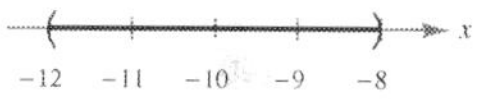

99. $R > C$

$89.95x > 35x + 2500$

$54.95x > 2500$

$x \ge 45$

To maximize the profit, the company should produce 45 or more units.

101. $x^2 - 1 < 0$

$(x - 1)(x + 1) < 0$

Critical numbers: $x = 1, x = -1$

Test intervals: $(-\infty, -1), (-1, 1), (1, \infty)$

103. $\dfrac{x+3}{x-1} \le 2$

$\dfrac{x+3}{x-1} - 2 \le 0$

$\dfrac{x+3}{x-1} - 2 \cdot \dfrac{x-1}{x-1} \le 0$

$\dfrac{x + 3 - 2x + 2}{x-1} \le 0$

$\dfrac{5 - x}{x - 1} \le 0$

Critical numbers: $x = 1, x = 5$

Test intervals: $(-\infty, 1), (1, 5), (5, \infty)$

105. $5(x + 1)(x - 3) < 0$

Critical numbers: $x = -1, x = 3$

Test intervals: $(-\infty, -1), (-1, 3), (3, \infty)$

Solution set: $(-1, 3)$

107. $x^3 - 9x < 0$

$x(x + 3)(x + 3) < 0$

Critical numbers: $x = 0, x = \pm 3$

Test intervals: $(-\infty, -3), (-3, 0), (0, 3), (3, \infty)$

Solution set: $(-\infty, -3) \cup (0, 3)$

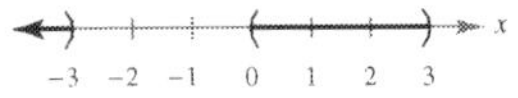

109.
$$\frac{2 + 3x}{4 - x} < 2$$
$$\frac{2 + 3x}{4 - x} - 2 < 0$$
$$\frac{2 + 3x - 2(4 - x)}{4 - x} < 0$$
$$\frac{5x - 6}{4 - x} < 0$$

Critical numbers: $x = \frac{6}{5}, x = 4$

Test intervals: $\left(-\infty, \frac{6}{5}\right), \left(\frac{6}{5}, 4\right), (4, \infty)$

Solution set: $\left(-\infty, \frac{6}{5}\right) \cup (4, \infty)$

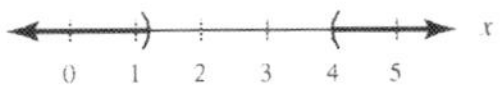

111. $-1.2x^2 + 4.76 > 1.32$
$$-1.2x^2 > -3.44$$
$$x^2 < 2.8667$$
$$-1.69 < x < 1.69$$

113.
$$\frac{1}{3.7x - 6.1} > 2.9$$
$$\frac{1}{3.7x - 6.1} - 2.9 > 0$$
$$\frac{1 - 2.9(3.7x - 6.1)}{3.7x - 6.1} > 0$$
$$\frac{-10.73x + 18.69}{3.7x - 6.1} > 0$$

Critical numbers: $x \approx 1.74, x \approx 1.65$

$1.65 < x < 1.74$

115. $x - 10 \geq 0$
$$x \geq 10$$

Domain: $[10, \infty)$

117. $x^2 - 15x + 54 \geq 0$

$(x - 6)(x - 9) \geq 0$

Critical numbers: $x = 6, x = 9$

Test intervals: $(-\infty, -6), (6, 9), (9, \infty)$

Domain: $(-\infty, -6] \cup [9, \infty)$

119. $S = -16t^2 + v_0t + s_0$

Let $v_0 = 100$ and $s_0 = 0$.
$$-16t^2 + 100t > 150$$
$$-16t^2 + 100t - 150 > 0$$
$$t = \frac{-100 \pm \sqrt{(100)^2 - 4(-16)(-150)}}{2(-16)}$$

$t = 2.5$ or $t = 3.75$

2.5 seconds $< t <$ 3.75 seconds

121.
$$P = 2l + 2w$$
$$80 = 2l + 2w$$
$$80 - 2l = 2w$$
$$40 - l = w$$
$$A \geq 380$$
$$lw \geq 380$$
$$l(40 - l) \geq 380$$
$$40l - l^2 \geq 380$$
$$l^2 - 40l + 380 \leq 0$$
$$l = \frac{40 \pm \sqrt{(-40)^2 - 4(1)(380)}}{2(1)}$$

$l \approx 15.5$ or $l \approx 24.5$

Solution set: $[15.5, 24.5]$

The length must be between 15.5 meters and 24.5 meters.

123. $2000\left(1 + \frac{r}{2}\right)^{16} > 2800$
$$\left(1 + \frac{r}{2}\right)^{16} > 1.4$$
$$1 + \frac{r}{2} > 1.021252193$$
$$\frac{r}{2} > 0.021252193$$
$$r > 0.0425043865$$

The interest rate must be greater than 4.25%.

125.
$$R - C \geq 1{,}000{,}000$$
$$x(75 - 0.0005x) - (25x + 100{,}000) \geq 1{,}000{,}000$$
$$75x - 0.0005x^2 - 25x - 100{,}000 \geq 1{,}000{,}000$$
$$-0.0005x^2 + 50x - 1{,}100{,}000 \geq 0$$
$$x = \frac{-50 \pm \sqrt{50^2 - 4(-0.0005)(-1{,}100{,}000)}}{2(-0.0005)}$$
$x \approx 32{,}679$ or $x \approx 67{,}321$

Solution set: $[32{,}679,\ 67{,}321]$

$$p = 75 - 0.0005(32{,}679) = \$58.66$$
$$p = 75 - 0.0005(67{,}321) = \$41.34$$

The company should set the price between \$41.34 and \$58.66 to obtain a profit of at least \$1,000,000.

Chapter 1 Test Yourself

1.
$$3(x + 2) - 8 = 4(2 - 5x) + 7$$
$$3x + 6 - 8 = 8 - 20x + 7$$
$$3x - 2 = -20x + 15$$
$$23x = 17$$
$$x = \tfrac{17}{23}$$

2. (a) Since you can take the cube root of any real number the domain is $(-\infty, \infty)$.

(b)
$$9 - x^2 \geq 0$$
$$(3 - x)(3 + x) \geq 0$$
Critical numbers: $x = \pm 3$

Test intervals: $(-\infty, -3), (-3, 3), (3, \infty)$

Test: Is $(3 - x)(3 + x) \geq 0$?

Domain: $[-3, 3]$

3. Total profit = April profit + May profit
$$625{,}509.12 = x + 0.92x$$
$$625{,}509.12 = 1.92x$$
$$325{,}786 = x$$
$$299{,}723.12 = 0.92x$$
The profit in April was \$325,786 and the profit in May was \$299,723.12.

4.
$$6x^2 + 7x = 5$$
$$6x^2 + 7x - 5 = 0$$
$$(3x + 5)(2x - 1) = 0$$
$3x + 5 = 0$ or $2x - 1 = 0$

$3x = -5$ or $2x = 1$

$x = -\tfrac{5}{3}$ or $x = \tfrac{1}{2}$

5.
$$12 + 5x - 2x^2 = 0$$
$$(3 + 2x)(4 - x) = 0$$
$3 + 2x = 0$ or $4 - x = 0$

$2x = -3$ or $-x = -4$

$x = -\tfrac{3}{2}$ or $x = 4$

6.
$$x^2 - 5 = 10$$
$$x^2 = 15$$
$$x = \pm\sqrt{15}$$

7.
$$(x + 5)^2 = -3x$$
$$x^2 + 10x + 25 = -3x$$
$$x^2 + 13x + 25 = 0$$
$$x = \frac{-13 \pm \sqrt{13^2 - 4(1)(25)}}{2(1)}$$
$$= \frac{-13 \pm \sqrt{169 - 100}}{2}$$
$$= \frac{-13 \pm \sqrt{69}}{2}$$

8.
$$3x^2 - 11x = 2$$
$$3x^2 - 11x - 2 = 0$$
$$x = \frac{-(-11) \pm \sqrt{(-11)^2 - 4(3)(-2)}}{2(3)}$$
$$= \frac{11 \pm \sqrt{121 + 24}}{6}$$
$$= \frac{11 \pm \sqrt{145}}{6}$$

9. $5.4x^2 - 3.2x - 2.5 = 0$

$$x = \frac{-(-3.2) \pm \sqrt{(-3.2)^2 - 4(5.4)(-2.5)}}{2(5.4)}$$

$$= \frac{3.2 \pm \sqrt{10.24 \pm 54}}{10.8}$$

$$= \frac{3.2 \pm \sqrt{64.24}}{10.8}$$

$$x = \frac{3.2 + \sqrt{64.24}}{10.8} \approx 1.038$$

$$x = \frac{3.2 - \sqrt{64.24}}{10.8} \approx -0.446$$

10. $|2x - 3| = 10$

$2x - 3 = -10$ or $2x - 3 = 10$

$2x = -7$ or $2x = 13$

$x = -\frac{7}{2}$ or $x = \frac{13}{2}$

11. $\sqrt{x - 3} + x = 5$

$\sqrt{x - 3} = 5 - x$

$x - 3 = 25 - 10x + x^2$

$0 = x^2 - 11x + 28$

$0 = (x + 4)(x - 7)$

$x - 4 = 0$ or $x - 7 = 0$

$x = 4$ or $x = 7$

$x = 4$ is the only solution.

(**Note:** $x = 7$ is extraneous.)

12. $x^4 - 10x^2 + 9 = 0$

$(x^2 - 1)(x^2 - 9) = 0$

$(x + 1)(x - 1)(x + 3)(x - 3) = 0$

$x = -1, 1, -3, 3$

13. $(x^2 - 9)^{2/3} = 9$

$(x^2 - 9)^2 = 9^3$

$x^2 - 9 = \pm\sqrt{729}$

$x^2 - 9 = \pm 27$

$x^2 = 9 \pm 27$

$x^2 = 36$ or $x^2 = -18$

$x = \pm 6$ No real solution

14. $\text{Revenue} = xp = x(40 - 0.0001x)$

$2{,}000{,}000 = 40x - 0.0001x^2$

$0.0001x^2 - 40x + 2{,}000{,}000 = 0$

$$x = \frac{-(-40) \pm \sqrt{(-40)^2 - 4(0.0001)(2{,}000{,}000)}}{2(0.0001)}$$

$$= \frac{40 \pm \sqrt{800}}{0.0002}$$

$$x = \frac{40 + 20\sqrt{2}}{0.0002} \approx 341{,}421 \text{ units}$$

or

$$x = \frac{40 - 20\sqrt{2}}{0.0002} \approx 58{,}579 \text{ units}$$

Since the equation is quadratic, it is possible to have two distinct solutions. In this case they would both have to be positive since x represents the number of units sold.

15. $\dfrac{3x + 1}{5} < 2$

$3x + 1 < 10$

$3x < 9$

$x < 3$

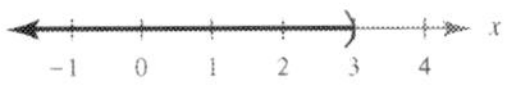

16. $|4 - 5x| \geq 24$

$4 - 5x \leq -24$ or $4 - 5x \geq 24$

$-5x \leq -28$ or $-5x \geq 20$

$x \geq \frac{28}{5}$ or $x \leq -4$

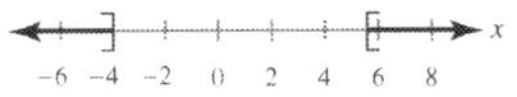

17. $\dfrac{x + 3}{x + 7} > 2$

$\dfrac{x + 3}{x + 7} - 2 > 0$

$\dfrac{x + 3 - 2(x + 7)}{x + 7} > 0$

$\dfrac{-x - 11}{x + 7} > 0$

$\dfrac{x + 3 - 2(x + 7)}{x + 7} > 0$

$\dfrac{-x - 11}{x + 7} > 0$

Critical numbers: $x = -11, -7$

Test intervals: $(-\infty, -11), (-11, -7), (-7, \infty)$

Solution set: $(-11, -7)$

$-11 < x < -7$

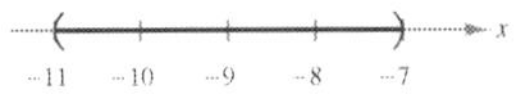

18. $3x^3 - 12x \le 0$

$3x(x^2 - 4) \le 0$

$3x(x + 2)(x - 2) \le 0$

Critical numbers: $x = 0, \pm 2$

Test intervals: $(-\infty, -2), (-2, 0), (0, 2), (2, \infty)$

Solution set: $(-\infty, -2] \cup [0, 2]$

$x \le -2,\ 0 \le x \le 2$

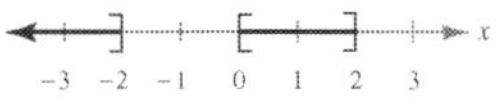

19. $R - C \ge 800{,}000$

$x(90 - 0.0004x) - (25x + 300{,}00) \ge 800{,}000$

$90x - 0.0004x^2 - 25x - 300{,}000 \ge 800{,}000$

$-0.0004x^2 + 65x - 1{,}100{,}000 \ge 0$

$$x = \frac{-65 \pm \sqrt{65^2 - 4(-0.0004)(-1{,}100{,}000)}}{2(-0.0004)}$$

$x \approx 19{,}189$ or $x \approx 143{,}311$

Solution set: $[19{,}189, 143{,}311]$

The company must sell between 19,189 units and 143,311 units to obtain a profit of at least $800,000.

20. $0.145t^2 + 0.79t + 107.4 > 190$

$0.145t^2 + 0.79t - 82.6 > 0$

The zeros are $t = \dfrac{-0.79 \pm \sqrt{(0.79)^2 - 4(0.145)(-82.6)}}{2(0.145)}$

Critical numbers: $t \approx 21.30, t \approx -26.75$

The total payroll for federal civilian employees will exceed $190 billion in 2012.

CHAPTER 2
Functions and Graphs

CHAPTER 2
Functions and Graphs

Section 2.1 Graphs of Equations

Skills Warm Up

1. $\sqrt{(2-6)^2 + [1-(-2)]^2} = \sqrt{(-4)^2 + 3^2}$
$= \sqrt{16+9} = \sqrt{25} = 5$

2. $\sqrt{(1-4)^2 + (-2-1)^2} = \sqrt{(-3)^2 + (-3)^2}$
$= \sqrt{9+9} = \sqrt{18} = 3\sqrt{2}$

3. $\dfrac{4+(-2)}{2} = \dfrac{2}{2} = 1$

4. $\dfrac{-1+(-3)}{2} = \dfrac{-4}{2} = -2$

5. $\sqrt{18} + \sqrt{45} = 3\sqrt{2} + 3\sqrt{5} = 3\left(\sqrt{2} + \sqrt{5}\right)$

6. $\sqrt{12} + \sqrt{44} = 2\sqrt{3} + 2\sqrt{11} = 2\left(\sqrt{3} + \sqrt{11}\right)$

7. $\sqrt{(4-x)^2 + (5-2)^2} = \sqrt{58}$

$$(4-x)^2 + (5-2)^2 = 58$$
$$(4-x)^2 + 9 = 58$$
$$(4-x)^2 = 49$$
$$4 - x = \pm\sqrt{49}$$
$$4 - x = 7 \Rightarrow x = -3$$
$$4 - x = -7 \Rightarrow x = 11$$

8. $\sqrt{(8-6)^2 + (y-5)^2} = 2\sqrt{5}$

$$(8-6)^2 + (y-5)^2 = 20$$
$$4 + (y-5)^2 = 20$$
$$(y-5)^2 = 16$$
$$y - 5 = \pm\sqrt{16}$$
$$y - 5 = 4 \Rightarrow y = 9$$
$$y - 5 = -4 \Rightarrow y = 1$$

9. $x^3 - 9x = 0$

$$x(x+3)(x-3) = 0$$
$$x = 0$$
$$x + 3 = 0 \Rightarrow x = -3$$
$$x - 3 = 0 \Rightarrow x = 3$$

10. $x^4 - 8x^2 + 16 = 0$

$$(x^2 - 4)(x^2 - 4) = 0$$
$$x^2 - 4 = 0 \Rightarrow x = \pm 2$$

1.

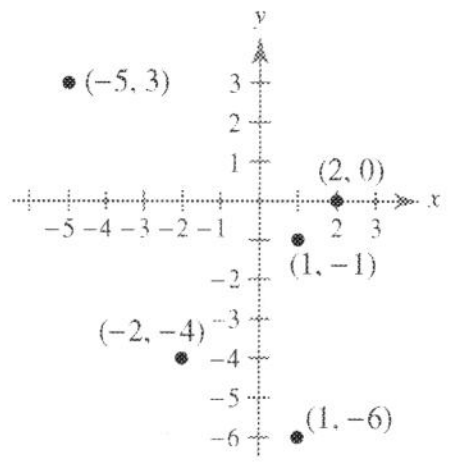

3. (a)

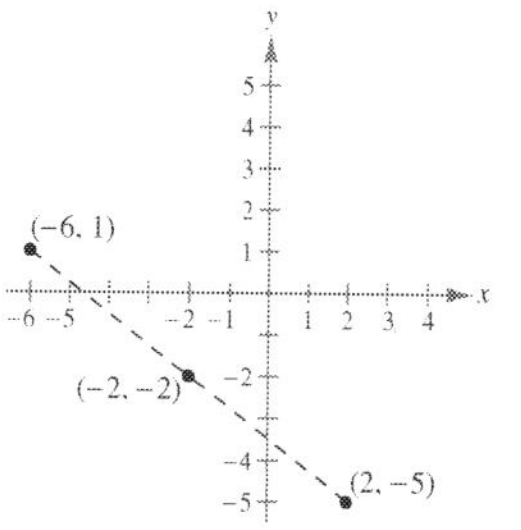

(b) $d = \sqrt{(2-(-6))^2 + (-5-1)^2}$
$= \sqrt{64+36} = 10$

(c) $\left(\dfrac{2+(-6)}{2}, \dfrac{(-5)+1}{2}\right) = (-2, -2)$

5. (a)

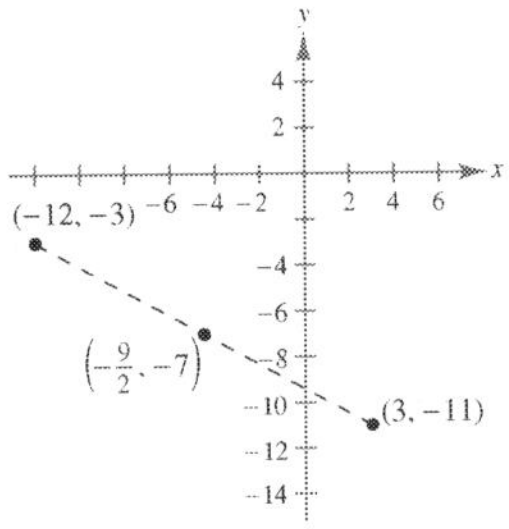

(b) $d = \sqrt{(-12-3)^2 + (-3-(-11))^2}$

$= \sqrt{225 + 64} = 17$

(c) $\left(\dfrac{-12+3}{2}, \dfrac{-3+(-11)}{2}\right) = \left(-\dfrac{9}{2}, -7\right)$

7. (a)

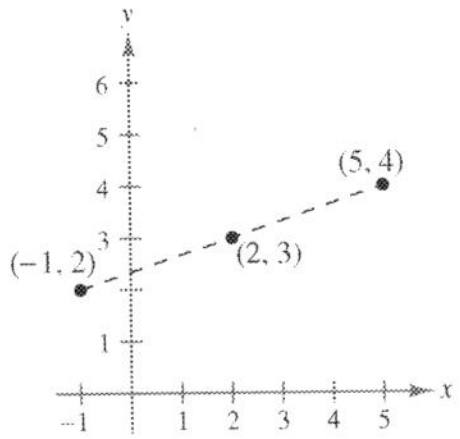

(b) $d = \sqrt{(5+1)^2 + (4-2)^2}$

$= \sqrt{36 + 4} = 2\sqrt{10}$

(c) $\left(\dfrac{-1+5}{2}, \dfrac{2+4}{2}\right) = (2, 3)$

9. $\left(\dfrac{1}{2}, 1\right), \left(-\dfrac{3}{2}, -5\right)$

(a)

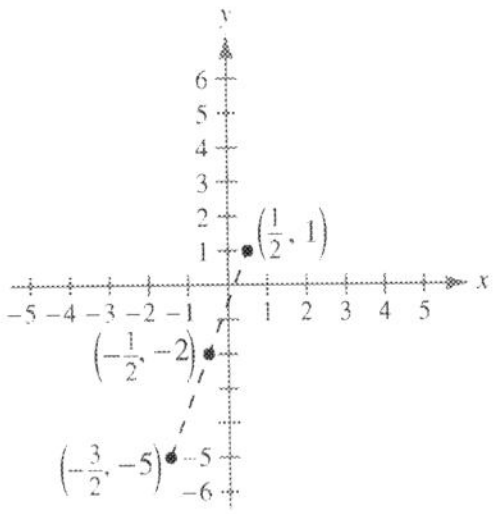

(b) $d = \sqrt{\left(-\dfrac{3}{2} - \dfrac{1}{2}\right)^2 + (-5-1)^2}$

$= \sqrt{(-2)^2 + (-6)^2} = \sqrt{40} = 2\sqrt{10}$

(c) Midpoint $= \left(\dfrac{\dfrac{1}{2} + \left(-\dfrac{3}{2}\right)}{2}, \dfrac{1+(-5)}{2}\right) = \left(-\dfrac{1}{2}, -2\right)$

11. (a)

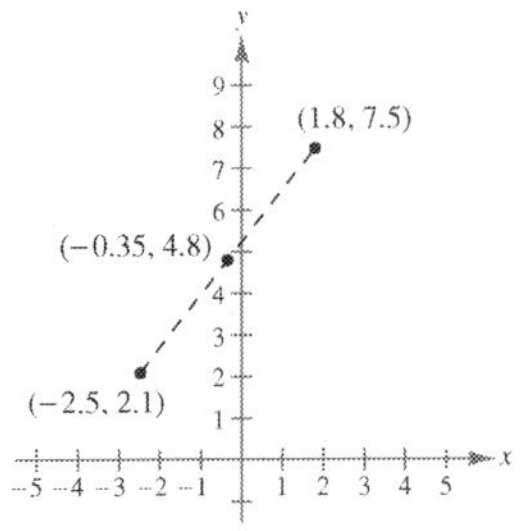

(b) $d = \sqrt{(2.1-7.5)^2 + (-2.5-1.8)^2}$

$= \sqrt{29.16 + 18.49} = \sqrt{47.65}$

(c) $\left(\dfrac{1.8-2.5}{2}, \dfrac{7.5+2.1}{2}\right) = (-0.35, 4.8)$

13. (a) $a = |4-0| = 4$

$b = |3-0| = 3$

$c = \sqrt{a^2 + b^2} = \sqrt{16+9} = 5$

(b) $c = \sqrt{(4-0)^2 + (3-0)^2} = \sqrt{16+9} = 5$

15. (a) $a = |7-(-3)| = 10$

$b = |4-1| = 3$

$c = \sqrt{a^2 + b^2} = \sqrt{100+9} = \sqrt{109}$

(b) $c = \sqrt{[7-(-3)]^2 + (4-1)^2}$

$= \sqrt{100+9} = \sqrt{109}$

17. $\sqrt{(x-3)^2 + (5-(-4))^2} = 15$

$(x-3)^2 + 81 = 225$

$(x-3)^2 = 144$

$x = 3 \pm 12 = -9, 15$

19. $\sqrt{(-15+3)^2 + (y+7)^2} = 20$

$144 + (y+7)^2 = 400$

$(y+7)^2 = 256$

$y = -7 \pm 16 = -23, 9$

21. $2x - 3y + 11 = 0$

(a) $2(2) - 3(5) + 11 = 0$, yes

(b) $2(3) - 3(2) + 11 = 11 \neq 0$, no

23. $y = \sqrt{x-5}$

(a) $\sqrt{9-5} = 2$, yes

(b) $\sqrt{21-5} = 4$, yes

25. $y = \frac{3}{4}x - 1$

x	−3	−2	−1	0	1	2	3
y	−3.25	−2.5	−1.75	−1	−0.25	0.5	1.25

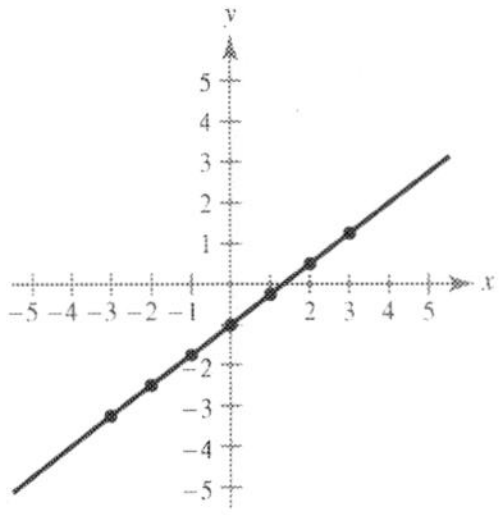

27.

x	−3	−2	−1	0	1	2	3
y	5	0	−3	−4	−3	0	5

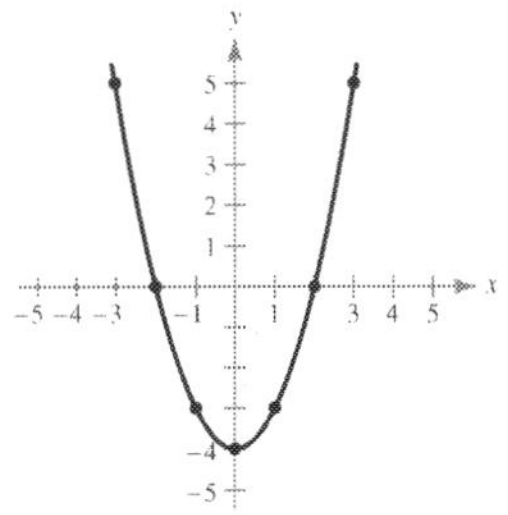

29. $y = 2x - 1 = 0$

$x = \frac{1}{2}$

x-intercept: $\left(\frac{1}{2}, 0\right)$

$y = 2(0) - 1 = -1$

y-intercept: $(0, -1)$

31. $y = x^2 + x - 2 = 0$

$(x + 2)(x - 1) = 0$

$x = -2, 1$

x-intercepts: $(-2, 0), (1, 0)$

$y = 0^2 + 0 - 2 = -2$

y-intercept: $(0, -2)$

33. $y = \sqrt{4 - x^2}$

$0 = \sqrt{4 - x^2}$

$0 = 4 - x^2$

$x^2 = 4$

$x = \pm 2$

x-intercepts: $(-2, 0), (2, 0)$

$y = \sqrt{4 - 0^2}$

$y = \sqrt{4}$

$y = 2$

y-intercept: $(0, 2)$

35. $2y - xy + 3x = 4$

$2(0) - x(0) + 3x = 4$

$x = \frac{4}{3}$

x-intercept: $\left(\frac{4}{3}, 0\right)$

$2y - 0(y) + 3(0) = 4$

$y = 2$

y-intercept: $(0, 2)$

37. All the ordered pairs on the x-axis have a y-coordinate of zero: $(x, 0)$. So, to find an x-intercept, we let $y = 0$ in an equation and solve for x. Similarly, all ordered pairs on the y-axis have an x-coordinate of zero: $(0, y)$. So, to find a y-intercept, we let $x = 0$ and solve for y. (Answers will vary.)

39. $(-x)^4 - 2y = 0 \Rightarrow x^4 - 2y = 0$

y-axis symmetry

41. $x - (-y)^2 = 0 \Rightarrow x - y^2 = 0$

x-axis symmetry

43. $y = \sqrt{16 - (-x)^2} \Rightarrow y = \sqrt{16 - x^2}$

y-axis symmetry

45. $-x(-y) = 2 \Rightarrow xy = 2$

Origin symmetry

47. $-y = \dfrac{-x}{(-x)^2 - 4} \Rightarrow y = \dfrac{x}{x^2 - 4}$

Origin symmetry

49. $(-x)^2 + y^2 = 25$

$x^2 + y^2 = 25$

y-axis symmetry

$x^2 + (-y)^2 = 25$

$x^2 + y^2 = 25$

x-axis symmetry

$(-x)^2 + (-y)^2 = 25$

$x^2 + y^2 = 25$

Origin symmetry

51. $y = -x^2 + 4$ y-axis symmetry

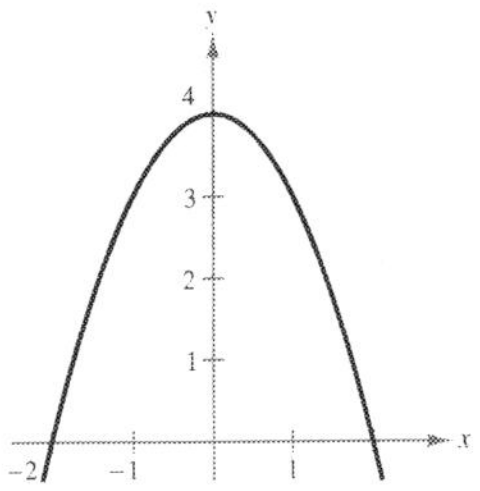

53. $y = -x^3 + x$

Origin symmetry

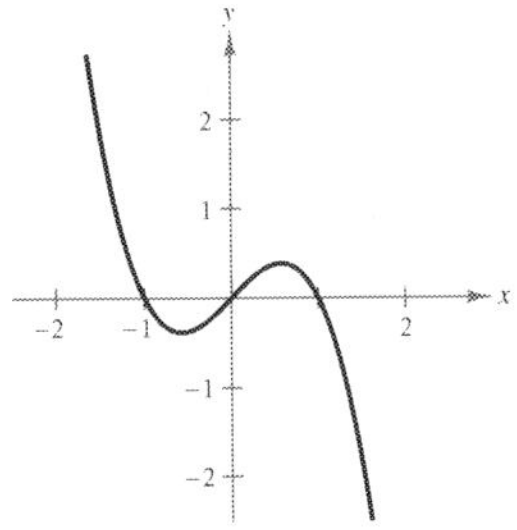

55. $y = 4 - x$ has intercepts $(0, 4)$ and $(4, 0)$; graph (c)

57. $y = x^2 + 2x$ has intercepts $(-2, 0)$ and $(0, 0)$; graph (f)

59. $y = x^3 - x$ has intercepts $(-1, 0)$, $(0, 0)$, and $(1, 0)$; graph (e)

61. $y = 5 - 3x$

Intercepts: $(0, 5)$, $\left(\frac{5}{3}, 0\right)$

No symmetry

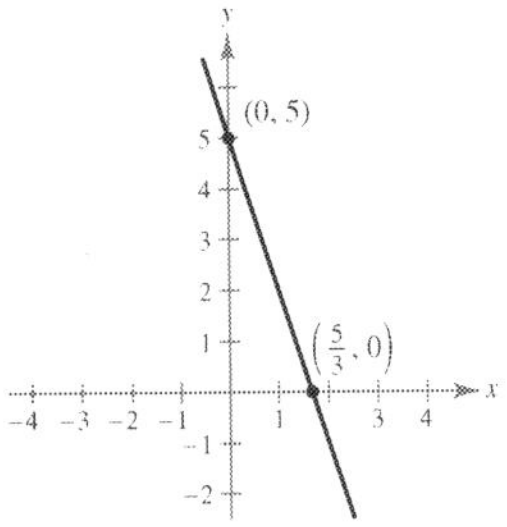

63. $y = 1 - x^2$

Intercepts: $(-1, 0)$, $(0, 1)$, $(1, 0)$

y-axis symmetry

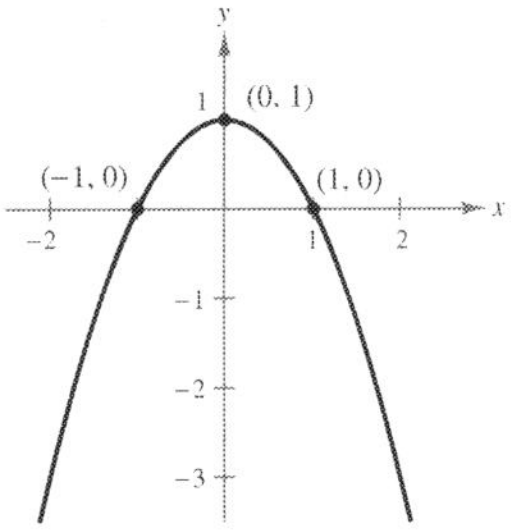

65. $y = x^2 - 4x + 3$

Intercepts: $(0, 3)$, $(1, 0)$, $(3, 0)$

No symmetry

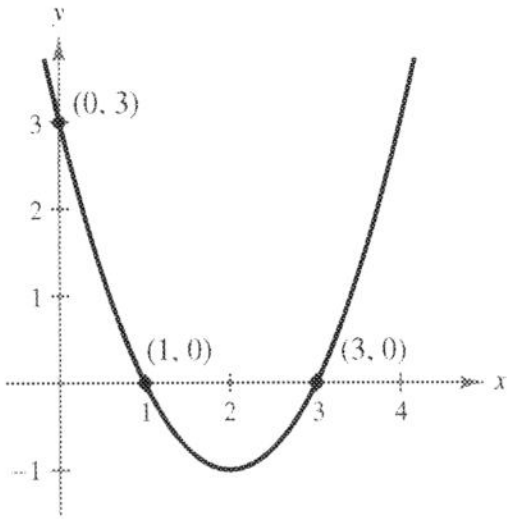

67. $y = x^3 + 2$

Intercepts: $\left(-\sqrt[3]{2}, 0\right)$, $(0, 2)$

No symmetry

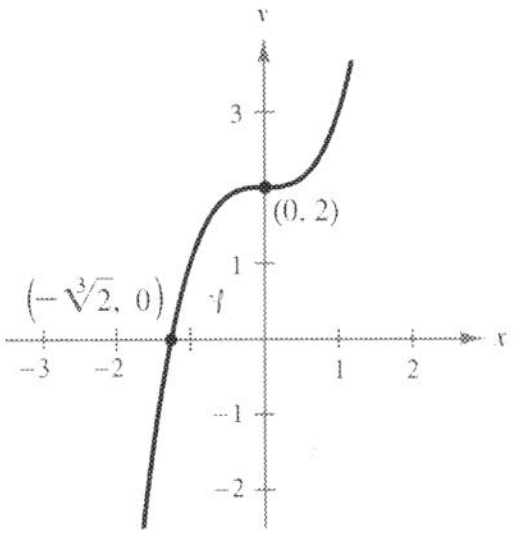

69. $y = \dfrac{8}{x^2 + 4}$

Intercept: $(0, 2)$

y-axis symmetry

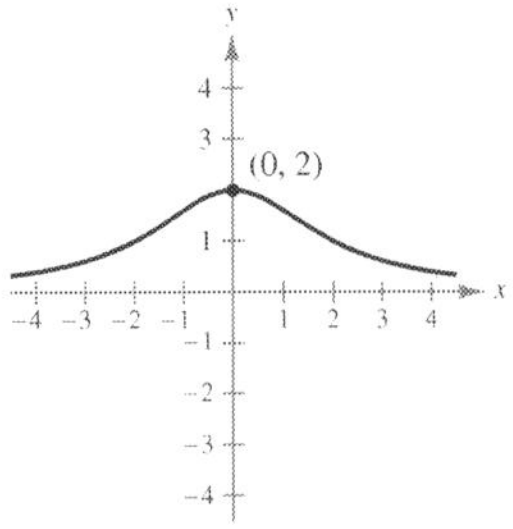

71. $y = \sqrt{x + 1}$

Intercepts: $(-1, 0), (0, 1)$

No symmetry

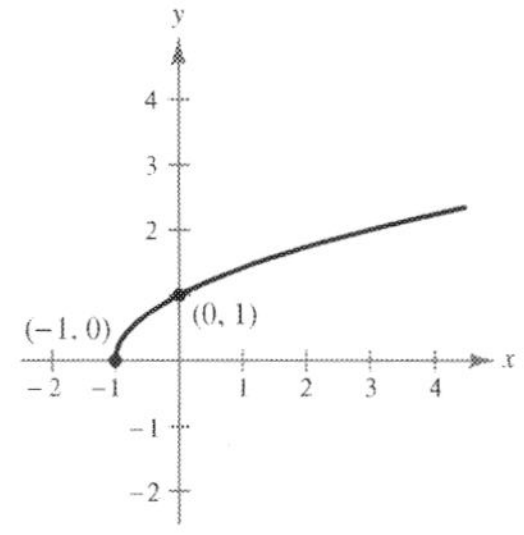

73. $y = \sqrt[3]{x}$

Intercept: $(0, 0)$

Origin symmetry

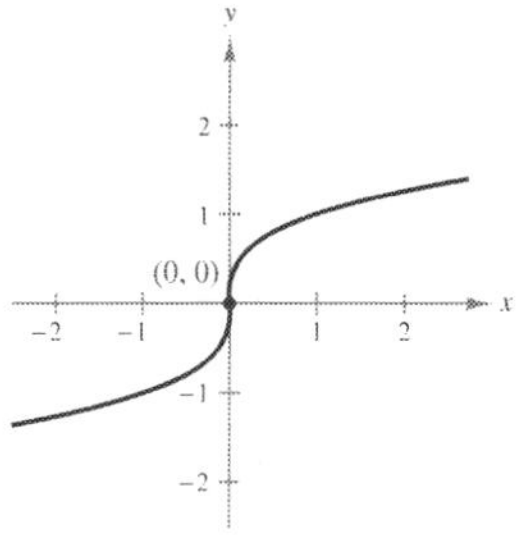

75. $y = |x - 4|$

Intercepts: $(0, 4), (4, 0)$

No symmetry

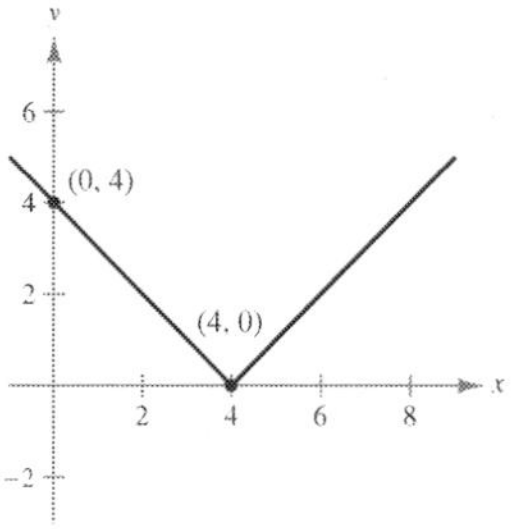

77. $x = y^2 - 1$

Intercepts: $(-1, 0), (0, -1), (0, 1)$

x-axis symmetry

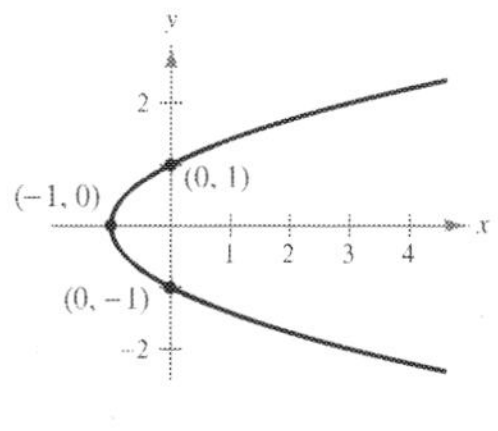

79. $x^2 + y^2 = 4$

Intercepts: $(-2, 0), (0, -2), (0, 2), (2, 0)$

x-axis, y-axis, and origin symmetry

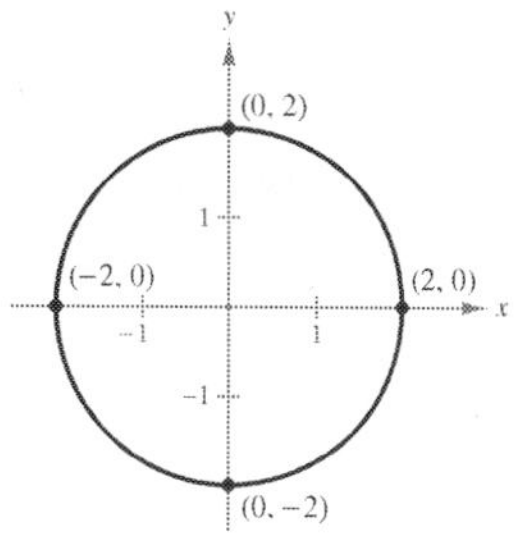

81. $r = \sqrt{(4 - 4)^2 + (2 - 0)^2} = \sqrt{0 + 4} = 2$

83. $r = \sqrt{(4 - 2)^2 + (3 - 2)^2} = \sqrt{4 + 1} = \sqrt{5}$

85. Center: $\left(\frac{5 + 1}{2}, \frac{3 + 1}{2}\right) = (3, 2)$

$$2r = \sqrt{(3 - 1)^2 + (5 - 1)^2}$$
$$2r = 2\sqrt{5}$$
$$r = \sqrt{5}$$

87. $x^2 + y^2 = 3^2$

$$x^2 + y^2 = 9$$

89. $[x - (-4)]^2 + (y - 1)^2 = (\sqrt{2})^2$

$$(x + 4)^2 + (y - 1)^2 = 2$$

91. $[x - (-1)]^2 + (y - 2)^2 = [0 - (-1)]^2 + (0 - 2)^2$

$$(x + 1)^2 + (y - 2)^2 = 5$$

93. $r = \frac{1}{2}\sqrt{(-3 - 5)^2 + [4 - (-2)]^2} = 5$

Center: $\left(\frac{-3 + 5}{2}, \frac{4 + (-2)}{2}\right) = (1, 1)$

$$(x - 1)^2 + (y - 1)^2 = 25$$

95. $x^2 + y^2 - 6x + 4y - 3 = 0$

$(x^2 - 6x + 9) + (y^2 + 4y + 4) = 3 + 9 + 4$

$(x - 3)^2 + (y + 2)^2 = 16$

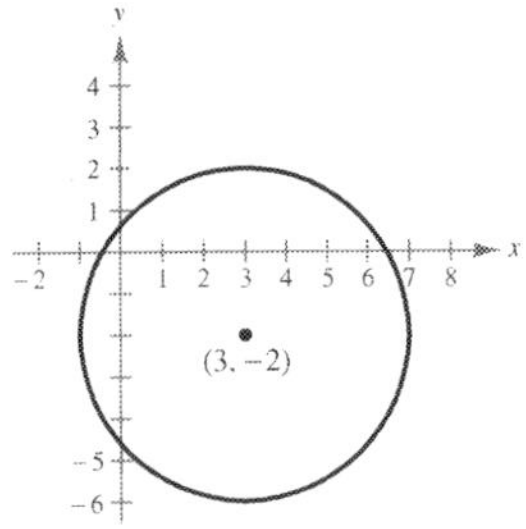

97. $x^2 + y^2 - 4x + 6y + 9 = 0$

$(x^2 - 4x + 4) + (y^2 + 6y + 9) = -9 + 9 + 4$

$(x - 2)^2 + (y + 3)^2 = 4$

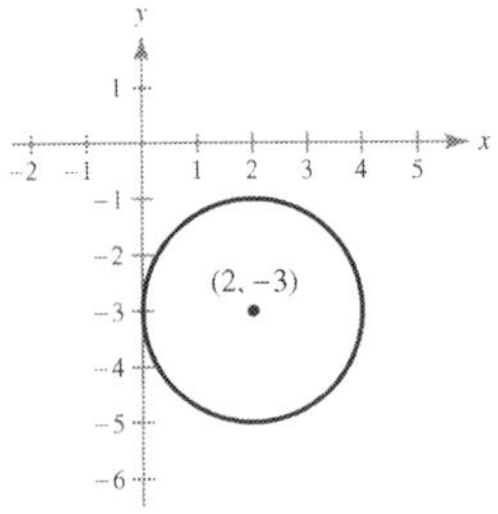

99. $2x^2 + 2y^2 - 2x - 2y - 3 = 0$

$\left(x^2 - x + \frac{1}{4}\right) + \left(y^2 - y + \frac{1}{4}\right) = \frac{3}{2} + \frac{1}{4} + \frac{1}{4}$

$\left(x - \frac{1}{2}\right)^2 + \left(y - \frac{1}{2}\right)^2 = 2$

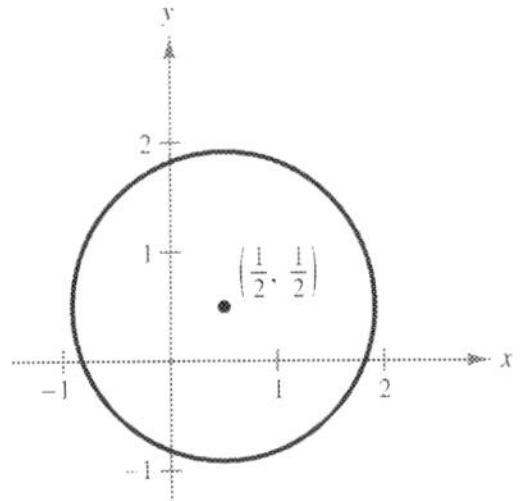

101. $16x^2 + 16y^2 + 16x + 40y - 7 = 0$

$\left(x^2 + x + \frac{1}{4}\right) + \left(y^2 + \frac{5}{2}y + \frac{25}{16}\right) = \frac{7}{16} + \frac{1}{4} + \frac{25}{16}$

$\left(x + \frac{1}{2}\right)^2 + \left(y + \frac{5}{4}\right)^2 = \frac{9}{4}$

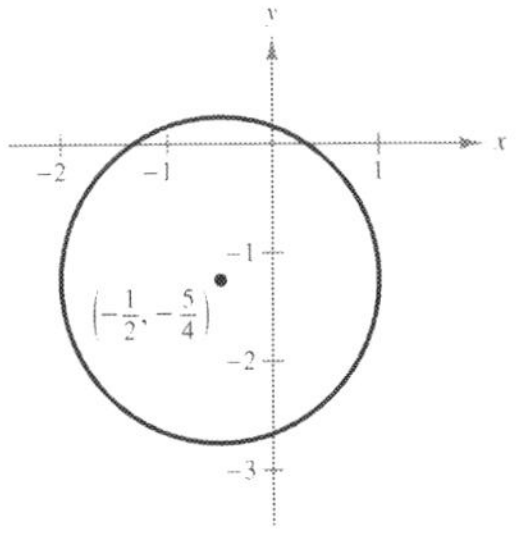

103. Center: $(3, -1)$, Radius: 5

$(x - 3)^2 + (y + 1)^2 = 25$

$x^2 - 6x + 9 + y^2 + 2y + 1 - 25 = 0$

$x^2 + y^2 - 6x + 2y - 15 = 0$

105. Answers will vary.

(a) From the graph, estimate the points:

1994: $(4, 2350)$

2000: $(10, 2750)$

Percent increase $= \dfrac{2750 - 2350}{2350} \approx 0.17$, or 17%

(b) From the graph, estimate the points:

2000: $(10, 2750)$

2008: $(18, 2975)$

Percent increase $= \dfrac{2975 - 2750}{2750} \approx 0.08$, or 8%

107. (a) $y = 5.3(0) + 482 = 482$

y-intercept: $(0, 482)$

The y-intercept represents the population (in millions of people) of North America in 2000.

(b)

x	−20	−10	0	10
y	376	429	482	535

x	20	30	40	50
y	588	641	694	747

(c)

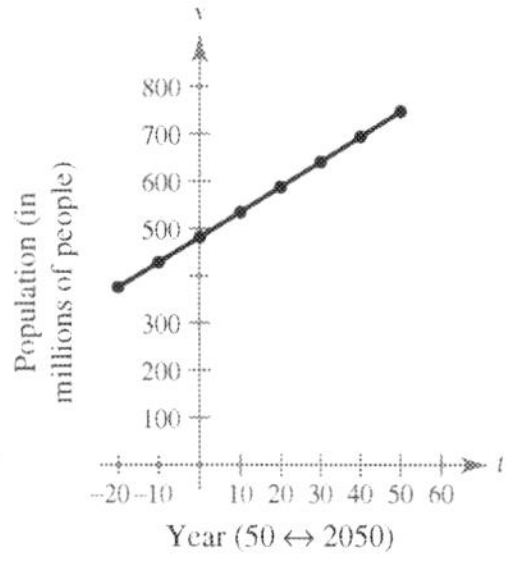

109. (a) $y = -24.91t^2 + 566.6t + 3624$

t	0	1	2	3	4
y	3624	4166	4658	5100	5492

t	5	6	7	8	9	10
y	5834	6127	6370	6563	6706	6799

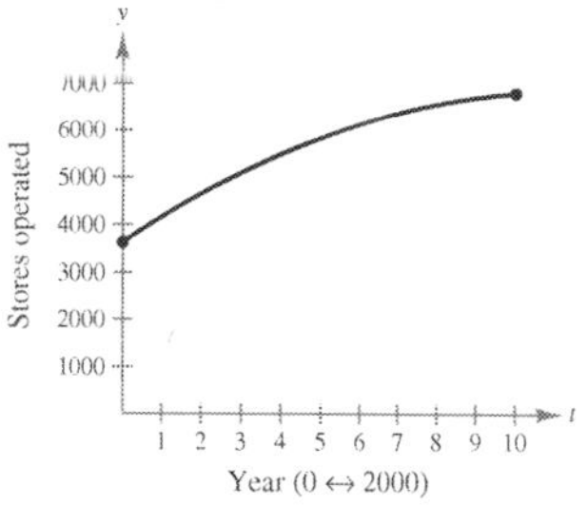

(b) When $t = 11$:

$$y = -24.91(11)^2 + 566.6(11) + 3624 = 6842.49$$

According to the model, in 2011 there will be about 6843 stores. The model's prediction for 2011 is not close to the plan of Family Dollar Stores.

(c)

t	12	13	14	15
y	6836.16	6780.01	6674.04	6518.25

(d)

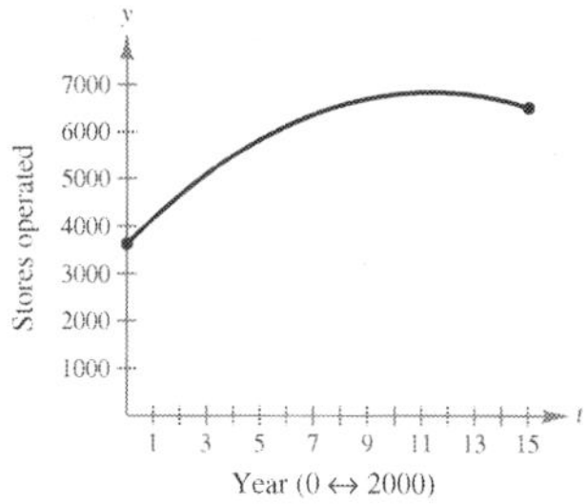

(e) The model does not support the company's expectations since the number of stores declines from 2013 to 2015. The graph begins to decrease in 2011.

111. (a) The center is at $(0, 0)$ and the radius is 87.5 cm so

$$x^2 + y^2 = (87.5)^2$$
$$x^2 + y^2 = 7656.25$$

(b) Quadrant II, Quadrant III, Quadrant II, Quadrant I, Quadrant IV

(c) $$d = \sqrt{[52 - (-40)]^2 + (-40 - 0)^2}$$
$$= \sqrt{92^2 + (-40)^2}$$
$$= \sqrt{10{,}064} \approx 100.3$$

The rat is about 100 cm from the platform.

Section 2.2 Lines in the Plane

Skills Warm Up

1. $\dfrac{4 - (-4)}{-3 - (-1)} = \dfrac{4 + 4}{-3 + 1} = \dfrac{8}{-2} = -4$

2. $\dfrac{-5 - 8}{0 - (-3)} = -\dfrac{13}{3}$

3. $\dfrac{-1}{\frac{4}{5}} = -\dfrac{5}{4}$

4. When $m = -3$:

$$-\frac{1}{m} = -\frac{1}{-3} = \frac{1}{3}.$$

Skills Warm Up —continued—

5. $2x - 3y = 6$
$$-3y = 6 - 2x$$
$$y = \tfrac{2}{3}x - 2$$

6. $4x + 2y = 0$
$$2y = -4x$$
$$y = -2x$$

7. $y - (-4) = 3[x - (-1)]$
$$y + 4 = 3(x + 1)$$
$$y + 4 = 3x + 3$$
$$y = 3x - 1$$

8. $y - 7 = \tfrac{2}{3}(x - 3)$
$$y - 7 = \tfrac{2}{3}x - 2$$
$$y = \tfrac{2}{3}x + 5$$

9. $y - (-1) = \dfrac{3 - (-1)}{2 - 4}(x - 4)$
$$y + 1 = -\frac{4}{2}(x - 4)$$
$$y + 1 = -2(x - 4)$$
$$y + 1 = -2x + 8$$
$$y = -2x + 7$$

10. $y - 5 = \dfrac{3 - 5}{0 - 2}(x - 2)$
$$y - 5 = \frac{-2}{-2}(x - 2)$$
$$y - 5 = x - 2$$
$$y = x + 3$$

1. $\text{slope} = \dfrac{\text{rise}}{\text{run}} = \dfrac{2}{4} = \dfrac{1}{2}$

3. $\text{slope} = \dfrac{\text{rise}}{\text{run}} = \dfrac{4}{-2} = -2$

5. (a) L_2

(b) L_3

(c) L_1

7.

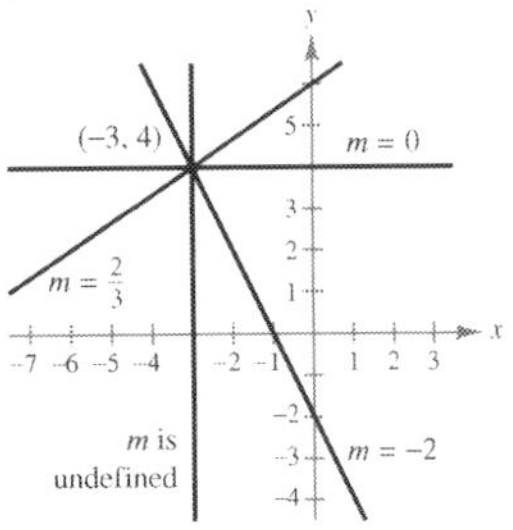

9.

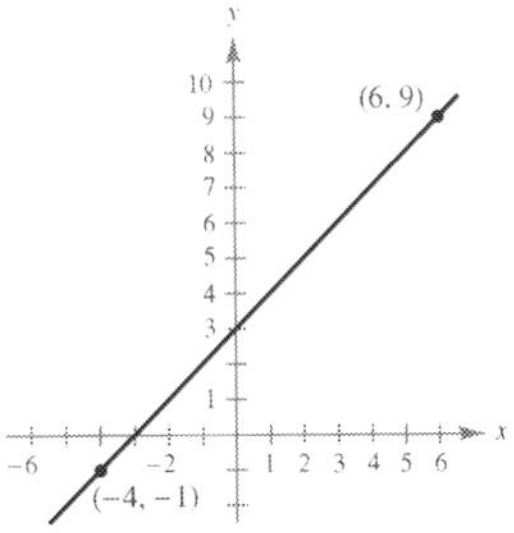

$\text{slope} = \dfrac{9 + 1}{6 + 4} = 1$

11.

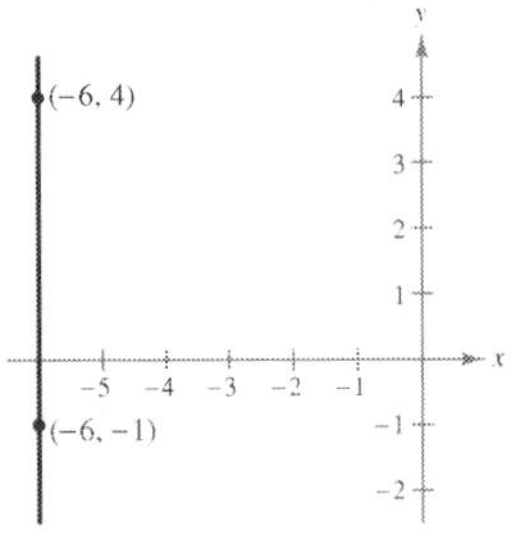

Slope is undefined.

13.

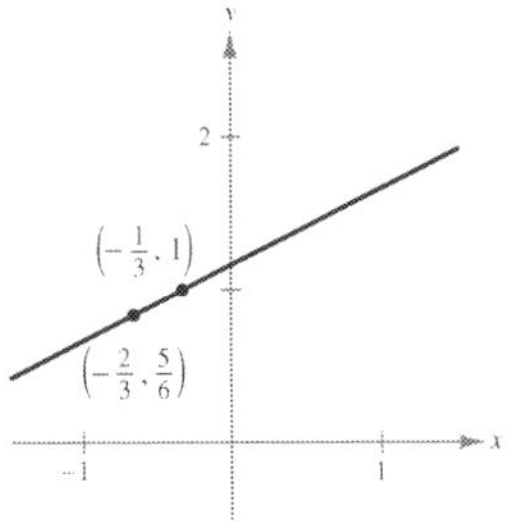

$$\text{slope} = \frac{1 - \frac{5}{6}}{-\frac{1}{3} - \left(-\frac{2}{3}\right)} = \frac{1}{2}$$

15. Since $m = 0$, y does not change. Three points are $(0, -2)$, $(-1, -2)$, and $(3, -2)$.

17. Since m is undefined, x does not change. Three points are $(2, 3)$, $(2, 0)$, and $(2, -1)$.

19. Since $m = -1$, y increases by 1 for every 1 unit increase in x. Three points are $(7, -4)$, $(8, -3)$, and $(6, -5)$.

21. Since $m = \frac{1}{2}$, y increases by 1 for every 2 unit increase in x. Three points are $(-4, 0)$, $(-2, 1)$, and $(0, 2)$.

23. $y = 1(x - 7)$

$y = x - 7 \Rightarrow x - y - 7 = 0$

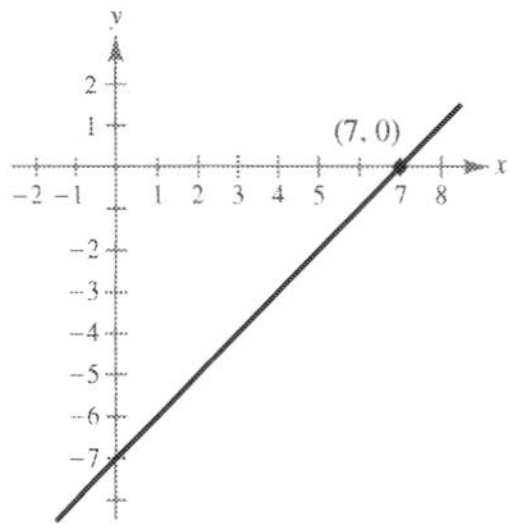

25. $y - 0 = -4(x + 2)$

$y = -4x - 8 \Rightarrow 4x + y + 8 = 0$

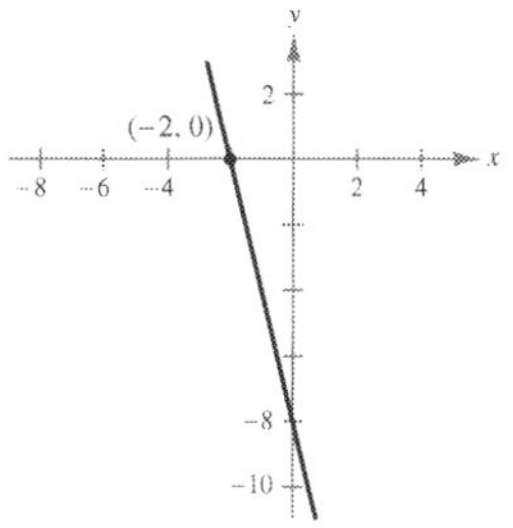

27. $y - 0 = -\frac{1}{3}(x - 4)$

$y = -\frac{1}{3}x + \frac{4}{3} \Rightarrow x + 3y - 4 = 0$

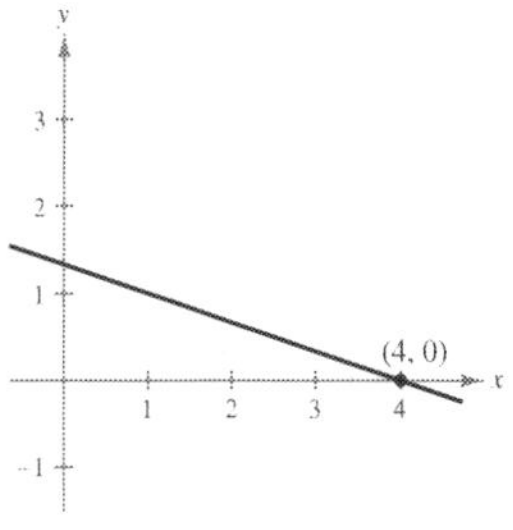

29. $y = -7$

$y + 7 = 0$

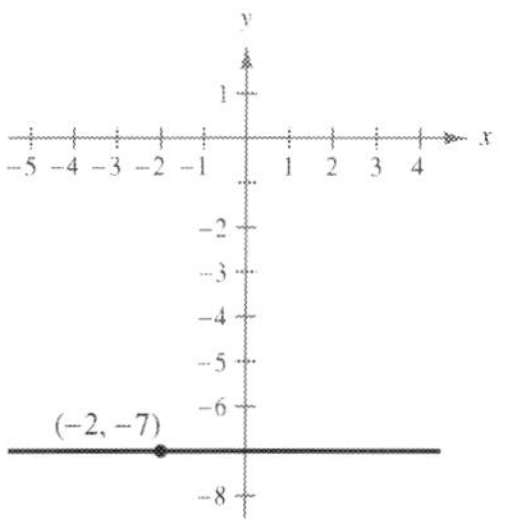

31. $y - \frac{5}{2} = \frac{4}{3}(x - 4)$

$y = \frac{4}{3}x - \frac{17}{6} \Rightarrow 8x - 6y - 17 = 0$

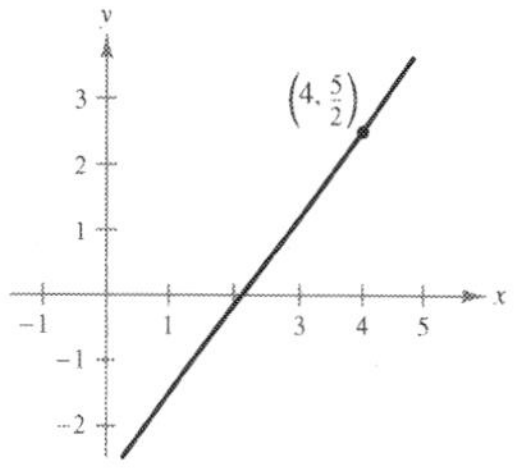

33. $y - 0 = \dfrac{-4 - 0}{-1 - 0}(x - 0)$

$y = 4x \Rightarrow 4x - y = 0$

35. $y + 4 = \left(\dfrac{3 + 4}{-7 - 7}\right)(x - 7)$

$y = -\dfrac{1}{2}(x - 7) - 4$

$y = -\dfrac{1}{2}x - \dfrac{1}{2} \Rightarrow x + 2y + 1 = 0$

37. Because both x-coordinates are -9, this is a vertical line with undefined slope.

$x = -9 \Rightarrow x + 9 = 0$

39. Because both y-coordinates are 7, this is a horizontal line with a slope of 0.

$y = 7 \Rightarrow y - 7 = 0$

41. $y - \dfrac{1}{2} = \dfrac{\frac{5}{4} - \frac{1}{2}}{\frac{1}{2} - 2}(x - 2)$

$y = -\dfrac{1}{2}(x - 2) + \dfrac{1}{2}$

$y = -\dfrac{1}{2}x + \dfrac{3}{2} \Rightarrow x + 2y - 3 = 0$

43. $y - 0.6 = \dfrac{-0.6 - 0.6}{-2 - 1}(x - 1)$

$y = 0.4(x - 1) + 0.6$

$y = 0.4x + 0.2 \Rightarrow 2x - 5y + 1 = 0$

45.

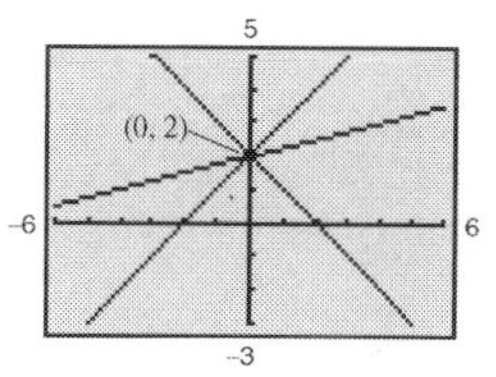

The y-intercept of each graph is $(0, 2)$.

47. $m = 2, (0, -1)$

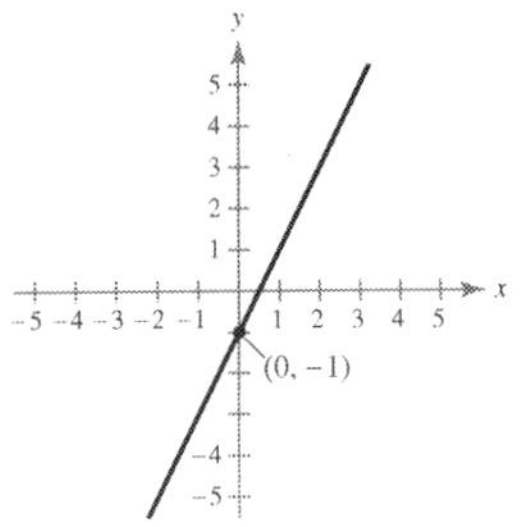

49. $4x - y - 6 = 0$

$y = 4x - 6$

Slope: $m = 4$

y-intercept: $(0, 6)$

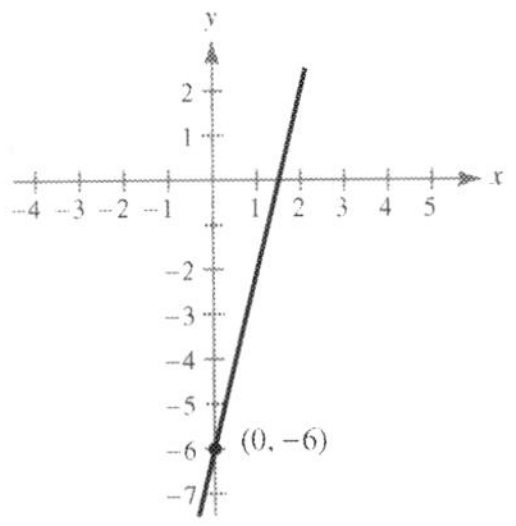

51. $8 - 3x = 0$

$x = \frac{8}{3}$

Slope: undefined

No y-intercept

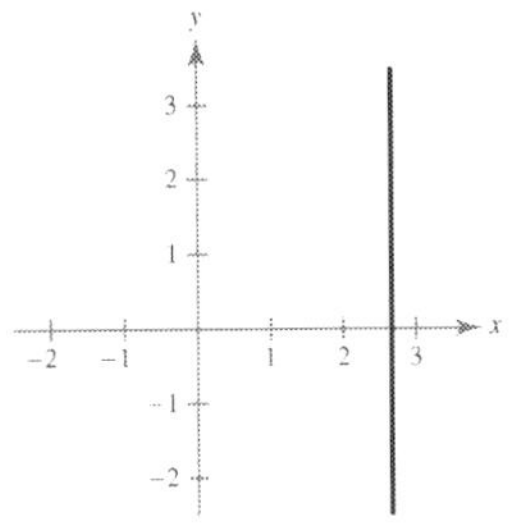

53. $7x + 6y - 30 = 0$

$y = -\frac{7}{6}x + 5$

Slope: $m = -\frac{7}{6}$

y-intercept: $(0, 5)$

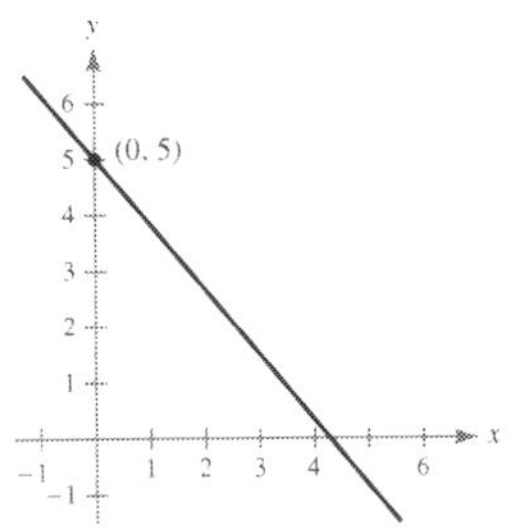

55. $2y - 7 = 0$

$2y = 7$

$y = \frac{7}{2}$

Slope: 0

y-intercept: $\left(0, \frac{7}{2}\right)$

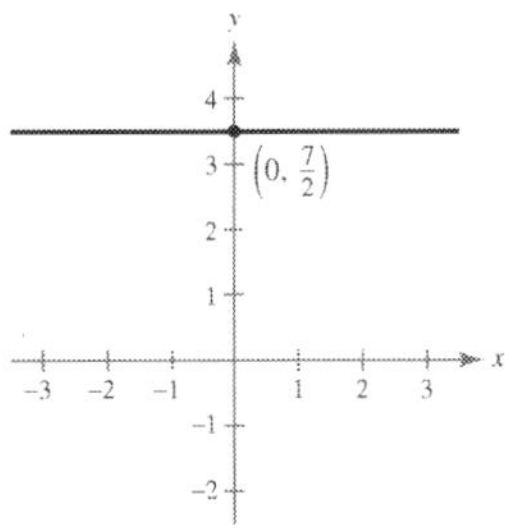

57. You could graph a vertical line and pick two convenient points on the vertical line. Find the slope. Regardless of the vertical line selected and the points selected, the slope formula would have zero in the denominator. Division by zero is not possible—so the slope does not exist. (Answers may vary.)

59. $\dfrac{x}{1} + \dfrac{y}{-4} = 1$

$4x - y - 4 = 0$

61. $\dfrac{x}{-2} + \dfrac{y}{-2} = 1$

$x + y + 2 = 0$

63. $\dfrac{x}{-1/6} + \dfrac{y}{-2/3} = 1$

$6x + \dfrac{3}{2}y = -1$

$12x + 3y + 2 = 0$

65. Changing the viewing window will affect the appearance of the slope of the line. Answers will vary.

67. $m_{L_1} = 3$

$m_{L_2} = 1$

L_1 and L_2 are neither parallel nor perpendicular.

69. $2x - y = 1$ $\qquad x + 2y = -1$

$-y = -2x + 1 \qquad 2y = -x - 1$

$y = 2x - 1 \qquad y = -\dfrac{1}{2}x - \dfrac{1}{2}$

L_1 and L_2 are perpendicular.

71. $x - 3y = -3 \qquad 2x - 6y = 6$

$-3y = -x - 3 \qquad -6y = -2x + 6$

$y = \frac{1}{3}x + 1 \qquad y = \frac{1}{3}x - 1$

L_1 and L_2 are parallel.

73. $2x - 3y - 15 = 0$ $\quad$ $3x + 2y + 8 = 0$

$-3y = -2x + 15$ $\quad$ $2y = -3x - 8$

$y = \frac{2}{3}x - 5$ $\quad$ $y = -\frac{3}{2}x - 4$

L_1 and L_2 are perpendicular.

75. $m_{L_1} = \dfrac{1 - 0}{-2 + 5} = \dfrac{1}{3}$

$m_{L_2} = \dfrac{2 - 1}{3 - 0} = \dfrac{1}{3} = m_{L_1}$

L_1 and L_2 are parallel.

77. $m_{L_1} = \dfrac{9 + 1}{5 - 0} = 2$

$m_{L_2} = \dfrac{1 - 3}{4 - 0} = -\dfrac{1}{2} = -\dfrac{1}{m_{L_1}}$

L_1 and L_2 are perpendicular.

79. $m_{L_1} = \dfrac{5 + 1}{1 + 2} = 2$

$m_{L_2} = \dfrac{-5 - 3}{5 - 1} = -2$

L_1 and L_2 are neither parallel nor perpendicular.

81. $m_{L_1} = \dfrac{4 - 7}{-6 + 1} = \dfrac{3}{5}$

$m_{L_2} = \dfrac{4 - 1}{5 - 0} = \dfrac{3}{5} = m_{L_1}$

L_1 and L_2 are parallel.

83. $y - 2x = -1$

$y = 2x - 1$

Slope: $m = 2$

(a) $y - 2 = 2(x - 6)$

$y = 2x - 10 \Rightarrow 2x - y - 10 = 0$

(b) $y - 2 = -\frac{1}{2}(x - 6)$

$y = -\frac{1}{2}x + 5 \Rightarrow x + 2y - 10 = 0$

85. $2x - 3y = 5$

$y = \frac{2}{3}x - \frac{5}{3}$

Slope: $m = \frac{2}{3}$

(a) $y + \frac{2}{3} = \frac{2}{3}\left(x - \frac{1}{4}\right)$

$y = \frac{2}{3}x - \frac{5}{6} \Rightarrow 4x - 6y - 5 = 0$

(b) $y + \frac{2}{3} = -\frac{3}{2}\left(x - \frac{1}{4}\right)$

$y = -\frac{3}{2}x - \frac{7}{24} \Rightarrow 36x + 24y + 7 = 0$

87. $y = -3$

Slope: $m = 0$

(a) $y = 0$

(b) $x = -1 \Rightarrow x + 1 = 0$

89. $m = \dfrac{34}{30(12)} = \dfrac{17}{180} > \dfrac{1}{12}$

Because $\frac{17}{180} > \frac{1}{12}$, the ramp is steeper than recommended.

91. $C - 0 = \dfrac{100 - 0}{212 - 32}(F - 32)$

$C = \dfrac{5}{9}(F - 32)$ or $F = \dfrac{9}{5}C + 32$

93. $y - 165{,}000 = \left(\dfrac{165{,}000 - 158{,}000}{2 - 1}\right)(x - 2)$

$y - 165{,}000 = 7000x - 14{,}000$

$y = 7000x + 151{,}000$

When $x = 4$, $y = 7000(4) + 151{,}000 = \$179{,}000$.

No, sales may not follow this linear pattern.

95. $m = \dfrac{4 - 1}{99 - 0} = \dfrac{1}{33}$ atmosphere per foot

$p - 4 = \dfrac{1}{33}(d - 99)$

$p - 4 = \dfrac{1}{33}d - 3$

$p = \dfrac{1}{33}d + 1$

Section 2.3 Linear Modeling and Direct Variation

Skills Warm Up

1.

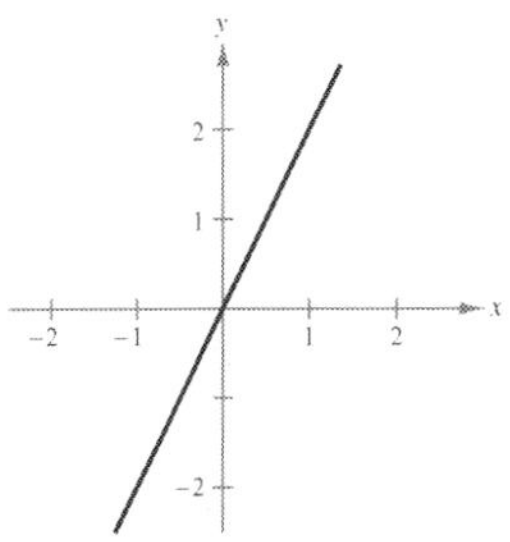

2.

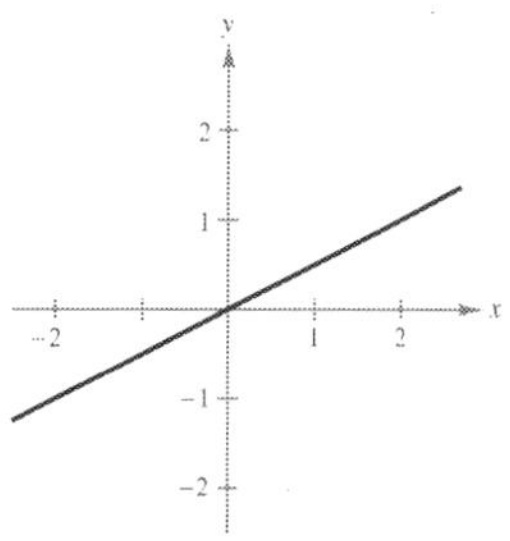

Skills Warm Up —*continued*—

3.

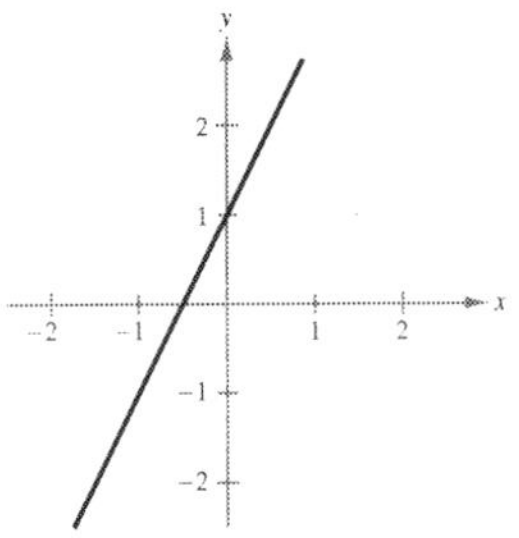

4.

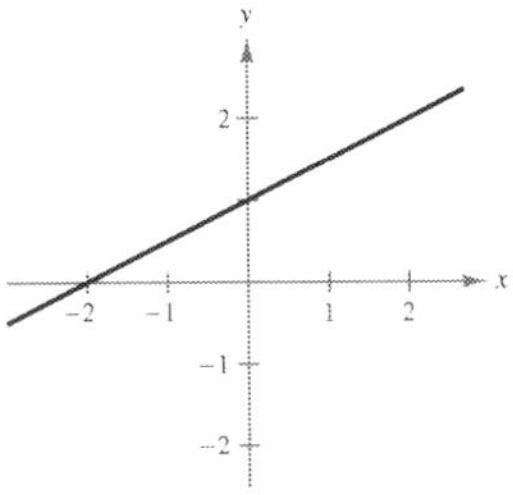

5. $y = x + 2$

6. $y = \frac{3}{2}x + 3$

7. $m = \dfrac{8 - 3}{6 - 1} = \dfrac{5}{5} = 1$

$y - 3 = x - 1$

$y = x + 2$

8. $m = \dfrac{10 - 4}{7 - 0} = \dfrac{6}{7}$

$y - 4 = \dfrac{6}{7}x$

$y = \dfrac{6}{7}x + 4$

9. $m = \dfrac{4.7 - 5.2}{5 - 1} = -\dfrac{1}{8}$

$y - 5.2 = -\dfrac{1}{8}(x - 1)$

$y - 5.2 = -\dfrac{1}{8}x + \dfrac{1}{8}$

$y = -\dfrac{1}{8}x + \dfrac{213}{40}$

$5x + 40y - 213 = 0$

10. $m = \dfrac{3.6 - 6.5}{8 - 2} = -\dfrac{29}{60}$

$y - 6.5 = -\dfrac{29}{60}(x - 2)$

$y - 6.5 = -\dfrac{29}{60}x + \dfrac{29}{30}$

$y = -\dfrac{29}{60}x + \dfrac{112}{15}$

$29x + 60y - 448 = 0$

1.

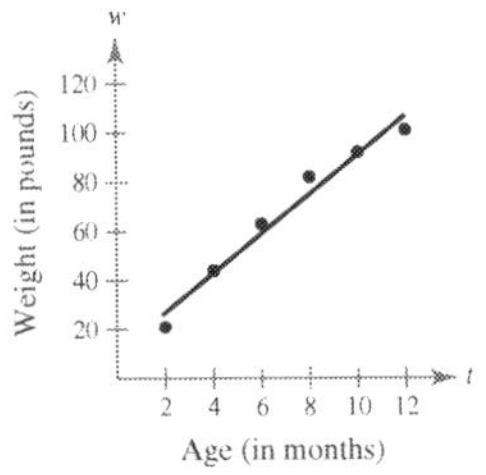

The model is a good fit for the actual data.

3.

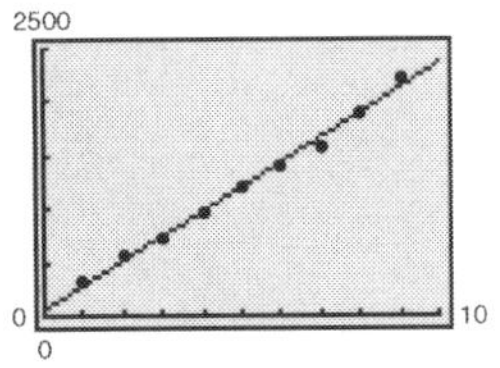

$y = 231.1x + 53.256$

The model is a good fit for the actual data.

5. $y = mx$

$3 = m(8)$

$\frac{3}{8} = m$

$y = \frac{3}{8}x$

7. $y = mx$

$300 = m(15)$

$20 = m$

$y = 20x$

9. $y = mx$

$3.2 = m(7)$

$\frac{16}{35} = m$

$y = \frac{16}{35}x$

11. $H = \frac{27}{9}p = 3p$

13. $c = \frac{12}{20}d = \frac{3}{5}d$

15. $I = kP$

$120 = k(2000)$

$0.06 = k$

$I = 0.06P$

17. (a) $y = kx$

$2211 = k(165{,}000)$

$0.0134 = k$

$y = 0.0134x$

(b) When $x = 185{,}000$,

$y = 0.0134(185{,}000) = 2479.$

The property tax on a house with assessed value of \$185,000 is \$2479.

19. (a) $M = kK$

$64 = k(103)$

$\frac{64}{103} = k$

$M = \frac{64}{103}K$

(b)

km/h	40	60	80	100	120
mi/h	24.85	37.28	49.71	62.14	74.56

21. $V = 2540 + 140(t - 12)$

$V = 2540 + 140t - 1680$

$V = 140t + 860,\ 12 \le t \le 17$

23. $V = 20{,}400 - 2142(t - 12)$

$V = 20{,}400 - 2142t + 25{,}704$

$V = -2142t + 46{,}104,\ 12 \le t \le 17$

25. $V = 154{,}000 + 10{,}780(t - 12)$

$V = 154{,}000 + 10{,}780t - 129{,}360$

$V = 10{,}780t + 24{,}640,\ 12 \le t \le 17$

27. (a) 2 minutes = 120 seconds

(0, 7000), (120, 4600)

$$h - 7000 = \frac{4600 - 7000}{120 - 0}(t - 0)$$

$$h = -20t + 7000$$

(b) When the parachutist reaches the ground, we have

$h = 0$

$0 = -20t + 7000$

$t = \frac{7000}{20} = 350$ seconds

≈ 5.833 minutes.

The time will be

(2 hours + 8 minutes) + (5 minutes + 50 seconds)

= 2:13:50 P.M.

29. $V = 875 - \left(\frac{875}{5}\right)t$

$= 875 - 175t,\ 0 \le t \le 5$

31. $W = 11.50 + 0.75x$

33. (a) $H = 0.5t + 4$

(b) 1.5 years is 18 months.

$H = 0.5(18) + 4 = 13$

The hair will be 13 inches long in 1.5 years.

35.

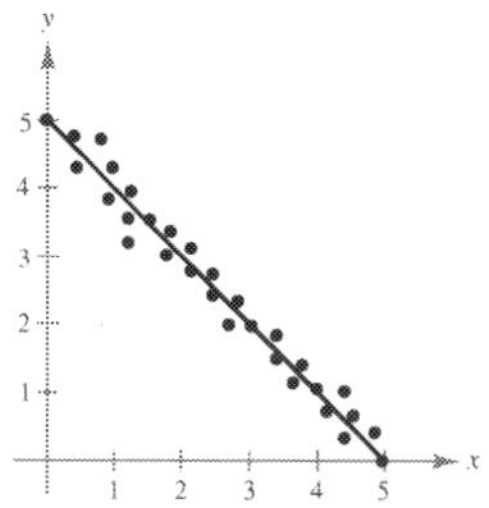

Two approximate points on the line are (0, 5) and (4.9, 0).

Thus, $y - 5 = \frac{0 - 5}{4.9 - 0}(x - 0)$

$y - 5 = -1.02x$

$y = -1.02x + 5$ or $y = -x + 5$.

37. No. The data cannot be approximated by a linear model.

39.

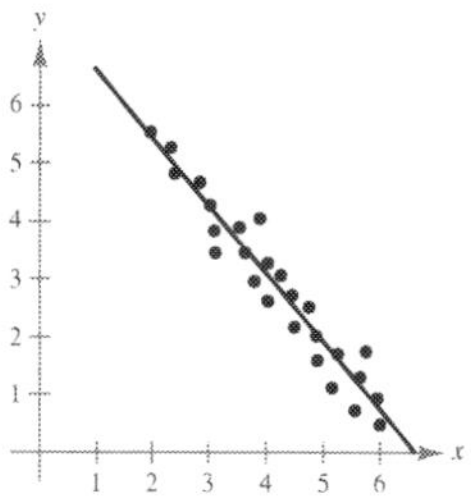

Two approximate points on the line are (2, 5.5) and (4, 3).

Thus, $y - 3 = \frac{3 - 5.5}{4 - 2}(x - 4)$

$y - 3 = -1.25x + 5$

$y = -1.25x + 8.$

41. (a)

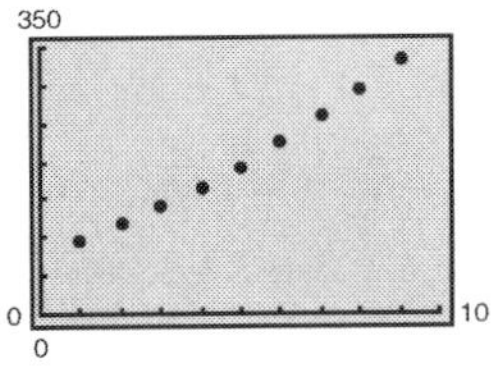

Yes, the data appear linear.

(b) $C = 29.19t + 57.1$

(c) The slope is 29.19. The number of autistic children receiving disability services is increasing by about 29 thousand children per year.

(d) For $t = 11, C = 29.19(11) + 57.1 = 378.19$.

In 2011, about 378 thousand autistic children will receive disability services.

For $t = 12, C = 29.19(12) + 57.1 = 407.38$.

In 2012, about 407 thousand autistic children will receive disability services.

The estimates are reasonable.

43. (a)

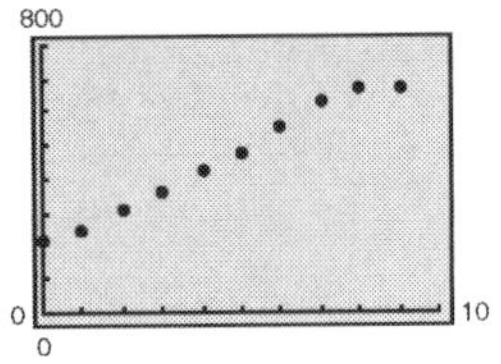

(b) Using $(0, 210.8)$ and $(8, 677.1)$:

$$y - 210.8 = \left(\frac{677.1 - 210.8}{8 - 0}\right)(t - 0)$$

$$y - 210.8 = 58.3t$$

$$y = 58.3t + 210.8$$

Answers will vary depending upon the selected ordered pairs.

(c) $y = 56.69t + 200.7$

Using the linear regression equation:

2010: $y = 56.69(10) + 200.7 = \$767.6$ million

2011: $y = 56.69(11) + 200.7 = \$824.29$ million

Using the equation from part (b):

2010: $y = 58.3(10) + 210.8 = \$793.8$ million

2011: $y = 58.3(11) + 210.8 = \$852.1$ million

(d) The projections by California Pizza Kitchen are lower than the predictions given by the models.

(e) Yes. Using the linear regression equation, the yearly revenue is expected to reach $y = 56.69(13) + 200.7 = \$937.67$ million by 2013. Using the equation from part (b), the yearly revenue is expected to be $y = 58.3(13) + 210.8 = \$968.7$ million by 2013.

45. (a)

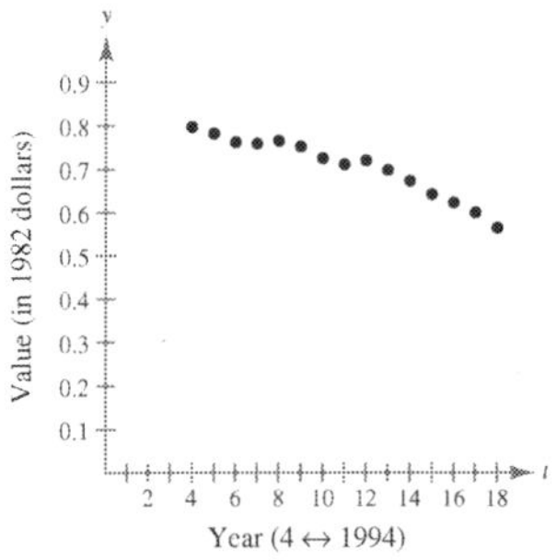

The data appear to be approximately linear.

(b) $y = -0.0152t + 0.873$

(c) 2010: $y = -0.0152(20) + 0.873 = \$0.569$

2011: $y = -0.0152(21) + 0.873 = \$0.5538$

Because the data follow a linear pattern from 1994 to 2008, you can assume the estimates for 2010 and 2011 are reliable.

47. (a)

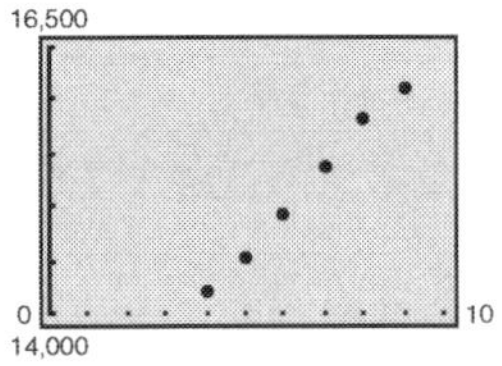

The data appear to be approximately linear.

(b) $E = 394.17t + 12{,}592.9$

(c) When $t = 11: E = 394.17(11) + 12{,}592.9 = 16{,}928.77$

When $t = 13: E = 394.17(13) + 12{,}592.9 = 17{,}717.11$

So, there will be about 16,928,770 employees in the health service industry in 2011 and about 17,717,110 employees in 2013.

(d)

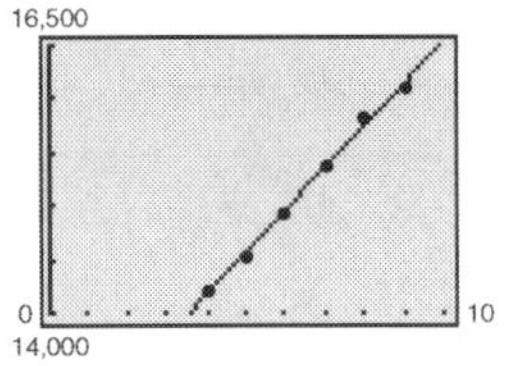

The predictions are most likely going to be right because the model is a good fit for the data.

49. The model for population because population tends to change at a consistent rate, whereas snowfall can be quite different from year to year. You should use more than three data points to find an accurate model.

51. Yes. When you start you have zero wrong moves $(x = 0)$ and 800 points $(y = 800)$. So, 800 is the y-intercept. Then you lose 50 points for each wrong move. This is a slope of -50.

53. A known rate of change corresponds to the slope of a linear model, so determine whether the rate of change of y with respect to x is constant. The data show a 10-unit increase in y for each 1-unit increase in x, so the rate of change is constant and the data is linear.

Section 2.4 Functions

Skills Warm Up

1. $2(-3)^3 + 4(-3) - 7 = 2(-27) - 12 - 7 = -73$

2. $4(-1)^2 - 5(-1) + 4 = 4 + 5 + 4 = 13$

3. $(x + 1)^2 + 3(x + 1) - 4 - (x^2 + 3x - 4)$

$= x^2 + 2x + 1 + 3x + 3 - 4 - x^2 - 3x + 4$

$= 2x + 4$

$= 2(x + 2)$

4. $(x - 2)^2 - 4(x - 2) - (x^2 - 4)$

$= x^2 - 4x + 4 - 4x + 8 - x^2 + 4$

$= -8x + 16$

$= -8(x - 2)$

5. $2x + 5y - 7 = 0$

$5y = 7 - 2x$

$y = \frac{7}{5} - \frac{2}{5}x$

6. $y^2 = x^2$

$y = \pm x$

7. $x^2 - 4 \ge 0$

$x^2 \ge 4$

$x \le -2, x \ge 2$

8. $9 - x^2 \ge 0$

$9 \ge x^2$

$-3 \le x \le 3$

9. $x^2 + 2x + 1 \ge 0$

All real numbers

10. $x^2 - 3x + 2 \ge 0$

$(x - 2)(x - 1) \ge 0$

$x \le 1, x \ge 2$

1. Because each element of A is matched with exactly one element of B, $\{(a, 1), (b, 1), (c, 1)\}$ represents a function from A to B.

3. Because $b \in A$ is not matched with an element of B, $\{(a, 1), (c, 3)\}$ does not represent a function from A to B.

5. Because each element of A is matched with exactly one element of B, $\{(0, 1), (1, -2,), (2, 0), (3, 2)\}$, represents a function from A to B.

7. Because each element of A is matched with exactly one element of B, $\{(0, 0), (1, 0), (2, 0), (3, 0)\}$ represents a function from A to B.

9. Because $(c, 2)$ and $(c, 3)$ are both in $\{(a, 1), (c, 2), (c, 3), (b, 3)\}$, this set does not represent a function from A to B.

11. Because the elements of A are the outputs rather than the inputs, $\{(0, a), (2, c), (3, b)\}$ does not represent a function from A to B.

13. Because each element of A is matched with exactly one element of B, $\{(a, 1), (b, 2), (c, 3)\}$ represents a function from A to B.

15. $x^2 + y^2 = 4 \Rightarrow y = \pm\sqrt{4 - x^2}$

Thus, y *is not* a function of x.

17. $x^2 + y = 4 \Rightarrow y = 4 - x^2$

Thus, y *is* a function of x.

19. $2x + 3y = 4 \Rightarrow y = \frac{1}{3}(4 - 2x)$

Thus, y *is* a function of x.

21. $y^2 = x^2 - 1 \Rightarrow y = \pm\sqrt{x^2 - 1}$

Thus, y *is not* a function of x.

23. $x^2y - x^2 + 4y = 0$

$y(x^2 + 4) = x^2$

$$y = \frac{x^2}{x^2 + 4}$$

Thus, y *is* a function of x.

25. $f(x) = 6 - 4x$

(a) $f(3) = 6 - 4(3) = -6$

(b) $f(-7) = 6 - 4(-7) = 34$

(c) $f(t) = 6 - 4(t) = 6 - 4t$

(d) $f(c + 1) = 6 - 4(c + 1) = 2 - 4c$

27. $g(x) = \dfrac{1}{x^2 - 2x}$

(a) $g(1) = \dfrac{1}{(1)^2 - 2(1)} = -1$

(b) $g(-3) = \dfrac{1}{(-3)^2 - 2(-3)} = \dfrac{1}{15}$

(c) $g(t) = \dfrac{1}{(t)^2 - 2(t)} = \dfrac{1}{t^2 - 2t}$

(d) $g(t + 1) = \dfrac{1}{(t + 1)^2 - 2(t + 1)} = \dfrac{1}{t^2 - 1}$

29. $f(x) = 2x - 3$

(a) $f(1) = 2(1) - 3 = -1$

(b) $f(-3) = 2(-3) - 3 = -9$

(c) $f(x - 1) = 2(x - 1) - 3 = 2x - 5$

(d) $f\left(\frac{1}{4}\right) = 2\left(\frac{1}{4}\right) - 3 = -\frac{5}{2}$

31. $h(t) = t^2 - 2t$

(a) $h(2) = 2^2 - 2(2) = 0$

(b) $h(-1) = (-1)^2 - 2(-1) = 3$

(c) $h(x + 2) = (x + 2)^2 - 2(x + 2) = x^2 + 2x$

(d) $h(1.5) = (1.5)^2 - 2(1.5) = -0.75$

33. $V(r) = \dfrac{4}{3}\pi r^3$

(a) $V(3) = \dfrac{4}{3}\pi(3)^3 = 36\pi$

(b) $V(0) = \dfrac{4}{3}\pi(0)^3 = 0$

(c) $V\left(\dfrac{3}{2}\right) = \dfrac{4}{3}\pi\left(\dfrac{3}{2}\right)^3 = \dfrac{9\pi}{2}$

(d) $V(2r) = \dfrac{4}{3}\pi(2r)^3 = \dfrac{32\pi r^3}{3}$

35. $f(y) = 3 - \sqrt{y}$

(a) $f(4) = 3 - \sqrt{4} = 1$

(b) $f(100) = 3 - \sqrt{100} = -7$

(c) $f(4x^2) = 3 - \sqrt{4x^2} = 3 - 2|x|$

(d) $f(0.25) = 3 - \sqrt{0.25} = 2.5$

37. $c(x) = \dfrac{1}{x^2 - 16}$

(a) $c(4) = \dfrac{1}{4^2 - 16}$ is undefined.

(b) $c(0) = \dfrac{1}{0^2 - 16} = -\dfrac{1}{16}$

(c) $c(y + 2) = \dfrac{1}{(y + 2)^2 - 16}$

$= \dfrac{1}{y^2 + 4y - 12}, \; y \neq -6, 2$

(d) $c(y - 2) = \dfrac{1}{(y - 2)^2 - 16}$

$= \dfrac{1}{y^2 - 4y - 12}, \; y \neq -2, 6$

39. $f(x) = \dfrac{|x|}{x}$

(a) $f(2) = \dfrac{|2|}{2} = 1$

(b) $f(-2) = \dfrac{|-2|}{-2} = -1$

(c) $f(x^2) = \dfrac{|x^2|}{x^2} = 1$

(d) $f(x - 1) = \dfrac{|x - 1|}{x - 1}$

41. $f(x) = \begin{cases} 3x - 1, & x < 0 \\ 2x + 3, & x \geq 0 \end{cases}$

(a) $f(-1) = 3(-1) - 1 = -4$

(b) $f(0) = 2(0) + 3 = 3$

(c) $f(-2) = 3(-2) - 1 = -7$

(d) $f(2) = 2(2) + 3 = 7$

43. $h(t) = \begin{cases} t^2 - 2, & t < 2 \\ 3, & t \ge 2 \end{cases}$

(a) $h(-2) = (-2)^2 - 2 = 2$

(b) $h(1) = 1^2 - 2 = -1$

(c) $h(2) = 3$

(d) $h(5) = 3$

45. $15 - 3x = 0$

$3x = 15$

$x = 5$

47. $x^2 - 9 = 0$

$x^2 = 9$

$x = \pm 3$

49. $x^3 - x = 0$

$x(x^2 - 1) = 0$

$x = 0 \quad x^2 - 1 = 0$

$x^2 = 1$

$x = \pm 1$

51. $\dfrac{3}{x-1} + \dfrac{4}{x-2} = 0$

$3(x - 2) + 4(x - 1) = 0$

$7x - 10 = 0$

$x = \dfrac{10}{7}$

53. $\{0, 3, 5, 8, 9\}$

55. $\{-2, -1, 0, 1, 2\}$

57. All real numbers x or $(-\infty, \infty)$

59. All real numbers $t \ne 0$ or $(-\infty, 0) \cup (0, \infty)$

61. All real numbers y or $(-\infty, \infty)$

63. $1 - x^2 \ge 0$

$x^2 \le 1$

$-1 \le x \le 1$

So the domain is $[-1, 1]$.

65. All real numbers $x \ne 0$ and $x \ne -2$ or $(-\infty, -2) \cup (-2, 0) \cup (0, \infty)$

67. $x + 1 \ge 0$ and $x - 2 \ne 0$

$x \ge -1 \qquad x \ne 2$

So the domain is $[-1, 2) \cup (2, \infty)$.

69. $x > 0$

The domain is $(0, \infty)$ or all real numbers greater than zero.

71. $f(x) = x^2$

$\{(-2, 4), (-1, 1), (0, 0), (1, 1), (2, 4)\}$

73. $f(x) = \sqrt{x + 2}$

$\{(-2, 0), (-1, 1), (0, \sqrt{2}), (1, \sqrt{3}), (2, 2)\}$

75. The domain of $f(x) = \sqrt{x - 2}$ is all real numbers $x \ge 2$ or $[2, \infty)$ since an even root of a negative number is not a real number.

The domain of $g(x) = \sqrt[3]{x - 2}$ is all real numbers or $(-\infty, \infty)$. The functions f and g have different domains because an odd root of a negative number is a real number but an even root of a negative number is not a real number.

77. There will be n ordered pairs because set A contains n elements and each element in the domain must be matched with exactly one element in the range.

79. (a) $V = x(18 - 2x)^2 = 4x(9 - x)^2$

(b) Because $V > 0$, the domain is $0 < x < 9$.

(c) $V = 4[18 - 2(4)]^2 = 4(10)^2 = 400 \text{ in.}^3$

81. $y = -\frac{1}{10}(30)^2 + 3(30) + 6 = 6$

Yes, at a distance of 30 feet, the height of the ball will be 6 feet. So, the teammate will be able to catch the ball without backing up.

83. 2000: $A = -3.4(10) + 979 = 945$ million acres

2009: $A = -3.4(19) + 979 = 914.4$ million acres

85. (a) $C = 57{,}000 + 2.05x$

(b) $\overline{C} = \dfrac{57{,}000 + 2.05x}{x}$

(c)

x	100	1000	10,000	100,000
$\overline{C}$	572.05	59.05	7.75	2.62

(d) Answers will vary. Sample answer: The average cost per unit decreases as x gets larger.

87. (a)

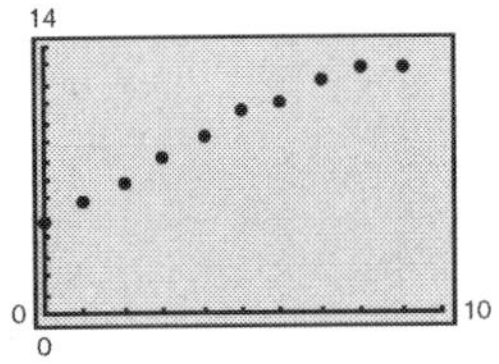

(b) Linear model: $P = 0.965t + 5.11$

Quadratic model: $P = -0.0456t^2 + 1.375t + 4.56$

(c)

Year	P (Actual)	P (Linear)	P (Quadratic)
2000	4.80	5.11	4.56
2001	5.89	6.08	5.89
2002	6.81	7.04	7.13
2003	8.10	8.01	8.27
2004	9.30	8.97	9.33
2005	10.55	9.94	10.30
2006	11.13	10.90	11.17
2007	12.09	11.87	11.95
2008	12.95	12.83	12.64
2009	12.91	13.80	13.24

The quadratic model is a better fit because its values for P tend to be closer to the actual values than those of the linear model.

89. (a) $r(t) = 0.75t$

(b) $A = 0.5625\pi t^2$

Time, t	1	2	3
Radius, r (in feet)	0.75	1.5	2.25
Area, r (in square feet)	1.767	7.069	15.904

Time, t	4	5
Radius, r (in feet)	3	3.75
Area, r (in square feet)	28.274	44.179

(c) $\dfrac{A(2)}{A(1)} = \dfrac{7.069}{1.767} = 4.000$

$\dfrac{A(4)}{A(2)} = \dfrac{28.274}{7.069} = 4.000$

$A(8) = 113.097$

91. (a) Correct; Each price has exactly one tax amount.

(b) Incorrect; For any given value of education there is not exactly one specific value of intelligence.

93. Answers will vary.

Chapter 2 Quiz Yourself

1. (a)

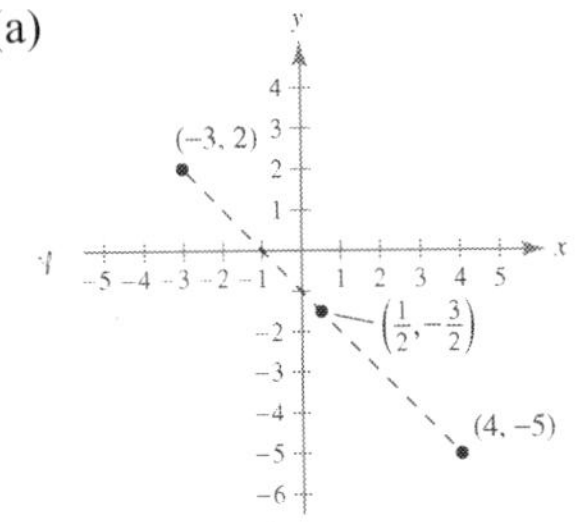

(b) $d = \sqrt{(4-(-3))^2 + (-5-2)^2}$

$= \sqrt{(7)^2 + (-7)^2}$

$= \sqrt{49 + 49}$

$= \sqrt{98}$

$= 7\sqrt{2}$

(c) $\left(\dfrac{-3+4}{2}, \dfrac{2+(-5)}{2}\right) = \left(\dfrac{1}{2}, -\dfrac{3}{2}\right)$

2. (a)

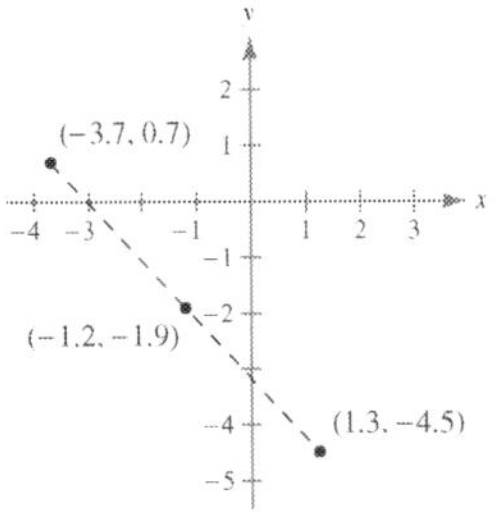

(b) $d = \sqrt{(-3.7-1.3)^2 + (0.7+4.5)^2}$

$= \sqrt{(5)^2 + (5.2)^2}$

$= \sqrt{52.04}$

≈ 7.214

(c) $\left[\dfrac{1.3+(-3.7)}{2}, \dfrac{-4.5+0.7}{2}\right] = (-1.2, -1.9)$

3. (a)

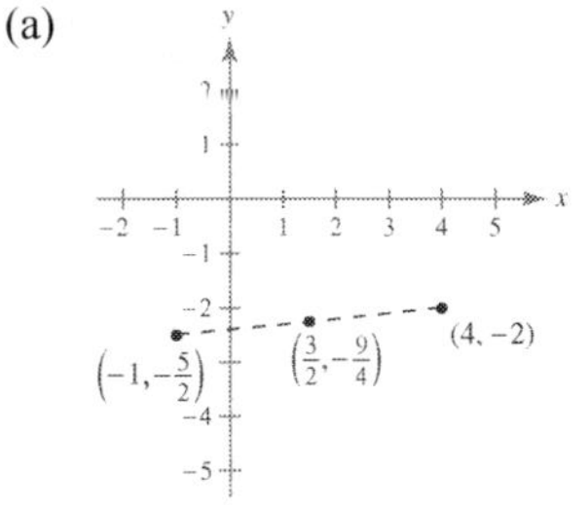

(b) $$d = \sqrt{(4 - (-1))^2 + \left(-2 - \frac{-5}{2}\right)^2}$$

$$= \sqrt{5^2 + \left(\frac{1}{2}\right)^2}$$

$$= \sqrt{25 + \frac{1}{4}}$$

$$= \sqrt{\frac{101}{4}}$$

$$= \frac{\sqrt{101}}{2}$$

$$\approx 5.025$$

(c) $$\left[\frac{4 + (-1)}{2}, \frac{-2 + \left(-\frac{5}{2}\right)}{2}\right] = \left(\frac{3}{2}, -\frac{9}{4}\right)$$

4.

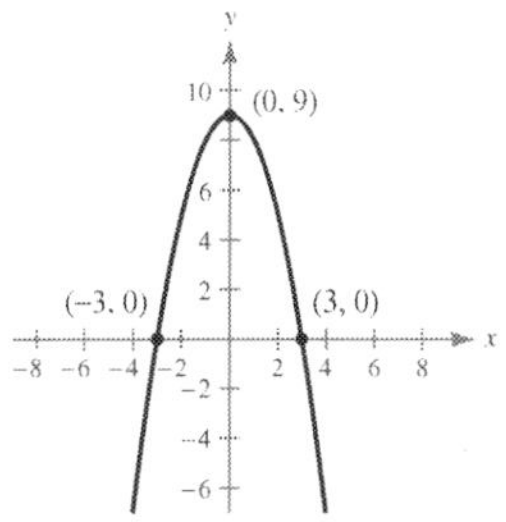

Intercepts: $(\pm 3, 0)$, $(0, 9)$

Symmetry: y-axis

5.

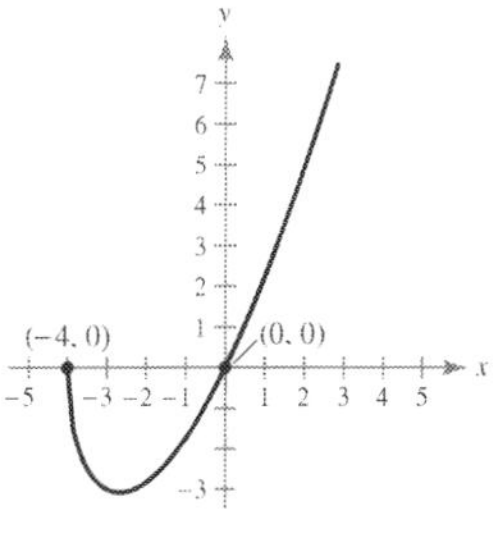

Intercepts: $(-4, 0)$, $(0, 0)$

Symmetry: None

6.

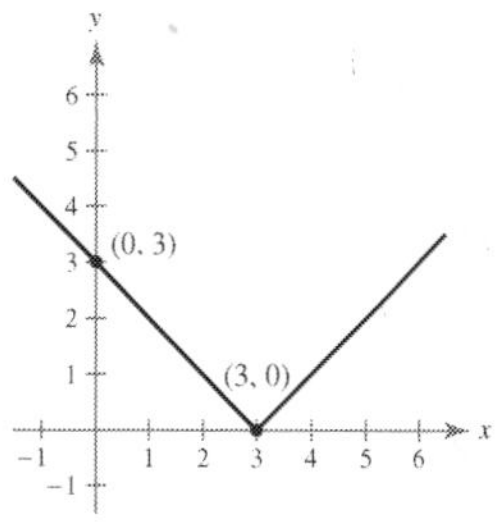

Intercepts: $(3, 0)$, $(0, 3)$

Symmetry: None

7. First use the Distance Formula to find the radius.

$$r = \sqrt{(4 - 0)^2 + (5 - 3)^2}$$

$$r = \sqrt{16 + 4}$$

$$r = \sqrt{20} = 2\sqrt{5}$$

So, the equation of the circle in standard form is:

$$(x - 4)^2 + (y - 5)^2 = 20$$

8.

$$x^2 + y^2 - 2x + 4y - 4 = 0$$

$$\left(x^2 - 2x + 1\right) + \left(y^2 + 4y + 4\right) = 4 + 1 + 4$$

$$(x - 1)^2 + (y + 2)^2 = 9$$

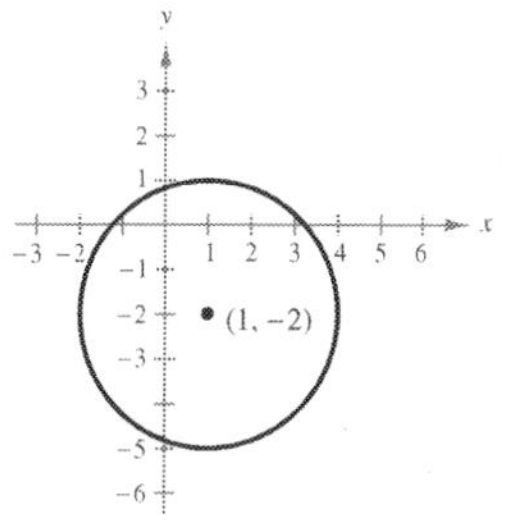

9. $y - 5 = \frac{2}{3}(x - 3)$

$y = \frac{2}{3}x + 3$

or $2x - 3y + 9 = 0$

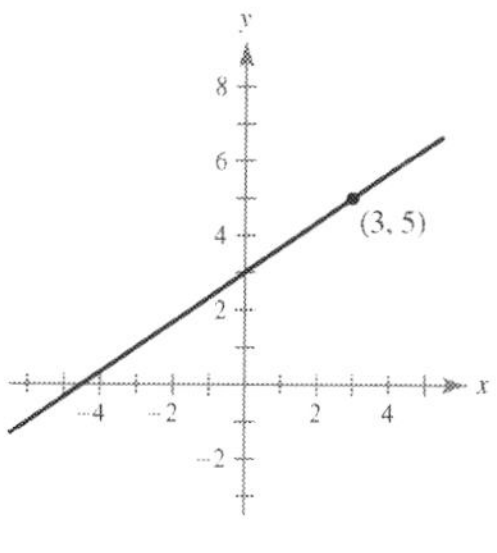

10. $y = 4$

$y - 4 = 0$

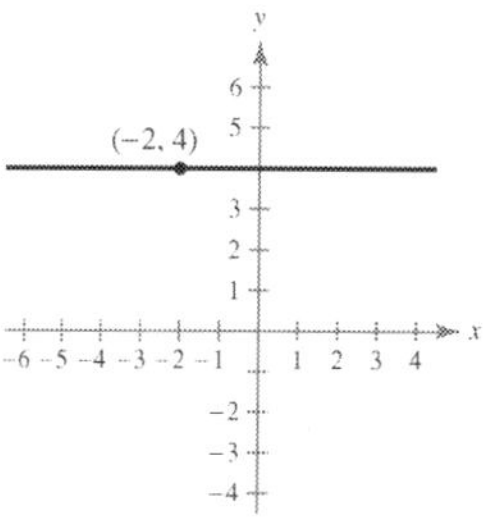

11. $x = 2$

$x - 2 = 0$

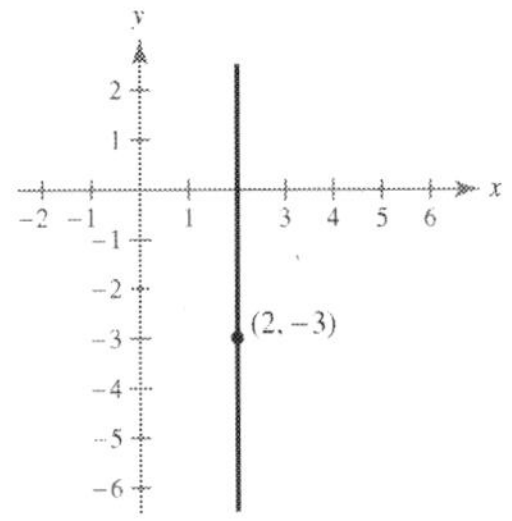

12. $y - (-5) = -2[x - (-2)]$

$y = -2x + 9$

or $2x + y + 9 = 0$

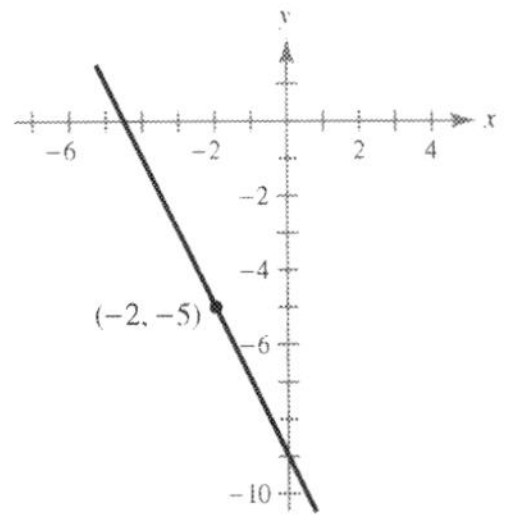

13. Let $t = 0$ represent 2010 and $t = 2$ represent 2012.

Ordered pairs: $(0, 233{,}134)$ and $(2, 244{,}288)$

Using the point-slope form of a line:

$$P - 233{,}134 = \left(\frac{244{,}288 - 233{,}134}{2 - 0}\right)(t - 0)$$

$$P - 233{,}134 = 5577t$$

$$P = 5577t + 233{,}134$$

When $t = 5$:

$$P = 5577(5) + 233{,}134 = 261{,}019$$

In 2015, the population of the city will be about 261,019 people.

14. $x^2 + y = 3$ can be rewritten as

$y = 3 - x^2$.

$y = 3 - x^2$ represents y as a function of x because each element in the domain corresponds to exactly one element in the range.

15. Yes; Each element of A is matched with exactly one element of B.

16. No; The element 8 in A has no corresponding value in B.

17. $f(x) = 3(x + 2) - 4$

(a) $f(0) = 3(2) - 4 = 2$

(b) $f(-3) = 3(-3 + 2) - 4 = 3(-1) - 4 = -7$

18. $g(t) = 2t^3 - t^2$

(a) $g(1) = 2(1)^3 - (1)^2 = 2 - 1 = 1$

(b) $g(-2) = 2(-2)^3 - (-2)^2 = 2(-8) - 4 = -20$

19. $x - 4 \ge 0$

$x \ge 4$

20. $x + 2 \ne 0$

$x \ne -2$

All real numbers $x \ne -2$

21.

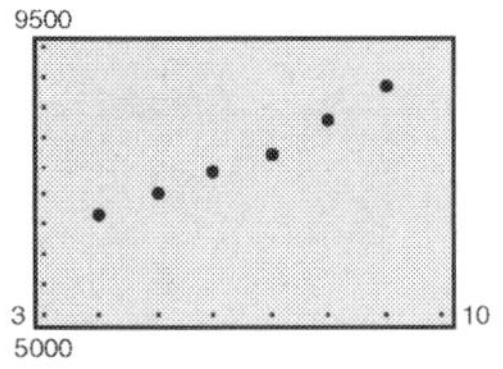

Linear model: $y = 428.5t + 4873$

Quadratic model: $y = 35.29t^2 - 30.2t + 6261$

22. Linear model:

When $t = 10$: $y = 428.5(10) + 4873 = 9158$

When $t = 11$: $y = 428.5(11) + 4873 = 9586.5$

Federal costs for school lunch programs will be \$9158 million in 2010 and \$9586.5 million in 2011.

Quadratic model:

When $t = 10$:

$y = 35.29(10)^2 - 30.2(10) + 6261 = 9488$

When $t = 11$:

$y = 35.29(11)^2 - 30.2(11) + 6261 = 10{,}198.89$

Federal costs for school lunch programs will be \$9488 million in 2010 and \$10,198.89 million in 2011.

23. $C = 2\pi r \Rightarrow r = \dfrac{C}{2\pi}$

Because the area of a circle is given by $A = \pi r^2$,

$$A = \pi\left(\frac{C}{2\pi}\right)^2.$$

$A = \dfrac{C^2}{4\pi}$ is the area of a circle as a function of its circumference.

Section 2.5 Graphs of Functions

Skills Warm Up

1. $f(2) = -(2)^3 + 5(2) = -8 + 10 = 2$

2. $f(6) = 6^2 - 6(6) = 0$

3. $f(-x) = -\dfrac{3}{x}$

4. $f(-x) = (-x)^2 + 3 = x^2 + 3$

5. $x^3 - 16x = 0$

   $x(x^2 - 16) = 0$

   $x = 0 \quad x = \pm 4$

6. $2x^2 - 3x + 1 = 0$

   $(2x - 1)(x - 1) = 0$

   $x = \frac{1}{2} \quad x = 1$

7. $g(x) = \dfrac{4}{x - 4}$

   $x - 4 \neq 0$

   All real numbers $x \neq 4$

8. $f(x) = \dfrac{2x}{(x - 4)(x - 5)}$

   $x - 4 \neq 0 \quad x - 5 \neq 0$

   $x \neq 4 \quad x \neq 5$

   All real numbers $x \neq 4$, $x \neq 5$

9. $5 - 3t \geq 0$

   $t \leq \frac{5}{3}$

10. All real numbers

1. $f(x) = \sqrt{x - 1}$

Domain: $[1, \infty)$

Range: $[0, \infty)$

$f(1) = \sqrt{1 - 1} = 0$

3. $f(x) = 4 - x^2$

Domain: $(-\infty, \infty)$

Range: $(-\infty, 4]$

$f(0) = 4 - 0^2 = 4$

5. $f(x) = x^3 - 1$

Domain: $(-\infty, \infty)$

Range: $(-\infty, \infty)$

$f(0) = 0^3 - 1 = -1$

7. $f(x) = \sqrt{25 - x^2}$

Domain: $[-5, 5]$

Range: $[0, 5]$

$f(0) = \sqrt{25 - 0^2} = 5$

9. $y = x^2$

A vertical line intersects the graph just once, so y is a function of x.

11. $x^2 + y^2 = 9$

Some vertical lines intersect the graph more than once, so y is not a function of x.

13. $y = \frac{1}{2}x^2$

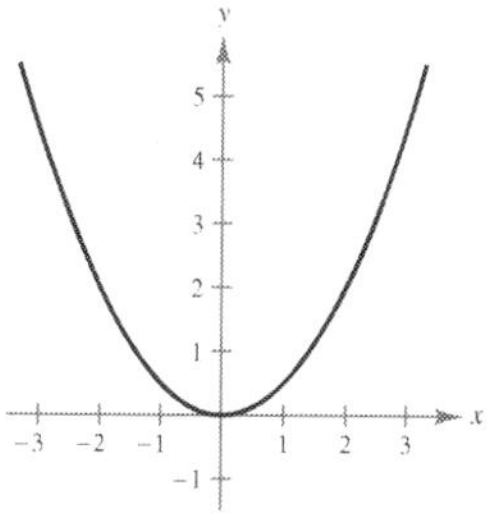

A vertical line intersects the graph just once, so y is a function of x.

15. $x^2 + y^2 = 25$

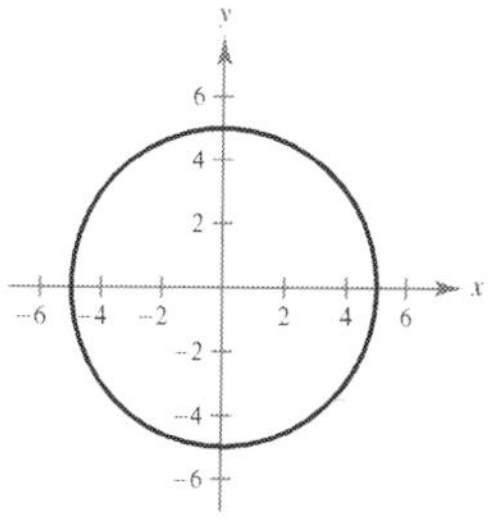

Some vertical lines intersect the graph more than once, so y is not a function of x.

17. $f(x) = 2x$

Increasing on $(-\infty, \infty)$

No change in the graph's behavior

19. $f(x) = x^3 - 3x^2$

Increasing on $(-\infty, 0)$ and $(2, \infty)$

Decreasing on $(0, 2)$

The graph's behavior changes at the points $(0, 0)$ and $(2, -4)$.

21. $f(x) = 3x^4 - 6x^2$

Decreasing on $(-\infty, -1)$ and $(0, 1)$

Increasing on $(-1, 0)$ and $(1, \infty)$

The graph's behavior changes at the points $(-1, -3)$, $(0, 0)$, and $(1, -3)$.

23. $f(x) = x\sqrt{x + 3}$

Decreasing on $(-3, -2)$

Increasing on $(-2, \infty)$

The graph's behavior changes at the point $(-2, -2)$.

25. $f(x) = x^2 - 4x$

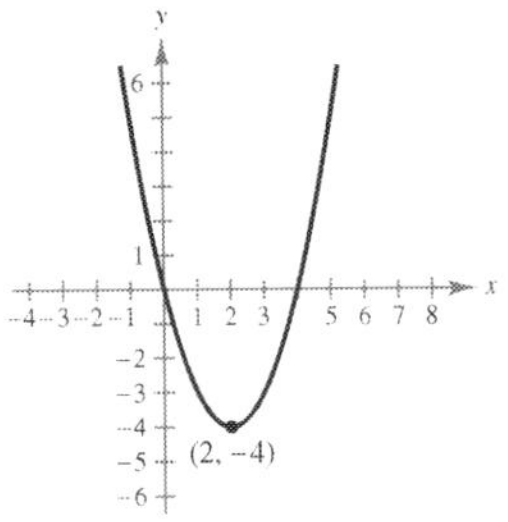

Increasing on $(2, \infty)$, decreasing on $(-\infty, 2)$

The behavior changes at $(2, -4)$.

27. $f(x) = x^3 - 3x^2 + 2$

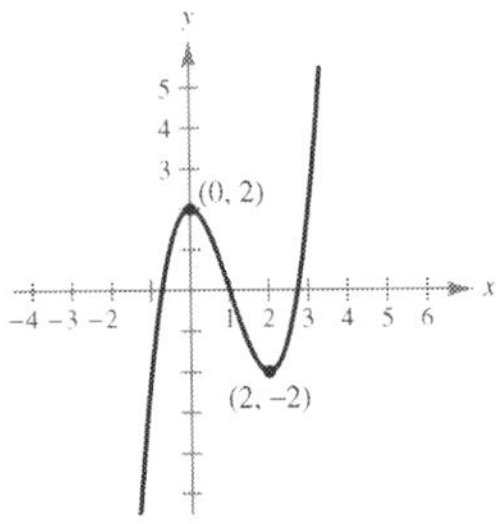

Increasing on $(-\infty, 0)$ and $(2, \infty)$, decreasing on $(0, 2)$

The behavior changes at $(0, 2)$ and $(2, -2)$.

29. $f(x) = x^2 - 4x + 1$

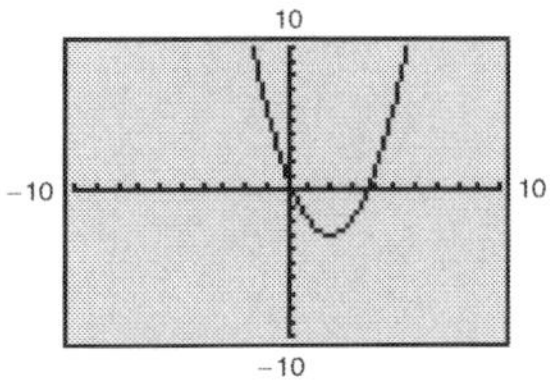

Decreasing on $(-\infty, 2)$

Increasing on $(2, \infty)$

Relative minimum: $(2, -3)$

31. $f(x) = x^3 - 3x^2$

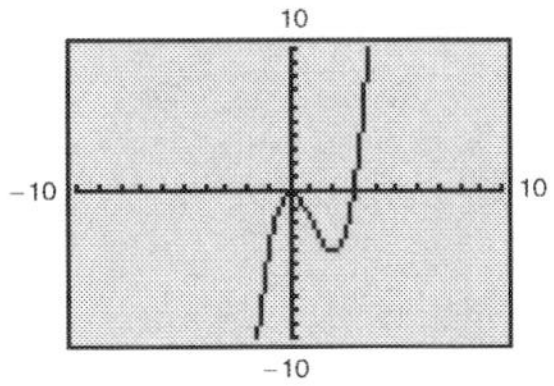

Increasing on $(-\infty, 0)$, $(2, \infty)$

Decreasing on $(0, 2)$

Relative maximum: $(0, 0)$

Relative minimum: $(2, -4)$

33. $f(x) = \frac{1}{4}(-4x^4 - 5x^3 + 10x^2 + 8x + 6)$

Relative maxima: $(-1.54, 3.29)$, $(0.95, 3.77)$

Relative minimum: $(-0.34, 1.14)$

Increasing: $(-\infty, -1.54)$, $(-0.34, 0.95)$

Decreasing: $(-1.54, -0.34)$, $(0.95, \infty)$

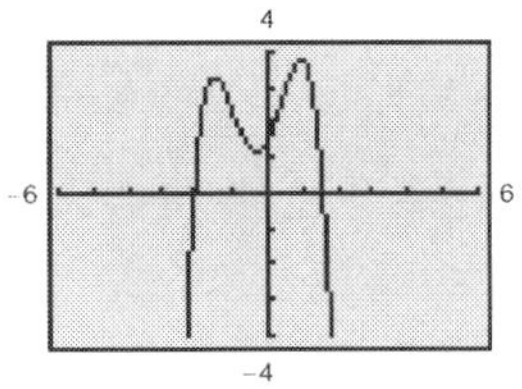

35. (a) $f(2) = [\![2]\!] = 2$

(b) $f(2.5) = [\![2.5]\!] = 2$

(c) $f(-2.5) = [\![-2.5]\!] = -3$

(d) $f(-4) = [\![-4]\!] = -4$

37. (a) $f(4) = [\![4 - 1.8]\!] = [\![2.2]\!] = 2$

(b) $f(3.7) = [\![3.7 - 1.8]\!] = [\![1.9]\!] = 1$

(c) $f(-5.8) = [\![-5.8 - 1.8]\!] = [\![-7.6]\!] = -8$

(d) $f(-6.3) = [\![-6.3 - 1.8]\!] = [\![-8.1]\!] = -9$

39. $f(-x) = (-x)^6 - 2(-x)^2 + 3 = x^6 - 2x^2 + 3 = f(x)$

f is even.

41. $g(-x) = (-x)^3 - 5(-x) = -x^3 + 5x = -g(x)$

g is odd.

43. $f(-x) = -x\sqrt{4 - (-x)^2} = -x\sqrt{4 - x^2}$

f is odd.

45. $f(x) = 3$, even

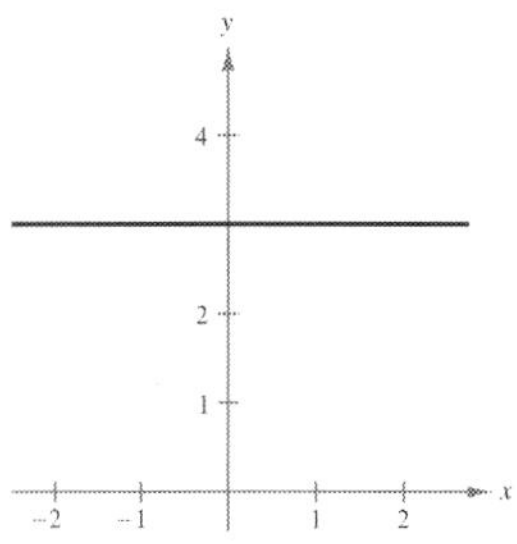

47. $f(x) = 5 - 3x$, neither even nor odd

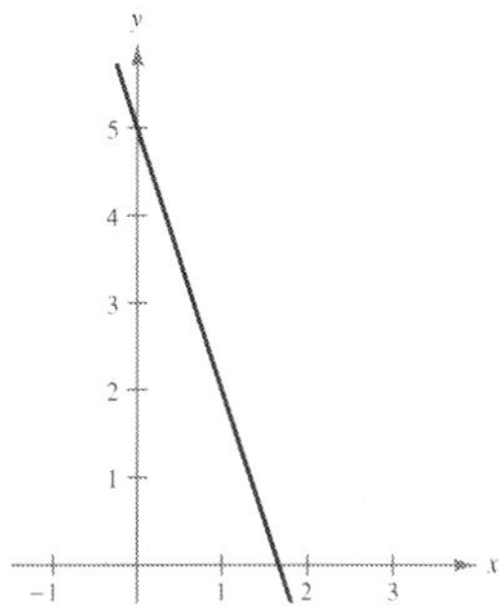

49. $g(s) = \frac{s^3}{4}$, odd

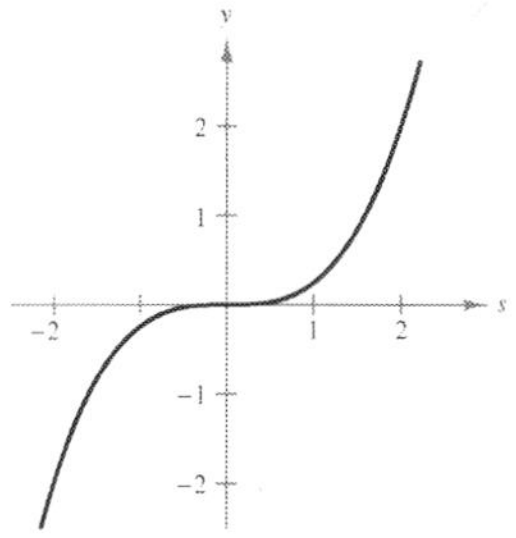

51. $f(x) = \sqrt{1 - x}$, neither even nor odd

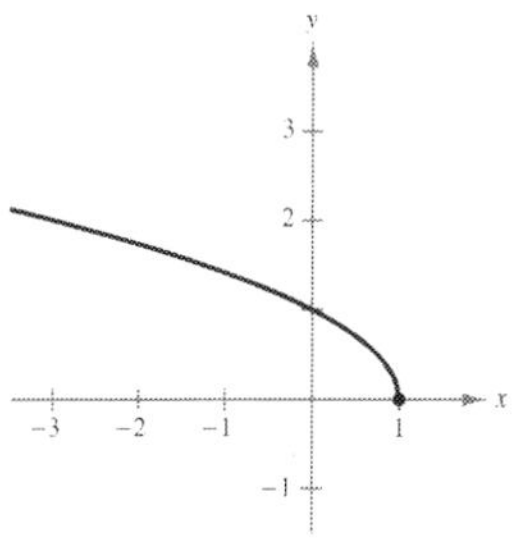

53. $f(x) = x^{3/2}$, neither even nor odd

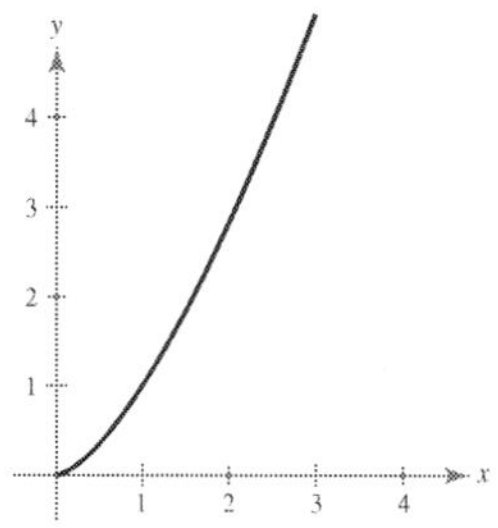

55.

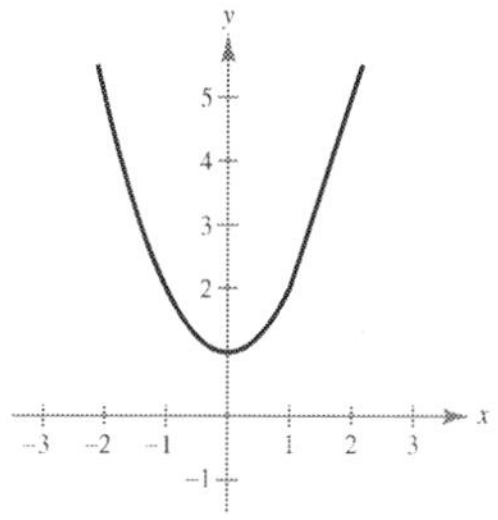

Neither even nor odd

57.

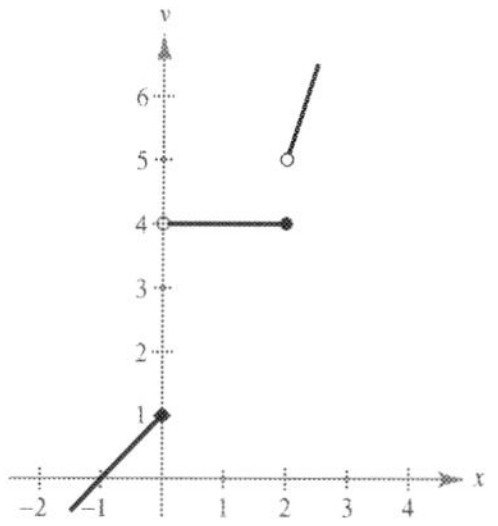

Neither even nor odd

59.

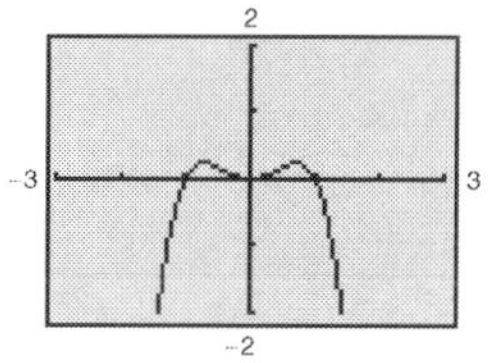

Even

61.

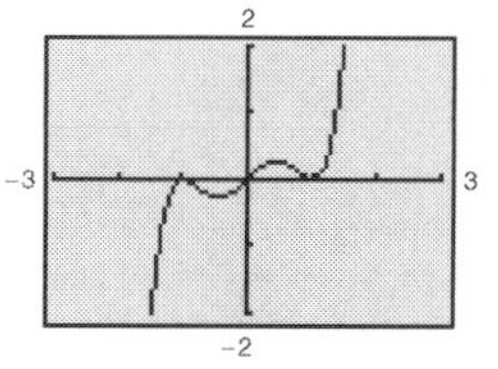

Odd

63.

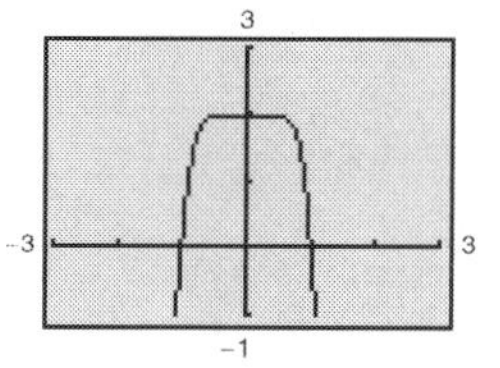

Even

65. $f(x) = 4 - x$

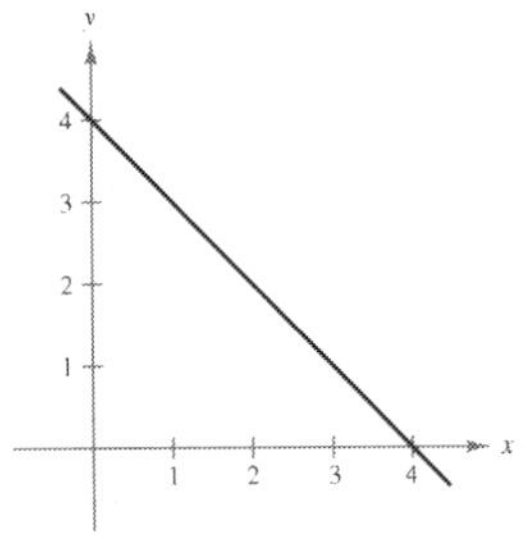

67. $f(x) = x^2 - 9$

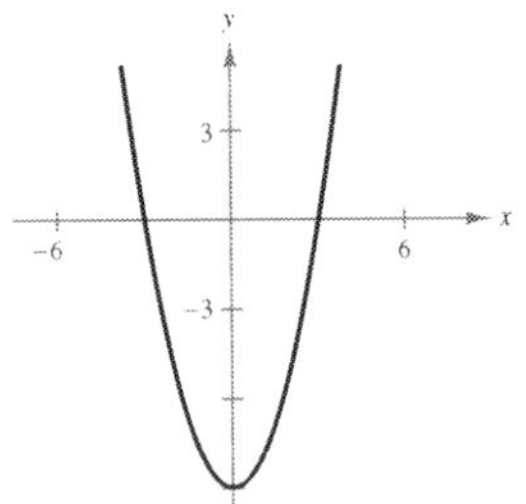

69. $f(x) = 1 - x^4$

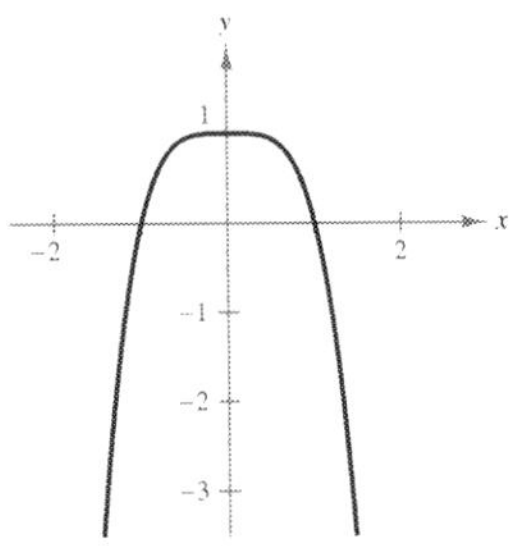

71. $f(x) = \frac{1}{3}(3 + |x|)$

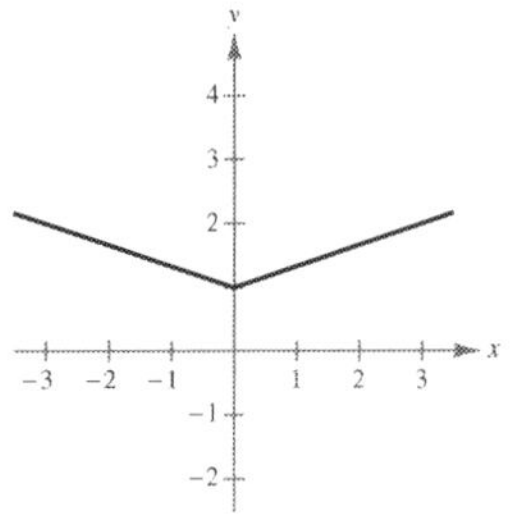

73. $f(x) = \sqrt{x + 3}$

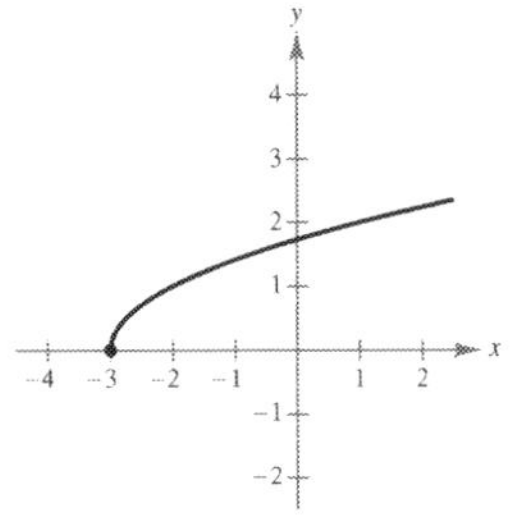

75. $f(x) = -[\![x]\!]$

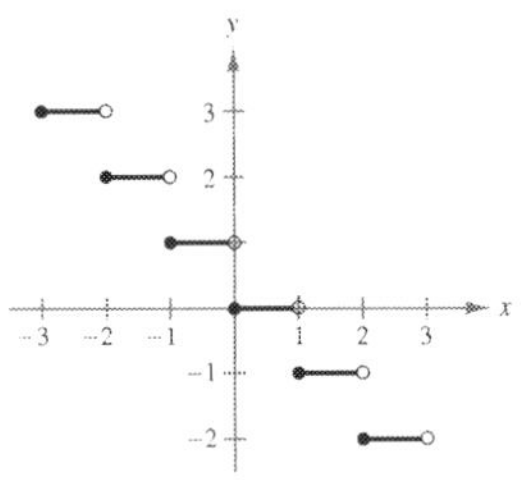

77. $f(x) = [\![x - 1]\!]$

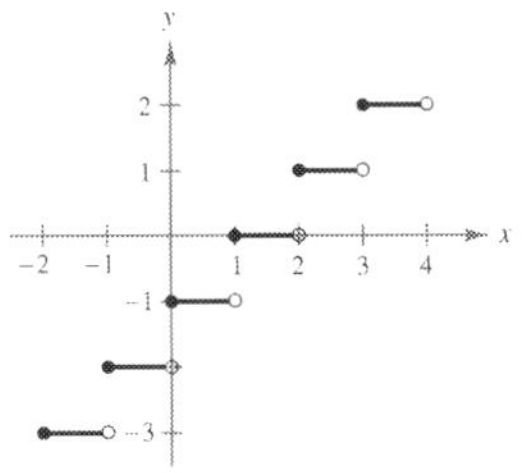

79. (a) From the graph, the maximum occurred at $t = 10$. The maximum price of gold in 2010 was

$P = 0.038214(10)^4 - 0.48703(10)^3 + 11.0120(10)^2 - 6.018(10) + 276.49 = 1212.62.$

(b) The price did not decrease during 2000–2010.
The price only increased during 2000–2010.

(c) The slope of the graph seems to increase quite rapidly in future years. The price of gold probably will not increase that quickly.

81. (a)

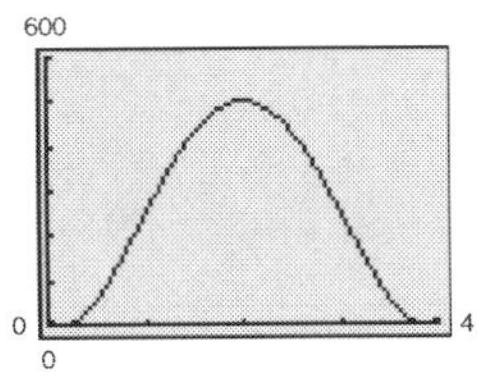

Increasing: 0 seconds to 2 seconds

Decreasing: 2 seconds to 4 seconds

(b) Maximum change in volume ≈ 501.9 milliliters

83.

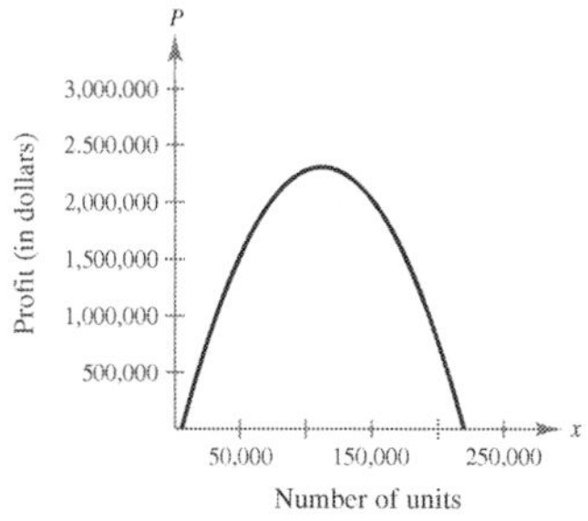

Approximately 112,500 units

85.

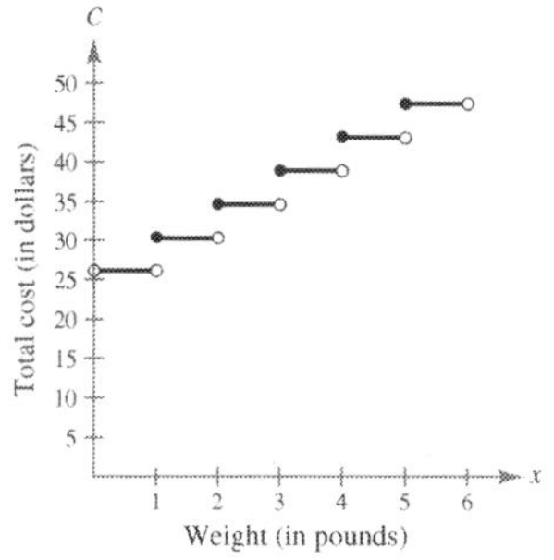

87.

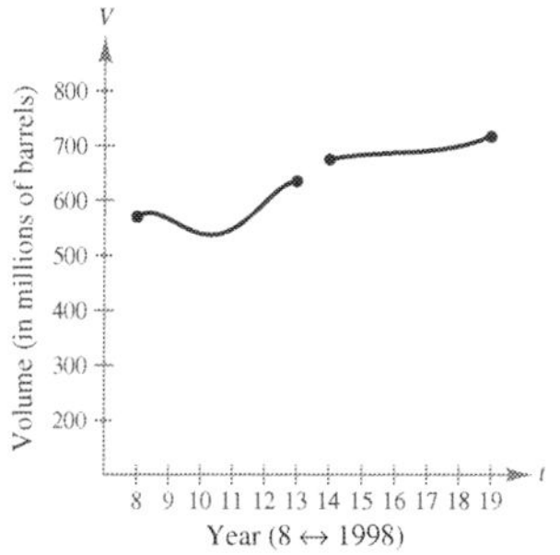

89. (a) $y = 425.71t + 705.54$

Domain: $(0, 4)$

Range: $(705.54, 2408.38)$

(b) $y = -202.950t^2 + 3304.67t - 7311.8$

Domain: $(5, 9)$

Range: $(4137.8, 6140.3)$

(c) $y = \begin{cases} 425.71t + 705.54, & 0 \le t \le 4 \\ -202.950t^2 + 3304.67t - 7311.8, & 5 \le t \le 9 \end{cases}$

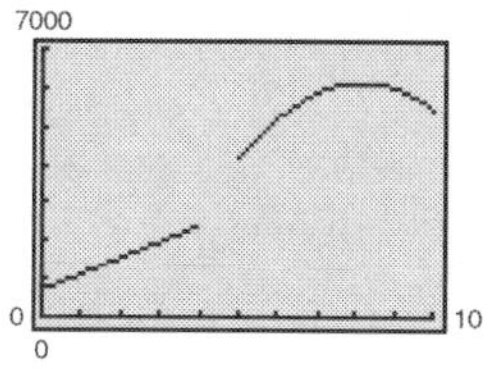

(d) The revenues increased from 2000 to 2008 and decreased from 2008 to 2009.

91. $\left(-\frac{3}{2}, 4\right)$

(a) If f is even, another point is $\left(\frac{3}{2}, 4\right)$.

(b) If f is odd, another point is $\left(\frac{3}{2}, -4\right)$.

93. $(4, 9)$

(a) If f is even, another point is $(-4, 9)$.

(b) If f is odd, another point is $(-4, -9)$.

Section 2.6 Transformations of Functions

Skills Warm Up

1. $f(3) = 3^2 - 4(3) + 15$

$= 9 - 12 + 15$

$= 12$

2. $f(-x) = \dfrac{2(-x)}{(-x - 3)}$

$= \dfrac{-2x}{-x - 3}$

3. $-x^3 + 10x = 0$

$-x(x^2 - 10) = 0$

$x = 0, x = \pm\sqrt{10}$

4. $3x^2 + 2x - 8 = 0$

$(3x - 4)(x + 2) = 0$

$x = \frac{4}{3}, x = -2$

5.

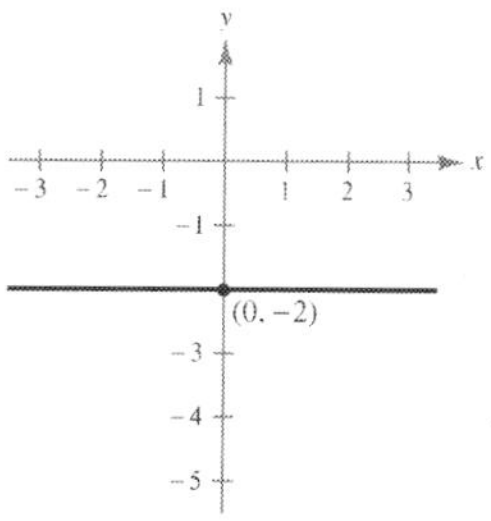

6.

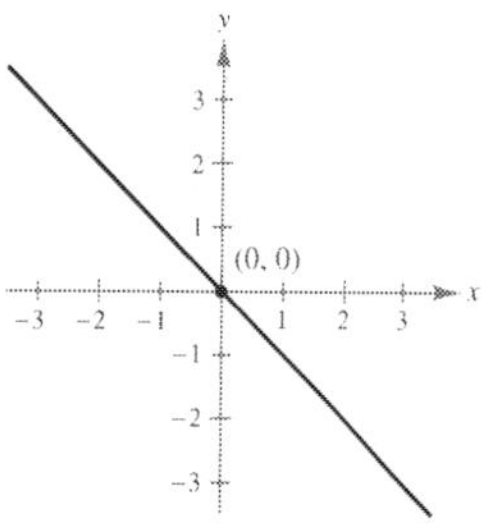

Skills Warm Up —*continued*—

7.

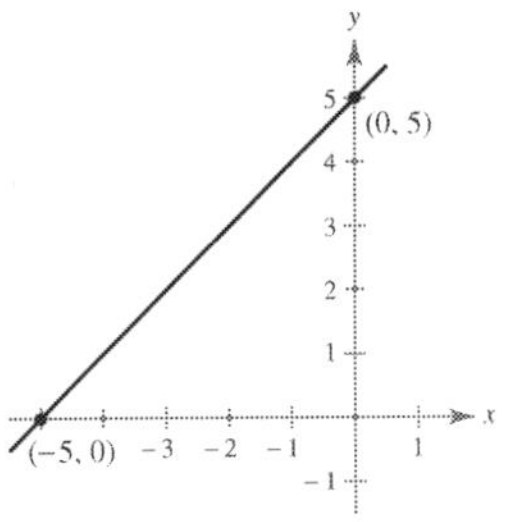

8.

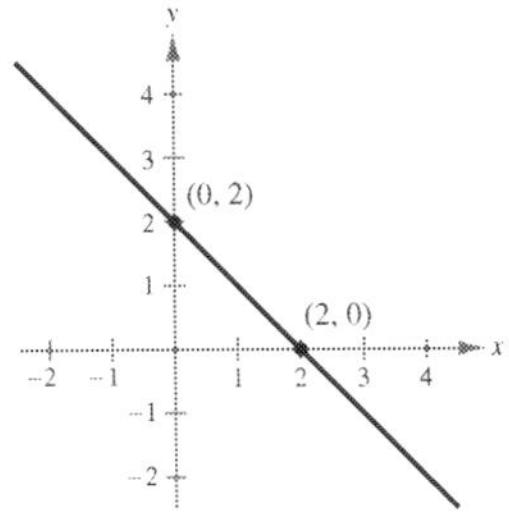

9.

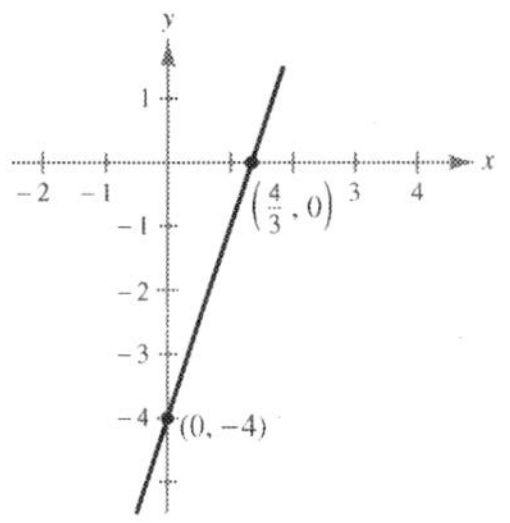

10.

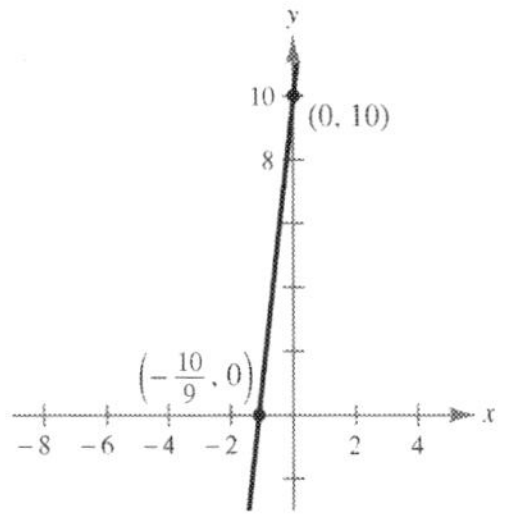

1. Shifted 4 units downward

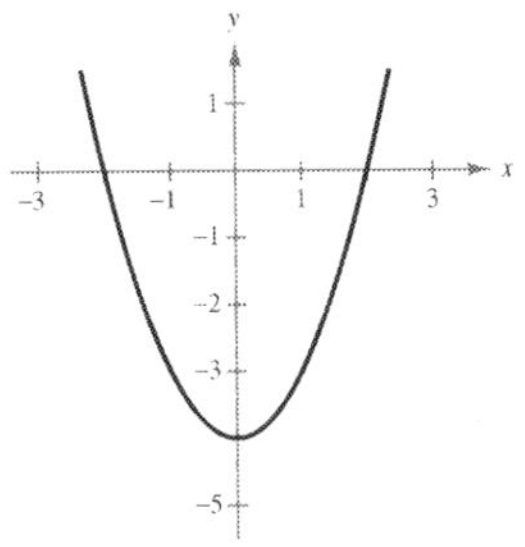

3. Shifted 2 units to the left

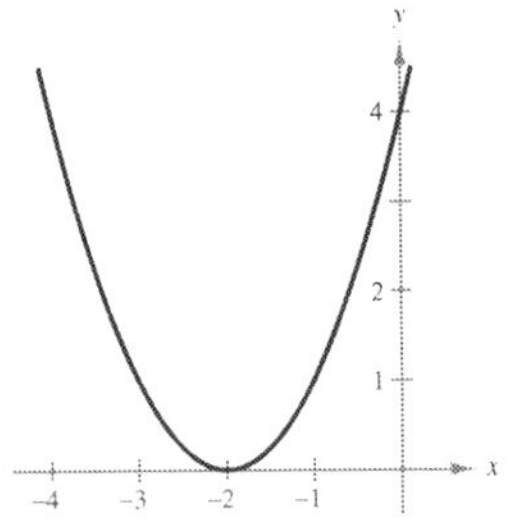

5. $g(x) = (x-4)^2 + 2 = f(x-4) + 2$

Horizontal shift 4 units to the right and vertical shift 2 units upward

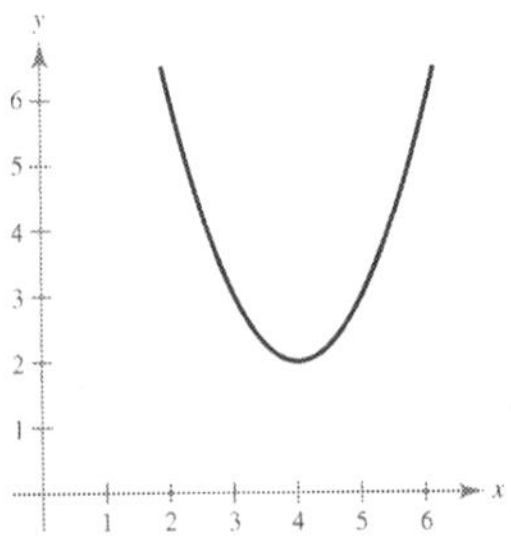

7. $g(x) = -x^2 + 1 = -f(x) + 1$

A reflection in the x-axis and a vertical shift 1 unit upward

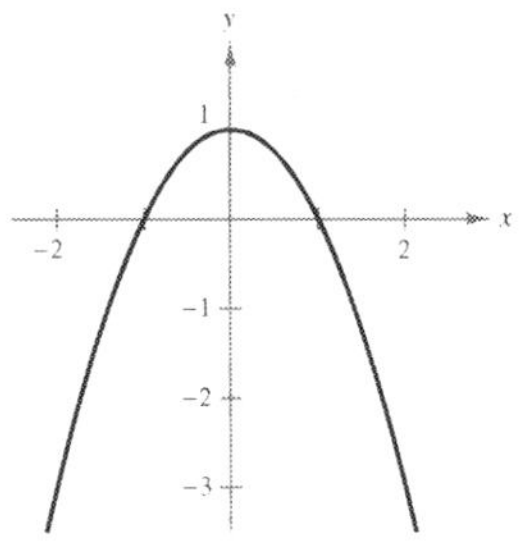

9. $g(x) = |x| + 2 = f(x) + 2$

Vertical shift 2 units upward

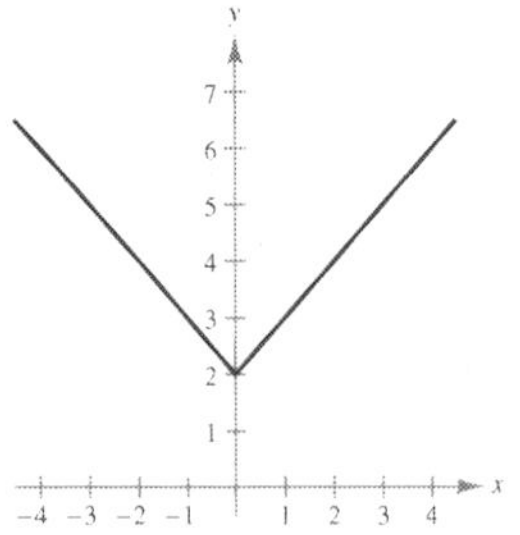

11. $g(x) = |x - 1| = f(x - 1)$

Horizontal shift 1 unit to the right

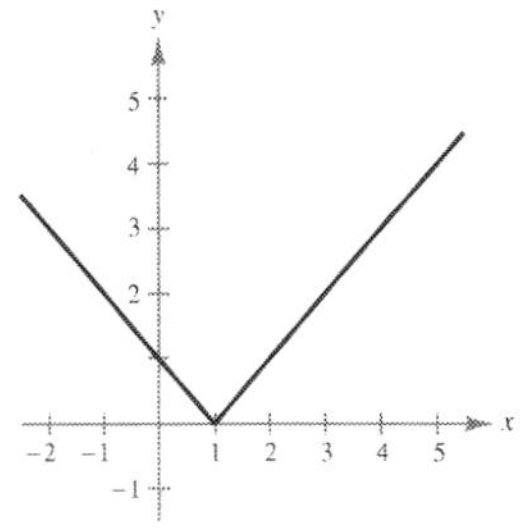

13. $g(x) = -|x| + 3 = -f(x) + 3$

A reflection in the x-axis and a vertical shift 3 units upward

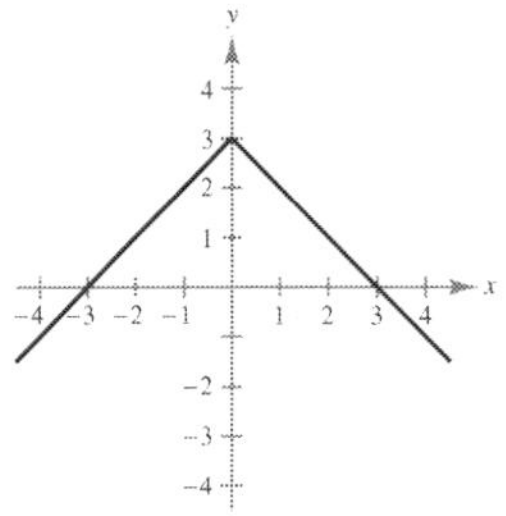

15. $g(x) = |x + 1| - 4 = f(x + 1) - 4$

Horizontal shift 1 unit to the left and vertical shift 4 units downward

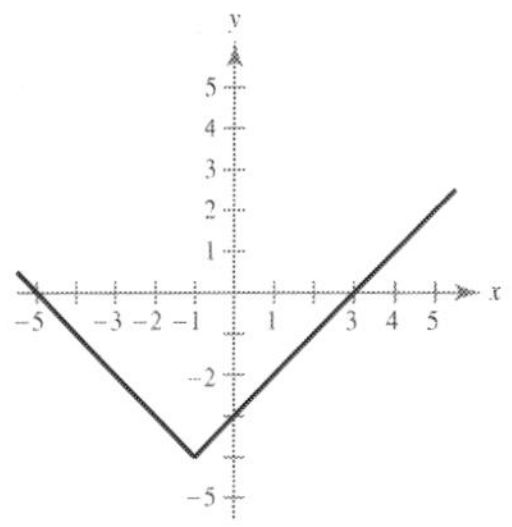

17. $g(x) = \sqrt{x - 3}$

Shifted 3 units to the right

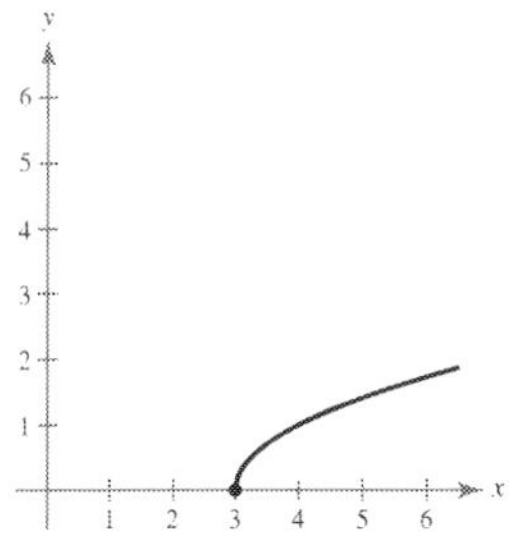

19. $y = \sqrt{x - 3} + 1 = f(x - 3) + 1$

Horizontal shift of 3 units to the right and a vertical shift 1 unit upward

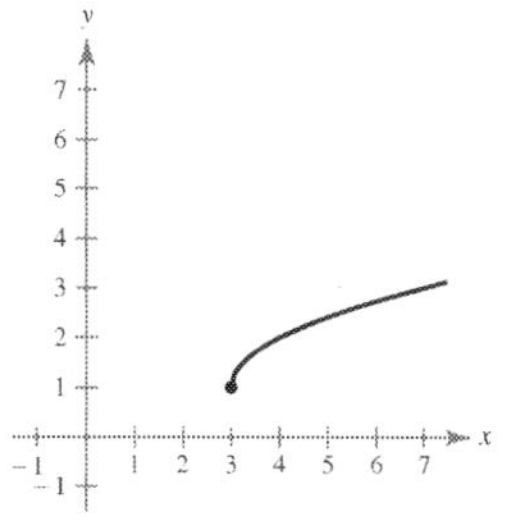

21. $y = \sqrt{2x} = \sqrt{2}\sqrt{x} = \sqrt{2}f(x)$

Vertical stretch by a factor of $\sqrt{2}$

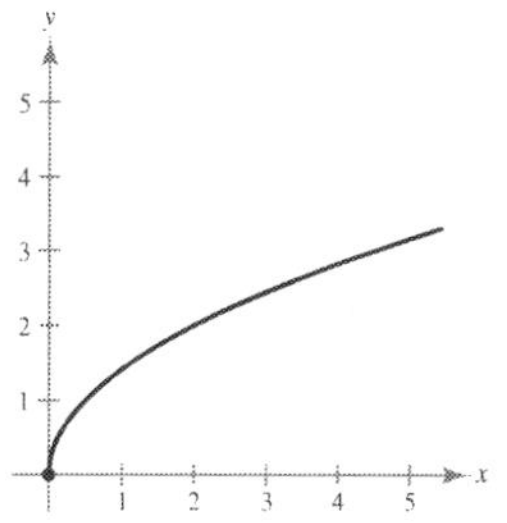

23. $g(x) = 2 + \sqrt{-x - 4} = f[(-1)(x + 4)] + 2$

The graph is reflected about the y-axis and shifted 4 units to the left and 2 units upward.

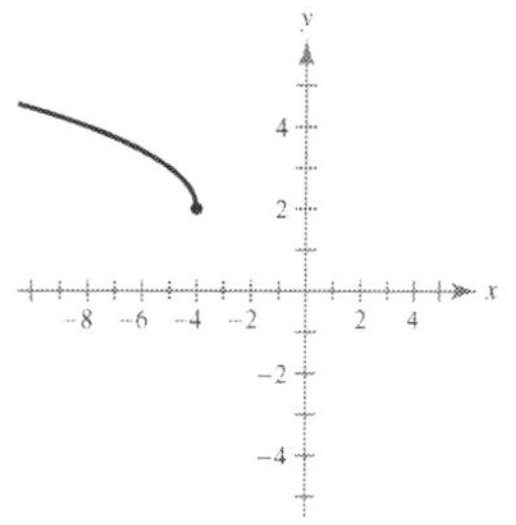

25. $y = \sqrt[3]{-x}$

Reflection about the y-axis

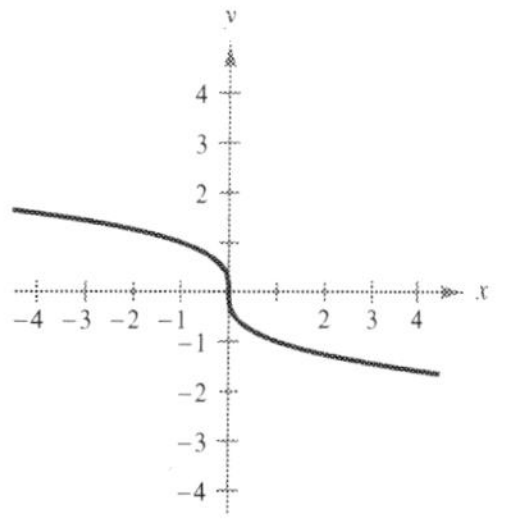

27. $y = \sqrt[3]{x} - 2$

The graph is shifted 2 units downward.

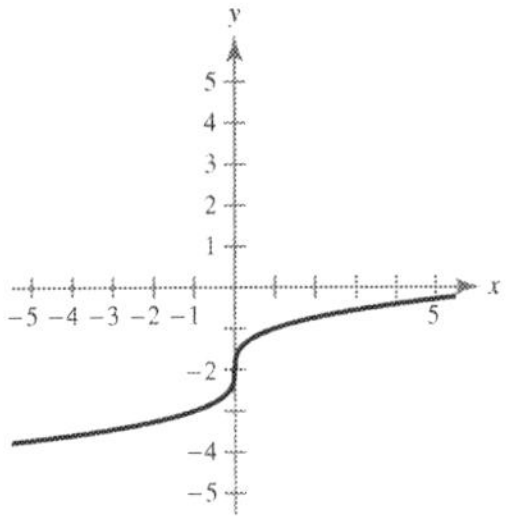

29. $y = 2 - \sqrt[3]{x + 1} = 2 - f(x + 1)$

Reflection about the x-axis, horizontal shift 1 unit to the left, and a vertical shift 2 units upward

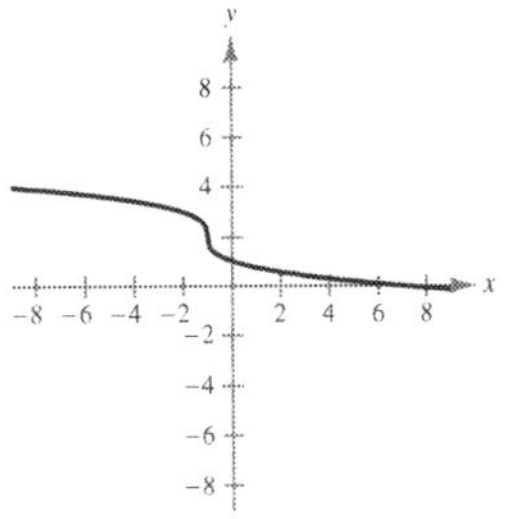

31. $y = 2\sqrt[3]{x + 1} - 1 = 2f(x + 1) - 1$

Vertical stretch by a factor of 2, horizontal shift 1 unit to the left, and a vertical shift 1 unit downward

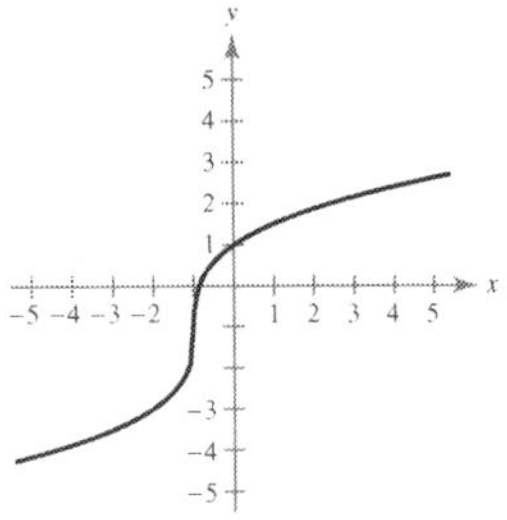

33. $y = \frac{1}{2}\sqrt[3]{x} = \frac{1}{2}f(x)$

Vertical shrink by a factor of $\frac{1}{2}$

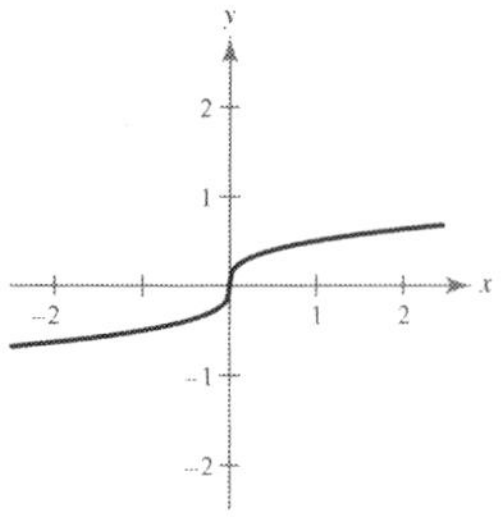

35. Common function: $y = x^3$

Transformation: horizontal shift 2 units to the right

Equation: $y = (x - 2)^3$

37. Common function: $y = x^2$

Transformation: reflection about the x-axis

Equation: $y = -x^2$

39. Common function: $y = \sqrt{x}$

Transformation: reflection about the x-axis and a vertical shift 1 unit upward

Equation: $y = -\sqrt{x} + 1$

41. Common function: $y = x$

Transformation: vertically stretched by a factor of 2

Equation: $y = 2x$

43. (a) $f(x) = \frac{1}{2}x + c$

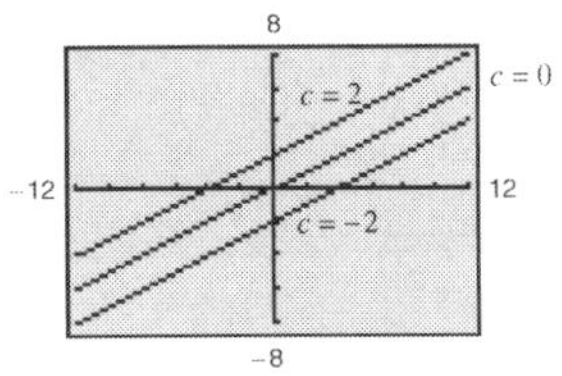

Vertical shift of 2 units

(b) $f(x) = \frac{1}{2}(x - c)$

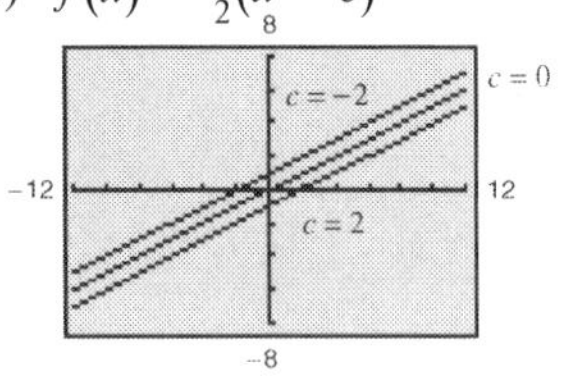

Horizontal shift of 2 units

(c) $f(x) = \frac{1}{2}(cx)$

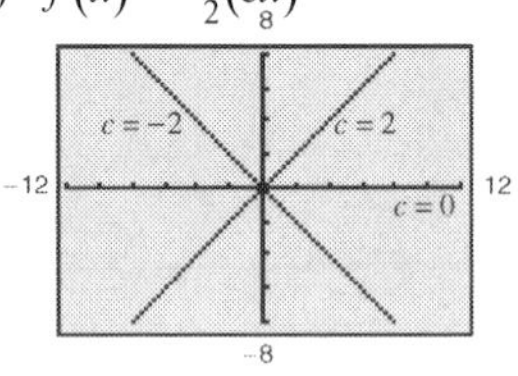

Slope of the function changes

45. (a) $g(x) = f(x - 1) + 1$

$g(x) = (x - 1)^2 + 1$

(b) $g(x) = -f(x + 1)$

$g(x) = -(x + 1)^2$

47. (a) Vertical shift 2 units upward

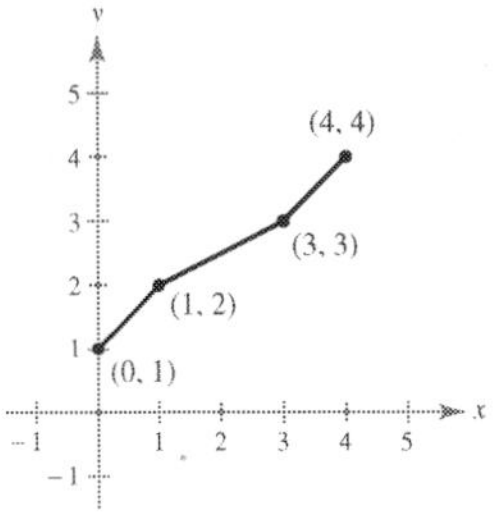

(b) Reflection in the x-axis

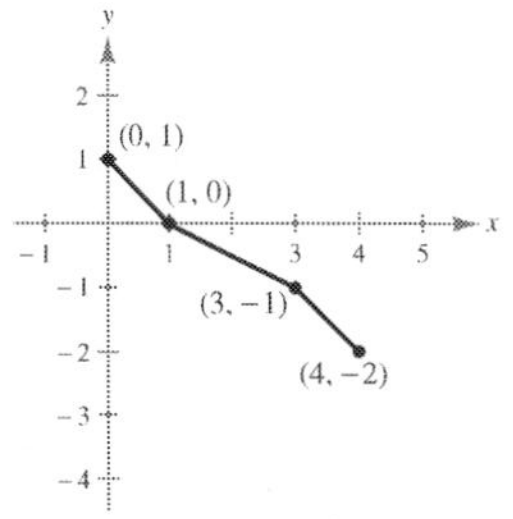

(c) Horizontal shift 2 units to the right

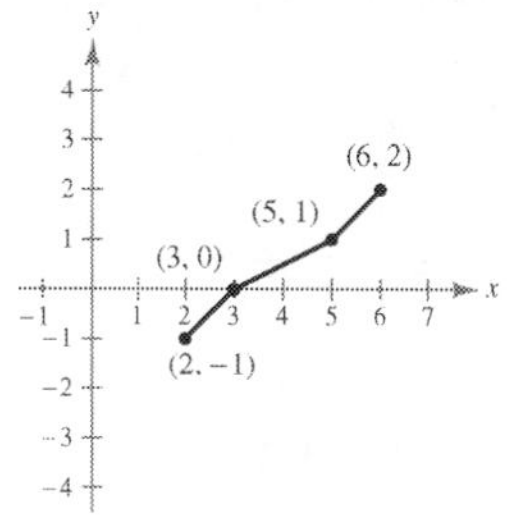

(d) Horizontal shift 3 units to the left

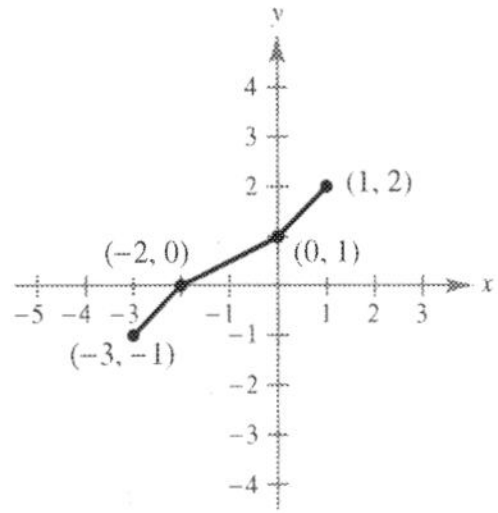

(e) Stretching by 2

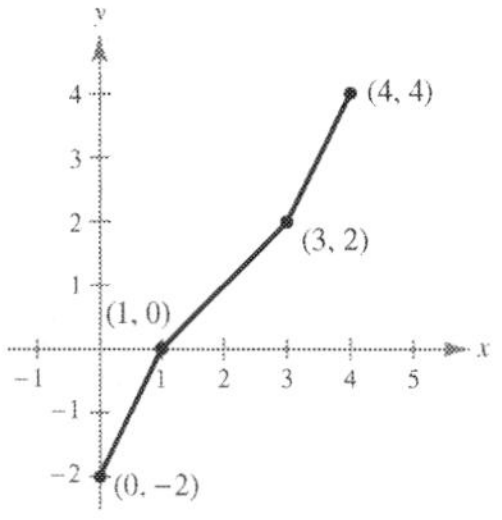

(f) Reflection in the y-axis

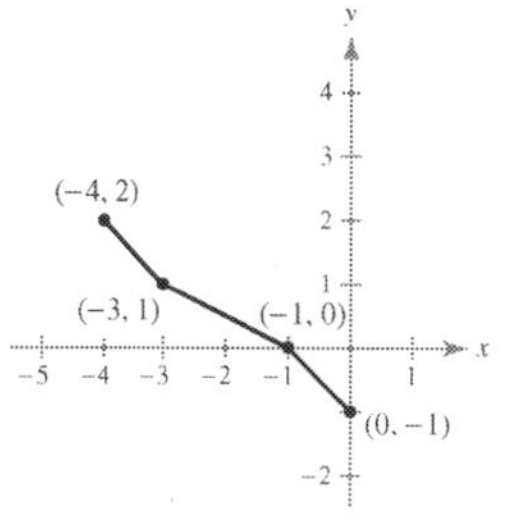

49. (a) Horizontal shift 5 units to the right

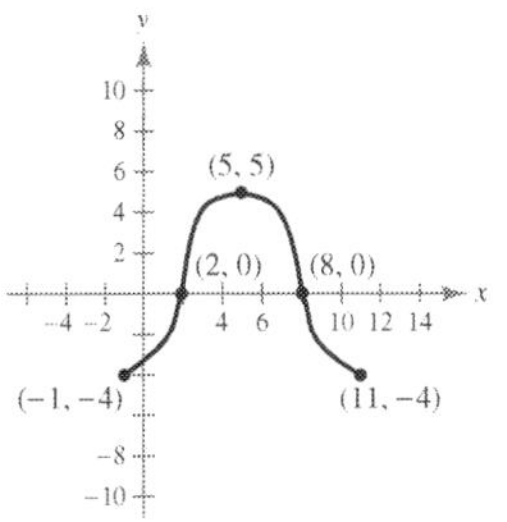

(b) Reflection about the x-axis and a vertical shift 3 units upward

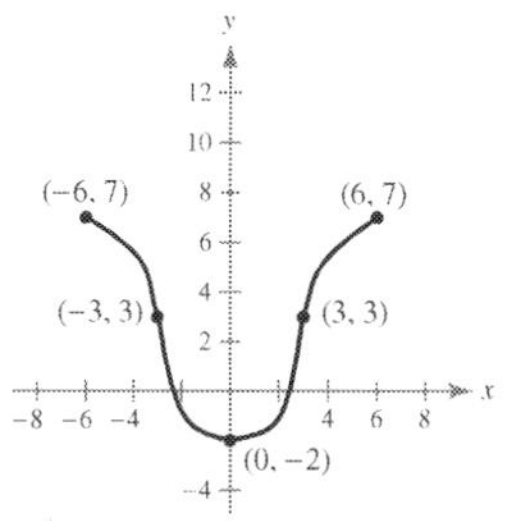

(c) Shrinking by $\frac{1}{3}$

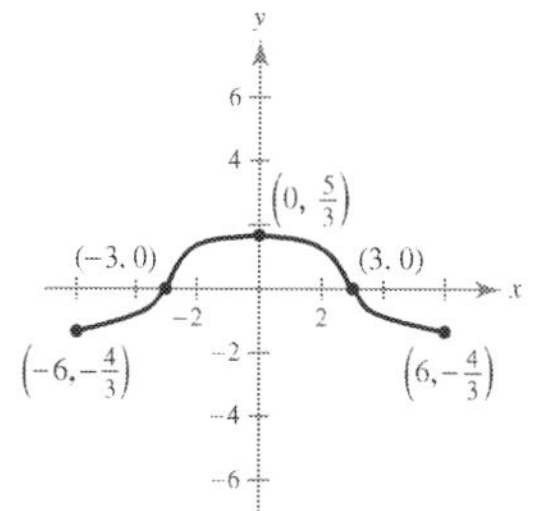

(d) Reflection about the x-axis and a horizontal shift of 1 unit to the left

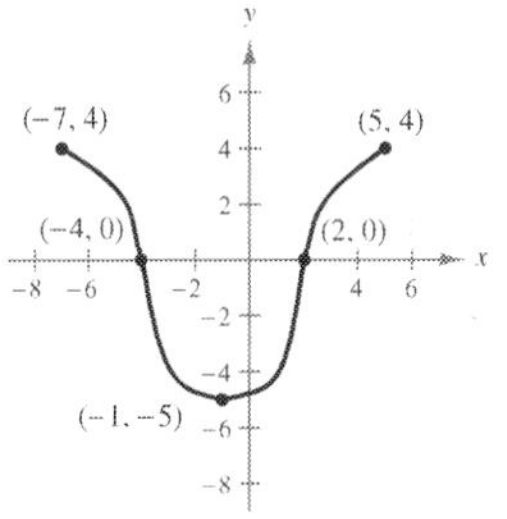

(e) Reflection about the y-axis

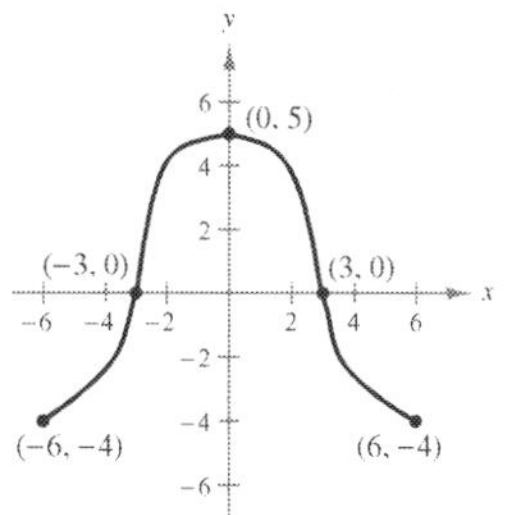

(f) Vertical shift 5 units downward

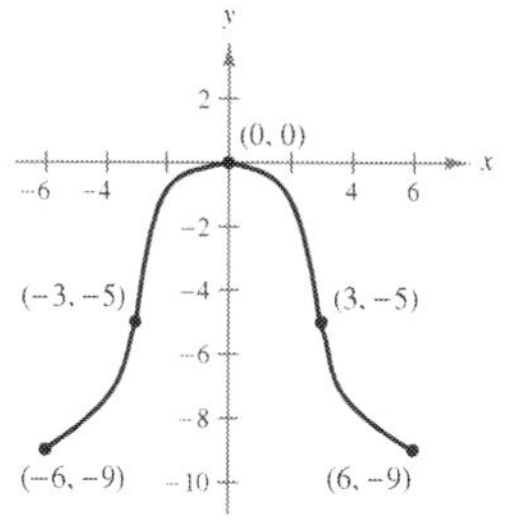

51. $y = x^3 - 2$

53. $y = 4x^3$

55. $h(x) = \sqrt{x - 4} - 3$

57. $h(x) = \frac{1}{2}\sqrt{x - 3}$

59. The graph is shifted 2 units upward, so
$g(x) = x^3 - 3x^2 + 2.$

61. The graph is shifted 1 unit to the left, so
$g(x) = (x + 1)^3 - 3(x + 1)^2 = x^3 - 3x - 2.$

63. Shift: horizontally 1 unit to the right and vertically 2 units downward

$h(x) = f(x - 1) - 2 = (x - 1)^2 - 2$

65. (a)

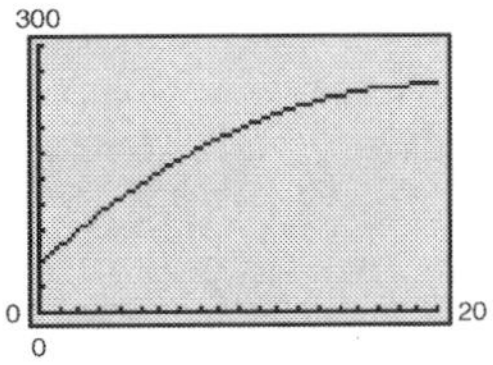

(b) $P(x) = 80 + 20x - 0.5x^2 - 25$

$P(x) = 55 + 20x - 0.5x^2, 0 \le x \le 20$

Vertical shift 25 units downward. Note the shift is 25 units since both x and P are measured in hundreds of dollars.

(c) $P\left(\frac{x}{100}\right) = 80 + \frac{x}{5} - 0.00005x^2,$

$0 \le x \le 2000$ (x in dollars)

Horizontal stretch

67. (a) The graph is stretched by a factor of 34 and shifted 3705 units upward.

(b)

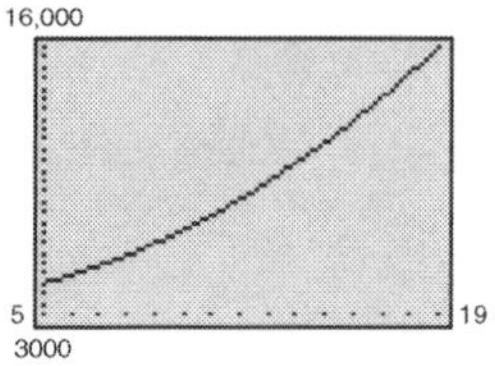

(c) $34t^2 + 3705 = 13{,}000$

$34t^2 = 9295$

$t^2 = \frac{9295}{34}$

$t = \sqrt{\frac{9295}{34}} \approx 16.5$

So, the amount of debt exceeded \$13 trillion in 2007.

(d) $M(t) = 34(t + 10)^2 + 3705$

To make a horizontal shift 10 years backward (10 units to the left), add 10 to t.

69. $y = x^7$ will be an odd function and its graph will be similar in shape to the graphs of $y = x^3$ and $y = x^5$.

$y = x^8$ will be an even function and its graph will be similar in shape to the graphs of $y = x^2$, $y = x^4$, and $y = x^6$.

Section 2.7 The Algebra of Functions

Skills Warm Up

1. $\frac{1}{x} + \frac{1}{1-x} = \frac{1-x}{x(1-x)} + \frac{x}{x(1-x)} = \frac{1}{x(1-x)}$

2. $\frac{2}{x+3} - \frac{2}{x-3} = \frac{2(x-3)}{(x+3)(x-3)} - \frac{2(x+3)}{(x+3)(x-3)}$

$= \frac{2x - 6 - 2x - 6}{(x+3)(x-3)}$

$= -\frac{12}{(x+3)(x-3)}$

3. $\frac{3}{x-2} - \frac{2}{x(x-2)} = \frac{3x-2}{x(x-2)}$

4. $\frac{x}{x-5} + \frac{1}{3} = \frac{3x}{3(x-5)} + \frac{x-5}{3(x-5)} = \frac{4x-5}{3(x-5)}$

5. $(x-1)\left(\frac{1}{\sqrt{x^2-1}}\right) = \frac{x-1}{\sqrt{x^2-1}} \cdot \frac{\sqrt{x^2-1}}{\sqrt{x^2-1}}$

$= \frac{(x-1)\left(\sqrt{x^2-1}\right)}{x^2-1}$

$= \frac{(x-1)\left(\sqrt{x^2-1}\right)}{(x-1)(x+1)}$

$= \frac{\sqrt{x^2-1}}{x+1}, \quad x \ne 1$

6. $\left(\frac{x}{x^2-4}\right)\left(\frac{x^2-x-2}{x^2}\right)$

$= \left(\frac{x}{(x+2)(x-2)}\right)\left(\frac{(x-2)(x+1)}{x^2}\right)$

$= \frac{x+1}{x(x+2)}, \ x \ne 2$

7. $\left(x^2 - 4\right) \div \left(\frac{x+2}{5}\right) = (x+2)(x-2)\left(\frac{5}{x+2}\right)$

$= 5(x-2)$

8. $\left(\frac{x}{x^2+3x-10}\right) \div \left(\frac{x^2+3x}{x^2+6x+5}\right)$

$= \left(\frac{x}{(x+5)(x-2)}\right)\left(\frac{(x+5)(x+1)}{x(x+3)}\right)$

$= \frac{x+1}{(x-2)(x+3)}, \quad x \ne -5, -1, 0$

Skills Warm Up —continued—

9. $$\frac{\left(\frac{1}{x}\right)+5}{3-\left(\frac{1}{x}\right)} = \frac{\frac{1+5x}{x}}{\frac{3x-1}{x}} = \left(\frac{1+5x}{x}\right)\left(\frac{x}{3x-1}\right) = \frac{1+5x}{3x-1},\ x \neq 0$$

10. $$\frac{\frac{x}{4}-\frac{4}{x}}{x-4} = \frac{\frac{x^2}{4x}-\frac{16}{4x}}{x-4} = \left(\frac{x^2-16}{4x}\right)\left(\frac{1}{x-4}\right) = \left(\frac{(x+4)(x-4)}{4x}\right)\left(\frac{1}{x-4}\right) = \frac{x+4}{4x},\ x \neq 4$$

1.

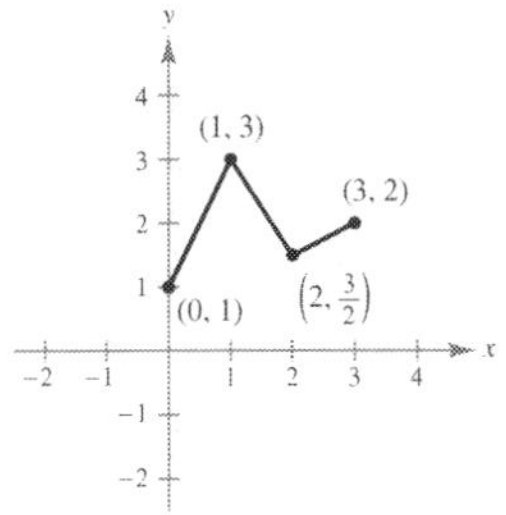

3.

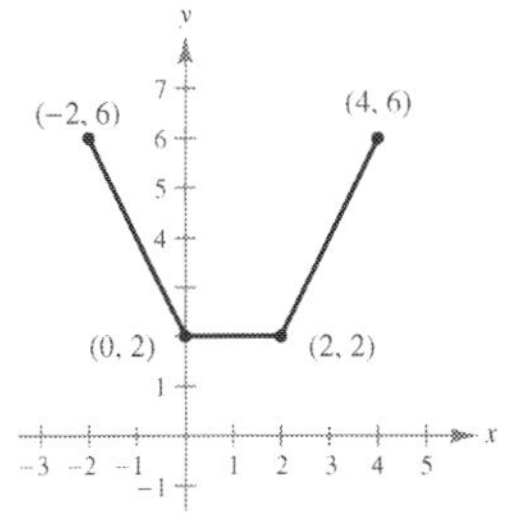

5. (a) $(f+g)(x) = f(x) + g(x) = (x+1) + (x-1) = 2x$

(b) $(f-g)(x) = f(x) - g(x) = (x+1) - (x-1) = 2$

(c) $(fg)(x) = f(x)\cdot g(x) = (x+1)(x-1) = x^2 - 1$

(d) $\left(\frac{f}{g}\right)(x) = \frac{f(x)}{g(x)} = \frac{x+1}{x-1}$

Domain: $(-\infty, 1) \cup (1, \infty)$

7. (a) $(f+g)(x) = f(x) + g(x) = 3x - 1 + 1 - x^2 = 3x - x^2$

(b) $(f-g)(x) = f(x) - g(x) = 3x - 1 - (1 - x^2) = x^2 + 3x - 2$

(c) $(fg)(x) = f(x)g(x) = (3x-1)(1-x^2) = -3x^3 + x^2 + 3x - 1$

(d) $\left(\frac{f}{g}\right)(x) = \frac{f(x)}{g(x)} = \frac{3x-1}{1-x^2}$

Domain: $(-\infty, -1) \cup (-1, 1) \cup (1, \infty)$

9. (a) $(f+g)(x) = f(x) + g(x) = (x^2+5) + \sqrt{1-x}$

(b) $(f-g)(x) = f(x) - g(x) = (x^2+5) - \sqrt{1-x}$

(c) $(fg)(x) = f(x)\cdot g(x) = (x^2+5)\sqrt{1-x}$

(d) $\left(\frac{f}{g}\right)(x) = \frac{f(x)}{g(x)} = \frac{x^2+5}{\sqrt{1-x}}$

Domain: $(-\infty, 1)$

11. (a) $(f + g)(x) = f(x) + g(x) = \frac{1}{x} + \frac{1}{x^2} = \frac{x + 1}{x^2}$

(b) $(f - g)(x) = f(x) - g(x) = \frac{1}{x} - \frac{1}{x^2} = \frac{x - 1}{x^2}$

(c) $(fg)(x) = f(x) \cdot g(x) = \frac{1}{x}\left(\frac{1}{x^2}\right) = \frac{1}{x^3}$

(d) $\left(\frac{f}{g}\right)(x) = \frac{f(x)}{g(x)} = \frac{\frac{1}{x}}{\frac{1}{x^2}} = \frac{x^2}{x} = x, \ x \neq 0$

Domain: $(-\infty, 0) \cup (0, \infty)$

13. $(f + g)(x) = 2x + 1 + x^2 - 2 = x^2 + 2x - 1$

$(f + g)(3) = 3^2 + 2(3) - 1 = 9 + 6 - 1 = 14$

15. $(f - g)(2t) = -(2t)^2 + 2(2t) + 3 = -4t^2 + 4t + 3$

17. $(fg)(x) = (2x + 1)(x^2 - 2) = 2x^3 + x^2 - 4x - 2$

$$\begin{aligned}(fg)(-2) &= 2(-2)^3 + (-2)^2 - 4(-2) - 2\\ &= 2(-8) + 4 + 8 - 2\\ &= -6\end{aligned}$$

19. $\left(\frac{f}{g}\right)(x) = \frac{2x + 1}{x^2 - 2}$

$\left(\frac{f}{g}\right)(5) = \frac{2(5) + 1}{5^2 - 2} = \frac{11}{23}$

21. $(f - g)(0) = -0^2 + 2(0) + 3 = 3$

23.

$$\begin{aligned}4f &= 4(2x + 1) = 8x + 4\\ 2g &= 2(x^2 - 2) = 2x^2 - 4\\ (4f)(2) - (2g)(4) &= [8(2) + 4] - [2(4)^2 - 4]\\ &= 20 - 28\\ &= -8\end{aligned}$$

25. (a) $(f \circ g)(x) = f(g(x)) = f(2x + 5) = 3(2x + 5) = 6x + 15$

(b) $(g \circ f)(x) = g(f(x)) = g(3x) = 2(3x) + 5 = 6x + 5$

(c) $(f \circ f)(x) = f(f(x)) = f(3x) = 3(3x) = 9x$

27. (a) $(f \circ g)(x) = f(g(x)) = f(3x + 1) = (3x + 1)^2 = 9x^2 + 6x + 1$

(b) $(g \circ f)(x) = g(f(x)) = g(x^2) = 3x^2 + 1$

(c) $(f \circ f)(x) = f(f(x)) = f(x^2) = (x^2)^2 = x^4$

29. (a) $(f \circ g)(x) = f(g(x)) = f(3x + 1) = \frac{1}{3}(3x + 1) - 3 = x - \frac{8}{3}$

(b) $(g \circ f)(x) = g(f(x)) = g(\frac{1}{3}x - 3) = 3(\frac{1}{3}x - 3) + 1 = x - 8$

31. (a) $(f \circ g)(x) = f(g(x)) = f(x^2 - 4) = \sqrt{x^2 - 4 + 4} = x$

(b) $(g \circ f)(x) = g(f(x)) = g(\sqrt{x + 4}) = (\sqrt{x + 4})^2 - 4 = x + 4 - 4 = x$

33. (a) $(f \circ g)(x) = f(g(x)) = f(\sqrt{x}) = (x^{1/2})^{1/2} = x^{1/4} = \sqrt[4]{x}$

(b) Since $f(x) = g(x)$, $(g \circ f)(x) = (f \circ g)(x) = \sqrt[4]{x}$.

35. (a) $(f \circ g)(x) = f(g(x)) = f(x + 6) = |x + 6|$

(b) $(g \circ f)(x) = g(f(x)) = g(|x|) = |x| + 6$

37. (a) All real numbers, or $(-\infty, \infty)$

(b) $x \geq 0$, or $[0, \infty)$

(c) $(f \circ g)(x) = f(g(x)) = f(\sqrt{x})$

$= (\sqrt{x})^2 + 3 = x + 3$

$x \geq 0$, or $[0, \infty)$

39. (a) All real numbers $x \neq 0$, or $(-\infty, 0) \cup (0, \infty)$

(b) All real numbers, or $(-\infty, \infty)$

(c) $(f \circ g)(x) = f(g(x)) = f(x - 2) = \dfrac{1}{(x - 2)^2}$

All real numbers $x \neq 2$, or $(-\infty, 2) \cup (2, \infty)$

41. (a) $(f + g)(3) = f(3) + g(3) = 2 + 1 = 3$

(b) $\left(\dfrac{f}{g}\right)(2) = \dfrac{f(2)}{g(2)} = \dfrac{0}{2} = 0$

43. (a) $(f \circ g)(2) = f(g(2)) = f(2) = 0$

(b) $(g \circ f)(2) = g(f(2)) = g(0) = 4$

45. $f(x) = x^2,\ g(x) = 2x + 1$ (Answers will vary.)

47. $f(x) = \sqrt[3]{x},\ g(x) = x^2 - 4$ (Answers will vary.)

49. $f(x) = \dfrac{1}{x},\ g(x) = x + 2$ (Answers will vary.)

51. $f(x) = x^2 + 2x,\ g(x) = x + 4$ (Answers will vary.)

53. $T = R + B = \frac{3}{4}x + \frac{1}{15}x^2,\ 0 \le x \le 60$

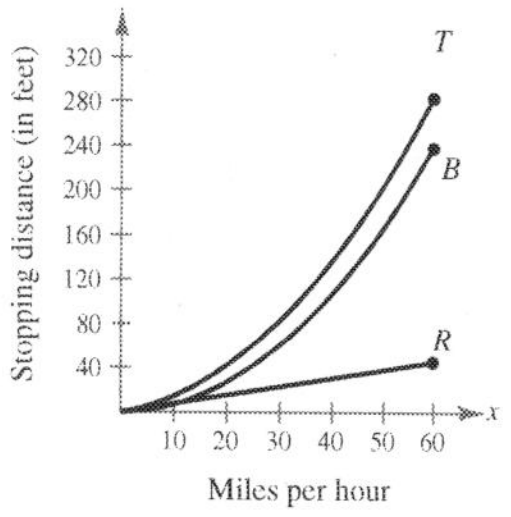

55. $(C \circ x)(t) = 50(30t) + 495 = 1500t + 495$

$C \circ x$ represents the cost of producing x units in t hours.

57. (a) Total yearly sales decreased from 2005 to 2012.

(b) Total yearly sales $= R_1 + R_2 = (525 - 15.2t) + (392 + 8.5t) = 917 - 6.7t$

The function is linear with a negative slope, so yearly sales decreased for increasing values of t.

59. (a)

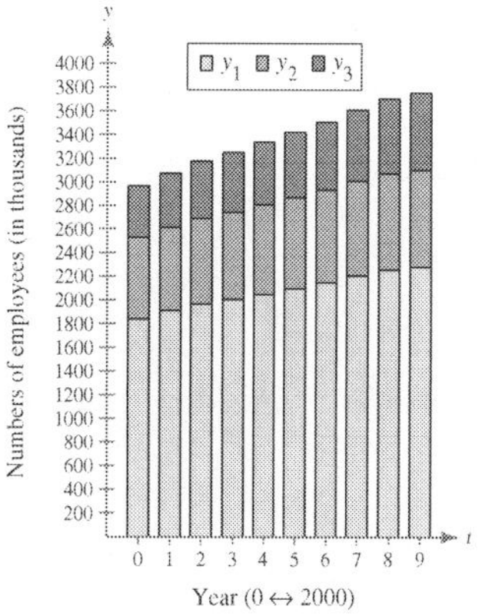

(b) $y_1 = 48.5t + 1856$

$y_2 = 15.2t + 694$

$y_3 = 23.7t + 434$

(c)

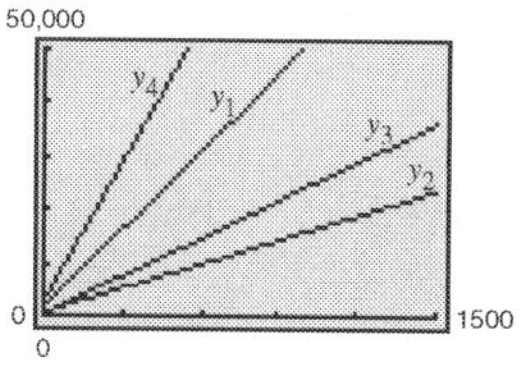

$y_4 = 87.4t + 2984$

When $t = 11$: $y_4 = 87.4(11) + 2984 = 3945.4$

When $t = 13$: $y_4 = 87.4(13) + 2984 = 4120.2$

In 2011, there were 3,945,400 employees.

In 2013, there were 4,120,200 employees.

61. $(A \circ r)(t) = A(r(t)) = \pi(0.6t)^2 = 0.36\pi t^2$

$A \circ r$ represents the area of the circle at time t.

63.

Year	2005	2006	2007	2008	2009
P/E	16.7	21.0	25.9	12.7	11.3

65. $\left(\dfrac{f}{g}\right)(x) = \dfrac{f(x)}{g(x)} = \dfrac{\sqrt{x}}{\sqrt{9 - x^2}}$

Domain: all real numbers $0 \le x < 3$, or $[0, 3)$

$\left(\dfrac{g}{f}\right)(x) = \dfrac{g(x)}{f(x)} = \dfrac{\sqrt{9 - x^2}}{\sqrt{x}}$

Domain: all real numbers $0 < x \le 3$ or $(0, 3]$

The domains differ because if $x = 3$, $\left(\dfrac{f}{g}\right)(x)$ would be undefined (division by zero), and if $x = 0$, $\left(\dfrac{g}{f}\right)(x)$ would be undefined (division by zero).

67. True. The range of g must be a subset of the domain of f for $(f \circ g)(x)$ to be defined.

Review Exercises for Chapter 2

1. (a)

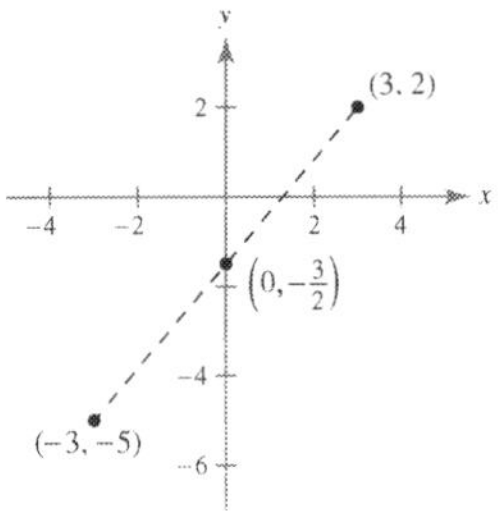

(b) $d = \sqrt{(3 - (-3))^2 + (2 - (-5))^2}$

$= \sqrt{6^2 + 7^2} = \sqrt{85} \approx 9.22$

(c) Midpoint: $\left(\dfrac{-3 + 3}{2}, \dfrac{-5 + 2}{2}\right) = \left(0, -\dfrac{3}{2}\right)$

3. (a)

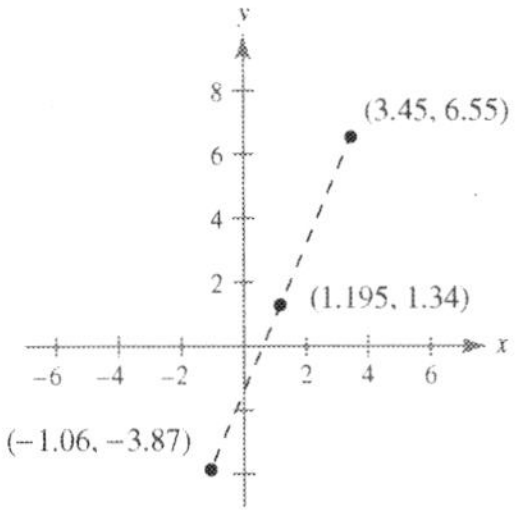

(b) $d = \sqrt{(3.45 + 1.06)^2 + (6.55 + 3.87)^2}$

$= \sqrt{128.9165} \approx 11.35$

(c) Midpoint:

$\left(\dfrac{-1.06 + 3.45}{2}, \dfrac{-3.87 + 6.55}{2}\right) = (1.195, 1.34)$

5. $\sqrt{(x - 10)^2 + (-5 - 10)^2} = 25$

$x^2 - 20x + 100 + 225 = 625$

$x^2 - 20x - 300 = 0$

$(x + 10)(x - 30) = 0$

$x = -10$ or $x = 30$

7. (a) $0 \stackrel{?}{=} 2(5)^2 - 7(5) - 15$

$0 \stackrel{?}{=} 50 - 35 - 15$

$0 = 0$ ✓

(5, 0) is a solution.

(b) $7 \stackrel{?}{=} 2(-2)^2 - 7(-2) - 15$

$7 \stackrel{?}{=} 8 + 14 - 15$

$7 = 7$ ✓

(−2, 7) is a solution.

9. $y = -\frac{1}{2}x + 2$

x	−2	0	2	3	4
y	3	2	1	$\frac{1}{2}$	0

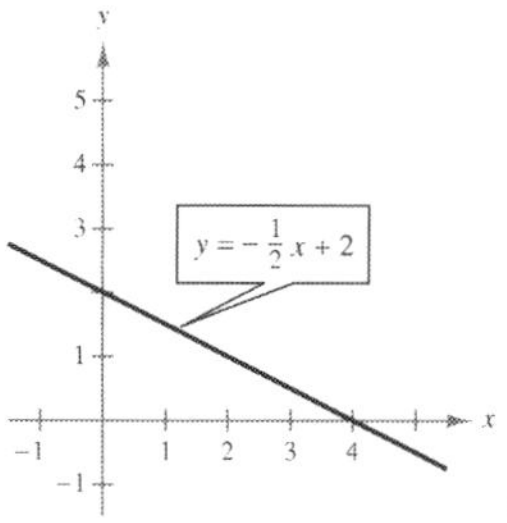

11. $y = x^2 + 3$

y-intercept: (0, 3)

y-axis symmetry

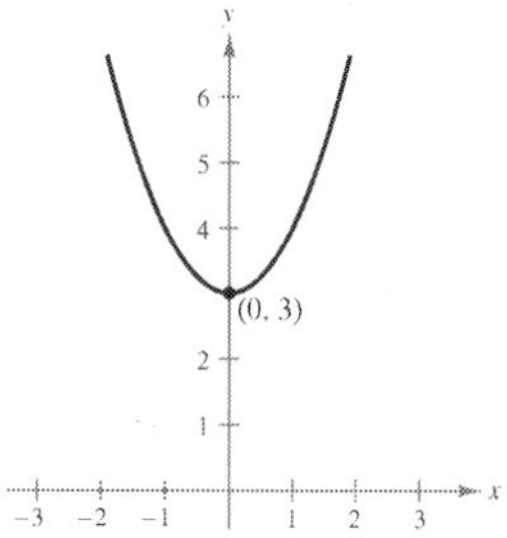

13. $y = 3x - 4$

x-intercept: $\left(\frac{4}{3}, 0\right)$

y-intercept: (0, −4)

Symmetry: none

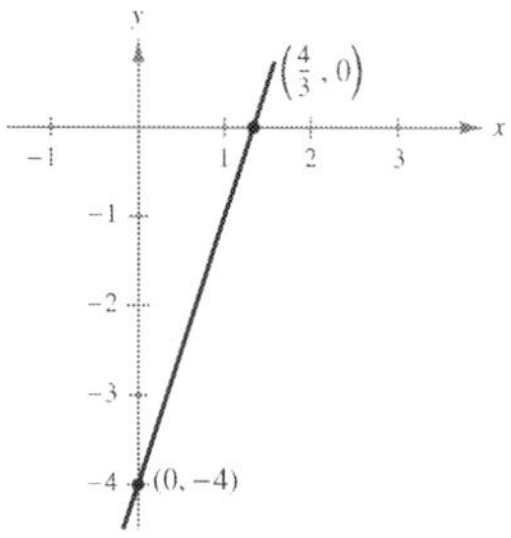

15. $y = x^3 + 1$

x-intercept: (−1, 0)

y-intercept: (0, 1)

Symmetry: none

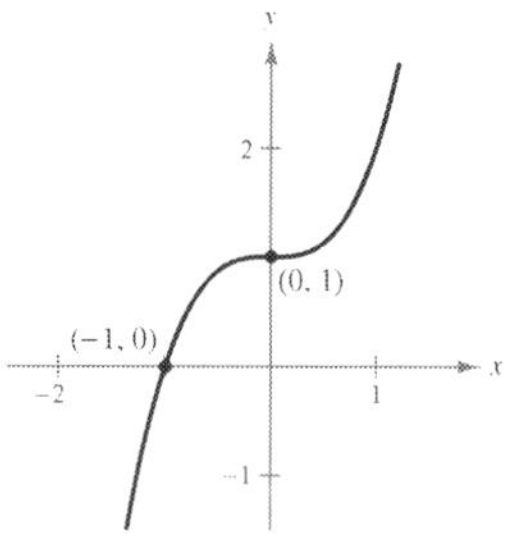

17. $(x + 1)^2 + (y - 2)^2 = 6^2$

$(x + 1)^2 + (y - 2)^2 = 36$

19. $x^2 + y^2 - 4x + 6y - 12 = 0$

$$(x^2 - 4x + 4) + (y^2 + 6y + 9) = 12 + 4 + 9$$

$$(x - 2)^2 + (y + 3)^2 = 25$$

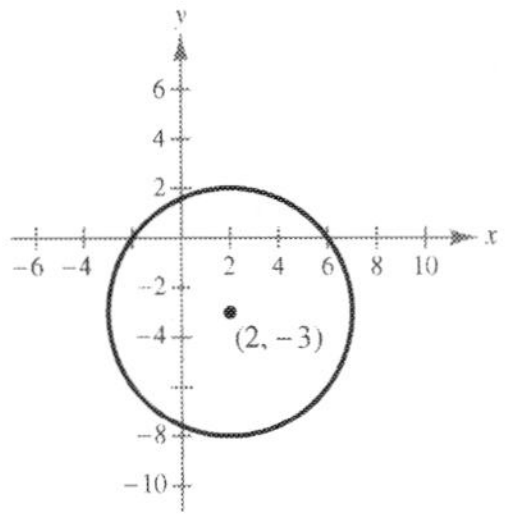

21. $m = \dfrac{-1 - 7}{2 - 3} = 8$

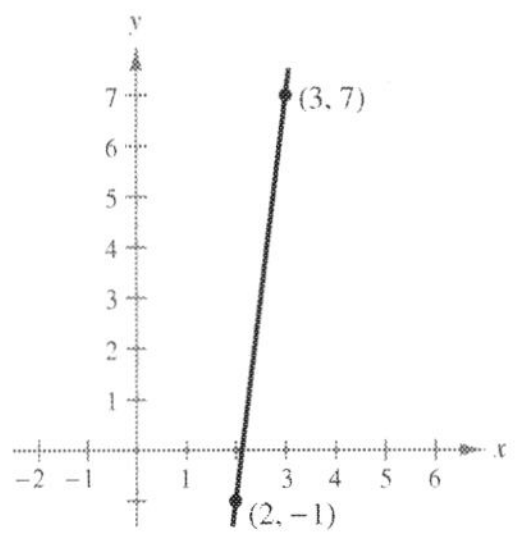

23. Since m is undefined, the line is vertical.

$x = 3 \Rightarrow x - 3 = 0$

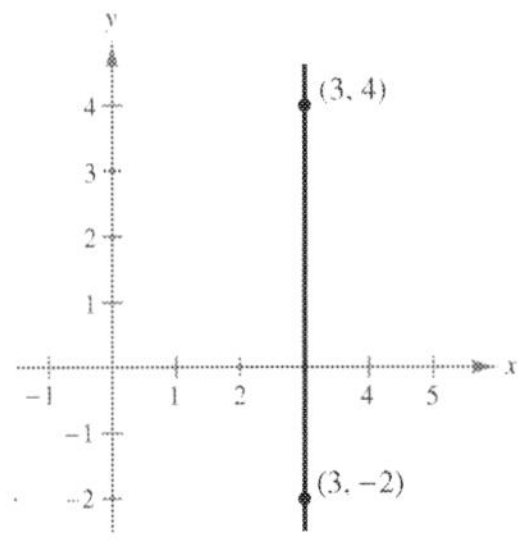

25. $y + 5 = \frac{3}{2}(x - 0)$

$y = \frac{3}{2}x - 5 \Rightarrow 3x - 2y - 10 = 0$

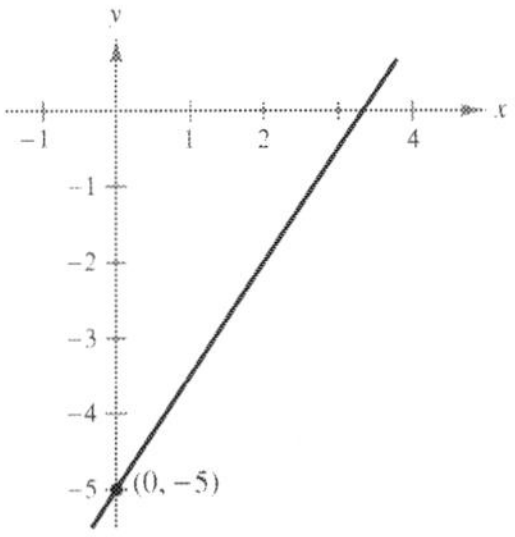

27. $y - 6 = 0(x + 2)$

$y - 6 = 0$

$y = 6$

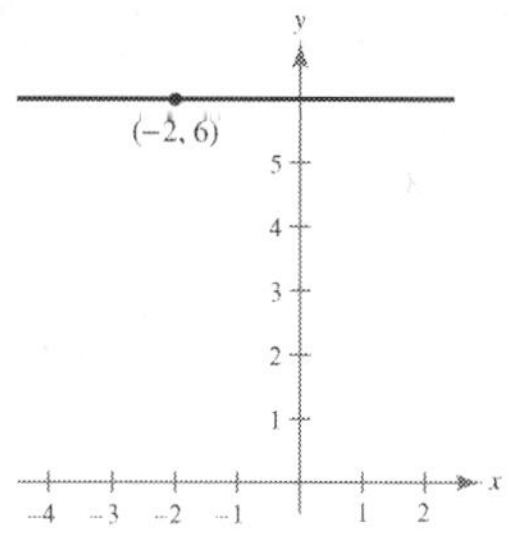

29. $8x - 4y + 16 = 0$

$-4y = -8x - 16$

$y = 2x + 4$

Slope: $m = 2$

y-intercept: $(0, 4)$

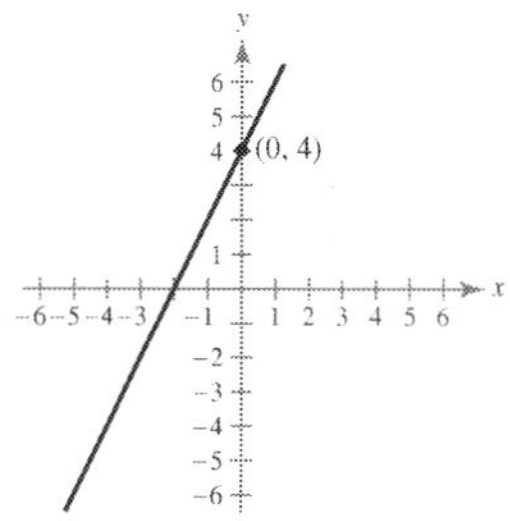

31. $17 - 5x = 10$

$-5x = -7$

$x = \frac{7}{5}$

Slope: undefined

y-intercept: none

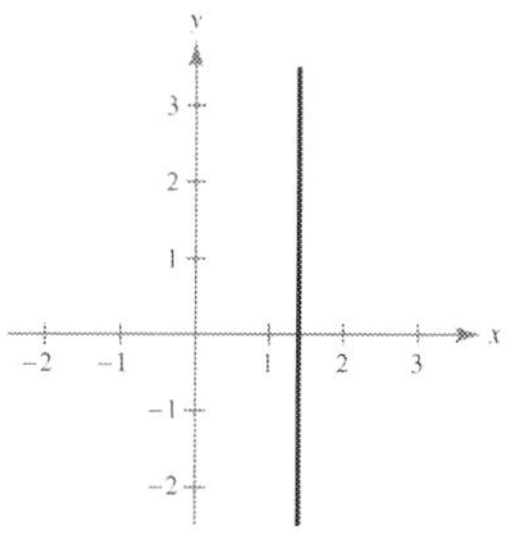

33. L_1: $(0, 3), (-2, 1)$

$m_1 = \dfrac{1 - 3}{-2 - 0} = 1$

L_2: $(-8, -3), (4, 9)$

$m_2 = \dfrac{9 + 3}{4 + 8} = 1$

Since $m_1 = m_2$, the lines are parallel.

35. L_1: $(3, 6), (-1, -5)$

$m_1 = \dfrac{-5 - 6}{-1 - 3} = \dfrac{11}{4}$

L_2: $(-2, 3), (4, 7)$

$m_2 = \dfrac{7 - 3}{4 + 2} = \dfrac{4}{6} = \dfrac{2}{3}$

The lines are neither parallel nor perpendicular.

37. $y = \frac{1}{2}x - 1$

$m_1 = \frac{1}{2}$

(a) $m_2 = \frac{1}{2}, (4, 3)$

$$y - 3 = \frac{1}{2}(x - 4)$$
$$y - 3 = \frac{1}{2}x - 2$$
$$y = \frac{1}{2}x + 1$$

(b) $m_2 = -\frac{1}{\frac{1}{2}} = -2$

$$y - 3 = -2(x - 4)$$
$$y - 3 = -2x + 8$$
$$y = -2x + 11$$

39. $5x - 4y = 8$

$$y = \frac{5}{4}x - 2$$
$$m_1 = \frac{5}{4}$$

(a) $m_2 = \frac{5}{4}, (3, -2)$

$$y + 2 = \frac{5}{4}(x - 3)$$
$$4y + 8 = 5x - 15$$
$$0 = 5x - 4y - 23$$

(b) $m_2 = -\frac{4}{5}, (3, -2)$

$$y + 2 = -\frac{4}{5}(x - 3)$$
$$5y + 10 = -4x + 12$$
$$4x + 5y - 2 = 0$$

41. (a) $y = -2$

(b) $x = -1$

43. $y = kx$

$$7 = k(3)$$
$$\frac{7}{3} = k$$
$$y = \frac{7}{3}x$$

45. $y = kx$

$$3480 = k(10)$$
$$348 = k$$
$$y = 348x$$

47. $A = \frac{30}{6}r = 5r$

49. $a = \frac{15}{20}b = \frac{3}{4}b$

51. $y = \frac{1260}{150{,}000}x$

Model: $y = 0.0084x$

When $x = 175{,}000$: $y = 0.0084(175{,}000) = 1470$

The property tax on a \$175,000 home is \$1470.

53. $\frac{3{,}150{,}000 - 2{,}950{,}000}{3 - 2} = 200{,}000$

$$y - 2{,}950{,}000 = 200{,}000(x - 2)$$
$$y = 200{,}000x + 2{,}550{,}000$$

When $x = 4$:

$y = 200{,}000(4) + 2{,}550{,}000 = 3{,}350{,}000$

In the fourth quarter, sales will be \$3,350,000.

55. Amount lost: $135{,}000 - 5500 = 129{,}500 \div 10$

$= 12{,}950$ lost per year

Model: $y = -12{,}950t + 135{,}000,\ 0 \le t \le 10$

57. $3x + y = 12$

$$y = -3x + 12$$

y is a function of x.

59. $y^2 = x + 3$

No, y is not a function of x because a value of x will be matched with two different values of y.

61. Because each element of A is paired with only one element of B, $\{(1, -3), (2, -7), (3, -3)\}$ represents a function from A to B.

63. (a) $f(5) = \sqrt{5 + 4} - 5 = \sqrt{9} - 5 = -2$

(b) $f(0) = \sqrt{0 + 4} - 5 = 2 - 5 = -3$

(c) $f(-4) = \sqrt{-4 + 4} - 5 = -5$

(d) $f(x + 3) = \sqrt{x + 3 + 4} - 5 = \sqrt{x + 7} - 5$

65. All real numbers x

67. $x \ge -5$

69. $1 \le t < 4,\ t > 4$

71. The domain of $h(x)$ is all real numbers except $x = 0$ because division by zero is undefined. The domain of $k(x)$ is all real numbers except $x = -2$ and $x = 2$ because if $x = -2$ or $x = 2$, then $x^2 - 4$ would equal zero and division by zero is undefined. Using a graphing utility and the table feature, $h(0)$ results in an error and $k(-2)$ and $k(2)$ also result in an error.

73. (a) $B = 6500\left(1 + \dfrac{0.0575}{4}\right)^{4t}$

(b) The domain is $t \geq 0$.

75. $f(x) = x^2 + 1$

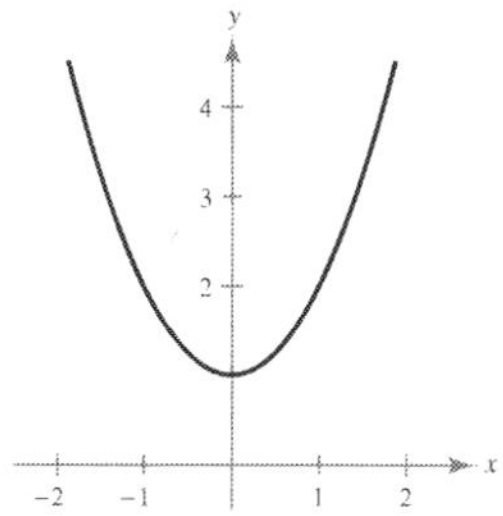

(a) Domain: all real numbers or $(-\infty, \infty)$

Range: $[1, \infty)$

(b) Increasing on $(0, \infty)$

Decreasing on $(-\infty, 0)$

(c) $f(-x) = (-x)^2 + 1 = x^2 + 1 = f(x)$

Even

(d) Relative minimum at $(0, 1)$

77. $f(x) = x^3 - 4x^2$

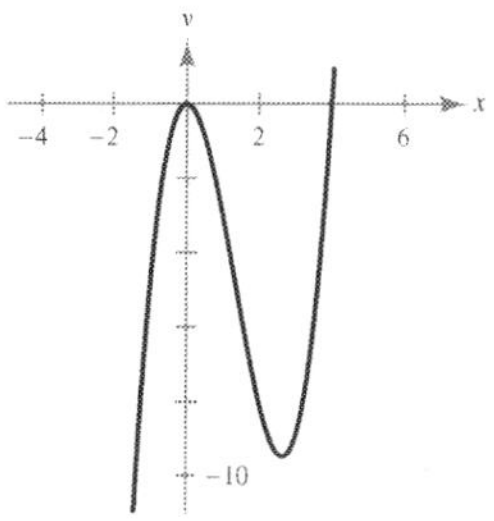

(a) Domain: All real numbers or $(-\infty, \infty)$

Range: All real numbers or $(-\infty, \infty)$

(b) Increasing on $(-\infty, 0)$ and $\left(\frac{8}{3}, \infty\right)$

Decreasing on $\left(0, \frac{8}{3}\right)$

(c) $f(-x) = (-x)^3 - 4(-x)^2 = -x^3 - 4x^2$

Neither odd nor even

(d) Relative maximum at $(0, 0)$

Relative minimum at $\left(\frac{8}{3}, -\frac{256}{27}\right)$

79. $y = \frac{1}{2}x^2$

This is a function because a vertical line intersects the graph only once.

81. $x - y^2 = 1$

This is not a function because a vertical line intersects the graph twice.

83. $f(x) = |x + 3|$

x-intercept: $(-3, 0)$

y-intercept: $(0, 3)$

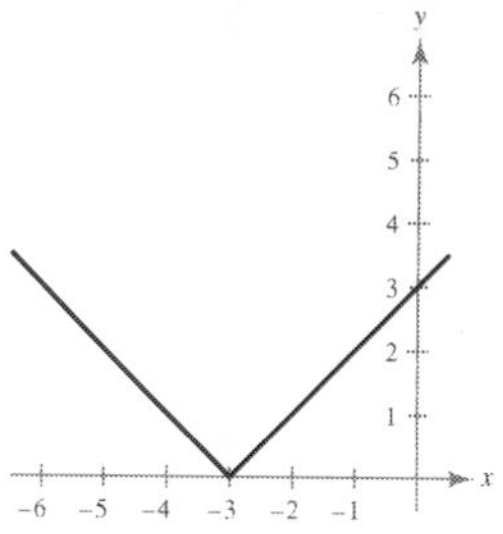

85. $h(x) = -[\![x]\!] + 1$

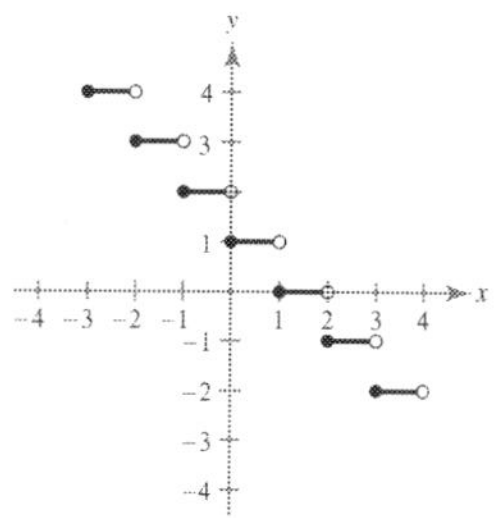

87. $g(x) = \begin{cases} x + 2, & x < 0 \\ 2, & x = 0 \\ x^2 + 2, & x > 0 \end{cases}$

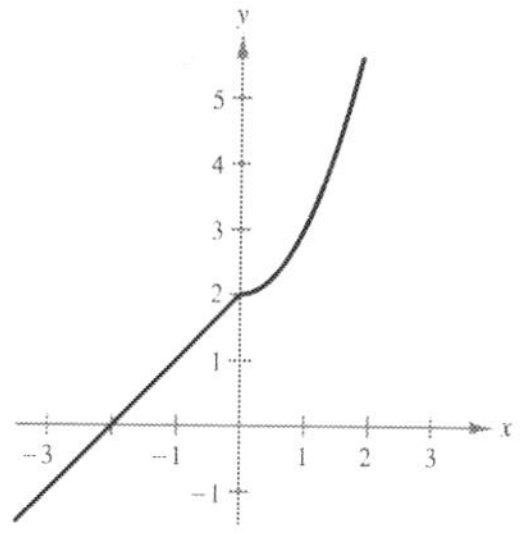

89. $h(x) = x^2 - 3x = x(x - 3)$

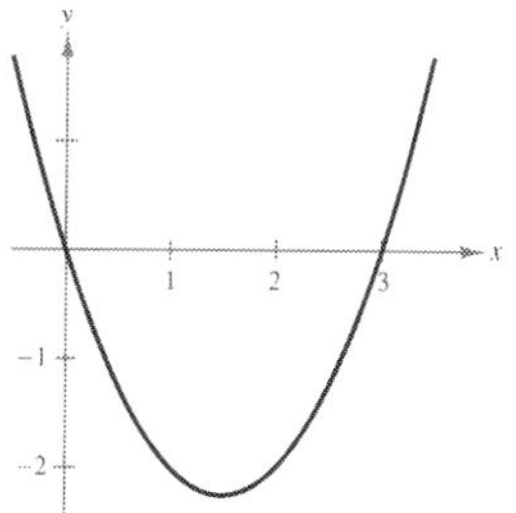

91. $g(x) = -f(x - 1) - 2$

Reflection in the x-axis, horizontal shift 1 unit to the right, and a vertical shift 2 units downward

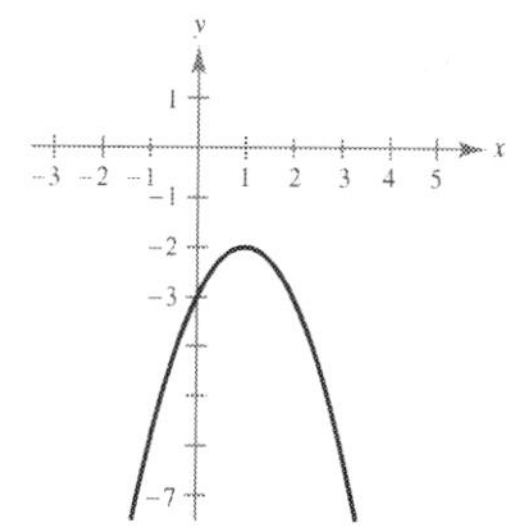

93. $g(x) = f(x - 2)$

Horizontal shift of 2 units to the right

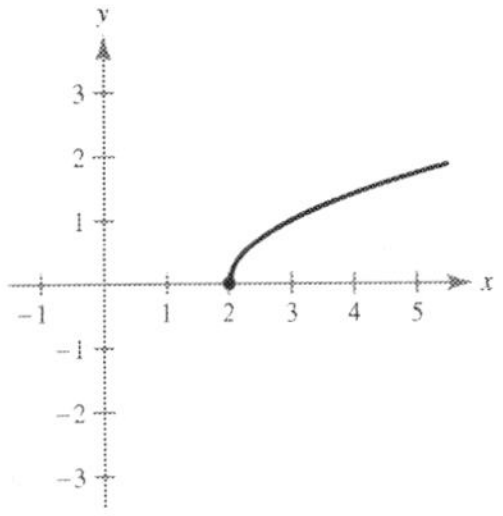

95. $y = f(x + 2)$

Horizontal shift 2 units to the left

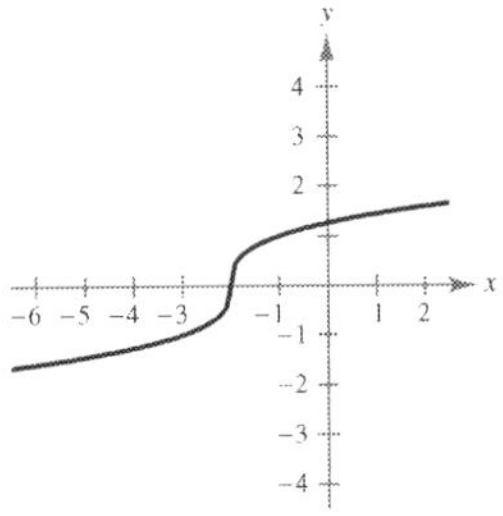

97. Common function: $y = \sqrt{x}$

Transformation: horizontal shift 3 units to the right and then reflected about the y-axis

Equation: $y = \sqrt{-(x - 3)} = \sqrt{-x + 3}$ or $\sqrt{3 - x}$

99. $(f + g)(x) = (3x - 1) + (x^2 + 2x) = x^2 + 5x - 1$

$(f - g)(x) = (3x - 1) - (x^2 + 2x) = -x^2 + x - 1$

$$(fg)(x) = (3x - 1)(x^2 + 2x)$$
$$= 3x^3 - x^2 + 6x^2 - 2x$$
$$= 3x^3 + 5x^2 - 2x$$

$$\left(\frac{f}{g}\right)(x) = \frac{3x + 1}{x^2 + 2x}, \; x \neq 0, \; x \neq -2$$

Domain of $\left(\frac{f}{g}\right)(x)$: all real numbers except

$x = 0$ and $x = -2$

101. $(f + g)(x) = x^2 + 3x + 2x - 5 = x^2 + 5x - 5$

$(f + g)(2) = 2^2 + 5(2) - 5 = 4 + 10 - 5 = 9$

103. $(fg)(x) = (x^2 + 3x)(2x - 5)$

$$= 2x^3 - 5x^2 + 6x^2 - 15x$$
$$= 2x^3 + x^2 - 15x$$

$(fg)(3) = 2(3)^3 + 3^2 - 15(3) = 54 + 9 - 45 = 18$

105. $f(x) = x^2, \; g(x) = x + 3$

(a) $(f \circ g)(x) = f(g(x))$
$$= f(x + 3)$$
$$= (x + 3)^2$$
$$= x^2 + 6x + 9$$

Domain: $(-\infty, \infty)$

(b) $(g \circ f)(x) = g(f(x)) = g(x^2) = x^2 + 3$

Domain: $(-\infty, \infty)$

107. $f(x) = \frac{1}{x}, \; g(x) = 3x + x^2$

(a) $(f \circ g)(x) = f(g(x)) = f(3x + x^2)$
$$= \frac{1}{3x + x^2}, \quad x \neq 0, x \neq -3$$

Domain: $(-\infty, -3) \cup (-3, 0) \cup (0, \infty)$

(b) $(g \circ f)(x) = g(f(x)) = g\left(\frac{1}{x}\right)$
$$= 3\left(\frac{1}{x}\right) + \left(\frac{1}{x}\right)^2 = \left(\frac{3}{x}\right) + \left(\frac{1}{x}\right)^2$$
$$= \frac{3x + 1}{x^2}, \; x \neq 0$$

Domain: $(-\infty, 0) \cup (0, \infty)$

109. $f(x) = x^2, g(x) = 6x - 5$; Answers will vary.

111. $f(x) = \frac{1}{x^2}, \; g(x) = x - 1$; Answers will vary.

113. (a) $r = \frac{x}{2}$

(b) $A = \pi r^2$

(c) $(A \circ r)(x) = \frac{\pi x^2}{4}$

$(A \circ r)(x)$ is the area of the base of the tank in terms of x, where x is the length of a side of the square base.

115. $(N \circ T)(t) = 8(2t + 2)^2 - 14(2t + 2) + 200$
$$= 8(4t^2 + 8t + 4) - 28t - 28 + 200$$
$$= 32t^2 + 64t + 32 - 28t - 28 + 200$$
$$= 32t^2 + 36t + 204$$

$(N \circ T)(t)$ represents the number of bacteria in the food at time t.

Chapter 2 Test Yourself

1. $d = \sqrt{(5-(-3))^2 + (-2-2)^2} = \sqrt{(8)^2 + (-4)^2} = \sqrt{80} = 4\sqrt{5} \approx 8.94$

Midpoint: $\left(\dfrac{-3+5}{2}, \dfrac{2+(-2)}{2}\right) = \left(\dfrac{2}{2}, \dfrac{0}{2}\right) = (1, 0)$

2. $d = \sqrt{(-2.37-3.25)^2 + (1.62-7.05)^2} = \sqrt{(-5.62)^2 + (-5.43)^2} = \sqrt{61.0693} \approx 7.81$

Midpoint: $\left(\dfrac{3.25+(-2.37)}{2}, \dfrac{7.05+1.62}{2}\right) = \left(\dfrac{0.88}{2}, \dfrac{8.67}{2}\right) = (0.44, 4.335)$

3. x-intercepts: $0 = (x+5)(x-3) \Rightarrow x = -5, 3$

$(-5, 0), (3, 0)$

y-intercept: $y = (0+5)(0-3) = -15$

$(0, -15)$

4. $(-y) = \dfrac{x}{x^2-4} \Rightarrow y = -\dfrac{x}{x^2-4}$ No x-axis symmetry

$y = \dfrac{(-x)}{(-x)^2-4} \Rightarrow y = -\dfrac{x}{x^2-4}$ No y-axis symmetry

$(-y) = \dfrac{(-x)}{(-x)^2-4} \Rightarrow y = \dfrac{x}{x^2-4}$ Origin symmetry

5.

$$x^2 + y^2 - 6x + 4y - 3 = 0$$
$$(x - 6x + 9) + (y^2 + 4y + 4) = 3 + 9 + 4$$
$$(x-3)^2 + (y+2)^2 = 16$$

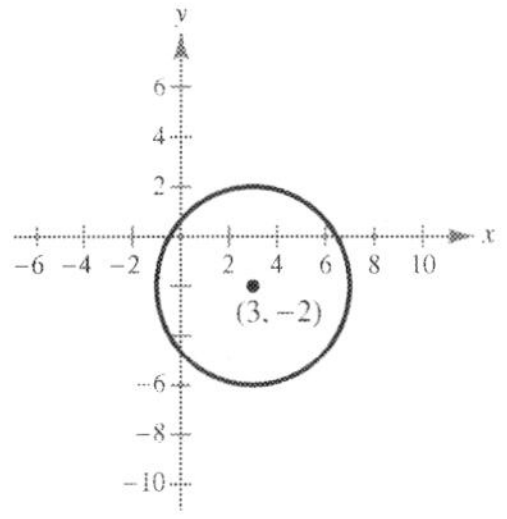

6.

$$y - 5 = \tfrac{2}{3}(x+3)$$
$$y - 5 = \tfrac{2}{3}x + 2$$
$$y = \tfrac{2}{3}x + 7$$
$$2x - 3y + 7 = 0$$

7.

$$2x - 3y = 5$$
$$-3y = -2x + 5$$
$$y = \tfrac{2}{3}x - \tfrac{5}{3}$$

Each value of x yields only one y-value, so it is a function of x. True.

8. $A = \{3, 4, 5\}, B = \{-1, -2, -3\}$

$C = \{(3, -9), (4, -2), (5, -3)\}$

Because -9 is not an element of B, C does **not** represent a function from A to B. False.

9. (a) Domain: $(-\infty, \infty)$

Range: $(-\infty, 2]$

(b) Increasing on $(-\infty, 0)$

Decreasing on $(0, \infty)$

(c) $f(-x) = 2 - (-x)^2$

$= 2 - x^2 = f(x)$

The function is even.

(d) Relative maximum at $(0, 2)$

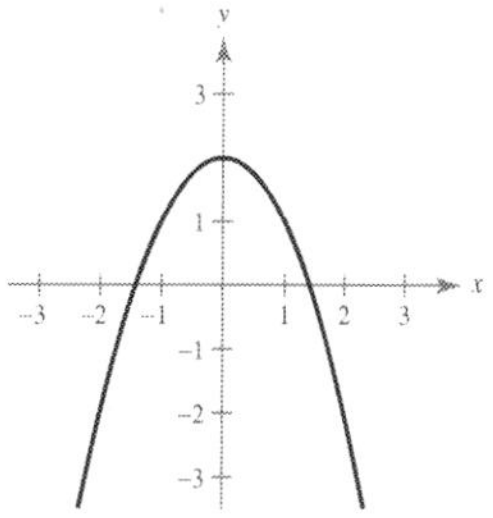

10. (a) Domain: $(-\infty, -2] \cup [2, \infty)$
Range: $[0, \infty)$

(b) Increasing on $(2, \infty)$
Decreasing on $(-\infty, -2)$

(c) $f(-x) = \sqrt{(-x)^2 - 4}$
$= \sqrt{x^2 - 4}$
$= f(x)$

The function is even.

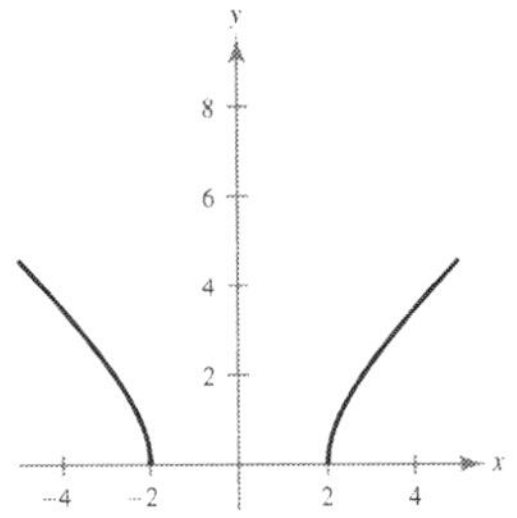

(d) Relative minimum at $(-2, 0)$ and $(2, 0)$

11. Reflected about the x-axis and shifted 2 units to the left.

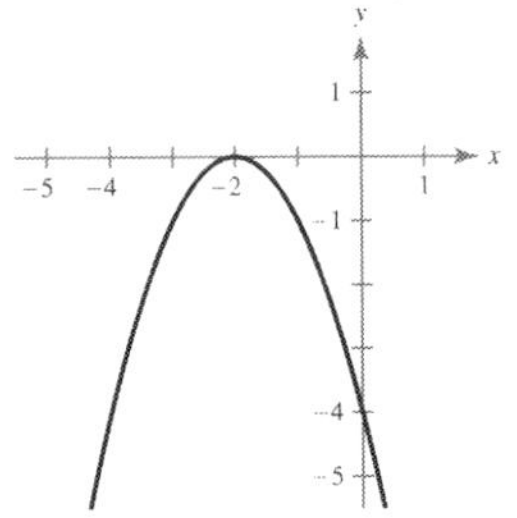

12. Shifted 3 units to the right and 4 units upward

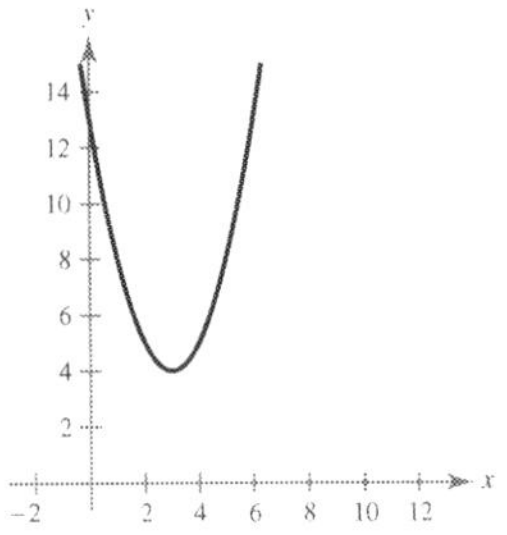

13. $(f - g)(x) = f(x) - g(x)$
$= (x^2 + 1) - (2x - 2)$
$= x^2 - 2x + 3$

14. $(fg)(x) = f(x)g(x)$
$= (x^2 + 1)(2x - 2)$
$= 2x^3 - 2x^2 + 2x - 2$

15. $(f \circ g)(x) = f(g(x))$
$= f(2x - 2)$
$= (2x - 2)^2 + 1$
$= 4x^2 - 8x + 5$

16. $(g \circ f)(x) = g(f(x))$
$= g(x^2 + 1)$
$= 2(x^2 + 1) - 2$
$= 2x^2$

17. When the equipment is new, $t = 0$, so $(0, 30{,}000)$ represents the value when the equipment is new and $(5, 3500)$ represents the value after 5 years.

$$V - 30{,}000 = \left(\frac{3500 - 30{,}000}{5 - 0}\right)(t - 0)$$
$$V - 30{,}000 = -5300t$$
$$V = 30{,}000 - 5300t$$

18.

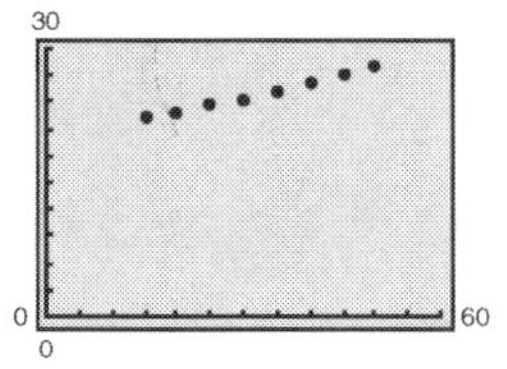

Linear model: $P = 0.17t + 19.3$

CHAPTER 3
Polynomial and Rational Functions

CHAPTER 3
Polynomial and Rational Functions

Section 3.1 Quadratic Functions and Models

Skills Warm Up

1. $2x^2 + 11x - 6 = 0$

$(x + 6)(2x - 1) = 0$

$x + 6 = 0 \rightarrow x = -6$

$2x - 1 = 0 \rightarrow x = \frac{1}{2}$

2. $5x^2 - 12x - 9 = 0$

$(5x + 3)(x - 3) = 0$

$5x + 3 = 0 \rightarrow x = -\frac{3}{5}$

$x - 3 = 0 \rightarrow x = 3$

3. $2x^2 - x - 3 = 0$

$(2x - 3)(x + 1) = 0$

$2x - 3 = 0 \rightarrow x = \frac{3}{2}$

$x + 1 = 0 \rightarrow x = -1$

4. $x^2 + 20x + 100 = 0$

$(x + 10)^2 = 0$

$x + 10 = 0$

$x = -10$

5. $x = \dfrac{6 \pm \sqrt{36 - 4(1)(4)}}{2(1)}$

$= \dfrac{6 \pm \sqrt{20}}{2}$

$= \dfrac{6 \pm 2\sqrt{5}}{2}$

$= 3 \pm \sqrt{5}$

6. $x = \dfrac{-4 \pm \sqrt{16 - 4(1)(1)}}{2(1)}$

$= \dfrac{-4 \pm \sqrt{12}}{2}$

$= \dfrac{-4 \pm 2\sqrt{3}}{2}$

$= -2 \pm \sqrt{3}$

7. $x = \dfrac{16 \pm \sqrt{(-16)^2 - 4(2)(25)}}{2(2)}$

$x = \dfrac{16 \pm \sqrt{56}}{4}$

$= \dfrac{16 \pm 2\sqrt{14}}{4}$

$= 4 \pm \dfrac{\sqrt{14}}{2}$

8. $x = \dfrac{-30 \pm \sqrt{900 - 4(3)(74)}}{6}$

$= \dfrac{-30 \pm \sqrt{12}}{6}$

$= \dfrac{-30 \pm 2\sqrt{3}}{6}$

$= -5 \pm \dfrac{\sqrt{3}}{3}$

9. $x = \dfrac{-3 \pm \sqrt{9 - 4(1)(1)}}{2}$

$= \dfrac{-3 \pm \sqrt{5}}{2}$

10. $x = \dfrac{-3 \pm \sqrt{9 - 4(-3)(1)}}{2}$

$= \dfrac{-3 \pm \sqrt{21}}{2}$

1. $f(x) = 5x^2$ vertically stretches the graph of $y = x^2$ by a factor of 5.

3. $f(x) = 3x^2$

Vertex: $(0, 0)$

Intercept: $(0, 0)$

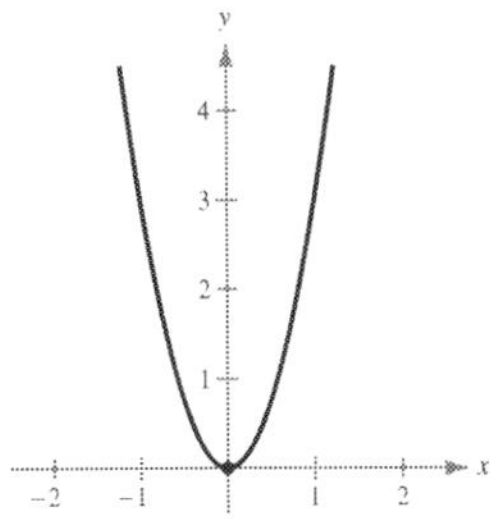

The graph of f is the graph of $y = x^2$ vertically stretched by a factor of 3.

5. $g(x) = -\frac{1}{2}x^2$

Vertex: $(0, 0)$

Intercept: $(0, 0)$

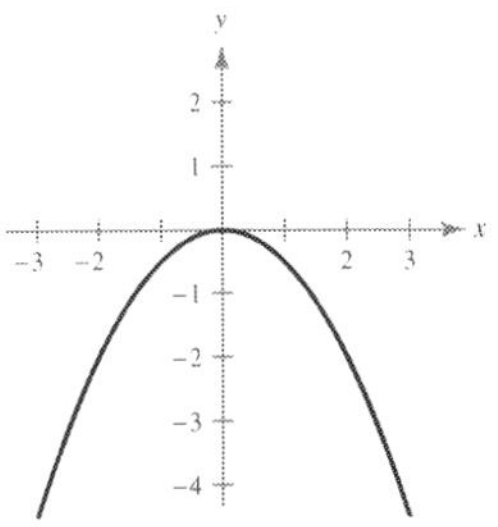

The graph of g is the graph of $y = x^2$ reflected over the x-axis and vertically shrunk by a factor of $\frac{1}{2}$.

7. The graph of f is the graph of $y = x^2$ reflected over the x-axis, shifted to the left one unit, and shifted upward one unit.

9. $f(x) = -(x - 3)^2$ opens downward and has vertex $(3, 0)$; graph (e)

11. $f(x) = x^2 - 4$ opens upward and has vertex $(0, -4)$; graph (a)

13. $f(x) = -(x + 1)^2 + 3$ opens downward and has vertex $(-1, 3)$; graph (f)

15. $f(x) = 16 - x^2$

Vertex: $(0, 16)$

Intercepts: $(-4, 0)$, $(0, 16)$, $(4, 0)$

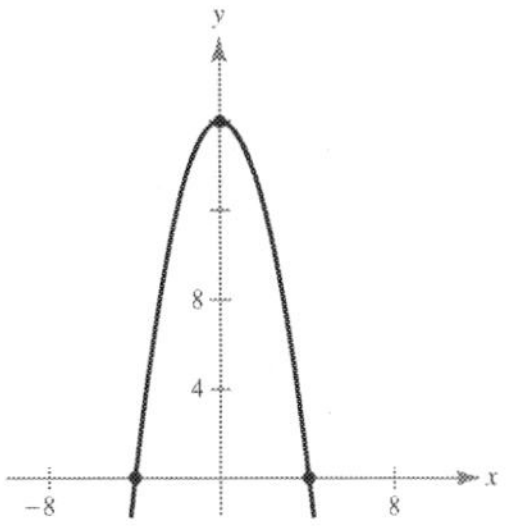

17. $f(x) = (x + 5)^2 - 6$

Vertex: $(-5, -6)$

Intercepts: $\left(-5 - \sqrt{6}, 0\right)$, $\left(-5 + \sqrt{6}, 0\right)$, $(0, 19)$

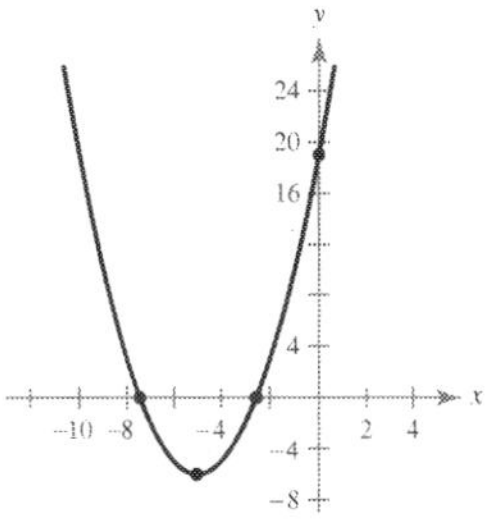

19. $g(x) = x^2 + 2x + 1 = (x + 1)^2$

Vertex: $(-1, 0)$

Intercepts: $(-1, 0)$, $(0, 1)$

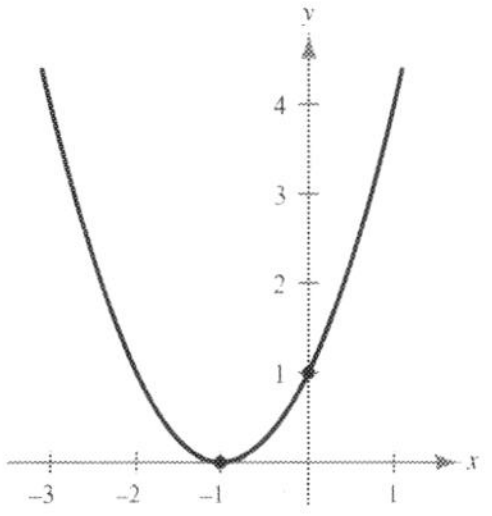

21. $f(x) = -(x^2 + 2x - 3) = -(x + 1)^2 + 4$

Vertex: $(-1, 4)$

Intercepts: $(-3, 0)$, $(0, 3)$, $(1, 0)$

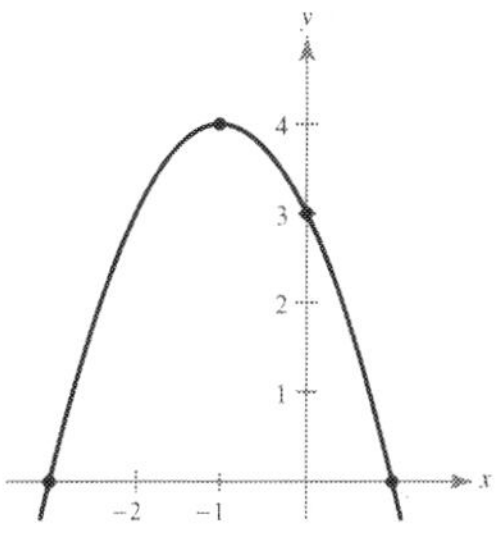

23. $f(x) = x^2 - x + \frac{5}{4} = \left(x - \frac{1}{2}\right)^2 + 1$

Vertex: $\left(\frac{1}{2}, 1\right)$

Intercept: $\left(0, \frac{5}{4}\right)$

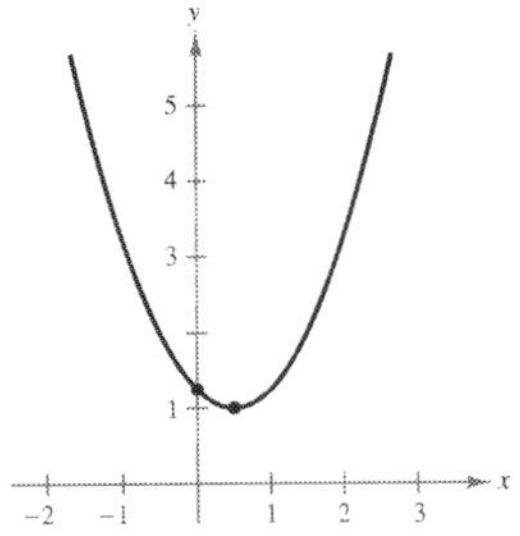

25. $f(x) = -x^2 + 2x + 5 = -(x - 1)^2 + 6$

Vertex: $(1, 6)$

Intercepts: $\left(1 \pm \sqrt{6}, 0\right), (0, 5)$

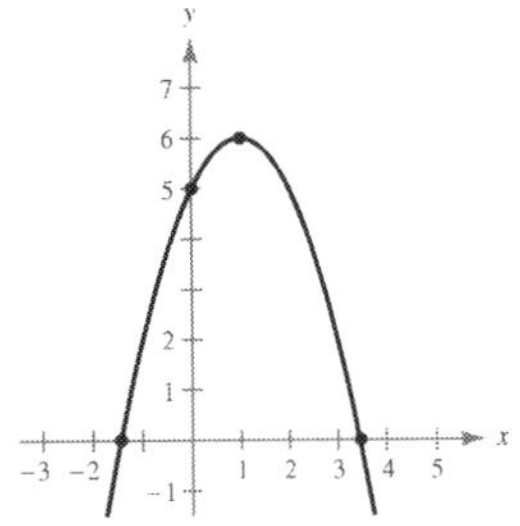

27. $h(x) = 4x^2 - 4x + 21 = 4\left(x - \frac{1}{2}\right)^2 + 20$

Vertex: $\left(\frac{1}{2}, 20\right)$

Intercept: $(0, 21)$

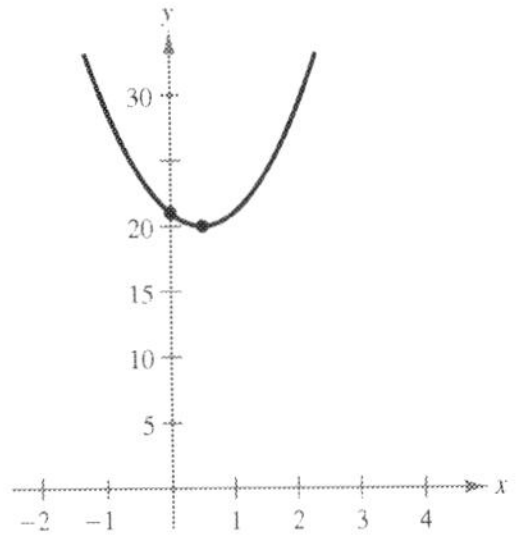

29. $f(x) = \frac{1}{4}\left(x^2 - 16x + 32\right) = \frac{1}{4}(x - 8)^2 - 8$

Vertex: $(8, -8)$

Intercepts: $(0, 8), \left(8 \pm 4\sqrt{2}, 0\right)$

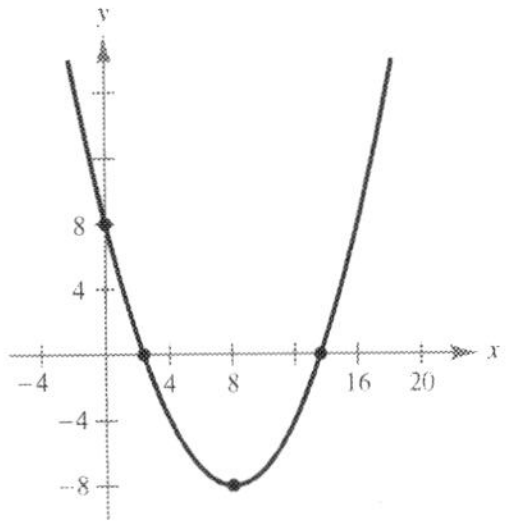

31. Vertex: $(-2, 0)$

$$\begin{aligned} f(x) &= a(x + 2)^2 + 0 \\ -4 &= a(0 + 2)^2 \\ a &= -1 \\ f(x) &= -(x + 2)^2 \end{aligned}$$

33. Vertex: $(3, -9)$

$$\begin{aligned} f(x) &= a(x - 3)^2 - 9 \\ 0 &= a(6 - 3)^2 - 9 \\ a &= 1 \\ f(x) &= (x - 3)^2 - 9 \end{aligned}$$

35. Vertex: $(-3, 3)$

$$\begin{aligned} f(x) &= a(x + 3)^2 + 3 \\ 1 &= a(-2 + 3)^2 + 3 \\ a &= -2 \\ f(x) &= -2(x + 3)^2 + 3 \end{aligned}$$

37.
$$\begin{aligned} y &= a(x - 2)^2 - 1 \\ -3 &= a(4 - 2)^2 - 1 \\ a &= -\frac{1}{2} \\ y &= -\frac{1}{2}(x - 2)^2 - 1 \end{aligned}$$

39.
$$\begin{aligned} y &= a(x - 5)^2 + 12 \\ 15 &= a(7 - 5)^2 + 12 \\ a &= \frac{3}{4} \\ y &= \frac{3}{4}(x - 5)^2 + 12 \end{aligned}$$

41. Answers will vary. Sample answer:

$f(x) = (x - 2)(x + 1) = x^2 - x - 2$

$g(x) = -(x - 2)(x + 1) = -x^2 + x + 2$

43. Answers will vary. Sample answer:

$f(x) = x(x - 10) = x^2 - 10x$

$g(x) = -x(x - 10) = -x^2 + 10x$

45. Answers will vary. Sample answer:

$f(x) = (x + 3)(2x + 1) = 2x^2 + 7x + 3$

$g(x) = -(x + 3)(2x + 1) = -2x^2 - 7x - 3$

47. Let x = the width of the rectangle. Then the length of the rectangle is $l = \frac{200 - 2x}{2} = 100 - x$.

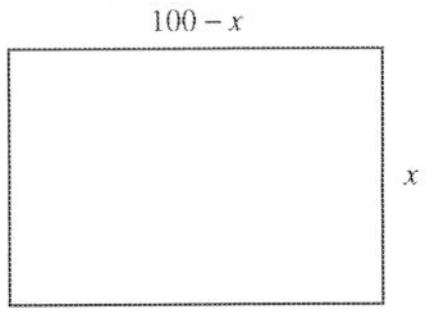

The area of the rectangle is:

$$A = l \cdot w = (100 - x)x$$
$$= 100x - x^2$$
$$= -(x - 50)^2 + 2500$$

The vertex of this parabola is at $x = 50$. At that point the area is as large as possible, 2500 square feet. The dimensions of the rectangle with optimal area are $x = 50$ feet by $l = 100 - x = 50$ feet (a square). Because the vertex is at (50, 2500), we know that the dimensions of the rectangle with optimal area are $x = 50$ feet by $l = 100 - x = 50$ feet and the optimal area is 2500 square feet. Additionally, the optimal area occurs when the rectangle is actually a square with sides of 50 feet.

49. $4x + 3y = 1200$

$$y = \frac{1200 - 4x}{3}$$

$$A = (2x)y$$
$$= 2x\left(\frac{1200 - 4x}{3}\right)$$
$$= \frac{2400}{3}x - \frac{8}{3}x^2$$
$$= -\frac{8}{3}(x - 150)^2 + \frac{180{,}000}{3}$$

The area will be maximum when $x = 150$. The dimensions for optimal area are $2x = 300$ feet by

$$y = \frac{1200 - 4x}{3} = \frac{600}{3} = 200 \text{ feet.}$$

51. $R = 1000x - 0.02x^2$

$$= -0.02(x - 25{,}000)^2 + 12{,}500{,}000$$

Vertex: (25,000, 12,500,000)

Maximum when $x = 25{,}000$ units

Optimal revenue occurs when 25,000 units are sold.

53. $C = 800 - 10x + 0.25x^2$

$$= 0.25(x - 20)^2 + 700$$

Vertex: (20, 700)

Minimum when $x = 20$ fixtures

Minimum cost occurs when 20 fixtures are produced.

55. $y = -\frac{4}{9}x^2 + \frac{24}{9}x + 10 = -\frac{4}{9}(x - 3)^2 + 14$

Vertex: (3, 14)

Maximum when $y = 14$ feet

The maximum height of the diver is 14 feet.

57. (a) and (b)

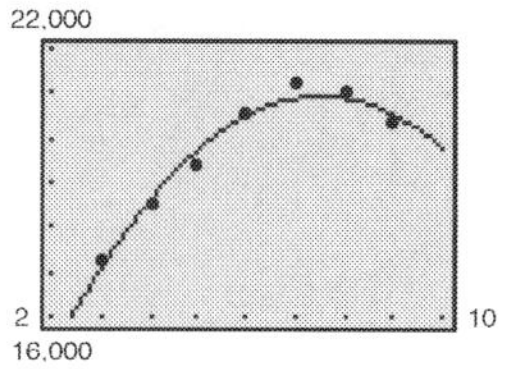

$$R = -186.43t^2 + 2811.9t + 10{,}329$$

(c) Yearly revenue was greatest in 2007 $(t \approx 7.5)$. Yes, this result agrees with the data.

59. (a)

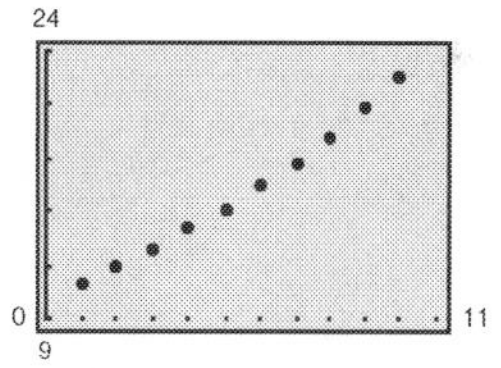

(b) $N = 0.047t^2 + 0.76t + 10.1$

(c)

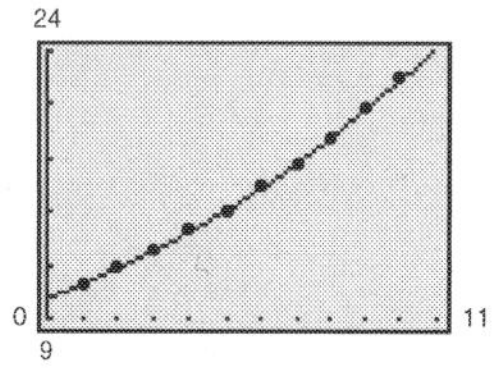

(d) In 2012, there will be about 25,988 female osteopathic physicians (when $t = 12$, $N = 25{,}988$).

61. (a)

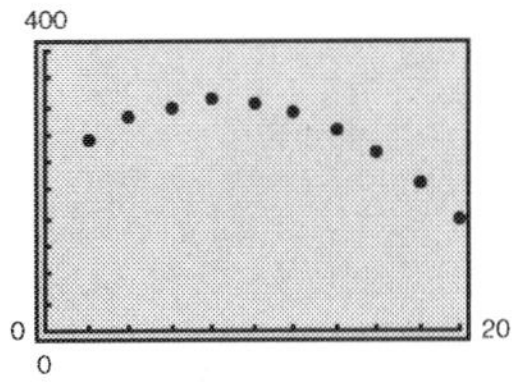

(b) $p(x) = -1.276x^2 + 21.63x + 236.5$

(c)

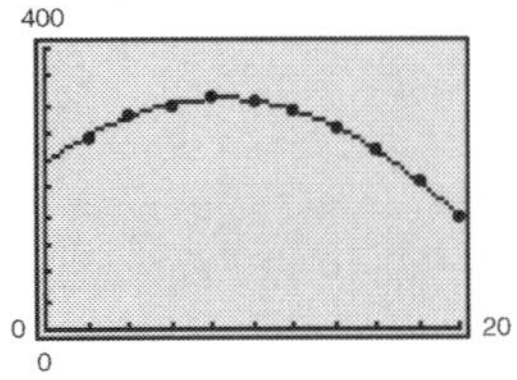

(d) The vertex is at about (8.5, 328.2). This means that producing about 85,000 units yields a maximum profit of about \$32,820,000.

(e) Answers will vary. Sample answer: Production costs may be growing faster than revenue, so profit decreases.

63.

$$\begin{aligned} f(x) &= ax^2 + bx + c \\ &= a\left(x^2 + \frac{b}{a}x\right) + c \\ &= a\left(x^2 + \frac{b}{a}x + \frac{b^2}{4a^2} - \frac{b^2}{4a^2}\right) + c \\ &= a\left(x + \frac{b}{2a}\right)^2 - \frac{b^2}{4a} + c \\ &= a\left(x - \left(-\frac{b}{2a}\right)\right)^2 + \frac{4ac - b^2}{4a} \end{aligned}$$

$$\begin{aligned} f\left(-\frac{b}{2a}\right) &= a\left(\frac{b^2}{4a^2}\right) + b\left(-\frac{b}{2a}\right) + c \\ &= \frac{b^2}{4a} - \frac{b^2}{2a} + c \\ &= \frac{b^2 - 2b^2 + 4ac}{4a} = \frac{4ac - b^2}{4a} \end{aligned}$$

So, the vertex occurs at

$$\left(-\frac{b}{2a}, \frac{4ac - b^2}{4a}\right) = \left(-\frac{b}{2a}, f\left(-\frac{b}{2a}\right)\right).$$

Section 3.2 Polynomial Functions of Higher Degree

Skills Warm Up

1. $12x^2 + 7x - 10 = (4x + 5)(3x - 2)$

2. $$\begin{aligned} 25x^3 - 60x^2 + 36x &= x(25x^2 - 60x + 36) \\ &= x(5x - 6)(5x - 6) \\ &= x(5x - 6)^2 \end{aligned}$$

3. $$\begin{aligned} 12z^4 + 17z^3 + 5z^2 &= z^2(12z^2 + 17z + 5) \\ &= z^2(z + 1)(12z + 5) \end{aligned}$$

4. $y^3 + 125 = (y + 5)(y^2 - 5y + 25)$

5. $$\begin{aligned} x^3 + 3x^2 - 4x - 12 &= x^2(x + 3) - 4(x + 3) \\ &= (x^2 - 4)(x + 3) \\ &= (x + 2)(x - 2)(x + 3) \end{aligned}$$

6. $$\begin{aligned} x^3 + 2x^2 + 3x + 6 &= x^2(x + 2) + 3(x + 2) \\ &= (x^2 + 3)(x + 2) \end{aligned}$$

7. $$\begin{aligned} 5x^2 + 8 &= 0 \\ 5x^2 &= -8 \\ x^2 &= -\frac{8}{5} \end{aligned}$$

No real solution

8. $$\begin{aligned} x^2 - 6x + 4 &= 0 \\ x &= \frac{6 \pm \sqrt{36 - 4(4)}}{2} \\ &= \frac{6 \pm \sqrt{20}}{2} \\ &= \frac{6 \pm 2\sqrt{5}}{2} \\ &= 3 \pm \sqrt{5} \end{aligned}$$

Skills Warm Up *—continued—*

9. $4x^2 + 4x - 11 = 0$

$$x = \frac{-4 \pm \sqrt{16 - 4(4)(-11)}}{2(4)}$$

$$= \frac{-4 \pm \sqrt{192}}{8}$$

$$= \frac{-4 \pm 8\sqrt{3}}{8}$$

$$= \frac{-1 \pm 2\sqrt{3}}{2}$$

10. $x^4 - 18x^2 + 81 = 0$

$$(x^2 - 9)(x^2 - 9) = 0$$

$$x = \pm 3$$

1. $f(x) = (x - 1)^4$

Because the leading coefficient is positive and the degree is even, the graph rises to the left and right; graph (b).

3. $f(x) = -x^4 + 1$

Because the leading coefficient is negative and the degree is even, the graph falls to the left and right; graph (c).

5. $f(x) = x^3 - 2$

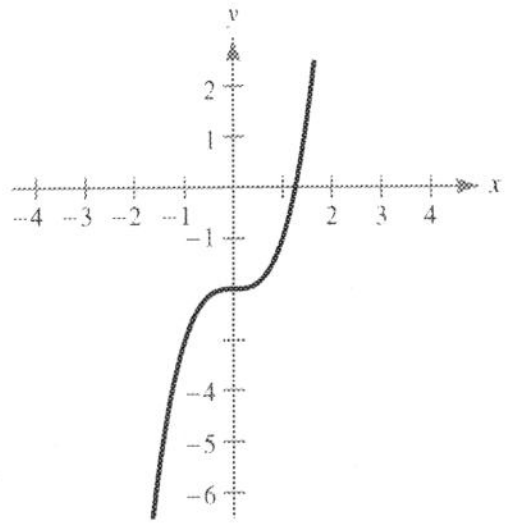

7. $f(x) = (x + 1)^3 - 4$

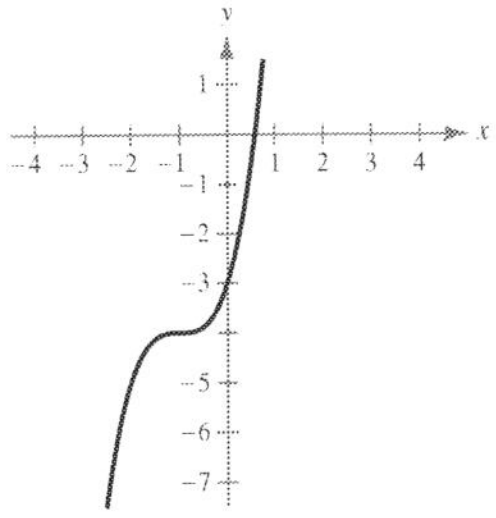

9. $f(x) = (x + 3)^4$

Horizontal shift 3 units to the left

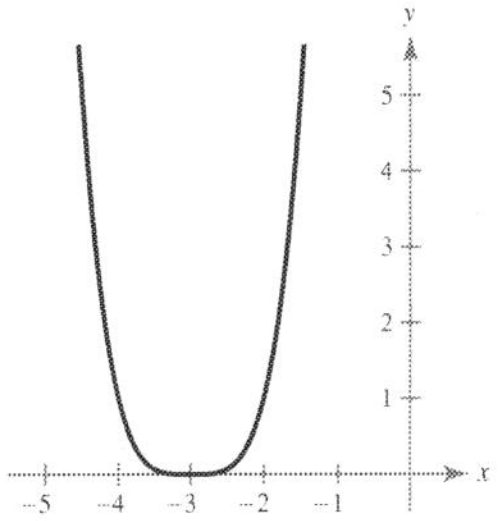

11. $f(x) = 3 - x^4$

Reflection in the x-axis and then a vertical shift 3 units upward

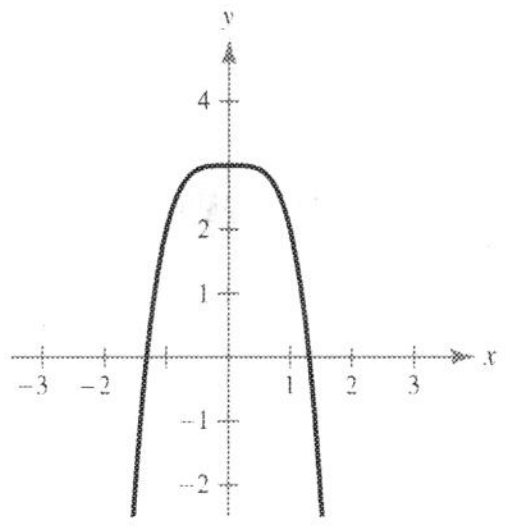

13. (a) 5; odd

(b) 2; positive

(c)

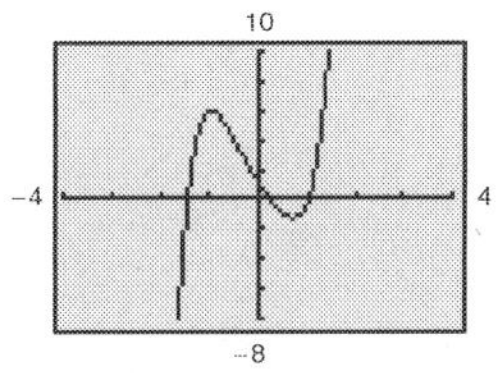

(d) The graph falls to the left and rises to the right.

15. (a) 5; odd

(b) −2; negative

(c)

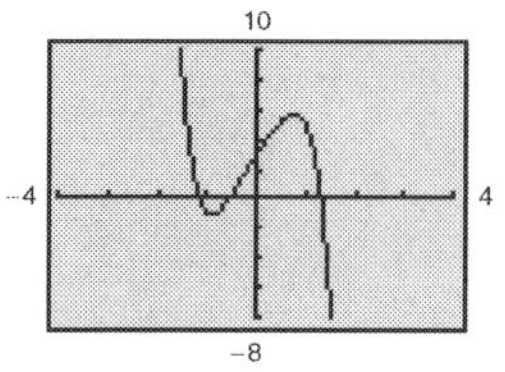

(d) The graph rises to the left and falls to the right.

17. (a) 2; even

(b) 2; positive

(c)

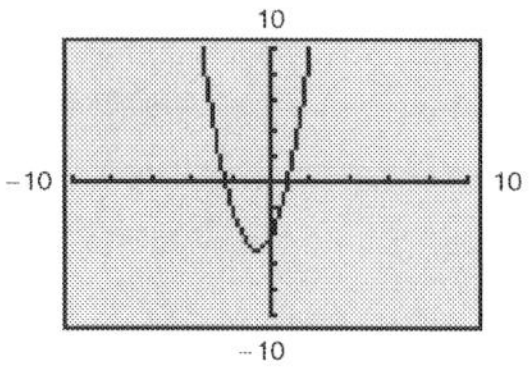

(d) The graph rises to the left and right.

19. (a) 6; even

(b) −1; negative

(c)

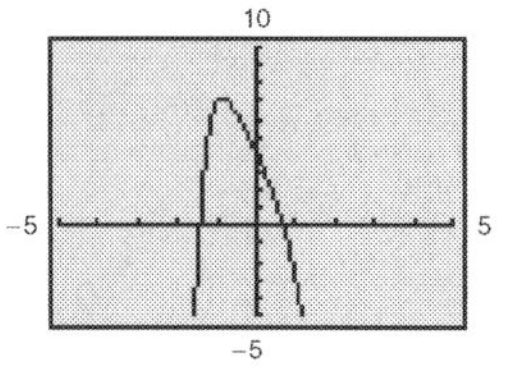

(d) The graph rises to the left and right.

21. When the leading coefficient is positive, the graph falls to the left and rises to the right. When the leading coefficient is negative, the graph rises to the left and falls to the right.

23. $f(x) = -x^3 + 1$

The degree is odd and the leading coefficient is negative. The graph rises to the left and falls to the right.

25. $g(x) = 6 - 4x^2 + x - 3x^5$

The degree is odd and the leading coefficient is negative. The graph rises to the left and falls to the right.

27. $f(x) = 4x^8 - 2$

The degree is even and the leading coefficient is positive. The graph rises to the left and right.

29. $f(x) = 2 + 5x - x^2 - x^3 + 2x^4$

The degree is even and the leading coefficient is positive. The graph rises to the left and right.

31. $h(t) = -\frac{2}{3}(t^2 - 5t + 3)$

The degree is even and the leading coefficient is negative. The graph falls to the left and right.

33. $f(x) = x^2 - 4x + 1$

Turning point: $n - 1 = 2 - 1 = 1$

Real zero: $n = 2$

35. $f(x) = -x^5 + x^4 - x$

Turning point: $n - 1 = 5 - 1 = 4$

Real zero: $n = 5$

37. $f(x) = 9 - x^2 = (x + 3)(x - 3)$

$x = \pm 3$

39. $h(t) = t^2 + 8t + 16 = (t + 4)^2$

$t = -4$

41. $f(x) = \frac{1}{3}x^2 + \frac{1}{3}x + \frac{2}{3}$

$= \frac{1}{3}(x^2 + x + 2)$

There are no real zeros.

43. $g(x) = -5x^2 - 10x + 20$

$x = -5(x^2 + 2x - 4)$

$= \dfrac{-2 \pm \sqrt{4 + 16}}{2}$

$= -1 \pm \sqrt{5}$

45. $f(t) = t^3 - 4t^2 + 4t = t(t - 2)^2$

$t = 0, 2$

47. $g(t) = \frac{1}{2}t^4 - \frac{1}{2} = \frac{1}{2}(t + 1)(t - 1)(t^2 + 1)$

$t = \pm 1$

49. $f(x) = 2x^4 - 2x^2 - 40$

$= 2(x^2 + 4)(x + \sqrt{5})(x - \sqrt{5})$

$x = \pm\sqrt{5}$

51. $f(x) = x^3 - 3x^2 + 2x - 6$

$= x^2(x - 3) + 2(x - 3)$

$= (x^2 + 2)(x - 3)$

$x = 3$

53. $f(x) = \frac{1}{2}(x^3 + 2x^2 - 3x)$ is a cubic with y-intercept (0, 0); graph (e).

55. $f(x) = \frac{1}{3}x^4 - x^2$ has intercepts $(0, 0)$ and $(\pm\sqrt{3}, 0)$; graph (g).

57. $f(x) = 3x^3 - 9x + 1$ is a cubic with y-intercept $(0, 1)$; graph (f).

59. $f(x) = -\frac{1}{3}x^3 + x - \frac{2}{3}$ is a cubic with y-intercept $(0, -\frac{2}{3})$; graph (a).

61. $f(x) = \frac{2}{3}x + 5$

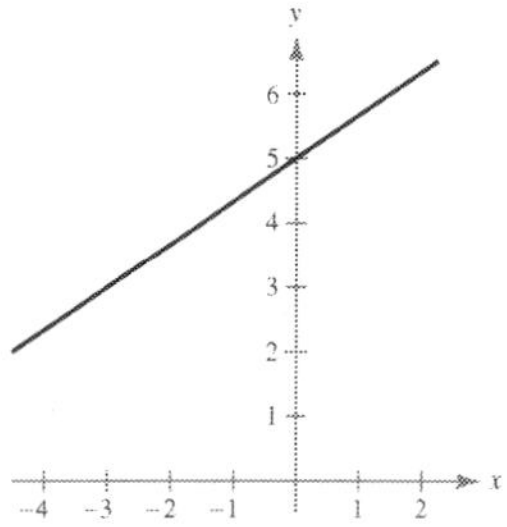

63. $f(t) = \frac{1}{2}(t^2 - 4t - 1) = \frac{1}{2}(t - 2)^2 + \frac{5}{2}$

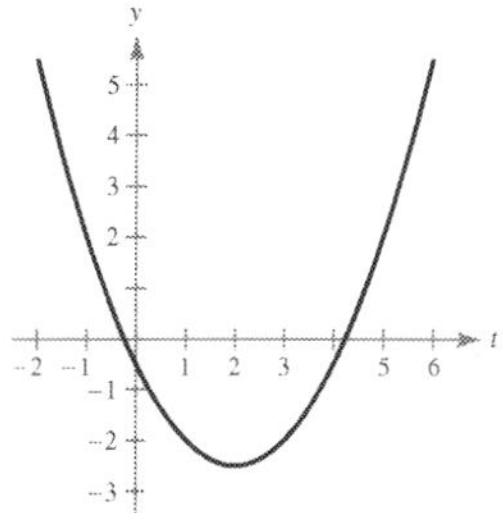

65. $f(x) = 4x^2 - x^3 = x^2(4 - x)$

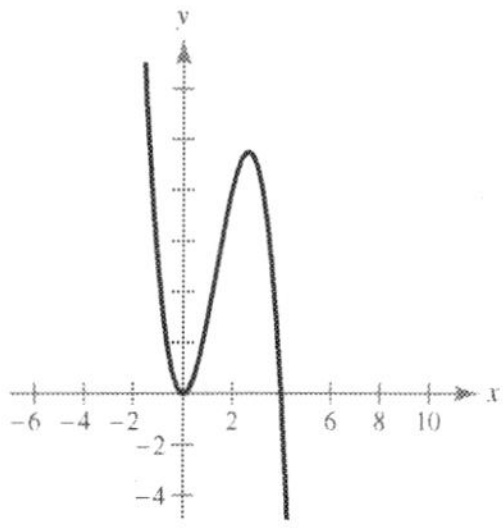

67. $f(x) = x^3 - 9x$

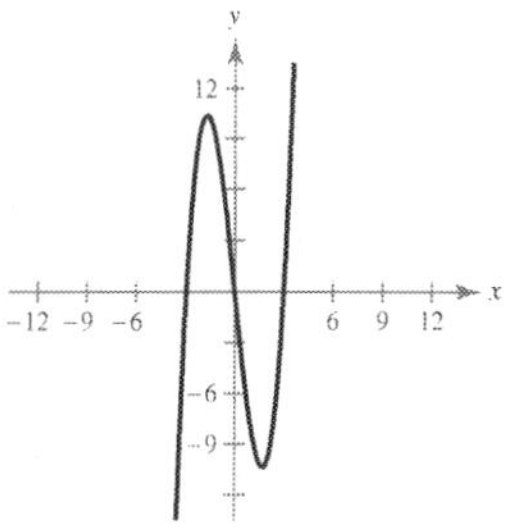

69. $g(t) = -\frac{1}{4}(t - 2)^2(t + 2)^2$

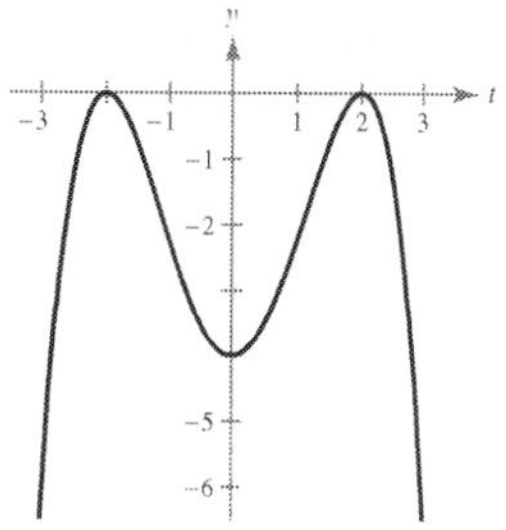

71. $f(x) = 1 - x^6$

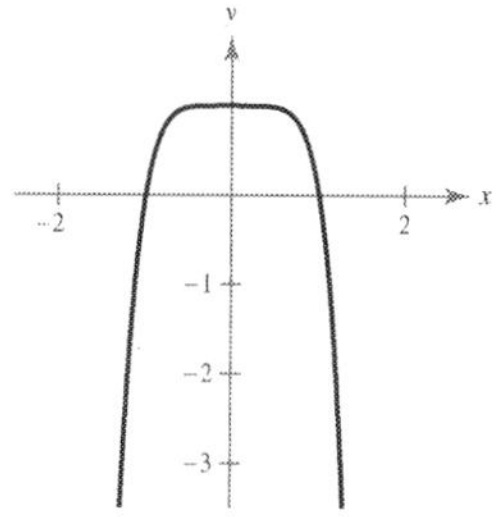

73. $a_n < 0$

Example: $f(x) = -(x - 1)^2(x + 1)(x - 3)$

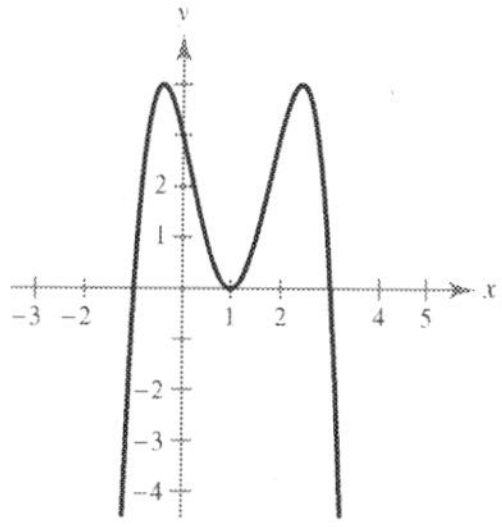

$a_n > 0$

Example: $f(x) = (x - 1)^2(x + 1)(x - 3)$

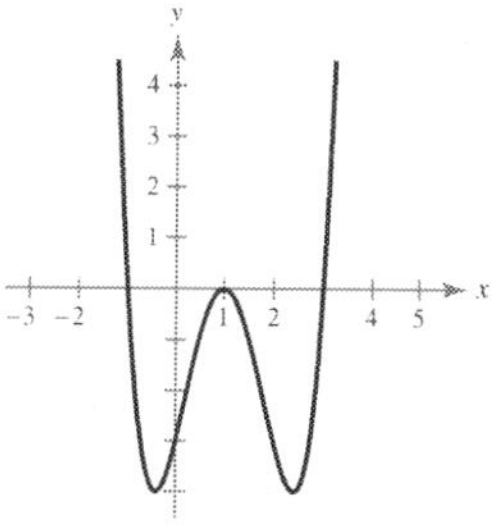

(Answers will vary.)

75. Because the graph intersects the x-axis at (4, 0) and (7, 0), $x - 4$ and $x - 7$ are factors of the fourth-degree polynomial. The graph just touches the x-axis at (0, 0), so x is a factor twice of the polynomial.

We know that

$$f(x) = a(x^2)(x - 4)(x - 7) = a(x^4 - 11x^3 + 28x^2).$$

Now the point (1, 9) is on the graph of $f(x)$, so

$$9 = a(1^2 - 11(1)^3 + 28(1)^2)$$
$$9 = 18a$$
$$a = \tfrac{1}{2}.$$

So, $f(x) = \tfrac{1}{2}(x^4 - 11x^3 + 28x^2)$.

77. (a)

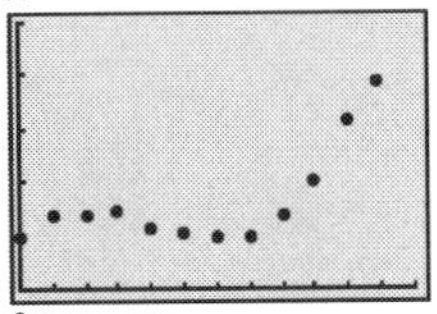

(b) Based on the scatter plot (data points fall to the left and rise to the right), the leading coefficient of the cubic model is positive.

(c) $P = 0.00011t^3 - 0.013t^2 + 0.37t + 10.8$

The model agrees with the prediction from part (b).

(d)

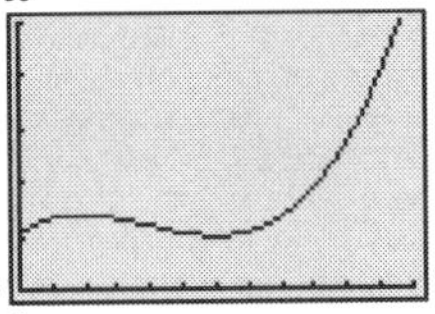

When t is about 112.8, or the year 2013, the immigrant population will be about 45 million. Based on the data, this prediction seems reasonable.

79.

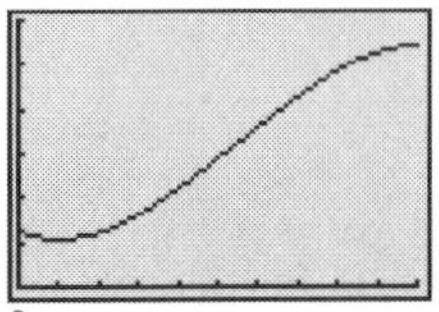

According to the graph, the point of diminishing returns is about (11, 324).

81. (a) Volume $= l \cdot w \cdot h = (24 - 2x)(24 - 4x)x$
$= 2(12 - x) \cdot 4(6 - x)x$
$= 8x(12 - x)(6 - x)$

(b) $x > 0$, $12 - x > 0$, $6 - x > 0$
$x < 12$ $x < 6$

Domain: $0 < x < 6$

(c)

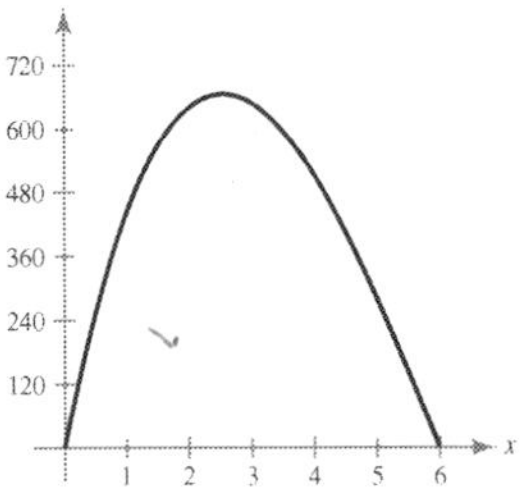

$x \approx 2.5$ corresponds to a maximum of about 665 cubic inches.

Section 3.3 Polynomial Division

Skills Warm Up

1. $(x - 1)(x^2 + 2) + 5 = x^3 + 2x - x^2 - 2 + 5$
$= x^3 - x^2 + 2x + 3$

2. $(x^2 - 3)(2x + 4) + 8 = 2x^3 + 4x^2 - 6x - 12 + 8$
$= 2x^3 + 4x^2 - 6x - 4$

3. $(x^2 + 1)(x^2 - 2x + 3) - 10 = x^4 + x^2 - 2x^3 - 2x + 3x^2 + 3 - 10$
$= x^4 - 2x^3 + 4x^2 - 2x - 7$

4. $(x + 6)(2x^3 - 3x) - 5 = 2x^4 - 3x^2 + 12x^3 - 18x - 5$
$= 2x^4 + 12x^3 - 3x^2 - 18x - 5$

Skills Warm Up —*continued*—

5. $x^2 - 4x + 3 = (x - 1)(x - 3)$

6. $8x^2 - 24x - 80 = 8(x^2 - 3x - 10)$
$= 8(x - 5)(x + 2)$

7. $3x^2 + 2x - 5 = (3x + 5)(x - 1)$

8. $9x^2 - 24x + 16 = (3x - 4)^2$

9. $4x^3 - 10x^2 + 6x = 2x(2x^2 - 5x + 3)$
$= 2x(x - 1)(2x - 3)$

10. $6x^3 + 7x^2 + 2x = x(6x^2 + 7x + 2)$
$= x(3x + 2)(2x + 1)$

1.
$$\begin{array}{r} x + 2 \\ x - 3 \overline{\big) \; x^2 - x - 6} \\ -(x^2 - 3x) \\ \hline 2x - 6 \\ -(2x - 6) \\ \hline 0 \end{array}$$

$$\frac{x^2 - x - 6}{x - 3} = x + 2$$

3.
$$\begin{array}{r} 3x - 4 \\ x - 1 \overline{\big) \; 3x^2 - 7x + 4} \\ -(3x^2 - 3x) \\ \hline -4x + 4 \\ -(-4x + 4) \\ \hline 0 \end{array}$$

$$\frac{3x^2 - 7x + 4}{x - 1} = 3x - 4$$

5.
$$\begin{array}{r} 2x + 4 \\ x + 3 \overline{\big) \; 2x^2 + 10x + 12} \\ -(2x^3 + 6x) \\ \hline 4x + 12 \\ -(4x + 12) \\ \hline 0 \end{array}$$

$$\frac{2x^2 + 10x + 12}{x + 3} = 2x + 4$$

7.
$$\begin{array}{r} x^2 + 2x + 3 \\ x + 1 \overline{\big) \; x^3 + 3x^2 + 5x + 3} \\ -(x^3 + x^2) \\ \hline 2x^2 + 5x \\ -(2x^2 + 2x) \\ \hline 3x + 3 \\ -(3x + 3) \\ \hline 0 \end{array}$$

$$\frac{x^3 + 3x^2 + 5x + 3}{x + 1} = x^2 + 2x + 3$$

9.
$$\begin{array}{r} x + 3 \\ 2x^2 - 1 \overline{\big) \; 2x^3 + 6x^2 - x - 3} \\ -(2x^3 \qquad - x) \\ \hline 6x^2 \qquad - 3 \\ -(6x^2 \qquad - 3) \\ \hline 0 \end{array}$$

$$\frac{2x^3 + 6x^2 - x - 3}{2x^2 - 1} = x + 3$$

11.
$$\begin{array}{r} x^3 + 3x^2 \qquad - 1 \\ x + 2 \overline{\big) \; x^4 + 5x^3 + 6x^2 - x - 2} \\ -(x^4 + 2x^3) \\ \hline 3x^3 + 6x^2 \\ -(3x^3 + 6x^2) \\ \hline -x - 2 \\ -(-x - 2) \\ \hline 0 \end{array}$$

$$\frac{x^4 + 5x^3 + 6x^2 - x - 2}{x + 2} = x^3 + 3x^2 - 1$$

13.
$$\begin{array}{r} 7 \\ x+4\overline{)\,7x+3} \\ -(7x+28) \\ \hline -25 \end{array}$$

$$\frac{7x+3}{x+4} = 7 - \frac{25}{x+4}$$

15.
$$\begin{array}{r} 3x+5 \\ 2x^2+0x+1\overline{)\,6x^3+10x^2+x+8} \\ -(6x^3+0x^2+3x) \\ \hline 10x^2-2x+8 \\ -(10x^2+0x+5) \\ \hline -2x+3 \end{array}$$

$$\frac{6x^3+10x^2+x+8}{2x^2+1} = 3x+5-\frac{2x-3}{2x^2+1}$$

17.
$$\begin{array}{r} x \\ x^2+0x-1\overline{)\,x^3+0x^2+0x-27} \\ -(x^3+0x^2-x) \\ \hline x-27 \end{array}$$

$$\frac{x^3-27}{x^2-1} = x + \frac{x-27}{x^2-1}$$

19.
$$\begin{array}{r} x^2-6x+17 \\ x+2\overline{)\,x^3-4x^2+5x-2} \\ -(x^3+2x^2) \\ \hline -6x^2+5x \\ -(-6x^2-12x) \\ \hline 17x-2 \\ -(17x+34) \\ \hline -36 \end{array}$$

$$\frac{x^3-4x^2+5x-2}{x+2} = x^2-6x+17-\frac{36}{x+2}$$

21.
$$\begin{array}{r} 2x^3+4x^2-2x-8 \\ x^2-2x+1\overline{)\,2x^5+0x^4-8x^3+0x^2+4x-1} \\ -(2x^5-4x^4+2x^3) \\ \hline 4x^4-10x^3+0x^2 \\ -(4x^4-8x^3+4x^2) \\ \hline -2x^3-4x^2+4x \\ -(-2x^3+4x^2-2x) \\ \hline -8x^2+6x-1 \\ -(-8x^2+16x-8) \\ \hline -10x+7 \end{array}$$

$$\frac{2x^5-8x^3+4x-1}{x^2-2x+1} = 2x^3+4x^2-2x-8-\frac{10x-7}{x^2-2x+1}$$

23.
$$\begin{array}{r|rrr} 1 & 1 & 6 & -7 \\ & & 1 & 7 \\ \hline & 1 & 7 & 0 \end{array}$$

Quotient: $x+7$

25.
$$\begin{array}{r|rrrr} -4 & 2 & 5 & -7 & 20 \\ & & -8 & 12 & -20 \\ \hline & 2 & -3 & 5 & 0 \end{array}$$

Quotient: $2x^2-3x+5$

27.
$$\begin{array}{r|rrrr} -2 & 4 & 8 & -9 & -18 \\ & & -8 & 0 & 18 \\ \hline & 4 & 0 & -9 & 0 \end{array}$$

Quotient: $4x^2-9$

29.
$$\begin{array}{r|rrrr} -10 & -1 & 0 & 75 & -250 \\ & & 10 & -100 & 250 \\ \hline & -1 & 10 & -25 & 0 \end{array}$$

Quotient: $-x^2+10x-25$

31.

−3	1	−4	−7	22	24
		−3	21	−42	60
	1	−7	14	−20	84

Quotient: $x^3 - 7x^2 + 14x - 20 + \dfrac{84}{x + 3}$

33.

6	10	−50	0	0	−800
		60	60	360	2160
	10	10	60	360	1360

Quotient: $10x^3 + 10x^2 + 60x + 360 + \dfrac{1360}{x - 6}$

35.

−1	1	0	0	0	0	0
		−1	1	−1	1	−1
	1	−1	1	−1	1	−1

Quotient: $x^4 - x^3 + x^2 - x + 1 - \dfrac{1}{x + 1}$

37.

2	−3	0	0	0	0
		−6	−12	−24	−48
	−3	−6	−12	−24	−48

Quotient: $-3x^3 - 6x^2 - 12x - 24 - \dfrac{48}{x - 2}$

39.

−1	−1	2	−3	5
		1	−3	6
	−1	3	−6	11

Quotient: $-x^2 + 3x - 6 + \dfrac{11}{x + 1}$

41.

$-\frac{1}{2}$	4	16	−23	−15
		−2	−7	15
	4	14	−30	0

Quotient: $4x^2 + 14x - 30$

43.

2	1	1	−12	20
		2	6	−12
	1	3	−6	8

$q(x) = x^2 + 3x - 6; r = 8$

$f(x) = (x - 2)(x^2 + 3x - 6) + 8$

$f(2) = (2 - 2)(2^2 + 3(2) - 6) + 8 = 0 + 8 = 8$

45.

$\frac{1}{3}$	3	2	5	−2
		1	1	2
	3	3	6	0

$q(x) = 3x^2 + 3x + 6; r = 0$

$f(x) = \left(x - \frac{1}{3}\right)(3x^2 + 3x + 6)$

$f\left(\frac{1}{3}\right) = \left(\frac{1}{3} - \frac{1}{3}\right)\left(3\left(\frac{1}{3}\right)^2 + 3\left(\frac{1}{3}\right) + 6\right) = 0$

47.

$\sqrt{3}$	1	2	−3	−12
		$\sqrt{3}$	$2\sqrt{3} + 3$	6
	1	$2 + \sqrt{3}$	$2\sqrt{3}$	−6

$q(x) = x^2 + (2 + \sqrt{3})x + 2\sqrt{3}; \quad r = -6$

$f(x) = (x - \sqrt{3})(x^2 + (2 + \sqrt{3})x + 2\sqrt{3}) - 6$

$f(\sqrt{3}) = (\sqrt{3} - \sqrt{3})\left((\sqrt{3})^2 + (2 + \sqrt{3})(\sqrt{3}) + 2\sqrt{3}\right) - 6 = -6$

49.

$1 + \sqrt{3}$	2	1	−14	−10
		$2 + 2\sqrt{3}$	$9 + 5\sqrt{3}$	10
	2	$3 + 2\sqrt{3}$	$-5 + 5\sqrt{3}$	0

$q(x) = 2x^2 + (3 + 2\sqrt{3})x + (-5 + 5\sqrt{3}); \quad r = 0$

$f(x) = \left[x - (1 + \sqrt{3})\right]\left[2x^2 + (3 + 2\sqrt{3})x + (-5 + 5\sqrt{3})\right]$

$f(1 + \sqrt{3}) = \left[1 + \sqrt{3} - 1 - \sqrt{3}\right]\left[2(1 + \sqrt{3})^2 + (3 + 2\sqrt{3})(1 + \sqrt{3}) + (-5 + 5\sqrt{3})\right] = 0$

51. $f(x) = 2x^5 - 3x^2 - 4x - 1$

(a) $\begin{array}{r|rrrrrr} -2 & 2 & 0 & 0 & -3 & -4 & -1 \\ & & -4 & 8 & -16 & 38 & -68 \\ \hline & 2 & -4 & 8 & -19 & 34 & \underline{-69} \end{array} = f(-2)$

(b) $\begin{array}{r|rrrrrr} -4 & 2 & 0 & 0 & -3 & -4 & -1 \\ & & -8 & 32 & -128 & 524 & -2080 \\ \hline & 2 & -8 & 32 & -131 & 520 & \underline{-2081} \end{array} = f(-4)$

(c) $\begin{array}{r|rrrrrr} 1 & 2 & 0 & 0 & -3 & -4 & -1 \\ & & 2 & 2 & 2 & -1 & -5 \\ \hline & 2 & 2 & 2 & -1 & -5 & \underline{-6} \end{array} = f(1)$

(d) $\begin{array}{r|rrrrrr} 3 & 2 & 0 & 0 & -3 & -4 & -1 \\ & & 6 & 18 & 54 & 153 & 447 \\ \hline & 2 & 6 & 18 & 51 & 149 & \underline{446} \end{array} = f(3)$

53. $f(x) = 2x^3 + 3x^2 + 8x - 14$

(a) $\begin{array}{r|rrrr} 2 & 2 & 3 & 8 & -14 \\ & & 4 & 14 & 44 \\ \hline & 2 & 7 & 22 & \underline{30} \end{array} = f(2)$

(b) $\begin{array}{r|rrrr} -1 & 2 & 3 & 8 & -14 \\ & & -2 & -1 & -7 \\ \hline & 2 & 1 & 7 & \underline{-21} \end{array} = f(-1)$

(c) $\begin{array}{r|rrrr} 1.1 & 2 & 3 & 8 & -14 \\ & & 2.2 & 5.72 & 15.092 \\ \hline & 2 & 5.2 & 13.72 & \underline{1.092} \end{array} = f(1.1)$

(d) $\begin{array}{r|rrrr} \frac{1}{2} & 2 & 3 & 8 & -14 \\ & & 1 & 2 & 5 \\ \hline & 2 & 4 & 10 & \underline{-9} \end{array} = f\left(\frac{1}{2}\right)$

55. (a) $\begin{array}{r|rrrr} -2 & 1 & 0 & -12 & -16 \\ & & -2 & 4 & 16 \\ \hline \end{array}$

$\begin{array}{r|rrrr} 4 & 1 & -2 & -8 & 0 \\ & & 4 & 8 & \\ \hline & 1 & 2 & 0 & \end{array}$

(b) The remaining factor is $x + 2$.

(c) $f(x) = x^3 - 12x - 16$
$= (x + 2)(x + 2)(x - 4)$
$= (x + 2)^2(x - 4)$

(d) The real zeros of $f(x)$ are -2 and 4.

(e)

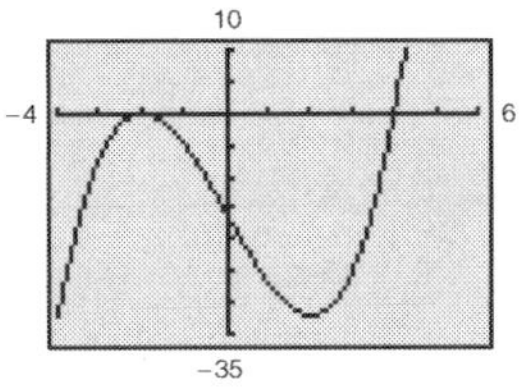

57. (a) $\begin{array}{r|rrrr} -\frac{1}{3} & 3 & 10 & -27 & -10 \\ & & -1 & -3 & 10 \\ \hline \end{array}$

$\begin{array}{r|rrrr} 2 & 3 & 9 & -30 & 0 \\ & & 6 & 30 & \\ \hline & 3 & 15 & 0 & \end{array}$

(b) The remaining factor is $x + 5$;

$$\frac{f(x)}{\left(x + \frac{1}{3}\right)(x - 2)} = 3(x + 5) \Rightarrow \frac{f(x)}{(3x + 1)(x - 2)} = x + 5.$$

(c) $f(x) = (3x + 1)(x - 2)(x + 5)$

(d) The real zeros of $f(x)$ are $-\frac{1}{3}$, 2, and -5.

(e)

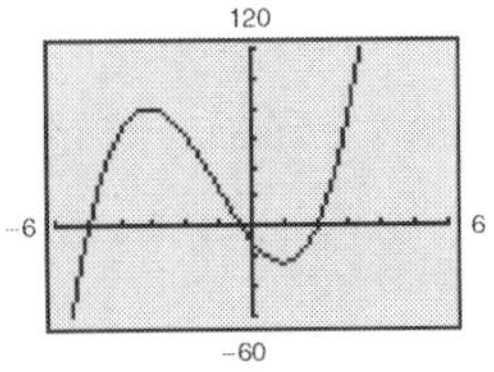

59. (a)
$$\begin{array}{r|rrrr} \sqrt{3} & 1 & 2 & -3 & -6 \\ & & \sqrt{3} & 3+2\sqrt{3} & 6 \\ \hline -2 & 1 & 2+\sqrt{3} & 2\sqrt{3} & 0 \\ & & -2 & -2\sqrt{3} & \\ \hline & 1 & \sqrt{3} & 0 & \end{array}$$

(b) The remaining factor of $f(x)$ is $\left(x + \sqrt{3}\right)$.

(c) $f(x) = x^3 + 2x^2 - 3x - 6$
$= \left(x - \sqrt{3}\right)\left(x + \sqrt{3}\right)(x + 2)$

(d) The real zeros of $f(x)$ are $-2, -\sqrt{3}$, and $\sqrt{3}$.

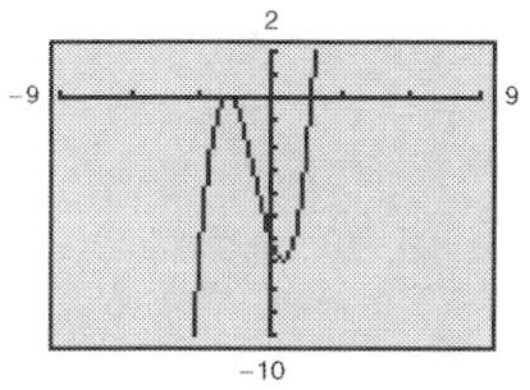

Note: To verify that -2 and $\sqrt{3}$ are real zeros of $f(x)$, you may want to enlarge that portion of the graph.

61. If the remainder is zero, then one of the polynomials is a factor or product of factors of the other polynomial.

For example, because $\dfrac{x^4 - 17x^2 + 16}{x^2 - 1} = x^2 - 16,$

then $x^4 - 17x^2 + 16$ can be written as $\left(x^2 - 16\right)\left(x^2 - 1\right)$ or $(x - 4)(x + 4)(x - 1)(x + 1)$.

63. $f(x) = x^3 - 2x^2 - 7x + 12$

Matches (e)

$x = 3$ is a real solution of $f(x) = 0$.

$$\begin{array}{r|rrrr} 3 & 1 & -2 & -7 & 12 \\ & & 3 & 3 & -12 \\ \hline & 1 & 1 & -4 & 0 \end{array}$$

$(x - 3)\left(x^2 + x - 4\right) = 0$

$x = 3$ or $x = \dfrac{-1 \pm \sqrt{17}}{2}$

(Use the Quadratic Formula.)

65. $f(x) = x^3 + 5x^2 + 6x + 2$

Matches (a)

$x = -1$ is a real solution of $f(x) = 0$.

$$\begin{array}{r|rrrr} -1 & 1 & 5 & 6 & 2 \\ & & -1 & -4 & -2 \\ \hline & 1 & 4 & 2 & 0 \end{array}$$

$(x + 1)\left(x^2 + 4x + 2\right) = 0$

$x = -1$ or $x = -2 \pm \sqrt{2}$

(Use the Quadratic Formula.)

67. $f(x) = x^3 + 3x^2 - 5x + 15$

Matches (b)

$x = -3$ is a real solution of $f(x) = 0$.

$$\begin{array}{r|rrrr} -3 & 1 & 3 & -5 & -15 \\ & & -3 & 0 & 15 \\ \hline & 1 & 0 & -5 & 0 \end{array}$$

$(x + 3)\left(x^2 - 5\right) = 0$

$x = -3$ or $x = \pm\sqrt{5}$

69. $a_n > 0$: $f(x) = 3\left(x - \frac{10}{3}\right)(x + 1)(x - 2)$
$= (3x - 10)(x + 1)(x - 2)$
$= 3x^3 - 13x^2 + 4x + 20$

$a_n < 0$: $f(x) = -3\left(x - \frac{10}{3}\right)(x + 1)(x - 2)$
$= (10 - 3x)(x + 1)(x - 2)$
$= -3x^2 + 13x^2 - 4x - 20$

There are infinitely many polynomials for $f(x)$. (Answers will vary.)

71.
$$\begin{array}{r} x^2 - 7x + 10 \\ x - 3 \overline{\big)\, x^3 - 10x^2 + 31x - 30} \\ \underline{-\left(x^3 - 3x^2\right)} \\ -7x^2 + 31x \\ \underline{-\left(-7x^2 + 21x\right)} \\ 10x - 30 \\ \underline{-(10x - 30)} \\ 0 \end{array}$$

So, $\dfrac{x^3 - 10x^2 + 31x - 30}{x - 3} = x^2 - 7x + 10.$

73.
$$\begin{array}{r} 3x^2 - x - 10 \\ 2x+1\overline{)\;6x^3 + x^2 - 21x - 10} \\ \underline{-(6x^3 + 3x^2)} \\ -2x^2 - 21x \\ \underline{-(-2x^2 - x)} \\ -20x - 10 \\ \underline{-(-20x - 10)} \\ 0 \end{array}$$

So, $\dfrac{6x^3 + x^2 - 21x - 10}{2x + 1} = 3x^2 - x - 10.$

75.
$$\begin{array}{r} x^2 - 6x \\ x^2 + x + 20\overline{)\;x^4 - 5x^3 + 14x^2 - 120x} \\ \underline{-(x^4 + x^3 + 20x^2)} \\ -6x^3 - 6x^2 - 120x \\ \underline{-(-6x^3 - 6x^2 - 120x)} \\ 0 \end{array}$$

So, $\dfrac{x^4 - 5x^3 + 14x^2 - 120x}{x^2 + x + 20} = x^2 - 6x.$

77.
$$\begin{array}{r} x^2 + 4x + 3 \\ x^2 - 9\overline{)\;x^4 + 4x^3 - 6x^2 - 36x - 27} \\ \underline{-(x^4 + 0x^3 - 9x^2)} \\ 4x^3 + 3x^2 - 36x \\ \underline{-(4x^3 + 0x^2 - 36x)} \\ 3x^2 - 27 \\ \underline{-(3x^2 + 0x - 27)} \\ 0 \end{array}$$

So, $\dfrac{x^4 + 4x^3 - 6x^2 - 36x - 27}{x^2 - 9} = x^2 + 4x + 3.$

79. $V = l \cdot w \cdot h$

$\dfrac{V}{h} = l \cdot w =$ square feet of floor space

−2	1	11	34	32
		−2	−18	−32
	1	9	16	0

$\dfrac{x^3 + 11x^2 + 34x + 32}{x + 2} = x^2 + 9x + 16$ square feet

81. Multiply the divisor and the quotient. Then add the remainder to obtain the dividend.

83. (a)

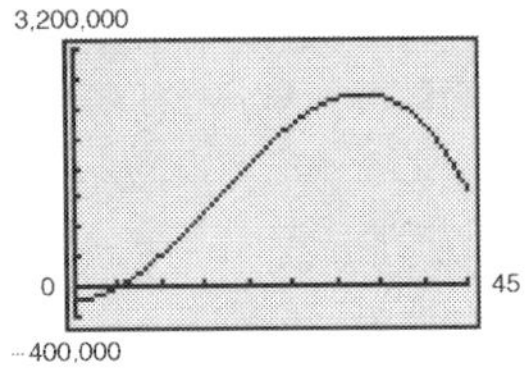

(b) Yes. If $x \approx 25.0366$, the advertising expense is \$250,366 and produces the same profit.

(c)

25.0366	−152	7545	0	−169,625
		−3805.5632	93,622.78339	2,343,996.179
	−152	3739.4368	93,622.78339	2,174,371.179

85.

−2	1	0	0	−2	1	c
		−2	4	−8	20	−42
	1	−2	4	−10	21	$c - 42$

To divide evenly, $c - 42$ must equal zero. Thus, c must equal 42.

Section 3.4 Real Zeros of Polynomial Functions

Skills Warm Up

1. $(x + 1)(x - \frac{2}{3})(x - 3) = 3x^3 - 8x^2 - 5x + 6$

2. $(x + 2)(x - \frac{3}{4})(x - 2)(x) = 4x^4 - 3x^3 - 16x^2 + 12x$

3. $$\begin{array}{r|rrrrrr} -3 & 1 & 0 & -9 & 0 & 5 & 18 \\ & & -3 & 9 & 0 & 0 & -15 \\ \hline & 1 & -3 & 0 & 0 & 5 & 3 \end{array}$$

$$x^4 - 3x^3 + 5 + \frac{3}{x + 3}$$

4. $$\begin{array}{r|rrrrr} -\frac{2}{3} & 3 & 17 & 10 & -9 & -8 \\ & & -2 & -10 & 0 & 6 \\ \hline & 3 & 15 & 0 & -9 & -2 \end{array}$$

$$3x^3 + 15x^2 - 9 - \frac{2}{x + \frac{2}{3}}$$

5. $f(x) = 2x^3 + 11x^2 + 2x - 4,\ x = \frac{1}{2}$

$$\begin{array}{r|rrrr} \frac{1}{2} & 2 & 11 & 2 & -4 \\ & & 1 & 6 & 4 \\ \hline & 2 & 12 & 8 & 0 \end{array}$$

$(x - \frac{1}{2})(2x^2 + 12x + 8) = 0$

$2(x - \frac{1}{2})(x^2 + 6x + 4) = 0$

$x = \frac{1}{2}$ or $x = -3 \pm \sqrt{5}$

(Use the Quadratic Formula.)

6. $f(x) = 6x^3 - 47x^2 - 124x - 60,\ x = 10$

$$\begin{array}{r|rrrr} 10 & 6 & -47 & -124 & -60 \\ & & 60 & 130 & 60 \\ \hline & 6 & 13 & 6 & 0 \end{array}$$

$(x - 10)(6x^2 + 13x + 6) = 0$

$(x - 10)(3x + 2)(2x + 3) = 0$

$x = 10$ or $x = -\frac{2}{3}$ or $x = -\frac{3}{2}$

7. $f(x) = 4x^3 - 13x^2 - 4x + 6,\ x = -\frac{3}{4}$

$$\begin{array}{r|rrrr} -\frac{3}{4} & 4 & -13 & -4 & 6 \\ & & -3 & 12 & -6 \\ \hline & 4 & -16 & 8 & 0 \end{array}$$

$(x - \frac{3}{4})(4x^2 - 16x + 8) = 0$

$4(x - \frac{3}{4})(x^2 - 4x + 2) = 0$

$x = \frac{3}{4}$ or $x = 2 \pm \sqrt{2}$

(Use the Quadratic Formula.)

8. $f(x) = 10x^3 + 51x^2 + 48x - 28,\ x = \frac{2}{5}$

$$\begin{array}{r|rrrr} \frac{2}{5} & 10 & 51 & 48 & -28 \\ & & 4 & 22 & 28 \\ \hline & 10 & 55 & 70 & 0 \end{array}$$

$(x - \frac{2}{5})(10x^2 + 55x + 70) = 0$

$5(x - \frac{2}{5})(2x^2 + 11x + 14) = 0$

$5(x - \frac{2}{5})(2x + 7)(x + 2) = 0$

$x = \frac{2}{5}$ or $x = -\frac{7}{2}$ or $x = -2$

9. $x^4 - 3x^2 + 2 = 0$

$(x^2 - 1)(x^2 - 2) = 0$

$x = \pm 1$ or $x = \pm\sqrt{2}$

10. $x^4 - 7x^2 + 12 = 0$

$(x^2 - 3)(x^2 - 4) = 0$

$x = \pm\sqrt{3}$ or $x = \pm 2$

1. $f(x) = x^3 + 2x^2 - 1$. Because the leading coefficient is 1, the possible rational zeros are ± 1.

   $f(1) = 2$ and $f(-1) = 0$. The actual rational zero is -1.

3. $f(x) = x^3 - 4x^2 + 1$. Because the leading coefficient is 1, the possible rational zeros are ± 1.

   $f(-1) = -4$, and $f(1) = -2$. There are no actual rational zeros.

5. $f(x) = x^3 - 7x - 6$. Because the leading coefficient is 1, the possible rational zeros are $\pm1, \pm2, \pm3, \pm6$.

$f(-1) = 0,\ f(1) = -12,\ f(-2) = 0,\ f(2) = -12,$

$f(-3) = -12,\ f(3) = 0,\ f(6) = 168,\ f(-6) = -180$

The actual rational zeros are $-2, -1$, and 3.

7. $f(x) = x^3 + x^2 - 4x - 4$

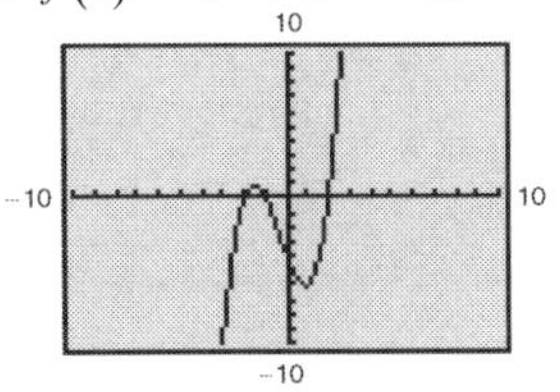

Possible rational zeros: $\pm 1, \pm 2, \pm 4$

From the graph, the rational zeros are ± 2 and -1.

9. $f(x) = x^3 - 9x^2 + 20x - 12$. Because the leading coefficient is 1, the possible rational zeros are $\pm1, \pm2, \pm3, \pm4, \pm6, \pm12$.

$$\begin{array}{r|rrrr} 1 & 1 & -9 & 20 & -12 \\ & & 1 & -8 & 12 \\ \hline & 1 & -8 & 12 & 0 \end{array}$$

$$\begin{aligned} f(x) &= (x - 1)(x^2 - 8x + 12) \\ &= (x - 1)(x - 2)(x - 6) \end{aligned}$$

The rational zeros are $x = 1, x = 2$, and $x = 6$.

11. $f(x) = 2x^3 - 3x^2 - 11x + 6$. Because the leading coefficient is 2, the possible rational zeros are:

$$\frac{\text{factors of } 6}{\text{factors of } 2} = \frac{\pm1, \pm2, \pm3, \pm6}{\pm1, \pm2}$$

$$= \pm1, \pm2, \pm3, \pm6, \pm\frac{1}{2}, \pm\frac{3}{2}$$

$$\begin{array}{r|rrrr} 3 & 2 & -3 & -11 & 6 \\ & & 6 & 9 & -6 \\ \hline & 2 & 3 & -2 & 0 \end{array}$$

$$\begin{aligned} f(x) &= (x - 3)(2x^2 + 3x - 2) \\ &= (x - 3)(2x - 1)(x + 2) \end{aligned}$$

The rational zeros are $x = 3, x = \frac{1}{2}$, and $x = -2$.

13. $f(x) = x^3 - \frac{3}{2}x^2 - \frac{23}{2}x + 6$

Consider $g(x) = 2x^3 - 3x^2 - 23x + 12$

(because all the coefficients, a_n, must be integers).

Possible rational zeros: $\pm1, \pm2, \pm3, \pm4, \pm6, \pm\frac{1}{2}, \pm\frac{3}{2}$

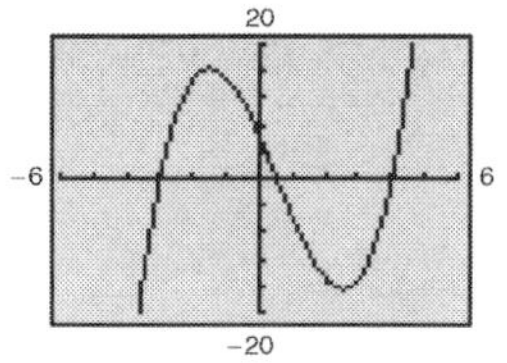

The actual rational zeros of $f(x)$ are $-3, \frac{1}{2}$, and 4.

15. $f(x) = 4x^4 - 17x^2 + 4$

Possible rational zeros: $\pm\frac{1}{4}, \pm\frac{1}{2}, \pm1, \pm2, \pm4$

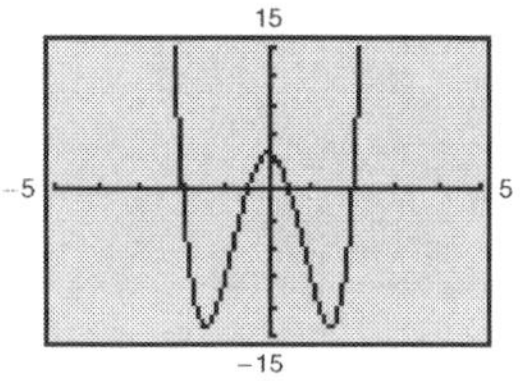

From the graph, the rational zeros are $\pm2, \pm\frac{1}{2}$.

17. $f(x) = x^3 - 6x^2 + 11x - 6$

Possible rational zeros: $\pm1, \pm2, \pm3, \pm6$

$$\begin{array}{r|rrrr} 1 & 1 & -6 & 11 & -6 \\ & & 1 & -5 & 6 \\ \hline & 1 & -5 & 6 & 0 \end{array}$$

$$\begin{aligned} x^2 - 5x + 6 &= 0 \\ (x - 3)(x - 2) &= 0 \end{aligned}$$

Real zeros: 1, 2, 3

19. $h(t) = t^3 + 12t^2 + 21t + 10$

Possible rational zeros: $\pm1, \pm2, \pm5, \pm10$

$$\begin{array}{r|rrrr} -1 & 1 & 12 & 21 & 10 \\ & & -1 & -11 & -10 \\ \hline & 1 & 11 & 10 & 0 \end{array}$$

$$\begin{aligned} t^2 + 11t + 10 &= 0 \\ (t + 1)(t + 10) &= 0 \end{aligned}$$

Real zeros: $-1, -10$

-1 is a repeated zero.

21. $C(x) = 2x^3 + 3x^2 - 1$

Possible rational zeros: $\pm\frac{1}{2}, \pm 1$

-1	2	3	0	-1
		-2	-1	1
	2	1	-1	0

$2x^2 + x - 1 = 0$

$(2x - 1)(x + 1) = 0$

Real zeros: $-1, \frac{1}{2}$

-1 is a repeated zero.

23. $f(x) = x^4 - 11x^2 + 18$

Possible rational zeros: $\pm 1, \pm 2, \pm 3, \pm 6, \pm 9, \pm 18$

3	1	0	-11	0	18
		3	9	-6	18
	1	3	-2	-6	0

-3	1	3	-2	-6
		-3	0	6
	1	0	-2	0

$x^2 - 2 = 0$

$x^2 = 2$

$x = \pm\sqrt{2}$

Real zeros: $\pm 3, \pm\sqrt{2}$

25. $z^4 - z^3 - 2z - 4 = 0$

Possible rational zeros: $\pm 1, \pm 2, \pm 4$

-1	1	-1	0	-2	-4
		-1	2	-2	4
	1	-2	2	-4	0

2	1	-2	2	-4
		2	0	4
	1	0	2	0

$z^2 + 2 = 0$

$b^2 - 4ac = 0^2 - 4(1)(2) = -8 < 0$

No other real zeros

Rational zeros: $-1, 2$

Note: The real numbers contain the rational numbers.

27. $2y^4 + 7y^3 - 26y^2 + 23y - 6 = 0$

Possible rational zeros: $\pm\frac{1}{2}, \pm\frac{3}{2}, \pm 1, \pm 2, \pm 3, \pm 6$

1	2	7	-26	23	-6
		2	9	-17	6
	2	9	-17	6	0

$\frac{1}{2}$	2	9	-17	6
		1	5	-6
	2	10	-12	0

$2y^2 + 10y - 12 = 0$

$2(y + 6)(y - 1) = 0$

Rational zeros: $-6, \frac{1}{2}, 1$

1 is a repeated zero.

29. $x^5 - x^4 - 3x^3 + 5x^2 - 2x = 0$

$x(x^4 - x^3 - 3x^2 + 5x - 2) = 0$

Possible rational zeros: $\pm 1, \pm 2$

1	1	-1	-3	5	-2
		1	0	-3	2
	1	0	-3	2	0

-2	1	0	-3	2
		-2	4	-2
	1	-2	1	0

$x^2 - 2x + 1 = 0$

$(x - 1)^2 = 0$

Rational zeros: $-2, 0, 1$

1 is a repeated zero.

31. $f(x) = 32x^3 - 52x^2 + 17x + 3$

(a) $\pm\frac{1}{32}, \pm\frac{1}{16}, \pm\frac{3}{32}, \pm\frac{1}{8}, \pm\frac{3}{16}, \pm\frac{1}{4}, \pm\frac{3}{8}, \pm\frac{1}{2}, \pm\frac{3}{4}, \pm 1, \pm\frac{3}{2}, \pm 3$

(b)

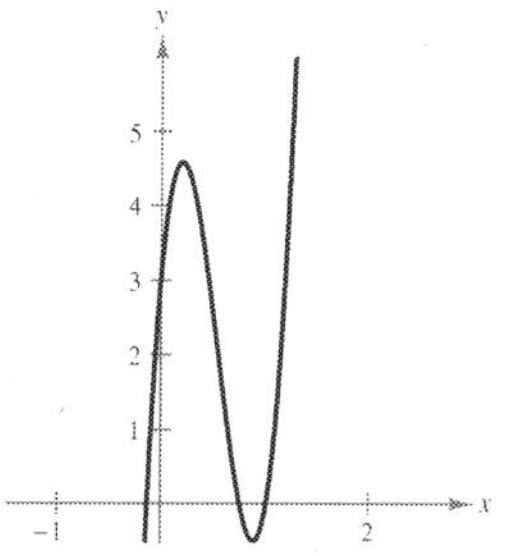

(c) Real zeros: $-\frac{1}{8}, \frac{3}{4}, 1$

33. $f(1) = 1 + 2 - 5 = -2$

$f(2) = 8 + 4 - 5 = 7$

Yes

35. $f(2) = 16 - 12 - 10 = -6$

$f(3) = 81 - 27 - 10 = 44$

Yes

37. $f(x) = x^3 + x - 1$, $[0, 1]$

x	0	0.1	0.2	0.3	0.4	0.5	0.6	0.7	0.8	0.9	1.0
$f(x)$	−1	−0.899	−0.792	−0.673	−0.536	−0.375	−0.184	0.043	0.312	0.629	1

Using a graphing utility, you can conclude that a real zero of $f(x)$ in the interval $[0, 1]$ is $x \approx 0.7$ (rounded to the nearest tenth).

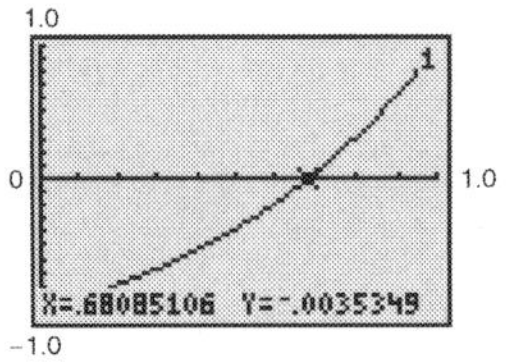

39. $f(x) = x^4 - 10x^2 - 11$, $[3, 4]$

x	3.0	3.1	3.2	3.3	3.4	3.5	3.6	3.7	3.8	3.9	4.0
$f(x)$	−20	−14.75	−8.542	−1.308	7.034	16.563	27.362	39.516	53.114	68.244	85

Using a graphing utility, you can conclude that a real zero of $f(x)$ in the interval $[3, 4]$ is $x \approx 3.3$ (rounded to the nearest tenth).

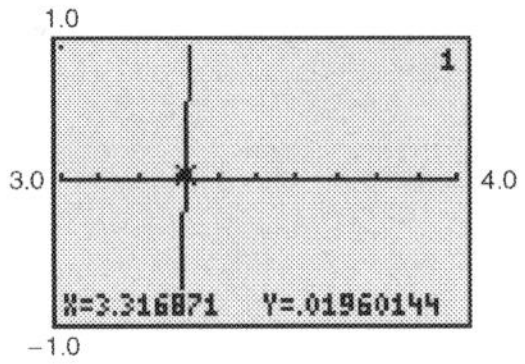

41. $f(x) = x^3 - 2x + 2$

Zero: $x \approx -1.769$

Matches (e)

43. $f(x) = 2x^3 - 6x^2 + 6x - 1$

Zero: $x \approx 0.206$

Matches (d)

45. $f(x) = -x^3 + 3x^2 - x + 1$

Zero: $x \approx 2.769$

Matches (f)

47. $f(x) = x^4 - x - 3$

Zeros: $x \approx -1.164, 1.453$

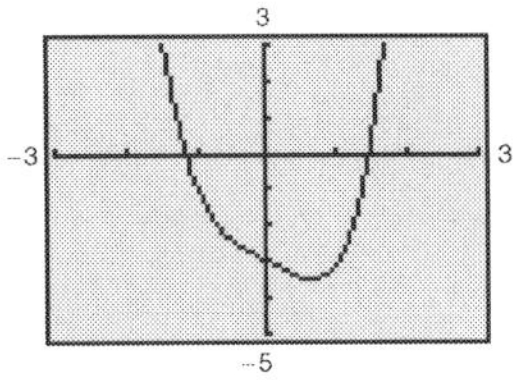

49. $f(x) = x^3 - 3.9x^2 + 4.79x - 1.881$

Zeros: $x \approx 0.900, 1.100, 1.900$

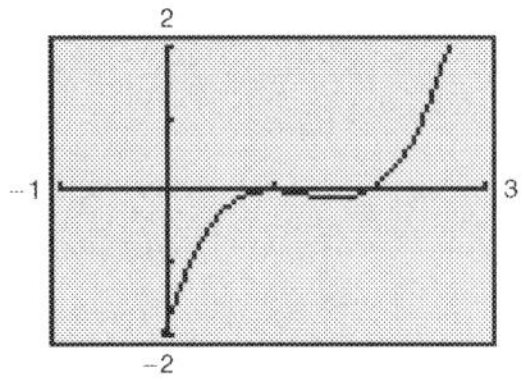

51. $f(x) = x^4 + x - 3$

Zeros: $x \approx -1.453, 1.164$

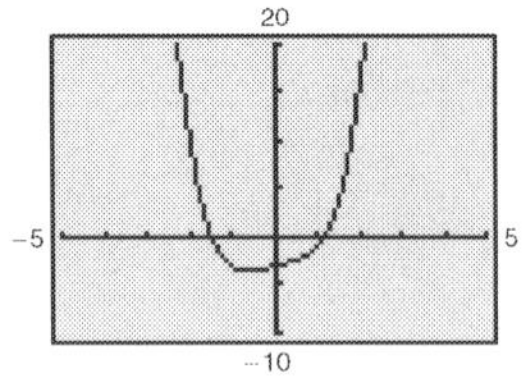

53. $f(x) = 7x^4 - 42x^3 + 43x^2 + 216x - 324$

Zeros: $x \approx -2.177, 1.563$

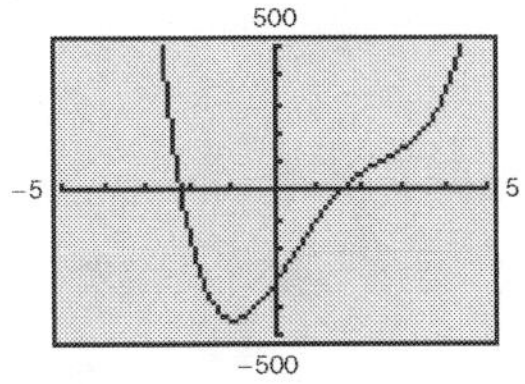

55. The function $f(x) = x^3 - 1$ has one rational zero and no irrational zeros so it matches (d).

57. The function $f(x) = x^3 - x$ has three rational zeros and no irrational zeros so it matches (b).

59. (a) $V = x(18 - 2x)(15 - 2x)$

Domain: $0 \le x \le 7.5$ or $[0, 7.5]$

(b)

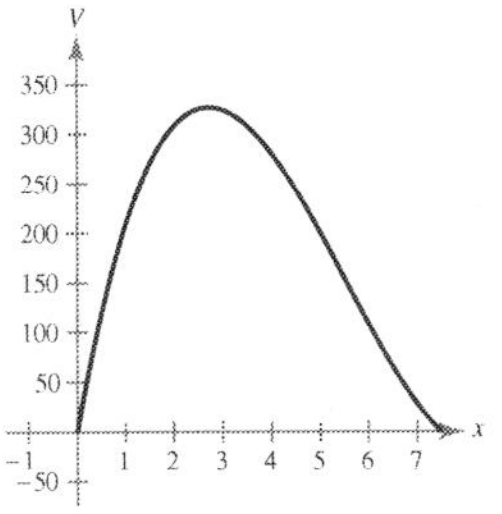

Approximate measurements: 2.72 inches × 12.56 inches × 9.56 inches

(c) $V = 108 = x(18 - 2x)(15 - 2x)$

$0 = 4x^3 - 66x^2 + 270x - 108$

$0 = (x - 6)(4x^2 - 42x + 18)$

$x = 6$ or $x = \dfrac{42\sqrt{1476}}{8} \approx 0.448$

A value of $x = \dfrac{42 + \sqrt{1476}}{8} \approx 10.052$ inches is impossible because it would yield a negative length and width.

(d) To make the tallest possible box with a volume of 108 cubic inches, let $x = 6$ inches.

61. $4x + y = 108 \Rightarrow y = 108 - 4x$

$V = x^2y = x^2(108 - 4x)$

$= -4x^3 + 108x^2 = 11{,}664$

$4x^3 - 108x^2 + 11{,}664 = 0$

$x^3 - 27x^2 + 2916 = 0$

$(x + 9)(x - 18)^2 = 0$

$x = 18$

$y = 36$

Box: 18 in. × 18 in. × 36 in.

63. (a) Current bin: $V = 2 \times 3 \times 4 = 24$ cubic feet

New bin: $V = 5(24) = 120$ cubic feet

$V = (2 + x)(3 + x)(4 + x) = 120$

(b) $x^3 + 9x^2 + 26x + 24 = 120$

$x^3 + 9x^2 + 26x - 96 = 0$

The only real zero of this polynomial is $x = 2$. All the dimensions should be increased by 2 feet, so the new bin will have dimensions of 4 feet by 5 feet by 6 feet.

65. $3t^4 + 2t^3 - 300t - 50 = 0$

Using the methods of this section, you have $t \approx 4.49$ hours.

67. (a)

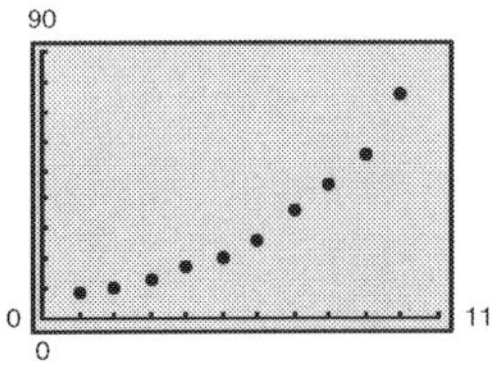

(b) Linear: $R = 6.923t - 7.43$

Quadratic: $R = 0.8884t^2 - 2.849t + 12.12$

Cubic: $R = 0.08806t^3 - 0.5646t^2 + 3.852t + 4.56$

Quartic: $R = 0.012630t^4 - 0.18979t^3 + 1.4688t^2 - 1.705t + 8.90$

(c) Linear

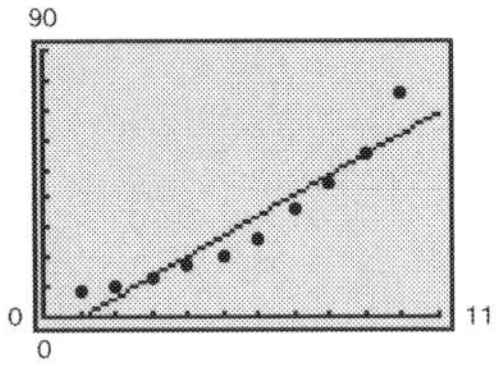

Quadratic

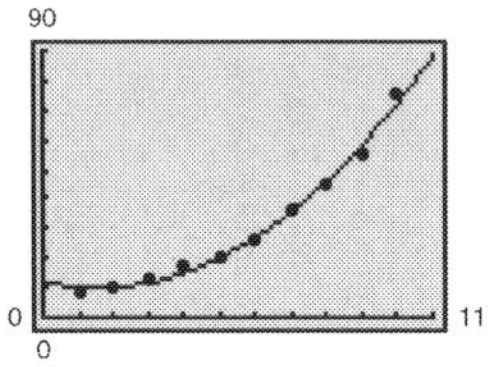

Cubic

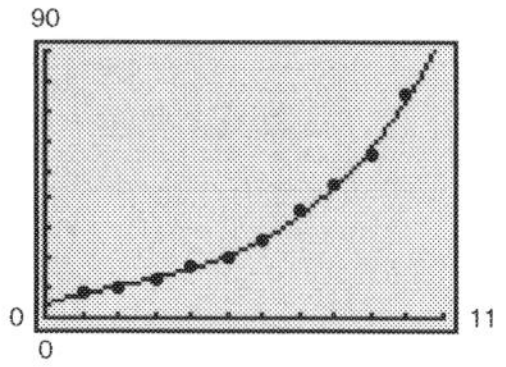

Quartic

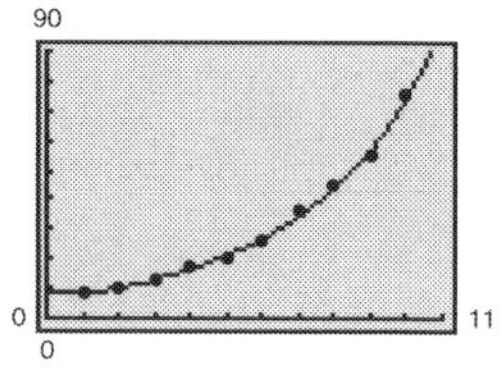

The nonlinear models fit the data well.

(d) Linear: $t \approx 15.52$ so revenue will be \$100 in 2015.

Quadratic: $t \approx 11.68$ so revenue will be \$100 in 2011.

Cubic: $t \approx 11.18$ so revenue will be \$100 in 2011.

Quartic: $t \approx 10.99$ so revenue will be \$100 in late 2010.

Answers will vary. Sample answer: The higher the degree of the model, the faster the graph climbs after 2010, so the quartic model makes the earliest prediction of when revenue per share will reach \$100. Based on the increasing trend of the last few data points, the year (2015) predicted by the linear model seems too late.

69. (a)

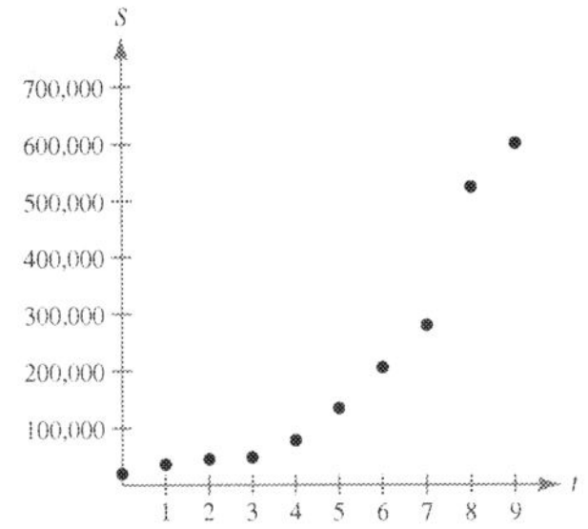

(b) Quadratic: $S = 11{,}001.49t^2 - 36{,}269.6t + 47{,}202$

Cubic: $S = 735.402t^3 + 1073.57t^2 - 2367.6t + 28{,}670$

(c) Quadratic: 2011 $(t \approx 11.1)$

Cubic: 2010 $(t \approx 10.6)$

Answers will vary. Sample answer: Based on the trend of the data, both models seem appropriate for predicting future values.

71. (a) $D = -x^3 + 54x^2 - 140x - 3000, \quad 10 \le x \le 50$

$D = 14{,}400$ when $x = 30$ and $x \approx 38.91$

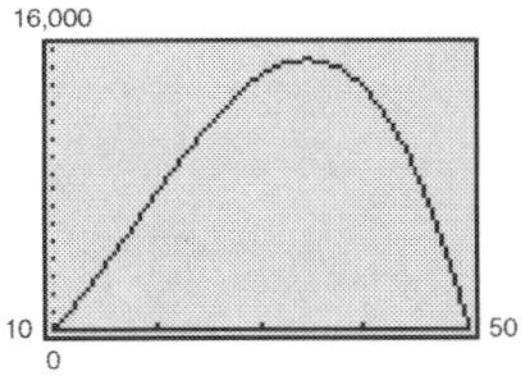

(b) $14{,}400 = -x^3 + 54x^2 - 140x - 3000, \; 10 \le x \le 50$

$x^3 - 54x^2 + 140x + 17{,}400 = 0$

Using a graphing utility, you can approximate the solutions.

Zeros: −14.91, 30, 38.91

(**Note:** −14.91 does not make sense in the context of the exercise.)

I would recommend that the company charge $38.91 for a cell phone because at that price, demand is 14,400 and the revenue is greater than it would be when $x = 30$.

73. $h(t) = -16t^2 + 48t$

Let $h = 64$ and solve for t.

$$64 = -16t^2 + 48t$$

$16t^2 - 48t + 64 = 0$

$16(t^2 - 3t + 4) = 0$

By the Quadratic Formula we have $t = \dfrac{3 \pm i\sqrt{7}}{2}$.

Because the equation yields only imaginary zeros, it is *not* possible for the ball to have reached a height of 64 feet.

75. No

77. Yes

Answers will vary.

Sample answer: $f(x) = (x + 1)(x^2 - 5)$

79. Answers will vary.

Chapter 3 Quiz Yourself

1. Vertex: $(-1, -2)$

Intercepts: $\left(-1 \pm \sqrt{2}, 0\right), (0, -1)$

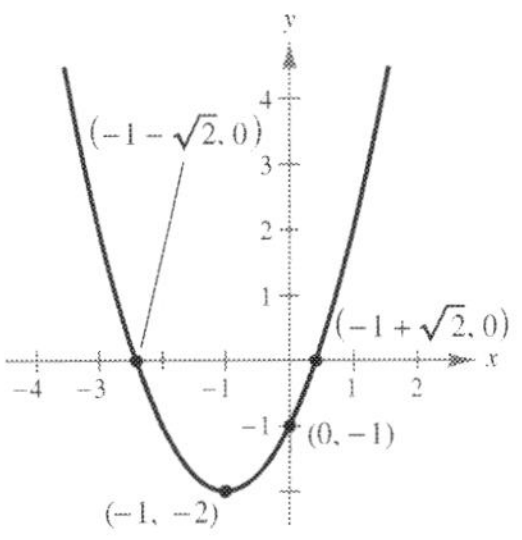

2. Vertex: $(0, 25)$

Intercepts: $(\pm 5, 0), (0, 25)$

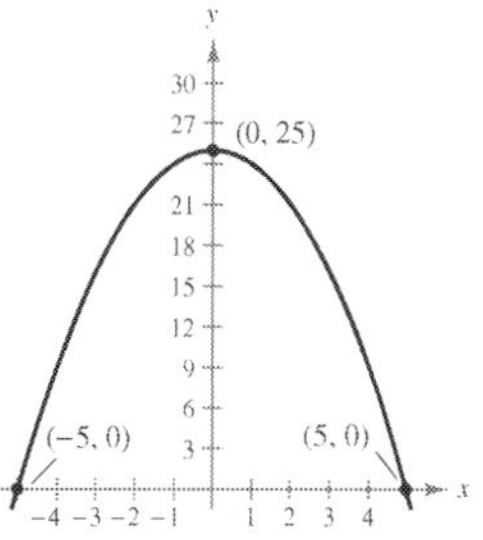

3. $f(x) = -2x^3 + 7x^2 - 9$

The degree is odd and the leading coefficient is negative. The graph rises to the left and falls to the right.

4. $f(x) = x^4 + 7x^2 - 8$

The degree is even and the leading coefficient is positive. The graph rises to the left and right.

5. $f(x) = 2x^4 + x^3 + 18x^2 - 4$ when $x = -3$

$$\begin{array}{r|rrrrr} -3 & 2 & 1 & 18 & 0 & -4 \\ & & -6 & 15 & -99 & 297 \\ \hline & 2 & -5 & 33 & -99 & 293 \end{array}$$

$f(-3) = 293$

6. $f(x) = x^4 - 5x^2 + 4,\ k = -1$

$$\begin{array}{r|rrrrr} -1 & 1 & 0 & -5 & 0 & 4 \\ & & -1 & 1 & 4 & -4 \\ \hline & 1 & -1 & -4 & 4 & 0 \end{array}$$

$$\begin{aligned} f(x) &= (x + 1)(x^3 - x^2 - 4x + 4) + 0 \\ &= x^4 - 5x^2 + 4 \end{aligned}$$

$$f(-1) = (-1)^4 - 5(-1)^2 + 4 = 0$$

7. $f(x) = x^3 + 5x^2 - 2x - 24,\ k = 2$

$$\begin{array}{r|rrrr} 2 & 1 & 5 & -2 & -24 \\ & & 2 & 14 & 24 \\ \hline & 1 & 7 & 12 & 0 \end{array}$$

$$\begin{aligned} f(x) &= (x - 2)(x^2 + 7x + 12) + 0 \\ &= x^3 + 5x^2 - 2x - 24 \end{aligned}$$

$$f(2) = (2)^3 + 5(2)^2 - 2(2) - 24 = 0$$

8.

$$\begin{array}{r} 2x^2 + 5x - 12 \\ x^2 + 2x - 15 \overline{\big)\, 2x^4 + 9x^3 - 32x^2 - 99x + 180} \\ \underline{-(2x^4 + 4x^3 - 30x^2)} \qquad\qquad \\ 5x^3 - 2x^2 - 99x \qquad \\ \underline{-(5x^3 + 10x^2 - 75x)} \qquad \\ -12x^2 - 24x + 180 \\ \underline{-12x^2 - 24x + 180} \\ 0 \end{array}$$

$$\frac{2x^4 + 9x^3 - 32x^2 - 99x + 180}{x^2 + 2x - 15} = 2x^2 + 5x - 12$$

9. Possible rational zeros: $\pm 1, \pm 5, \pm 7, \pm 35, \pm \frac{1}{2}, \pm \frac{5}{2}, \pm \frac{7}{2}, \pm \frac{35}{2}$

By testing we see that $x = -\frac{7}{2}$ is a zero.

$$\begin{array}{r|rrrr} -\frac{7}{2} & -2 & -7 & 10 & 35 \\ & & 7 & 0 & -35 \\ \hline & -2 & 0 & 10 & 0 \end{array}$$

$$\begin{aligned} \text{Thus, } f(x) &= \left(x + \tfrac{7}{2}\right)(-2x^2 + 10) \\ &= \left(x + \tfrac{7}{2}\right)(-2)(x^2 - 5) \\ &= -(2x + 7)(x^2 - 5). \end{aligned}$$

The real zeros of $f(x)$ are $x = -\frac{7}{2}$ and $x = \pm\sqrt{5}$.

10.

$$\begin{aligned} f(x) &= 4x^4 - 37x^2 + 9 \\ &= (4x^2 - 1)(x^2 - 9) \\ &= (2x + 1)(2x - 1)(x + 3)(x - 3) \end{aligned}$$

The real zeros of $f(x)$ are $x = \pm\frac{1}{2}$ and $x = \pm 3$.

11.

$$\begin{aligned} f(x) &= 3x^4 + 4x^3 - 3x - 4 \\ &= x^3(3x + 4) - (3x + 4) \\ &= (3x + 4)(x^3 - 1) \end{aligned}$$

The real zeros of $f(x)$ are $x = -\frac{4}{3}$ and $x = 1$.

12.

$$\begin{aligned} f(x) &= 2x^3 - 3x^2 + 2x - 3 \\ &= x^2(2x - 3) + (2x - 3) \\ &= (2x - 3)(x^2 + 1) \end{aligned}$$

The only real zero of $f(x)$ is $x = \frac{3}{2}$.

13. When the advertising expense is \$450,000,

$$P = -95(45)^3 + 5650(45)^2 - 250{,}000 = \$2{,}534{,}375$$

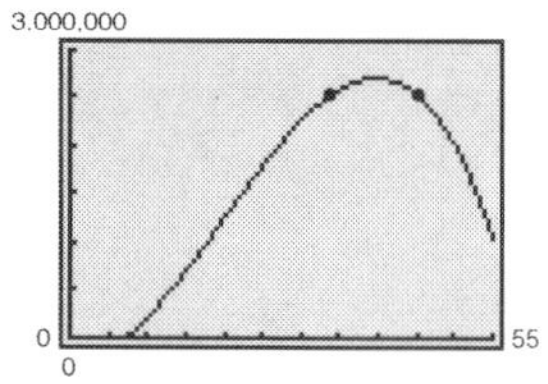

Using methods of this chapter, $x \approx 33.76$. The advertising expense of \$337,600 would yield the same profit.

14. (a)

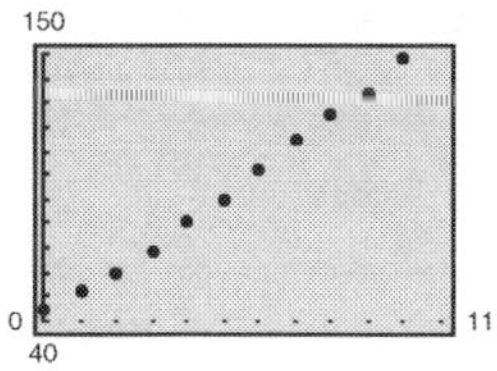

(b) Linear: $A = 10.53t + 39.9$

Quadratic: $A = 0.210t^2 + 8.43t + 43.0$

Cubic: $A = -0.0386t^3 + 0.789t^2 + 6.22t + 44.4$

Quartic: $A = 0.0062t^4 - 0.1633t^3 + 1.568t^2 + 4.66t + 44.9$

(c) Linear, Quadratic, Cubic, Quartic

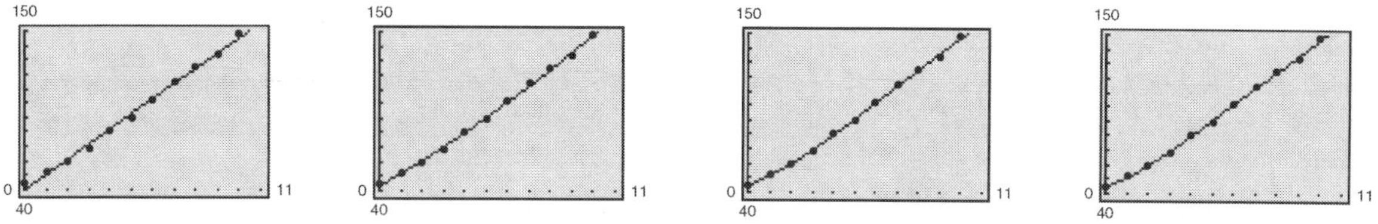

Each model could be considered a good fit, but the cubic and quartic models appear to fit the data a little better than the linear and quadratic models.

(d) Linear: 2012 $(t \approx 12.36)$

Quadratic: 2011 $(t \approx 11.67)$

Cubic: 2012 $(t \approx 12.49)$

Quartic: 2011 $(t \approx 11.79)$

Answers will vary. Sample answer: The linear and cubic predictions are close to each other and the quadratic and quartic predictions are close to each other. The quadratic and quartic models give earlier predictions because their graphs rise faster in the years after 2010 than the graphs of the linear and cubic models.

Section 3.5 Complex Numbers

Skills Warm Up

1. $\sqrt{12} = 2\sqrt{3}$

2. $\sqrt{500} = 10\sqrt{5}$

3. $\sqrt{20} - \sqrt{5} = 2\sqrt{5} - \sqrt{5} = \sqrt{5}$

4. $\sqrt{27} - \sqrt{243} = 3\sqrt{3} - 9\sqrt{3} = -6\sqrt{3}$

5. $\sqrt{24}\sqrt{6} = \sqrt{144} = 12$

6. $2\sqrt{18}\sqrt{32} = 2\sqrt{576} = 2(24) = 48$

7. $\dfrac{1}{\sqrt{7}} \cdot \dfrac{\sqrt{7}}{\sqrt{7}} = \dfrac{\sqrt{7}}{7}$

8. $\dfrac{5}{\sqrt{5}} \cdot \dfrac{\sqrt{5}}{\sqrt{5}} = \dfrac{5\sqrt{5}}{5} = \sqrt{5}$

9. $x = \dfrac{-1 \pm \sqrt{1 + 4}}{2} = \dfrac{-1 \pm \sqrt{5}}{2}$

10. $x = \dfrac{-2 \pm \sqrt{4 + 4}}{2} = \dfrac{-2 \pm 2\sqrt{2}}{2} = -1 \pm \sqrt{2}$

1. $i = i \quad i^5 = i \quad i^9 = i \quad i^{13} = i$

$i^2 = -1 \quad i^6 = -1 \quad i^{10} = -1 \quad i^{14} = -1$

$i^3 = -i \quad i^7 = -i \quad i^{11} = -i \quad i^{15} = -i$

$i^4 = 1 \quad i^8 = 1 \quad i^{12} = 1 \quad i^{16} = 1$

Let n be a positive integer. The patterns are:

$i^{4n+1} = i \qquad i^{4n+3} = -i$

$i^{4n+2} = -1 \qquad i^{4n} = 1$

(Answers will vary.)

3. $a + bi = 7 + 12i$ when $a = 7$ and $b = 12$.

5. $(a + 3) + 2bi = 7 - 4i$ when $a + 3 = 7 \Rightarrow a = 4$ and $2b = -4 \Rightarrow b = -2$.

7. $(-4 + 3i) + (6 - 2i) = (-4 + 6) + (3 - 2)i = 2 + i$

9. $(12 + 5i) - (7 - i) = (12 - 7) + (5 + 1)i$
$= 5 + 6i$

11. $-\frac{5}{2}i + \left(\frac{5}{3} + \frac{11}{3}i\right) = \frac{5}{3} + \left(-\frac{5}{2} + \frac{11}{3}\right)i$
$= \frac{5}{3} + \left(-\frac{15}{6} + \frac{22}{6}\right)i$
$= \frac{5}{3} + \frac{7}{6}i$

13. $(2.7 + 4.3i) + 6.0 - (7.4 - 4.7i)$
$= 2.7 + 4.3i + 6.0 - 7.4 - 4.7i$
$= (2.7 + 6.0 - 7.4) + (4.3 + 4.7)i$
$= 1.3 + 9i$

15. $(3 + 4i)(3 - 4i) = 9 - 12i + 12i - 16i^2$
$= 9 + 16$
$= 25$

17. $(2 + 3i)(1 - i) = 2 - 2i + 3i - 3i^2$
$= 2 + i + 3$
$= 5 + i$

19. $5i(4 - 6i) = 20i - 30i^2 = 30 + 20i$

21. $(5 + 6i)(5 + 6i) = 25 + 30i + 30i + 36i^2$
$= 25 + 60i - 36$
$= -11 + 60i$

23. $(3 + 4i)(3 + 4i) + (3 - 4i)(3 - 4i)$

$9 + 12i + 12i + 16i^2 + 9 - 12i - 12i + 16i^2 = 18 - 32 = -14$

25. $\left(\sqrt{5} - \sqrt{3}i\right)\left(\sqrt{5} + \sqrt{3}i\right) = \left(\sqrt{5}\right)^2 - \left(\sqrt{3}\right)^2 i^2 = 5 + 3 = 8$

27. $9 + \sqrt{-16} = 9 + \sqrt{16}i = 9 + 4i$

The complex conjugate is $9 - 4i$.

29. $-3 - \sqrt{-12} = -3 - \sqrt{12}i = -3 - 2\sqrt{3}i$

The complex conjugate is $-3 + 2\sqrt{3}i$.

31. -21 is in standard form. The imaginary part is 0. The complex conjugate of -21 is -21.

33. $-6i + i^2 = -6i - 1 = -1 - 6i$

The complex conjugate is $-1 + 6i$.

35. $-5i^5 = -5i$

The complex conjugate is $5i$.

37. $\left(\sqrt{-6}\right)^2 + 3 = \left(\sqrt{6}i\right)^2 + 3 = -6 + 3 = -3$

The imaginary part is 0. The complex conjugate is -3.

39. $\dfrac{3 - i}{3 + i} \cdot \dfrac{3 - i}{3 - i} = \dfrac{8 - 6i}{10} = \dfrac{4 - 3i}{5} = \dfrac{4}{5} - \dfrac{3}{5}i$

41. $\dfrac{5}{4 - 2i} = \dfrac{5}{4 - 2i} \cdot \dfrac{4 + 2i}{4 + 2i}$
$= \dfrac{20 + 10i}{16 + 4}$
$= \dfrac{20 + 10i}{20}$
$= \dfrac{2 + i}{2}$
$= 1 + \dfrac{1}{2}i$

43. $\dfrac{7 + 10i}{2i} \cdot \dfrac{2i}{2i} = \dfrac{-20 + 14i}{-4} = \dfrac{10 - 7i}{2} = 5 - \dfrac{7}{2}i$

45. $\dfrac{1}{(2i)^3} = \dfrac{1}{8i^3} = \dfrac{1}{8i^3} \cdot \dfrac{i}{i} = \dfrac{i}{8} = \dfrac{1}{8}i$

47. $\dfrac{4}{(1-2i)^3} = \dfrac{4}{-11+2i}\cdot\dfrac{-11-2i}{-11-2i}$

$= \dfrac{-44-8i}{125} = -\dfrac{44}{125} - \dfrac{8}{125}i$

49. $\dfrac{(21-7i)(4+3i)}{2-5i} = \dfrac{(21-7i)(4+3i)}{2-5i}\cdot\dfrac{2+5i}{2+5i}$

$= \dfrac{(105+35i)(2+5i)}{4+25}$

$= \dfrac{35+595i}{29} = \dfrac{35}{29} + \dfrac{595}{29}i$

51. $\left(-2+\sqrt{-8}\right) - \left(5-\sqrt{-50}\right) = -2 + 2\sqrt{2}i - 5 + 5\sqrt{2}i = -7 + 7\sqrt{2}i$

53. $\sqrt{-3}\cdot\sqrt{-8} = \left(\sqrt{3}i\right)\left(2\sqrt{2}i\right) = 2\sqrt{6}i^2 = -2\sqrt{6}$

55. $\left(\sqrt{-10}\right)^2 = \left(\sqrt{10}i\right)^2 = \left(\sqrt{10}\right)^2\left(i^2\right) = -10$

57. $\left(2-\sqrt{-8}\right)\left(8+\sqrt{-6}\right) = \left(2-2\sqrt{2}i\right)\left(8+\sqrt{6}i\right)$

$= 16 - 16\sqrt{2}i + 2\sqrt{6}i - 4\sqrt{3}i^2$

$= 16 - 16\sqrt{2}i + 2\sqrt{6}i + 4\sqrt{3}$

$= \left(16+4\sqrt{3}\right) + \left(2\sqrt{6} - 16\sqrt{2}\right)i$

59. Error: $(3-2i)(3+2i) = 9 - 4i^2$

$= 9 + 4 \text{ (not } 9-4)$

$= 13$

The mistake came from not multiplying $(3-2i)(3+2i)$ correctly. Remind the student that

$(3-2i)(3+2i) = 9 + 6i - 6i - 4i^2$

$= 9 - 4i^2$

$= 9 - 4(-1)$

$= 9 + 4 = 13.$

61. $x^2 + 9 = 0$

$x^2 = -9$

$x = \pm\sqrt{-9}$

$x = \pm 3i$

63. $x^2 - 2x + 2 = 0$

$x = \dfrac{-(-2) \pm \sqrt{(-2)^2 - 4(1)(2)}}{2(1)}$

$= \dfrac{2 \pm \sqrt{-4}}{2}$

$= \dfrac{2 \pm 2i}{2}$

$= 1 \pm i$

65. $4x^2 + 16x + 17 = 0$

$x = \dfrac{-16 \pm \sqrt{16^2 - 4(4)(17)}}{2(4)}$

$= \dfrac{-16 \pm \sqrt{-16}}{8}$

$= \dfrac{-16 \pm 4i}{8}$

$= -2 \pm \dfrac{1}{2}i$

67. $4x^2 + 16x + 20 = 0$

$4\left(x^2 + 4x + 5\right) = 0$

$x^2 + 4x + 5 = 0$

$x = \dfrac{-4 \pm \sqrt{4^2 - 4(1)(5)}}{2(1)}$

$= \dfrac{-4 \pm \sqrt{-4}}{2}$

$= \dfrac{-4 \pm 2i}{2}$

$= -2 \pm i$

69. $16t^2 - 4t + 3 = 0$

$$t = \frac{-(-4) \pm \sqrt{(-4)^2 - 4(16)(3)}}{2(16)}$$
$$= \frac{4 \pm \sqrt{-176}}{32}$$
$$= \frac{4 \pm 4\sqrt{11}i}{32}$$
$$= \frac{1}{8} \pm \frac{\sqrt{11}}{8}i$$

71. No. Sample answer: A quadratic equation will have either two real solutions or two imaginary solutions. In the Quadratic Formula, the discriminant is either positive (yielding two real solutions) or negative (yielding two imaginary solutions).

73.

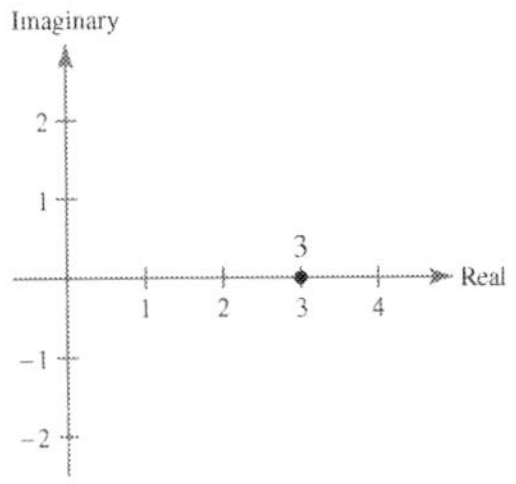

75.

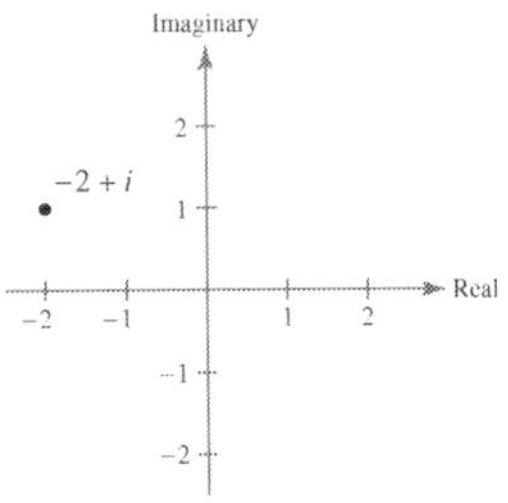

77.

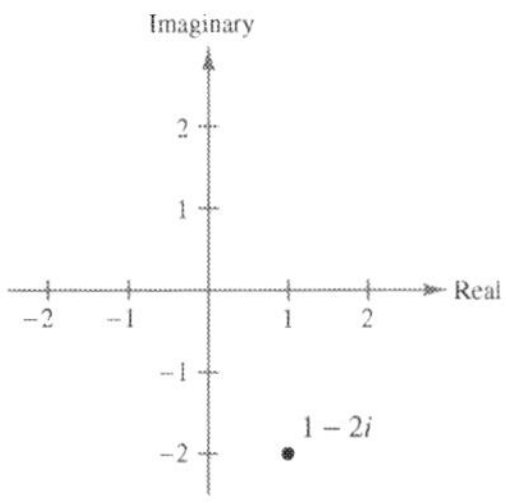

79. The complex number 0 is in the Mandelbrot Set because for $c = 0$, the corresponding Mandelbrot sequence is 0, 0, 0, 0, 0, 0, ... , which is bounded.

81. The complex number 1 is not in the Mandelbrot Set because for $c = 1$, the corresponding Mandelbrot sequence is 1, 2, 5, 26, 677, 458,330, ... , which is unbounded.

83. The complex number $\frac{1}{2}i$ is in the Mandelbrot Set because for $c = \frac{1}{2}i$, the corresponding Mandelbrot sequence is

$$\frac{1}{2}i,\ -\frac{1}{4} + \frac{1}{2}i,\ -\frac{3}{16} + \frac{1}{4}i,\ -\frac{7}{256} + \frac{13}{32}i,$$
$$-\frac{10{,}767}{65{,}636} + \frac{1957}{4096}i,\ -\frac{864{,}513{,}055}{4{,}294{,}967{,}296} + \frac{46{,}037{,}845}{134{,}217{,}728}i,\ \ldots,$$

which is bounded.

85. False, if $b = 0$ then $a + bi = a - bi = a$. That is, if the complex number is real, the number equals its conjugate.

87. Answers will vary.

Section 3.6 The Fundamental Theorem of Algebra

Skills Warm Up

1. $4 - \sqrt{29}i;\ 4 + \sqrt{29}i$
2. $-5 - 12i;\ -5 + 12i$
3. $-1 + 4\sqrt{2}i;\ -1 - 4\sqrt{2}i$
4. $6 + \frac{1}{2}i;\ 6 - \frac{1}{2}i$
5. $(-3 + 6i) - (10 - 3i) = -13 + 9i$
6. $(12 - 4i) + 20i = 12 + 16i$
7. $(4 - 2i)(3 + 7i) = 12 + 28i - 6i - 14i^2 = 26 + 22i$
8. $(2 - 5i)(2 + 5i) = 4 + 25 = 29$
9. $\frac{1 + i}{1 - i} \cdot \frac{1 + i}{1 + i} = \frac{2i}{2} = i$
10. $(3 + 2i)^3 = (5 + 12i)(3 + 2i)$
$$= 15 + 46i + 24i^2$$
$$= -9 + 46i$$

1. 1

3. 3

5. 4

7. $f(x) = x^2 + 25$

Zeros: $x = \pm 5i$

$f(x) = (x + 5i)(x - 5i)$

9. $f(x) = x^2 - 12x + 26$

Zeros: $x = \dfrac{12 \pm 2\sqrt{10}}{2} = 6 \pm \sqrt{10}$

$f(x) = \left(x - 6 - \sqrt{10}\right)\left(x - 6 + \sqrt{10}\right)$

11. $h(x) = x^2 - 4x + 1$

Zeros: $x = \dfrac{4 \pm 2\sqrt{3}}{2} = 2 \pm \sqrt{3}$

$h(x) = \left(x - 2 - \sqrt{3}\right)\left(x - 2 + \sqrt{3}\right)$

13. $f(x) = x^4 - 81 = \left(x^2 - 9\right)\left(x^2 + 9\right)$

Zeros: $x = \pm 3, \pm 3i$

$f(x) = (x + 3)(x - 3)(x + 3i)(x - 3i)$

15. $g(x) = x^3 + 5x = x\left(x^2 + 5\right)$

Zeros: $x = 0, \pm\sqrt{5}i$

$g(x) = x\left(x - \sqrt{5}i\right)\left(x + \sqrt{5}i\right)$

17. $h(x) = x^3 - 11x^2 - 15x + 325$

$= (x + 5)\left(x^2 - 16x + 65\right)$

$x = \dfrac{16 \pm \sqrt{4}i}{2} = 8 \pm i$

Zeros: $x = -5, x = 8 \pm i$

$h(x) = (x + 5)(x - 8 - i)(x - 8 + i)$

19. $g(x) = x^3 - 6x^2 + 13x - 10$

Zeros: $2 \begin{array}{|rrrr} 1 & -6 & 13 & -10 \\ & 2 & -8 & 10 \\ \hline 1 & -4 & 5 & 0 \end{array}$

$x = \dfrac{4 \pm \sqrt{4}i}{2} = 2 \pm i$

$g(x) = (x - 2)(x - 2 + i)(x - 2 - i)$

21. $f(t) = t^3 - 3t^2 - 15t + 125$

Zeros: $-5 \begin{array}{|rrrr} 1 & -3 & -15 & 125 \\ & -5 & 40 & -125 \\ \hline 1 & -8 & 25 & 0 \end{array}$

$t = \dfrac{8 \pm \sqrt{36}i}{2} = 4 \pm 3i$

$f(t) = (t + 5)(t - 4 + 3i)(t - 4 - 3i)$

23. $f(x) = x^3 + 24x^2 + 214x + 740$

Zeros: $-10 \begin{array}{|rrrr} 1 & 24 & 214 & 740 \\ & -10 & -140 & -740 \\ \hline 1 & 14 & 74 & 0 \end{array}$

$x = \dfrac{-14 \pm \sqrt{100}i}{2} = -7 \pm 5i$

$f(x) = (x + 10)(x + 7 + 5i)(x + 7 - 5i)$

25. $h(x) = x^3 + 9x^2 + 27x + 35$

Zeros: $-5 \begin{array}{|rrrr} 1 & 9 & 27 & 35 \\ & -5 & -20 & -35 \\ \hline 1 & 4 & 7 & 0 \end{array}$

$x = \dfrac{-4 \pm \sqrt{12}i}{2} = -2 \pm \sqrt{3}i$

$h(x) = (x + 5)\left(x + 2 + \sqrt{3}i\right)\left(x + 2 - \sqrt{3}i\right)$

27. $f(x) = 16x^3 - 20x^2 - 4x + 15$

Zeros: $-\frac{3}{4} \begin{array}{|rrrr} 16 & -20 & -4 & 15 \\ & -12 & 24 & -15 \\ \hline 16 & -32 & 20 & 0 \end{array}$

$x = \dfrac{32 \pm \sqrt{256}i}{32} = 1 \pm \dfrac{1}{2}i$

$f(x) = (4x + 3)(2x - 2 + i)(2x - 2 - i)$

29. $f(x) = 5x^3 - 9x^2 + 28x + 6$

Zeros: $-\frac{1}{5} \begin{array}{|rrrr} 5 & -9 & 28 & 6 \\ & -1 & 2 & -6 \\ \hline 5 & -10 & 30 & 0 \end{array}$

$x = \dfrac{10 \pm \sqrt{500}i}{10} = 1 \pm \sqrt{5}i$

$f(x) = (5x + 1)\left(x - 1 + \sqrt{5}i\right)\left(x - 1 - \sqrt{5}i\right)$

31. $g(x) = x^4 - 4x^3 + 8x^2 - 16x + 16$

Zeros:

$$\begin{array}{r|rrrrr} 2 & 1 & -4 & 8 & -16 & 16 \\ & & 2 & -4 & 8 & -16 \\ \hline 2 & 1 & -2 & 4 & -8 & 0 \\ & & 2 & 0 & 8 & \\ \hline & 1 & 0 & 4 & 0 & \end{array}$$

$x^2 + 4 = 0 \Rightarrow x^2 = -4 \Rightarrow x = \pm 2i$

$g(x) = (x - 2)^2(x + 2i)(x - 2i)$

33. $f(x) = x^4 + 10x^2 + 9 = (x^2 + 9)(x^2 + 1)$

Zeros: $x = \pm i, \pm 3i$

$f(x) = (x + i)(x - i)(x + 3i)(x - 3i)$

35. $f(t) = t^5 + 5t^4 - 7t^3 - 43t^2 - 8t - 48$

Zeros:

$$\begin{array}{r|rrrrrr} -4 & 1 & 5 & -7 & -43 & -8 & -48 \\ & & -4 & -4 & 44 & -4 & 48 \\ \hline -4 & 1 & 1 & -11 & 1 & -12 & 0 \\ & & -4 & 12 & -4 & 12 & \\ \hline 3 & 1 & -3 & 1 & -3 & 0 & \\ & & 3 & 0 & 3 & & \\ \hline & 1 & 0 & 1 & 0 & & \end{array}$$

$t^2 + 1 = 0 \Rightarrow t = \pm i$

$f(t) = (t + 4)^2(t - 3)(t + i)(t - i)$

37. Answers will vary. Sample answer:

$f(x) = (x - 2)(x + i)(x - i)$
$= (x - 2)(x^2 + 1)$
$= x^3 - 2x^2 + x - 2$

39. Answers will vary. Sample answer:

$f(x) = (x - 1)(x - 2 - i)(x - 2 + i)$
$= (x - 1)(x^2 - 4x + 5)$
$= x^3 - 5x^2 + 9x - 5$

41. Answers will vary. Sample answer:

$f(x) = (x + 4)(x - 3i)(x + 3i)(x - 2i)(x + 2i)$
$= (x + 4)(x^2 + 9)(x^2 + 4)$
$= x^5 + 4x^4 + 13x^3 + 52x^2 + 36x + 144$

43. Answers will vary. Sample answer:

$f(x) = (x + 5)^2(x - 1 - \sqrt{3}i)(x - 1 + \sqrt{3}i)$
$= (x^2 + 10x + 25)(x^2 - 2x + 4)$
$= x^4 + 8x^3 + 9x^2 - 10x + 100$

45. Answers will vary. Sample answer:

$f(x) = (3x - 2)(x + 1)(x - 3 - \sqrt{2}i)(x - 3 + \sqrt{2}i)$
$= (3x^2 + x - 2)(x^2 - 6x + 11)$
$= 3x^4 - 17x^3 + 25x^2 + 23x - 22$

47. $x^4 - 7x^2 - 8$

(a) $(x^2 - 8)(x^2 + 1)$

(b) $(x - 2\sqrt{2})(x + 2\sqrt{2})(x^2 + 1)$

(c) $(x - 2\sqrt{2})(x + 2\sqrt{2})(x - i)(x + i)$

49. $x^4 - 5x^3 + 4x^2 + x - 15$

(a) $(x^2 - 2x + 3)(x^2 - 3x - 5)$

(b) $x = \dfrac{3 \pm \sqrt{9 + 20}}{2} = \dfrac{3 \pm \sqrt{29}}{2}$

$$(x^2 - 2x + 3)\left(x - \frac{3 + \sqrt{29}}{2}\right)\left(x - \frac{3 - \sqrt{29}}{2}\right)$$

(c) $x = \dfrac{2 \pm \sqrt{4 - 12}}{2} = \dfrac{2 \pm \sqrt{-8}}{2} = 1 \pm \sqrt{2}i$

$$(x - 1 + \sqrt{2}i)(x - 1 - \sqrt{2}i)\left(x - \frac{3 + \sqrt{29}}{2}\right)\left(x - \frac{3 - \sqrt{29}}{2}\right)$$

51. $f(x) = 2x^3 + 3x^2 + 50x + 75$

$$\begin{array}{r|rrrr} 5i & 2 & 3 & 50 & 75 \\ & & 10i & -50+15i & -75 \\ \hline -5i & 2 & 3+10i & 15i & 0 \\ & & -10i & -15i & \\ \hline & 2 & 3 & 0 & \end{array}$$

$2x + 3 = 0 \Rightarrow x = -\frac{3}{2}$

Zeros: $\pm 5i, -\frac{3}{2}$

53. $f(x) = x^4 - 2x^3 + 37x^2 - 72x + 36$

$$\begin{array}{r|rrrrr} 6i & 1 & -2 & 37 & -72 & 36 \\ & & 6i & -36-12i & 72+6i & -36 \\ \hline -6i & 1 & -2+6i & 1-12i & 6i & 0 \\ & & -6i & 12i & -6i & \\ \hline & 1 & -2 & 1 & 0 & \end{array}$$

$x^2 - 2x + 1 = 0 \Rightarrow (x-1)^2 = 0 \Rightarrow x = 1$

Zeros: $1, \pm 6i$

55. $f(x) = 4x^3 + 23x^2 + 34x - 10$

$$\begin{array}{r|rrrr} -3+i & 4 & 23 & 34 & -10 \\ & & -12+4i & -37-i & 10 \\ \hline -3-i & 4 & 11+4i & -3-i & 0 \\ & & -12-4i & 3+i & \\ \hline & 4 & -1 & 0 & \end{array}$$

$4x - 1 = 0 \Rightarrow x = \frac{1}{4}$

Zeros: $\frac{1}{4}, -3 \pm i$

57. $f(x) = x^4 + 3x^3 - 5x^2 - 21x + 22$

$$\begin{array}{r|rrrrr} -3+\sqrt{2}i & 1 & 3 & -5 & -21 & 22 \\ & & -3+\sqrt{2}i & -2-3\sqrt{2}i & 27+2\sqrt{2}i & -22 \\ \hline -3-\sqrt{2}i & 1 & \sqrt{2}i & -7-3\sqrt{2}i & 6+2\sqrt{2}i & 0 \\ & & -3-\sqrt{2}i & 9+3\sqrt{2}i & -6-2\sqrt{2}i & \\ \hline & 1 & -3 & 2 & 0 & \end{array}$$

$x^2 - 3x + 2 = 0 \Rightarrow (x-2)(x-1) = 0 \Rightarrow x = 2, x = 1$

Zeros: $1, 2, -3 \pm \sqrt{2}i$

59. $f(x) = 8x^3 - 14x^2 + 18x - 9$

$$\left(x - \frac{1-\sqrt{5}i}{2}\right)\left(x - \frac{1+\sqrt{5}i}{2}\right) = \left(x^2 - x + \frac{3}{2}\right)$$

$$\begin{array}{r} 8x - 6 \\ x^2 - x + \frac{3}{2}\overline{\big)\; 8x^3 - 14x^2 + 18x - 9} \\ -(8x^3 - 8x^2 + 12x) \\ \hline -6x^2 + 6x - 9 \\ -(-6x^2 + 6x - 9) \\ \hline 0 \end{array}$$

$8x - 6 = 0 \Rightarrow x = \frac{3}{4}$

Zeros: $\frac{3}{4}, \frac{1 \pm \sqrt{5}i}{2}$

61. $f(x) = x^4 + 3x^3 - 5x^2 - 21x + 22$

(a) Real zeros: 1, 2

(b)
$$\begin{array}{r|rrrrr} 1 & 1 & 3 & -5 & -21 & 22 \\ & & 1 & 4 & -1 & -22 \\ \hline 2 & 1 & 4 & -1 & -22 & 0 \\ & & 2 & 12 & 22 & \\ \hline & 1 & 6 & 11 & 0 & \end{array}$$

$x^2 + 6x + 11 = 0$

$$x = \frac{-6 \pm \sqrt{6^2 - 4(1)(11)}}{2(1)} = \frac{-6 \pm \sqrt{-8}}{2}$$

$$= \frac{-6 \pm 2\sqrt{2}i}{2} = -3 \pm \sqrt{2}i$$

63. $f(x) = 8x^3 - 14x^2 + 18x - 9$

(a) Real zero: $\dfrac{3}{4}$

(b)

$\frac{3}{4}$	8	−14	18	−9
		6	−6	9
	8	−8	12	0

$$8x^2 - 8x + 12 = 0$$

$$4(2x^2 - 2x + 3) = 0 \Rightarrow 2x^2 - 2x + 3 = 0$$

$$x = \frac{-(-2) \pm \sqrt{(-2)^2 - 4(2)(3)}}{2(2)}$$

$$= \frac{2 \pm \sqrt{-20}}{4} = \frac{2 \pm 2\sqrt{5}i}{4}$$

$$= \frac{1 \pm \sqrt{5}i}{2}$$

65. $x^4 - 5x^2 + 4 = 0$

$(x^2 - 4)(x^2 - 1) = 0$

$x = \pm 2, \pm 1$

The x-intercepts of $y = x^4 - 5x^2 + 4$ are $(\pm 2, 0)$ and $(\pm 1, 0)$. The connection between the x-intercepts of $y = x^4 - 5x^2 + 4$ and the solutions of the equation $x^4 - 5x^2 + 4 = 0$ is that an x-intercept of the graph of a function tells us that the intercept value is a real zero of the related equation.

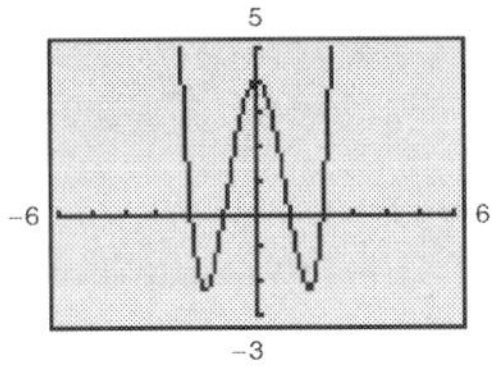

67. Answers will vary. Sample answer:

(a) Four real zeros: 1 (repeated), 2, 3

$$f(x) = (x + 1)(x + 1)(x - 2)(x - 3)$$
$$= x^4 - 7x^3 + 17x^2 - 17x + 6$$

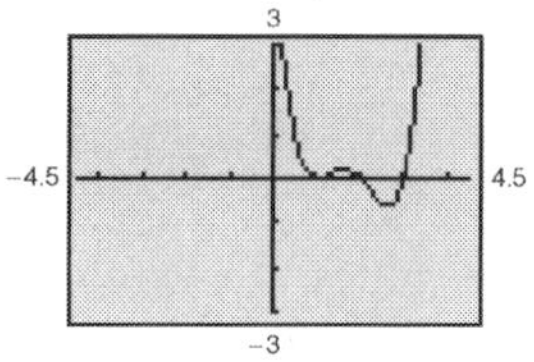

(b) Two real zeros: 1, 2, ($\pm i$ complex)

$$g(x) = (x - 1)(x - 2)(x^2 + 1)$$
$$= x^4 - 3x^3 + 3x^2 - 3x + 2$$

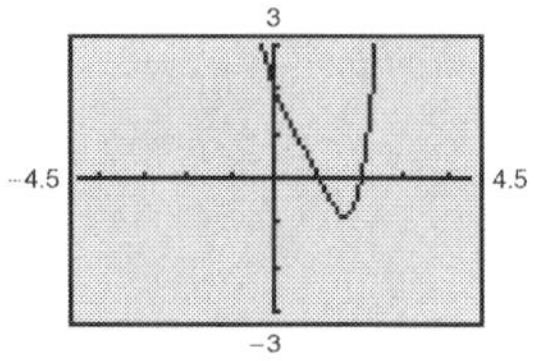

(c) No real zeros: (four complex zeros: $\pm i$, $\pm 2i$)

$$h(x) = x^4 + 5x^2 + 4$$

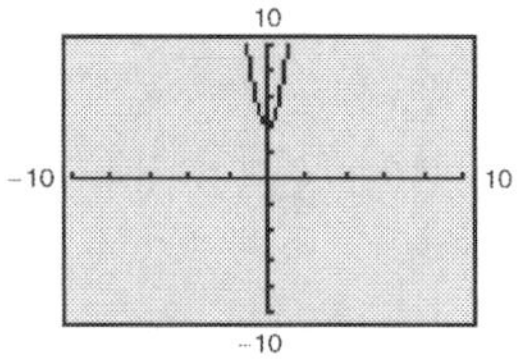

Similarities: The graphs of f, g, and h rise to the left and rise to the right.

Differences: f intersects the x-axis three times—one is a "touch" because $x = 1$ is a repeated zero.

g intersects the x-axis twice because g has two real zeros.

h does not intersect the x-axis because all zeros are complex.

69. $-0.0001x^2 + 140x = 50{,}000{,}000$

$$-0.0001x^2 + 140x - 50{,}000{,}000 = 0$$

$$x = \frac{-140 \pm \sqrt{140^2 - 4(-0.0001)(-50{,}000{,}000)}}{2(-0.0001)}$$

$$= 700{,}000 \pm 100{,}000i$$

Because the solutions are imaginary, there is no quantity of units x or price p that will yield a revenue of \$50 million.

71. Imaginary zeros of polynomials occur in conjugate pairs. So, a fourth-degree polynomial with real coefficients cannot have exactly three imaginary zeros.

73. Imaginary zeros can only occur in conjugate pairs. Because f has three zeros, one or all of them must be real numbers.

Section 3.7 Rational Functions

Skills Warm Up

1. $x^2 - 4x = x(x - 4)$

2. $2x^3 - 6x = 2x(x^2 - 3)$

3. $x^2 - 3x - 10 = (x - 5)(x + 2)$

4. $x^2 - 7x + 10 = (x - 5)(x - 2)$

5. $x^3 + 4x^2 + 3x = x(x^2 + 4x + 3)$
$= x(x + 3)(x + 1)$

6. $x^3 - 4x^2 - 2x + 8 = x^2(x - 4) - 2(x - 4)$
$= (x^2 - 2)(x - 4)$

7.

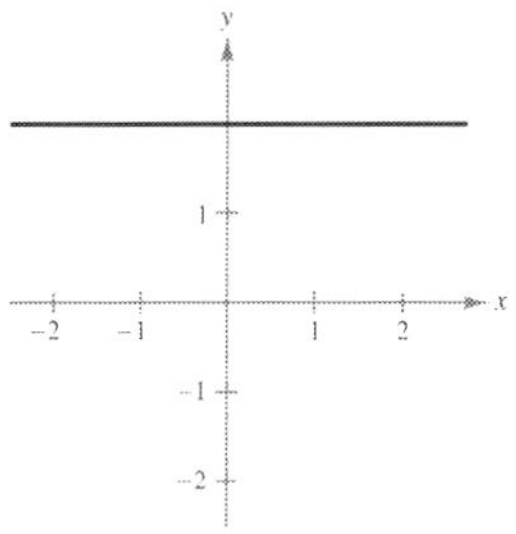

8.

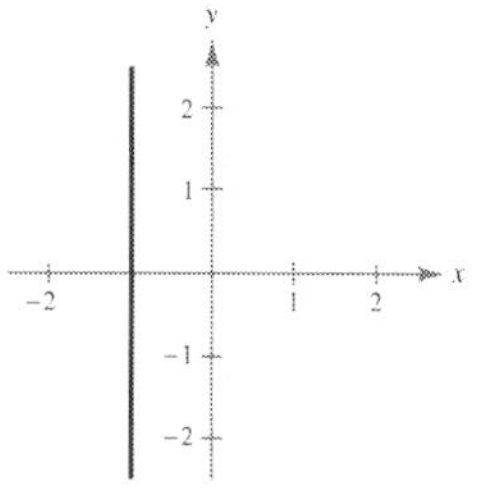

9.

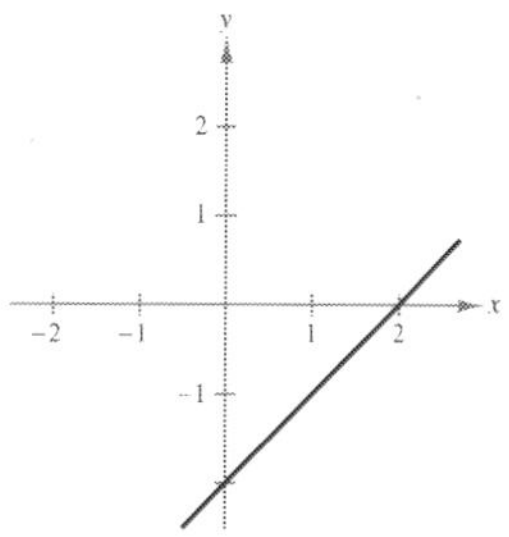

10.

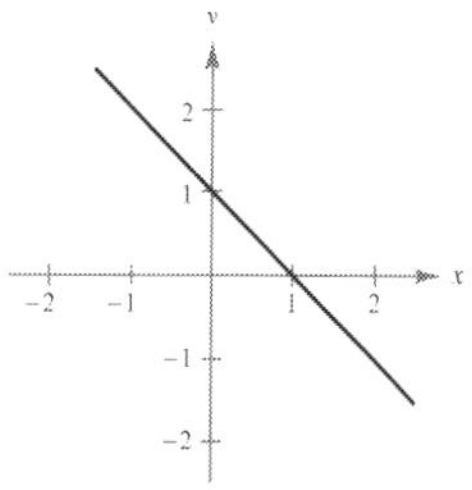

1. $f(x) = \frac{1}{x - 2}$

Domain: All real numbers except $x = 2$

$f(x)$ approaches $-\infty$ as x approaches 2 from the left.

$f(x)$ approaches ∞ as x approaches 2 from the right.

3. $f(x) = \frac{4x}{x + 1}$

Domain: All real numbers except $x = -1$

$f(x)$ approaches $-\infty$ as x approaches -1 from the left.

$f(x)$ approaches ∞ as x approaches -1 from the right.

5. $f(x) = \frac{3x^2}{x^2 - 1}$

Domain: All real numbers except $x = -1$ and $x = 1$

$f(x)$ approaches ∞ as x approaches -1 from the left.

$f(x)$ approaches $-\infty$ as x approaches -1 from the right.

$f(x)$ approaches $-\infty$ as x approaches 1 from the left.

$f(x)$ approaches ∞ as x approaches 1 from the right.

7. Horizontal asymptote: $y = 3$

Vertical asymptote: $x = -1$

9. Horizontal asymptote: $y = -1$

Vertical asymptote: $x = 5$

11. Horizontal asymptote: $y = 3$

Vertical asymptote: none

13. Horizontal asymptote: $y = 0$

Vertical asymptote: $x = -4$

15. $f(x) = -\frac{4}{x - 3}$

Vertical asymptote: $x = 3$

Horizontal asymptote: $y = 0$

Intercept: $\left(0, \frac{4}{3}\right)$

Graph (f)

17. $f(x) = \dfrac{x + 1}{x}$

Vertical asymptote: $x = 0$

Horizontal asymptote: $y = \dfrac{1}{1} = 1$

Intercept: $(-1, 0)$

Graph (a)

19. $f(x) = \dfrac{x + 4}{x - 2}$

Vertical asymptote: $x = 2$

Horizontal asymptote: $y = 1$

Intercepts: $(-4, 0), (0, -2)$

Graph (c)

21. $g(x) = f(x) - 2 = \dfrac{1}{x} - 2 = \dfrac{1 - 2x}{x}$

Vertical shift 2 units downward

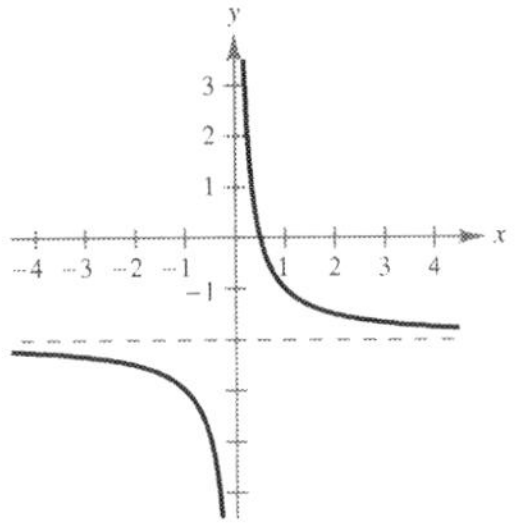

23. $g(x) = -f(x) = -\dfrac{1}{x}$

Reflection about the x-axis

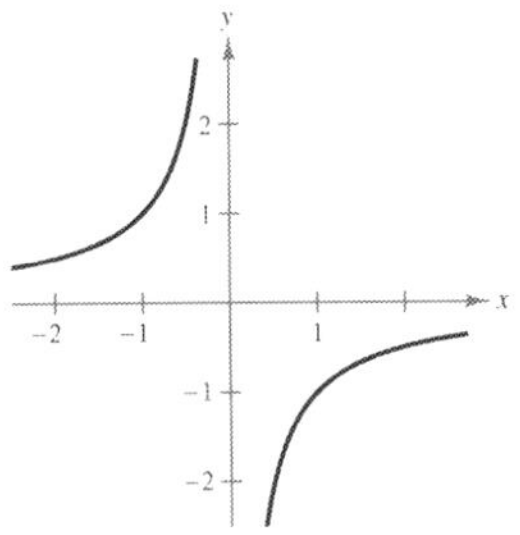

25. $g(x) = f(x) + 3 = \dfrac{4}{x^2} + 3 = \dfrac{4 + 3x^2}{x^2}$

Vertical shift 3 units upward

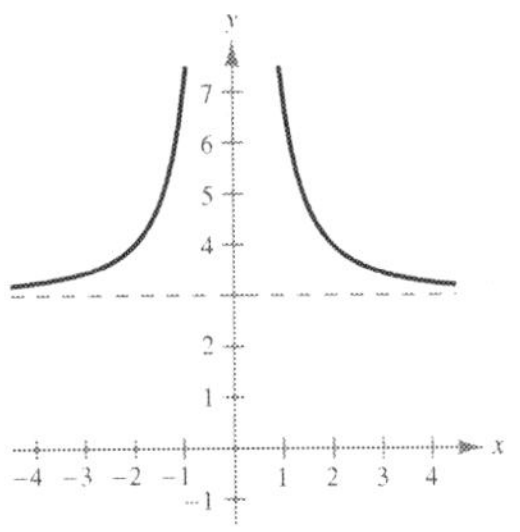

27. $g(x) = -f(x) = -\dfrac{4}{x^2}$

Reflection about the x-axis

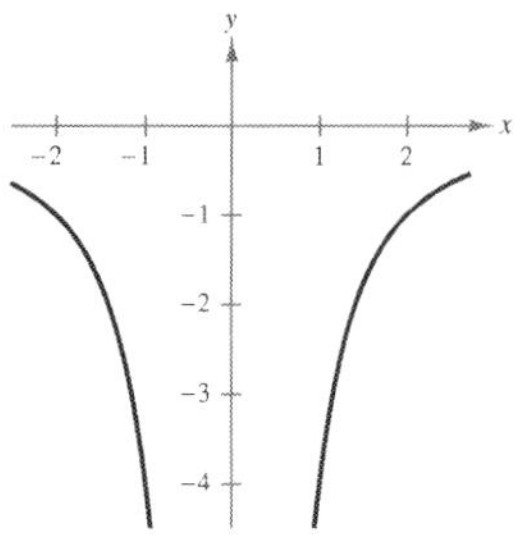

29. $g(x) = f(x) + 5 = \dfrac{8}{x^3} + 5 = \dfrac{8 + 5x^3}{x^3}$

Vertical shift 5 units upward

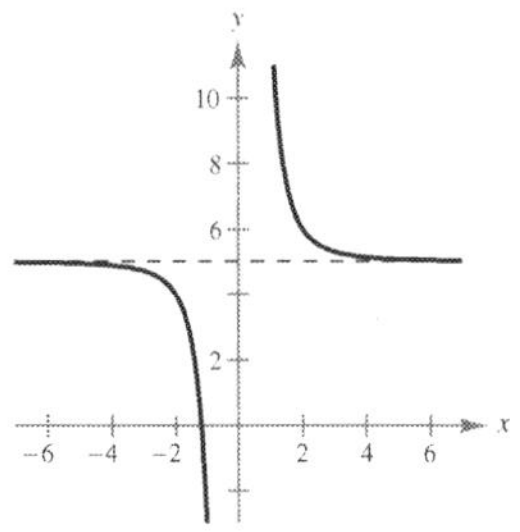

31. $g(x) = -f(x) = -\dfrac{8}{x^3}$

Reflection about the x-axis

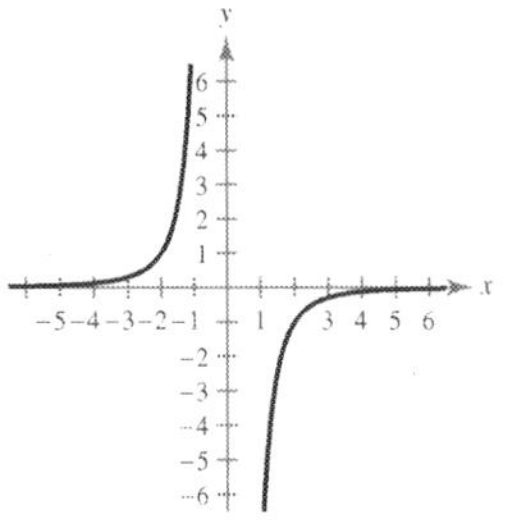

33. $f(x) = \dfrac{1}{x + 3}$

Intercept: $\left(0, \dfrac{1}{3}\right)$

Vertical asymptote: $x = -3$

Horizontal asymptote: $y = 0$

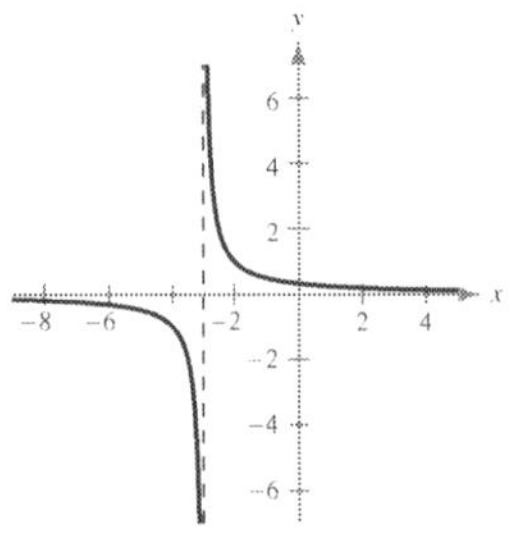

35. $f(x) = \dfrac{-1}{x + 1}$

Intercept: $(0, -1)$

Vertical asymptote: $x = -1$

Horizontal asymptote: $y = 0$

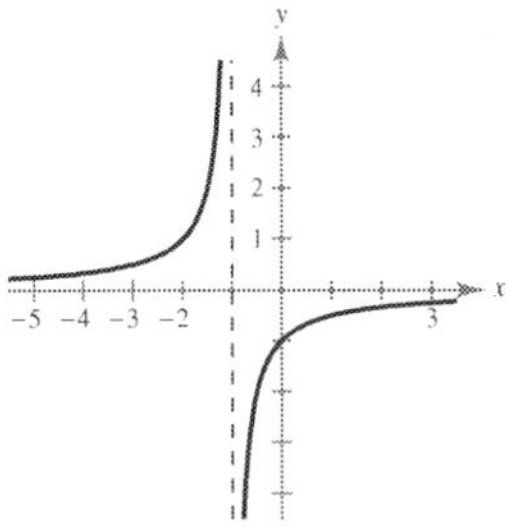

37. $f(x) = \dfrac{x + 4}{x - 5}$

Intercepts: $(-4, 0), \left(0, -\dfrac{4}{5}\right)$

Vertical asymptote: $x = 5$

Horizontal asymptote: $y = 1$

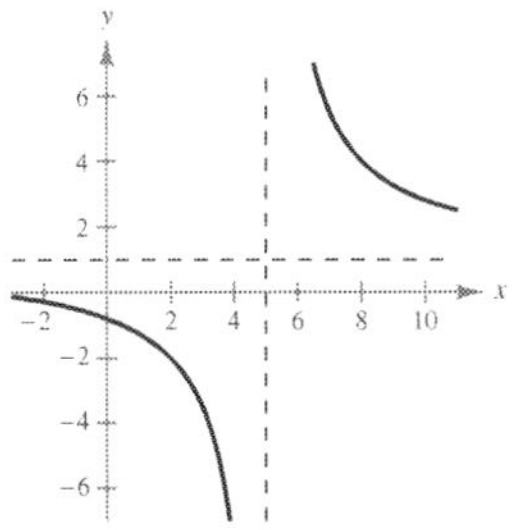

39. $f(x) = \dfrac{2 + x}{1 - x}$

Intercepts: $(-2, 0), (0, 2)$

Vertical asymptote: $x = 1$

Horizontal asymptote: $y = -1$

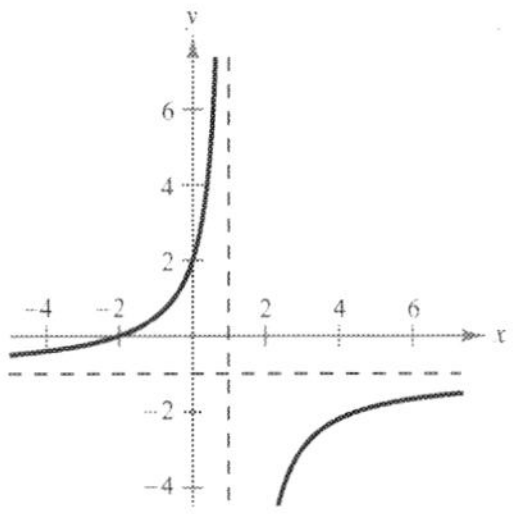

41. $f(t) = \dfrac{3t + 1}{t}$

Intercept: $\left(-\dfrac{1}{3}, 0\right)$

Vertical asymptote: $t = 0$

Horizontal asymptote: $y = 3$

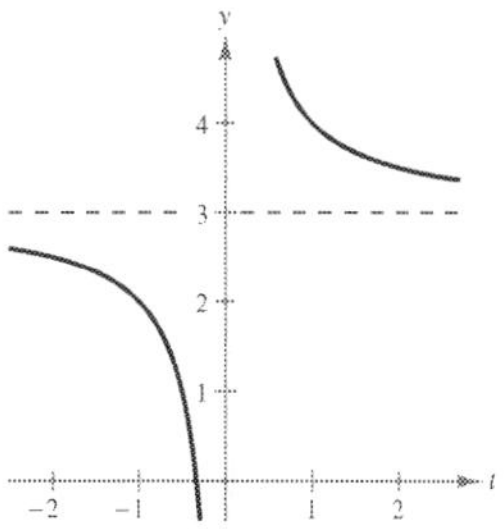

43. $C(x) = \dfrac{5 + 2x}{1 + x}$

Intercepts: $\left(-\dfrac{5}{2}, 0\right), (0, 5)$

Vertical asymptote: $x = -1$

Horizontal asymptote: $y = 2$

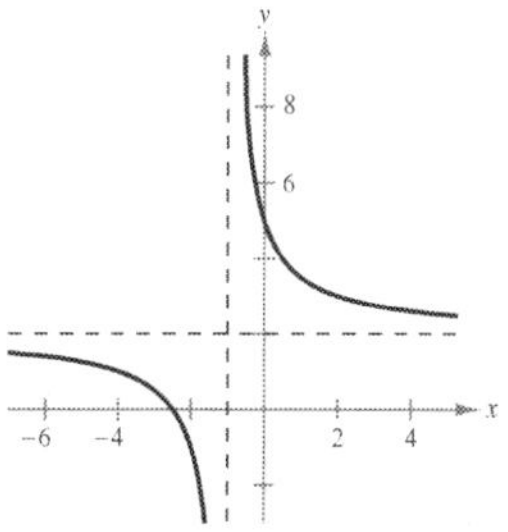

45. $g(x) = \dfrac{1}{x + 2} + 2 = \dfrac{2x + 5}{x + 2}$

Intercepts: $\left(-\dfrac{5}{2}, 0\right), \left(0, \dfrac{5}{2}\right)$

Vertical asymptote: $x = -2$

Horizontal asymptote: $y = 2$

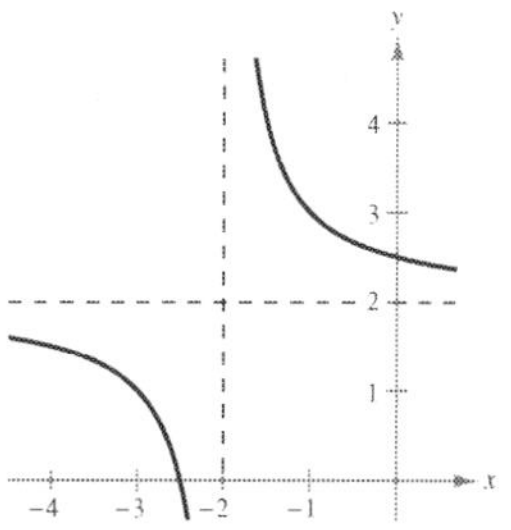

47. $f(x) = \dfrac{1}{x^2} + 2 = \dfrac{2x^2 + 1}{x^2}$

Intercept: none

Vertical asymptote: $x = 0$

Horizontal asymptote: $y = 2$

y-axis symmetry

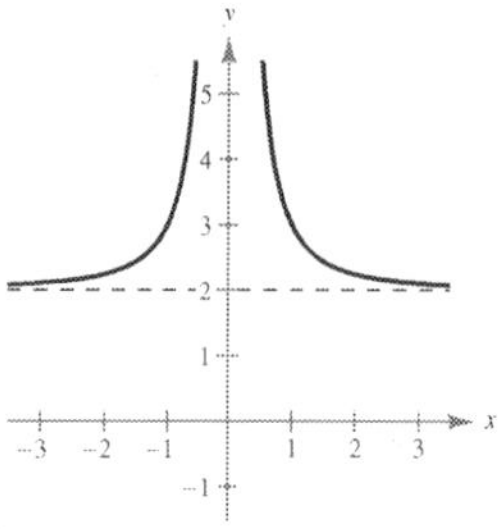

49. $h(x) = \dfrac{x^2}{x^2 - 9}$

Intercept: $(0, 0)$

Vertical asymptotes: $x = \pm 3$

Horizontal asymptote: $y = 1$

y-axis symmetry

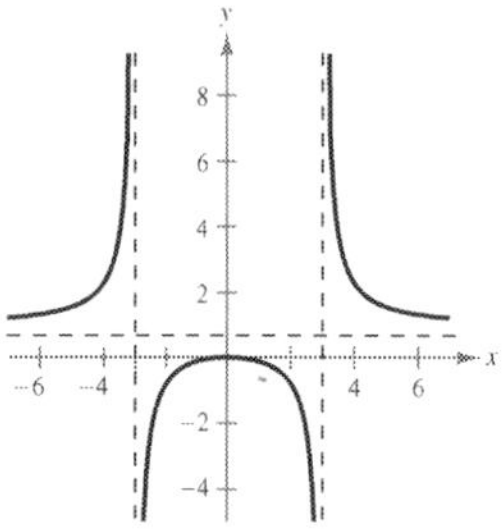

51. $g(s) = \dfrac{s}{s^2 + 1}$

Intercept: $(0, 0)$

Horizontal asymptote: $y = 0$

Origin symmetry

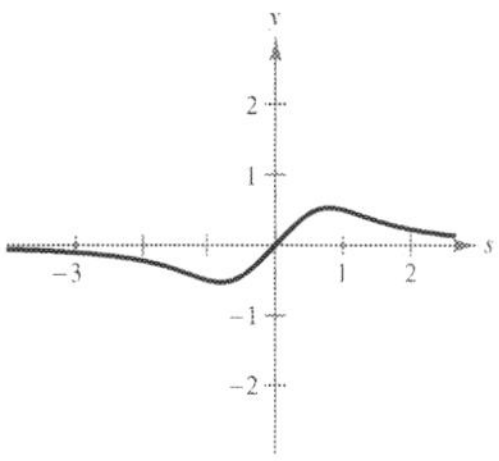

53. $f(x) = \dfrac{x}{x^2 - 3x - 4} = \dfrac{x}{(x - 4)(x + 1)}$

Intercept: $(0, 0)$

Vertical asymptotes: $x = -1, 4$

Horizontal asymptote: $y = 0$

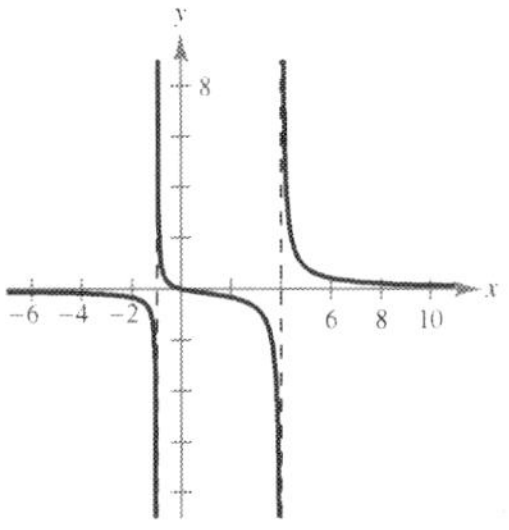

55. $f(x) = \dfrac{3x}{x^2 - x - 2} = \dfrac{3x}{(x + 1)(x - 2)}$

Intercept: $(0, 0)$

Vertical asymptotes: $x = -1, 2$

Horizontal asymptote: $y = 0$

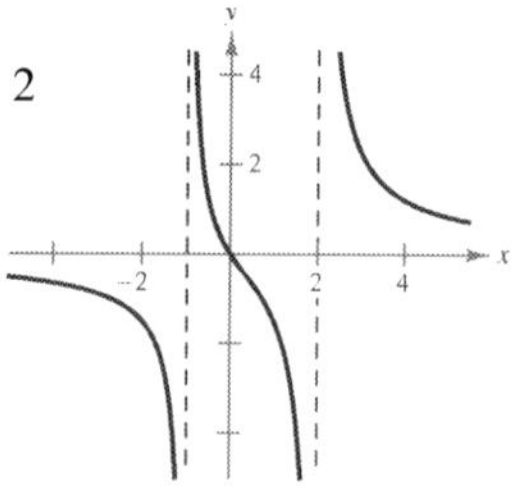

57. $g(x) = \dfrac{4(x^2 - 4)}{x^2 - 16}$

Intercepts: $(0, 1), (-2, 0), (2, 0)$

Horizontal asymptote: $y = 4$

Vertical asymptotes: $x = \pm 4$

y-axis symmetry

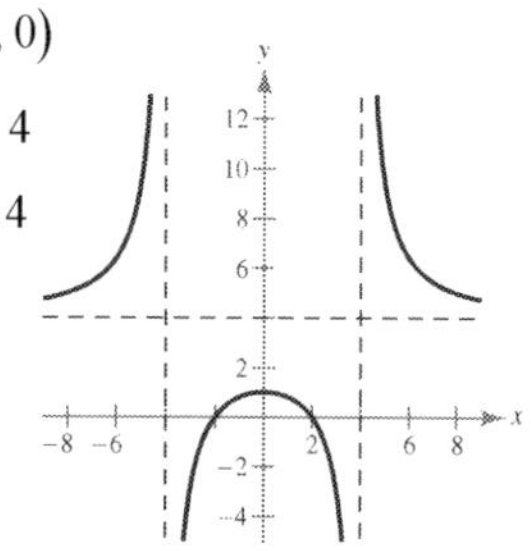

59. $f(x) = \dfrac{2x^2 + 1}{x}$

No intercepts

Vertical asymptote: $x = 0$

Slant asymptote: $y = 2x$

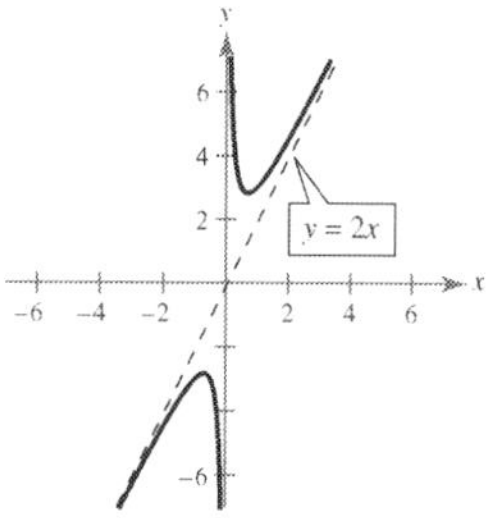

61. $h(x) = \dfrac{x^2}{x - 1}$

Intercept: $(0, 0)$

Vertical asymptote: $x = 1$

Slant asymptote:

$y = x + 1$

No symmetry

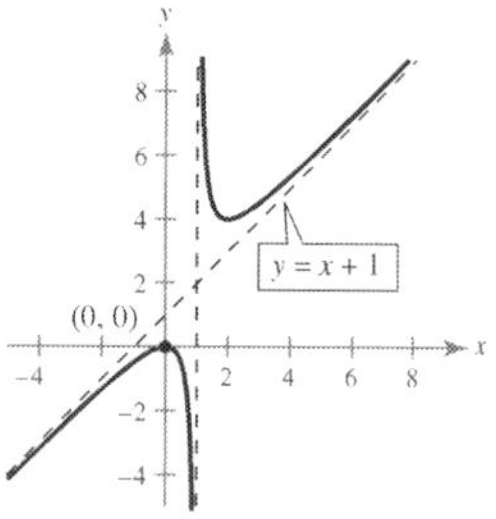

63. $g(x) = \dfrac{x^3}{2x^2 - 8}$

Intercept: $(0, 0)$

Vertical asymptotes:

$x = \pm 2$

Slant asymptote: $y = \dfrac{1}{2}x$

Origin symmetry

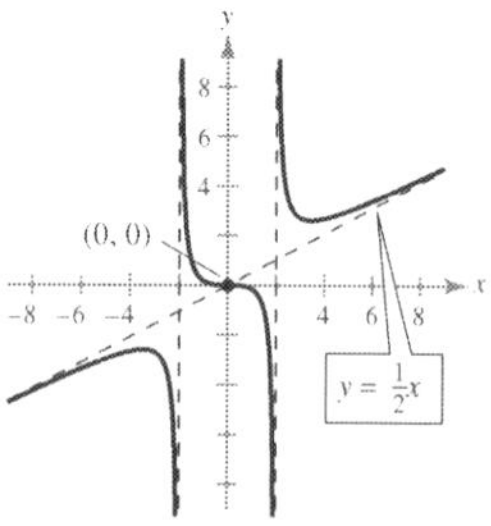

65. Vertical asymptote: None $\Rightarrow$ The denominator is not zero for any value of x (unless the numerator is also zero there).

Horizontal asymptote: $y = 2 \Rightarrow$ The degree of the numerator equals the degree of the denominator.

$f(x) = \dfrac{2x^2}{x^2 + 1}$ is one possible function. There are many correct answers.

67. Vertical asymptotes: $x = -2, x = 1 \Rightarrow (x + 2)(x - 1)$ are factors of the denominator.

Horizontal asymptotes: None $\Rightarrow$ The degree of the numerator is greater than the degree of the denominator.

$f(x) = \dfrac{x^3}{(x + 2)(x - 1)}$ is one possible function. There are many correct answers.

69.

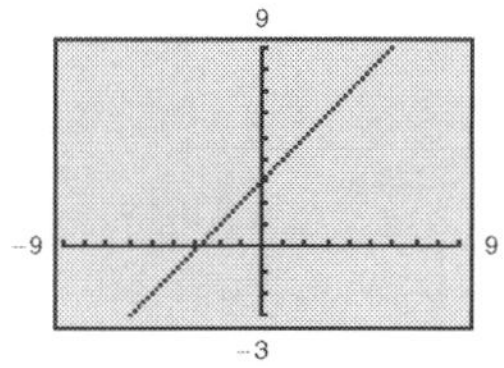

The graph of $f(x)$ has a hole at $x = 3$.

71.

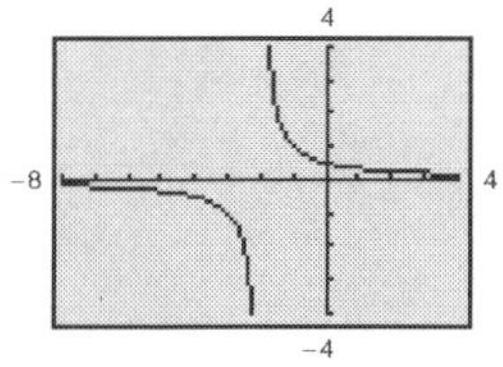

The graph of $f(x)$ has a hole at $x = 2$.

73. The domain of the new function should exclude $x = 7$.

75. No, given $f(x) = \dfrac{a_nx^n + \cdots + a_0}{b_mx^m + \cdots + b_0}$, if $n > m$, there is no horizontal asymptote and n must be greater than m for a slant asymptote to occur.

77. $C = \dfrac{528p}{100 - p}, \ 0 \le p < 100$

(a) $C(25) = \dfrac{528(25)}{100 - 25} = \$176$ million

(b) $C(50) = \dfrac{528(50)}{100 - 50} = \$528$ million

(c) $C(75) = \dfrac{528(75)}{100 - 75} = \$1584$ million

(d) $C \Rightarrow \infty$ as $x \Rightarrow 100^-$

No, the model would generate a zero in the denominator.

79. (a)

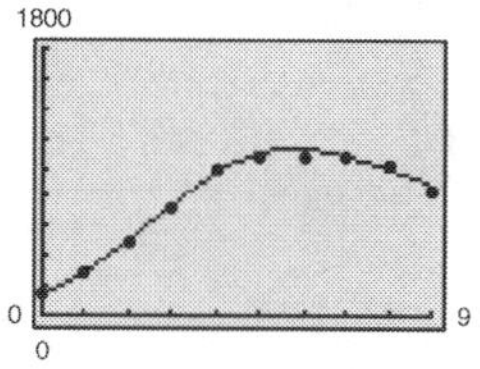

The model fits the data well.

(b) For 2012, $t = 12$:

$$N = \frac{104.05(12) + 140.3}{0.026(12)^2 - 0.21(12) + 1} \approx 624.5$$

For 2015, $t = 15$:

$$N = \frac{104.05(15) + 140.3}{0.026(15)^2 - 0.21(15) + 1} \approx 459.7$$

For 2020, $t = 20$:

$$N = \frac{104.05(20) + 140.3}{0.026(20)^2 - 0.21(20) + 1} \approx 308.5$$

In 2012, 624.5 million DVDs will be sold.

In 2015, 459.7 million DVDs will be sold.

In 2020, 308.5 million DVDs will be sold.

Answers will vary.

(c) The horizontal asymptote is $N = 0$. The model predicts that, over time, the number of DVDs sold will decrease and approach 0.

81. (a)

n	1	2	3	4	5	6	7	8	9	10
P	0.60	0.79	0.86	0.90	0.92	0.93	0.94	0.95	0.95	0.96

(b) 100%

83. (a)

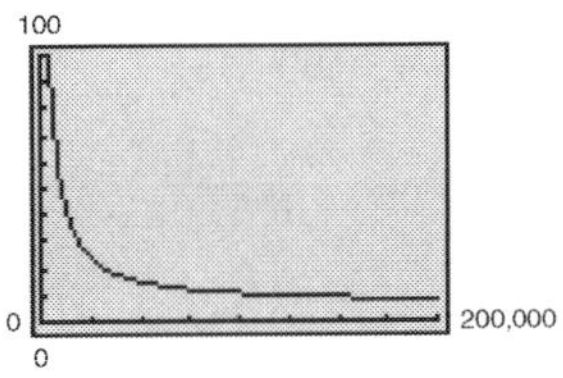

(b) $\overline{C}(10{,}000) = \$51$

$\overline{C}(100{,}000) = \$10.50$

$\overline{C}(1{,}000{,}000) = \$6.45$

$\overline{C}(10{,}000{,}000) \approx \$6.05$

Eventually the average recycling price per pound will approach $6.

85. (a) $E = \dfrac{27{,}904t + 906{,}950}{3.023t + 250.90}$

(b)

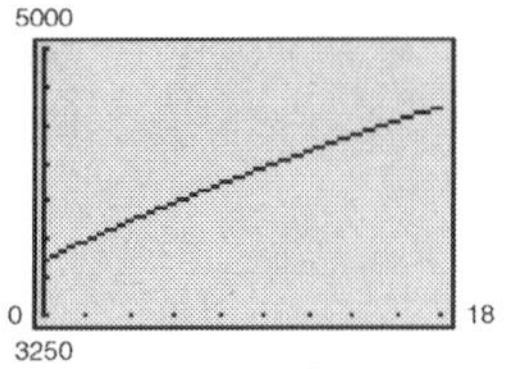

(c) For 2008, let $t = 18$:

$$E = \frac{27{,}904(18) + 906{,}950}{3.023(18) + 250.90} \approx 4615.6$$

In 2008, per capita domestic sales of electricity was about 4615.6 kWh per person.

87. (a)

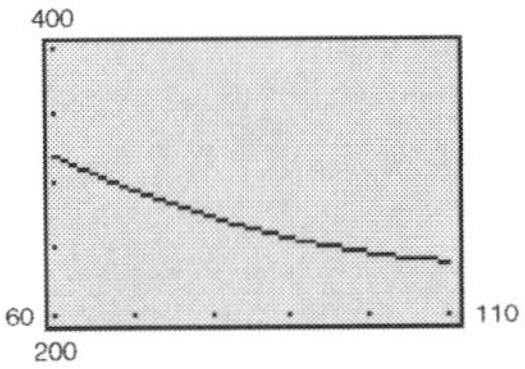

(b) For 2014, $t = 114$:

$$y = 0.0237(114)^2 - 5.624(114) + 571.67 \approx 238.54$$

For 2018, $t = 118$:

$$y = 0.0237(118)^2 - 5.624(118) + 571.67 \approx 238.04$$

In 2014, the winning time will be 238.54 seconds and in 2018, the winning time will be 238.04 seconds.

(c) No. Because y is a quadratic function, it does not have a horizontal asymptote. Answers will vary.

Sample answer: Because winning times cannot decrease to zero, it seems reasonable that a horizontal asymptote should not exist.

Review Exercises for Chapter 3

1. $f(x) = (x + 3)^2 - 5$

Vertex: $(-3, -5)$

y-intercept: $(0, 4)$

x-intercepts: $\left(-3 \pm \sqrt{5}, 0\right)$

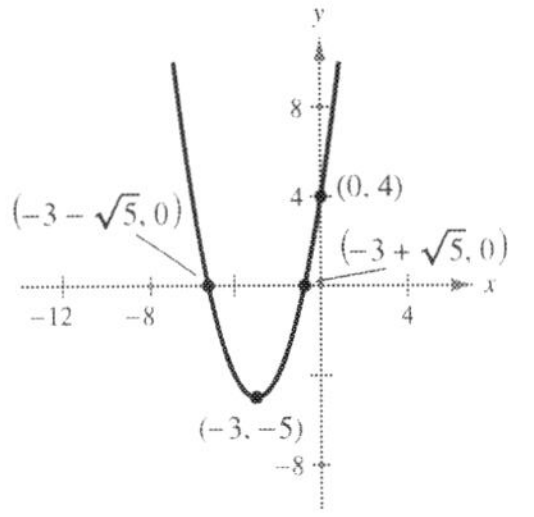

3. $h(x) = 3x^2 - 12x + 11 = 3(x - 2)^2 - 1$

Vertex: $(2, -1)$

y-intercept: $(0, 11)$

x-intercepts: $\left(\dfrac{6 \pm \sqrt{3}}{3}, 0\right)$

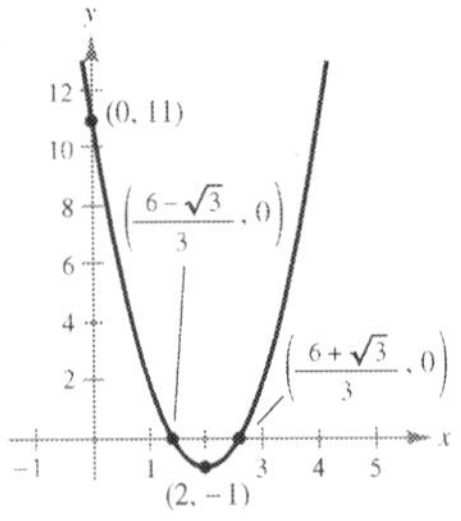

5. $f(x) = a(x + 5)^2 - 1$

$6 = a(-2 + 5)^2 - 1$

$\frac{7}{9} = a$

$f(x) = \frac{7}{9}(x + 5)^2 - 1$

7. $2x + 2y = 500$

$y = 250 - x$

$A = xy$

$= x(250 - x)$

$= 250x - x^2$

$= -(x^2 - 125)^2 + 15{,}625$

Vertex: $(125, 15{,}625)$

The dimensions that produce the maximum area are 125 feet × 125 feet.

9. (a)

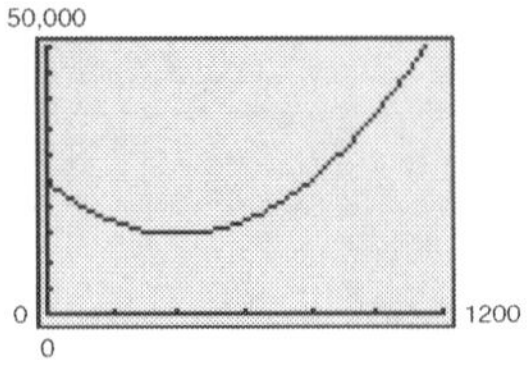

(b) About 385 units

(c) Write the equation in standard form. The vertex is the minimum cost.

11. (a)

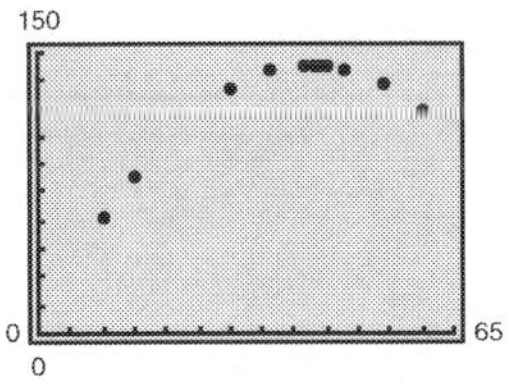

(b) $d(x) = -0.077x^2 + 6.59x + 2.4$

(c)

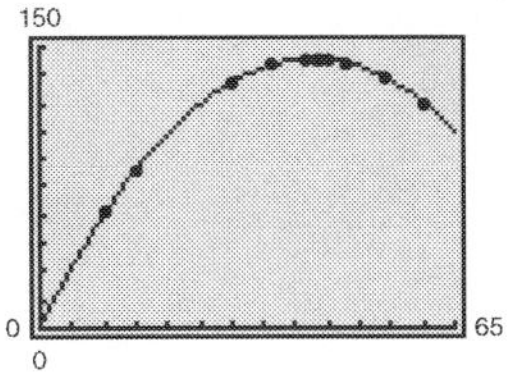

(d) (42.8, 143.4)

The vertex gives the angle $42.8°$, which results in the greatest distance of 143.4 meters.

13. $f(x) = (x + 4)^5$

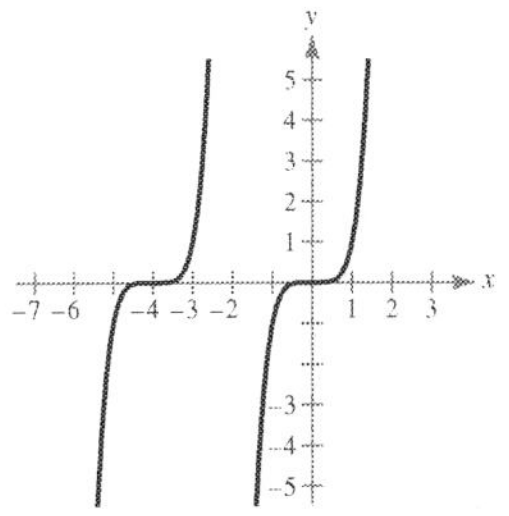

15. $f(x) = 3 - \frac{1}{2}x^5$

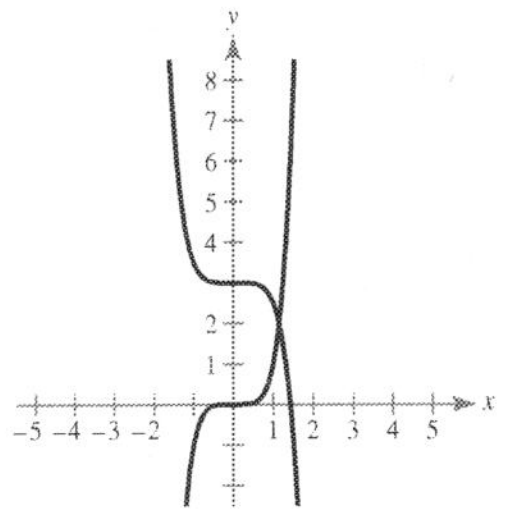

17. $f(x) = x^6 - 2$

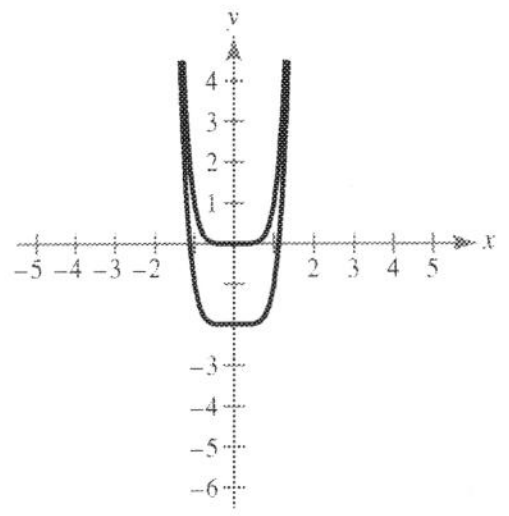

19. $f(x) = -\frac{1}{2}x^6 - 5$

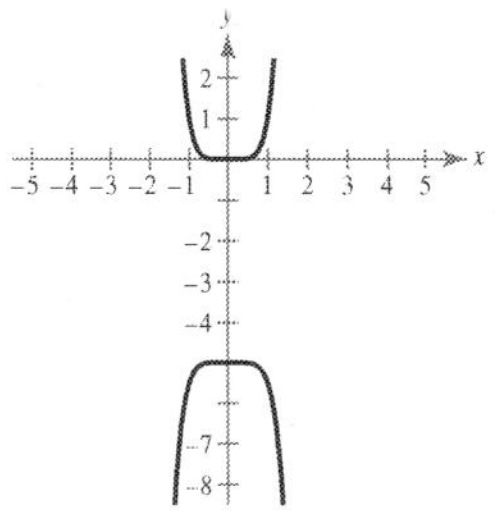

21. $f(x) = \frac{1}{2}x^3 + 2x$

The degree is odd and the leading coefficient is positive. The graph falls to the left and rises to the right.

23. $f(x) = -x^6 + 3x^4 - x^2 + 6$

The degree is even and the leading coefficient is negative. The graph falls to the left and falls to the right.

25. $f(x) = 16 - x^2 = (4 + x)(4 - x)$

Real zeros: $x = \pm 4$

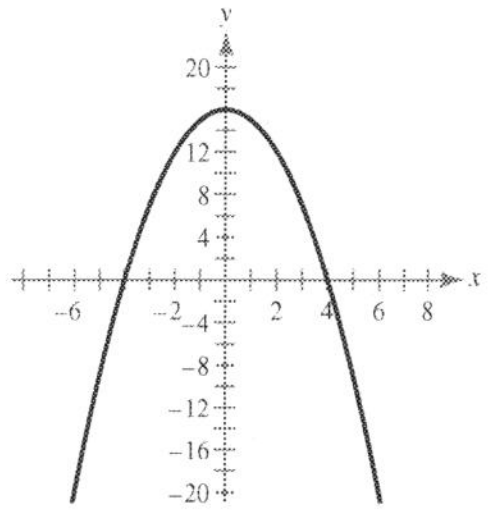

27. $f(x) = x^3 - 7x^2 + 10x$

$= x(x^2 - 7x + 10)$

$= x(x - 5)(x - 2)$

Real zeros: $x = 0, 2, 5$

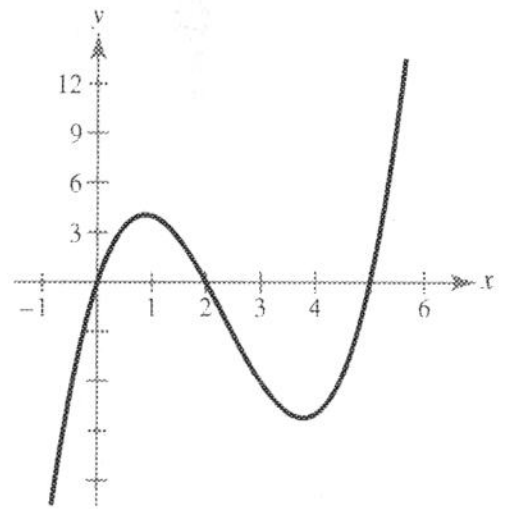

29.
$$
\begin{array}{r}
x^2 - 3x + 1 \\
2x + 1 \overline{\smash{)}\ 2x^3 - 5x^2 - x} \\
\underline{-(2x^3 + x^2)} \\
-6x^2 - x \\
\underline{-(-6x^2 - 3x)} \\
2x \\
\underline{-(2x + 1)} \\
-1
\end{array}
$$

$$\frac{2x^3 - 5x^2 - x}{2x + 1} = x^2 - 3x + 1 - \frac{1}{2x + 1}$$

31. $$\begin{array}{r|rrrr} -3 & 1 & 0 & -6 & 9 \\ & & -3 & 9 & -9 \\ \hline & 1 & -3 & 3 & 0 \end{array}$$

$$\frac{x^3 - 6x + 9}{x + 3} = x^2 - 3x + 3$$

33. $f(x) = 6 + 2x^2 - 3x^3$

(a) $$\begin{array}{r|rrrr} 2 & -3 & 2 & 0 & 6 \\ & & -6 & -8 & -16 \\ \hline & -3 & -4 & -8 & -10 \end{array}$$

$f(2) = -10$

(b) $$\begin{array}{r|rrrr} -1 & -3 & 2 & 0 & 6 \\ & & 3 & -5 & 5 \\ \hline & -3 & 5 & -5 & 11 \end{array}$$

$f(-1) = 11$

35. (a) $$\begin{array}{r|rrrr} 5 & 1 & -4 & -11 & 30 \\ & & 5 & 5 & -30 \\ \hline \end{array}$$

$$\begin{array}{r|rrrr} -3 & 1 & 1 & -6 & 0 \\ & & -3 & 6 & \\ \hline & 1 & -2 & 0 & \end{array}$$

(b) The remaining factor is $x - 2$.

(c) $f(x) = x^3 - 4x^2 - 11x + 30$
$= (x + 3)(x - 2)(x - 5)$

(d) Real zeros: $x = -3, 2, 5$

(e)

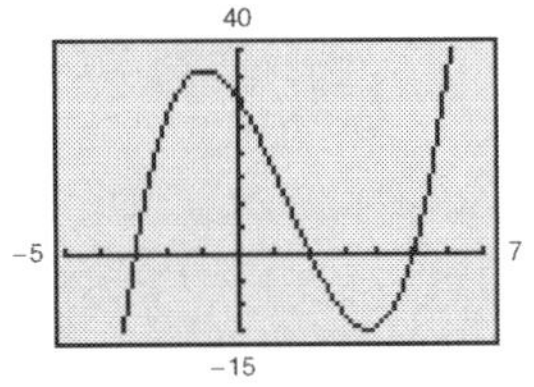

37. $$\begin{array}{r|rrrr} 2 & 1 & 9 & 2 & -48 \\ & & 2 & 22 & 48 \\ \hline & 1 & 11 & 24 & 0 \end{array}$$

$$\frac{x^3 + 9x^2 + 2x - 48}{x - 2} = x^2 + 11x + 24$$

39. (a)

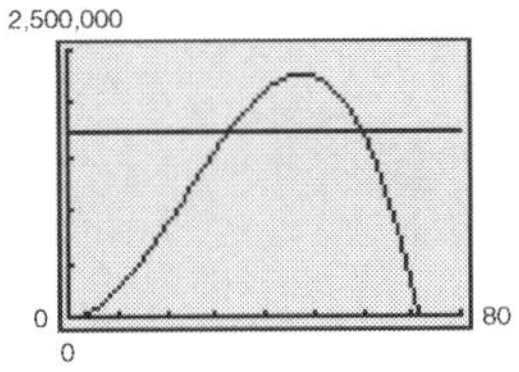

$\approx \$325,167$

(b) Answers will vary.

41. $f(x) = -4x^3 + 8x^2 - 3x + 15$

Possible rational numbers: $\pm 1, \pm 3, \pm 5, \pm 15,$

$\pm\frac{1}{2}, \pm\frac{3}{2}, \pm\frac{5}{2}, \pm\frac{15}{2}, \pm\frac{1}{4}, \pm\frac{3}{4}, \pm\frac{5}{4}, \pm\frac{15}{4}$

The only real root of $f(x)$ is $x \approx 2.357$, which is not one of the possible rational zeros.

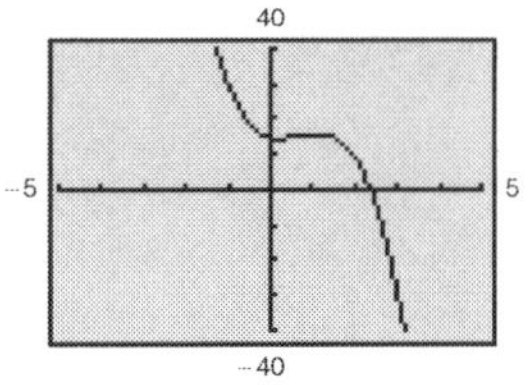

43. $f(x) = x^3 + 2x^2 - 5x - 6$

$$\begin{array}{r|rrrr} 1 & 1 & 2 & -5 & -6 \\ & & -1 & -1 & 6 \\ \hline & 1 & 1 & -6 & 0 \end{array}$$

$f(x) = (x + 1)(x^2 + x - 6)$
$= (x + 1)(x + 3)(x - 2)$

Zeros: $-3, -1, 2$

45. $h(x) = 3x^4 - 27x^2 + 60$
$= 3(x^4 - 9x^2 + 20)$
$= 3(x^2 - 4)(x^2 - 5)$
$= 3(x + 2)(x - 2)(x + \sqrt{5})(x - \sqrt{5})$

Zeros: $\pm 2, \pm\sqrt{5}$

47. $B(x) = 6x^3 - 19x^2 + 11x + 6$

$$\begin{array}{r|rrrr} 2 & 6 & -19 & 11 & 6 \\ & & 12 & -14 & -6 \\ \hline & 6 & -7 & -3 & 0 \end{array}$$

$B(x) = (x - 2)(6x^2 - 7x - 3)$
$= (x - 2)(3x + 1)(2x - 3)$

Zeros: $2, -\frac{1}{3}, \frac{3}{2}$

49. $p(x) = x^4 - x^3 - 2x - 4$

$$\begin{array}{r|rrrrr} -1 & 1 & -1 & 0 & -2 & -4 \\ & & -1 & 2 & -2 & 4 \\ \hline 2 & 1 & -2 & 2 & -4 & 0 \\ & & 2 & 0 & 4 & \\ \hline & 1 & 0 & 2 & 0 & \end{array}$$

$p(x) = x^4 - x^3 - 2x - 4 = (x + 1)(x - 2)(x^2 + 2)$

Zeros: $-1, 2$

51. $f(x) = x^3 - 4x + 3, [-3, -2], x \approx -2.3$

x	–3	–2.9	–2.8	–2.7	–2.6	–2.5	–2.4	–2.3	–2.2	–2.1	–2
–3	–12	–9.789	–7.752	–5.883	–4.176	–2.625	–1.224	0.033	1.152	2.139	3

53. $f(x) = 5x^3 - 11x - 3$

Using the ZOOM and TRACE keys, you can approximate the real zeros as $x \approx -1.321$, $x \approx -0.283$, and $x \approx 1.604$.

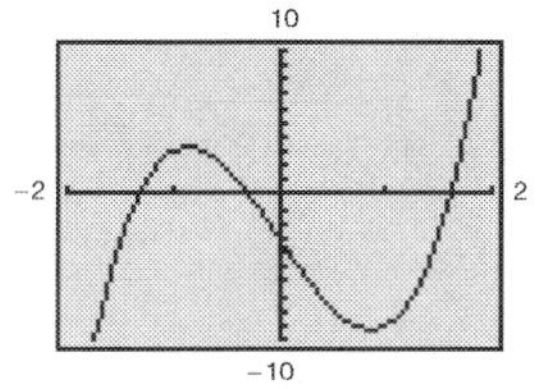

55. (a)

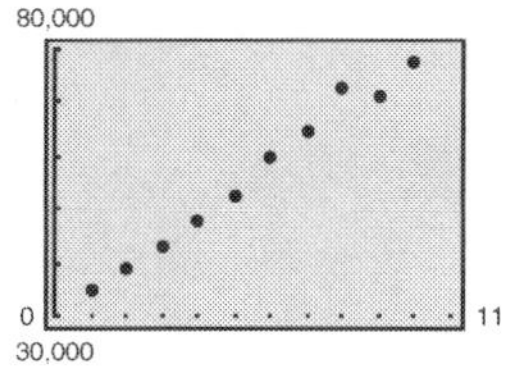

(b) Linear: $R = 4996.7t + 28{,}927$

Quadratic: $R = -36.14t^2 + 5394.3t + 28{,}132$

Cubic: $R = -58.393t^3 + 927.35t^2 + 950.5t + 33{,}142$

Quartic: $R = 3.4725t^4 - 134.788t^3 + 1486.42t^2 - 577.4t + 34{,}333$

(c) Linear

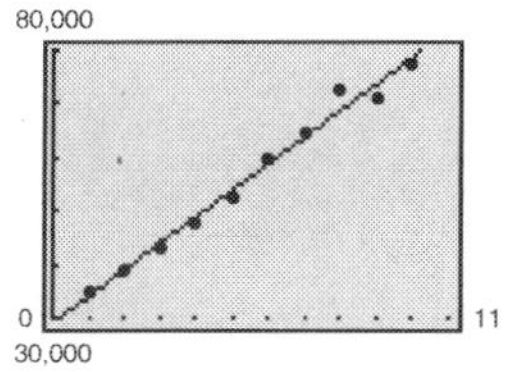

Quadratic

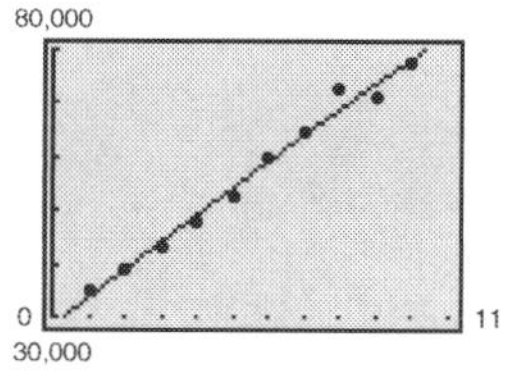

Cubic

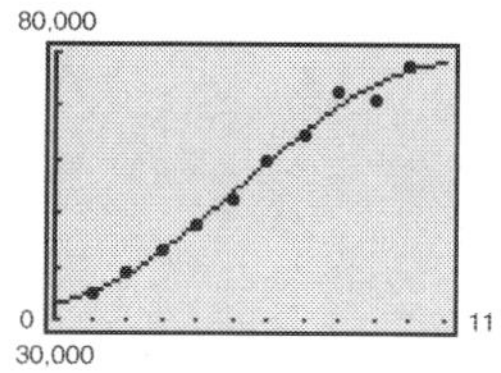

Quartic

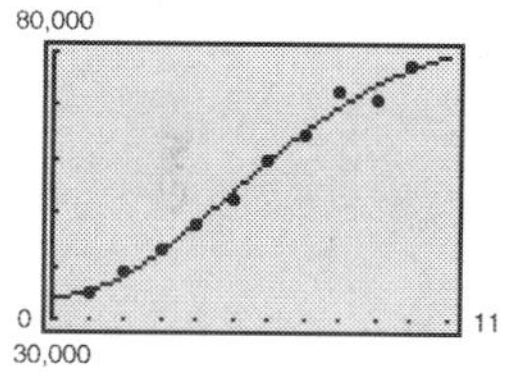

Each model could be considered a good fit for the data.

(d) Linear: 2011 $(t \approx 11.6)$

Quadratic: 2011 $(t \approx 11.9)$

Cubic: never

Quartic: 2018 $(t \approx 18.8)$

Answers will vary. Sample answer: The linear and quadratic predictions are around 2011. The cubic model predicts a drop after 2010 while the quartic model levels off for a few years before increasing.

57. $\sqrt{-32} = \sqrt{32}i = 4\sqrt{2}i$

The complex conjugate: $-4\sqrt{2}i$

59. $-3 + \sqrt{-16} = -3 + 4i$

The complex conjugate: $-3 - 4i$

61. $(7 - 3i) + (-2 + 5i) = 7 - 3i - 2 + 5i$

$= (7 - 2) + (-3 + 5)i$

$= 5 + 2i$

63. $(5 + 8i)(5 - 8i) = 25 - 64i^2$

$= 25 + 64$

$= 89$

65. $-2i(4 - 5i) = -8i + 10i^2 = -10 - 8i$

67. $(3 + 4i)^2 = 9 + 24i + 16i^2$

$= 9 + 24i - 16$

$= -7 + 24i$

69. $(3 + 2i)^2 + (3 - 2i)^2 = 9 + 12i + 4i^2 + 9 - 12i + 4i^2$
$= 18 + 8i^2 = 18 - 8 = 10$

71. $\dfrac{8 - i}{2 + i} = \dfrac{8 - i}{2 + i} \cdot \dfrac{2 - i}{2 - i}$
$= \dfrac{16 - 10i + i^2}{4 + 1} = \dfrac{15 - 10i}{5} = 3 - 2i$

73. $\dfrac{4 - 3i}{i} = \dfrac{4 - 3i}{i} \cdot \dfrac{-i}{-i} = \dfrac{-4i + 3i^2}{1} = -3 - 4i$

75. $2x^2 - x + 3 = 0$
$a = 2, b = -1, c = 3$
$x = \dfrac{-(-1) \pm \sqrt{(-1)^2 - 4(2)(3)}}{2(2)}$
$= \dfrac{1 \pm \sqrt{1 - 24}}{4}$
$= \dfrac{1 \pm \sqrt{-23}}{4}$
$= \dfrac{1 \pm \sqrt{23}i}{4}$
$= \dfrac{1}{4} \pm \dfrac{\sqrt{23}}{4}i$

77. $4x^2 + 11x + 3 = 0$
$a = 4, b = 11, c = 3$
$x = \dfrac{-11 \pm \sqrt{(11)^2 - 4(4)(3)}}{2(4)} = \dfrac{-11 \pm \sqrt{73}}{8}$

79.

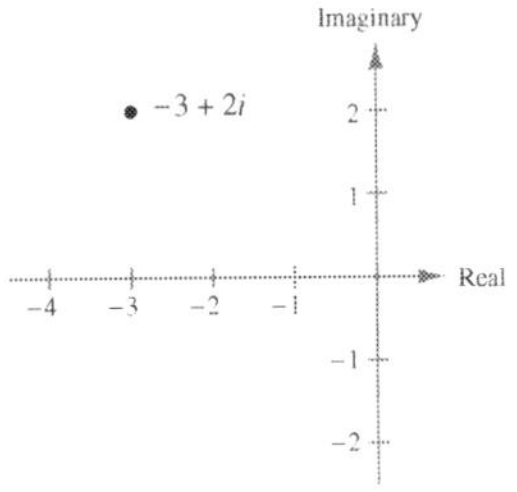

81. $f(x) = x^4 - 81$
$= (x^2 - 9)(x^2 + 9)$
$= (x - 3)(x + 3)(x - 3i)(x + 3i)$
$\pm 3, \pm 3i$

83. $f(t) = t^3 + 5t^2 + 3t + 15$
$= t^2(t + 5) + 3(t + 5)$
$= (t + 5)(t^2 + 3)$
$= (t + 5)(t + \sqrt{3}i)(t - \sqrt{3}i)$
$-5, \pm\sqrt{3}i$

85. $g(x) = 4x^3 - 8x^2 + 9x - 18$
$= 4x^2(x - 2) + 9(x - 2)$
$= (x - 2)(4x^2 + 9)$
$= (x - 2)(2x + 3i)(2x - 3i)$
$2, \pm\dfrac{3i}{2}$

87. Answers will vary. Sample answer: $x^3 - x^2 + 9x - 9$

89. (a) $f(x) = x^4 + 5x^2 - 24 = (x^2 - 3)(x^2 + 8)$
(b) $f(x) = (x - \sqrt{3})(x + \sqrt{3})(x^2 + 8)$
(c) $f(x) = (x - \sqrt{3})(x + \sqrt{3})(x - 2\sqrt{2}i)(x + 2\sqrt{2}i)$

91.

$-4i$	4	-1	64	-16
		$-16i$	$-64 + 4i$	16
$4i$	4	$-1 - 16i$	$4i$	0
		$16i$	$-4i$	
	4	-1	0	

$f(x) = 4x^3 - x^2 + 64x - 16$
$= (x + 4i)(x - 4i)(4x - 1)$
Zeros: $\pm 4i, \frac{1}{4}$

93.

$-1 + 3i$	1	7	24	58	40
		$-1 + 3i$	$-15 + 15i$	$-54 + 12i$	-40
$-1 - 3i$	1	$6 + 3i$	$9 + 15i$	$4 + 12i$	0
		$-1 - 3i$	$-5 - 15i$	$-4 - 12i$	
	1	5	4	0	

$f(x) = x^4 + 7x^3 + 24x^2 + 58x + 40$
$= [x - (-1 + 3i)][x - (-1 - 3i)](x^2 + 5x + 4)$
$= (x + 1 - 3i)(x + 1 + 3i)(x + 1)(x + 4)$
Zeros: $-1 \pm 3i, -1, -4$

95. Domain: All real numbers except $x = -2$

Vertical asymptote: $x = -2$

Horizontal asymptote: $y = 0$

97. Domain: All real numbers except $x = -3$ and $x = 3$

Vertical asymptotes: $x = 3, x = -3$

Horizontal asymptote: $y = 2$

99. $P(x) = \dfrac{3 - x}{x + 2}$

x-intercept: $(3, 0)$

y-intercept: $\left(0, \dfrac{3}{2}\right)$

Vertical asymptote:

$x = -2$

Horizontal asymptote:

$y = -1$

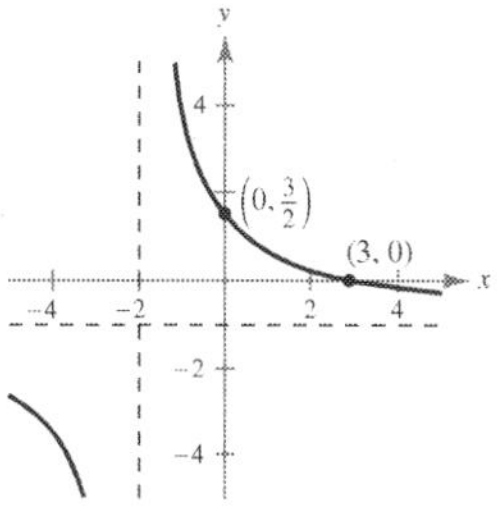

101. $g(x) = \dfrac{1}{x^2 - 4} + 2 = \dfrac{2x^2 - 7}{x^2 - 4}$

x-intercepts: $\left(\pm\dfrac{\sqrt{14}}{2}, 0\right)$

y-intercept: $\left(0, \dfrac{7}{4}\right)$

Vertical asymptotes:

$x = \pm 2$

Horizontal asymptote: $y = 2$

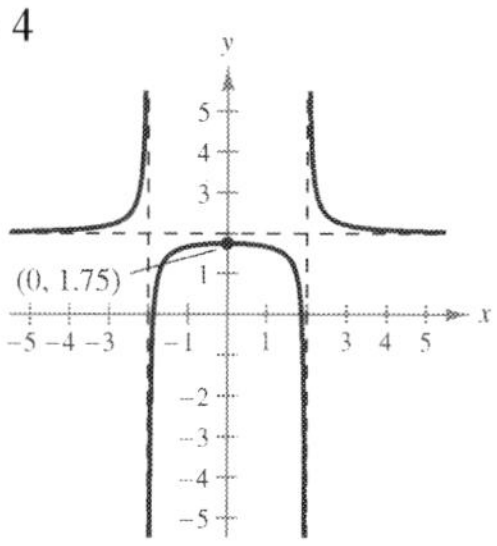

103. None

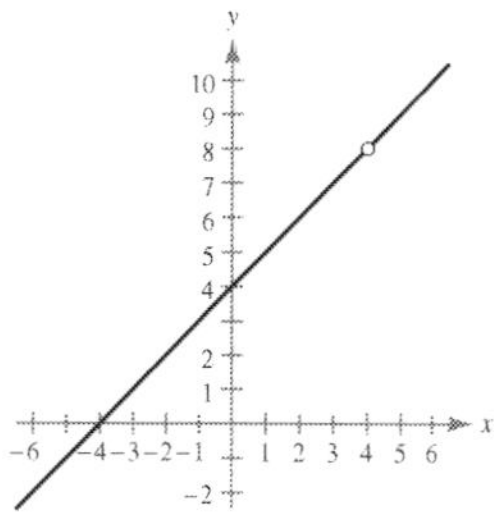

105. (a) $N(5) = \dfrac{20(3 + 5(5))}{1 + 0.06(5)} = 430{,}769$ fish

$N(10) = \dfrac{20(3 + 5(10))}{1 + 0.06(10)} = 662{,}500$ fish

$N(25) = \dfrac{20(3 + 5(25))}{1 + 0.06(25)} = 1{,}024{,}000$ fish

(b) $\dfrac{100}{0.06} = 1{,}666{,}667$

107. (a) $C = \dfrac{105{,}000(25)}{100 - 25} = \$35{,}000$

(b) $C = \dfrac{105{,}000(60)}{100 - 60} = \$157{,}500$

Chapter 3 Test Yourself

1.

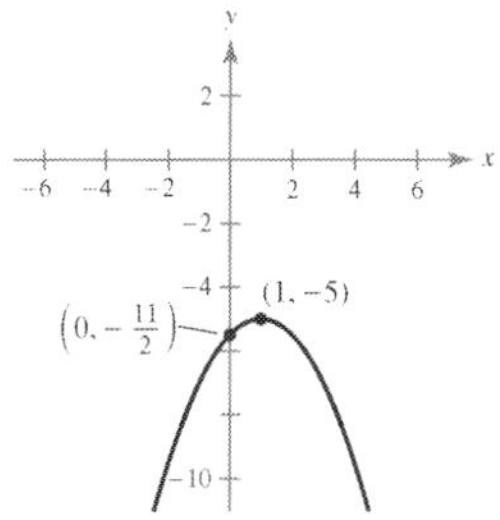

Vertex: $(1, -5)$

Intercept: $\left(0, -\frac{11}{2}\right)$

2. (a) $f(x) = 12x^3 - 5x^2 - 49x + 15$

The degree is odd and the leading coefficient is positive. The graph falls to the left and rises to the right.

(b) $f(x) = 5x^4 - 3x^3 + 2x^2 + 11x + 12$

The degree is even and the leading coefficient is positive. The graph rises to the left and rises to the right.

3.

$$\begin{array}{r}
x^2 + 7x + 12 \\
x^2 - 3x - 10 \overline{\big)\, x^4 + 4x^3 - 19x^2 - 106x - 120} \\
-(x^4 - 3x^3 - 10x^2) \\
\hline
7x^3 - 9x^2 - 106x \\
-(7x^3 - 21x^2 - 70x) \\
\hline
12x^2 - 36x - 120 \\
-(12x^2 - 36x - 120) \\
\hline
0
\end{array}$$

$$\frac{x^4 + 4x^3 - 19x^2 - 106x - 120}{x^2 - 3x - 10} = x^2 + 7x + 12$$

$$= (x + 3)(x + 4)$$

4. $\dfrac{\pm 1, \pm 3, \pm 9, \pm 27}{\pm 1, \pm 2, \pm 4} = \pm 1, \pm 3, \pm 9, \pm 27, \pm\dfrac{1}{2},$

$\pm\dfrac{3}{2}, \pm\dfrac{9}{2}, \pm\dfrac{27}{2}, \pm\dfrac{3}{4}, \pm\dfrac{9}{4}, \pm\dfrac{1}{4}, \pm\dfrac{27}{4}$

$$\begin{array}{r|rrrrr} -\frac{3}{2} & 4 & -16 & 3 & 36 & -27 \\ & & -6 & 33 & -54 & 27 \\ \hline & 4 & -22 & 36 & -18 & 0 \end{array} \checkmark$$

$$\begin{array}{r|rrrr} \frac{3}{2} & 4 & -22 & 36 & -18 \\ & & 6 & -24 & 18 \\ \hline & 4 & -16 & 12 & 0 \end{array} \checkmark$$

$4x^2 - 16x + 12 = 0$

$4(x^2 - 4x + 3) = 0$

$4(x - 1)(x - 3) = 0$

$f(x) = (x - 1)(x - 3)\left(x + \frac{3}{2}\right)\left(x - \frac{3}{2}\right)$

5. (a)

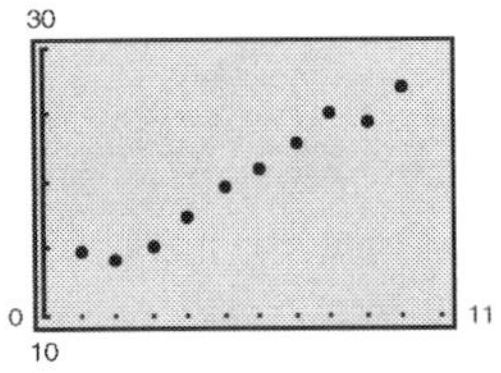

(b) Linear: $S = 1.517t + 11.81$

Quadratic: $S = 0.0240t^2 + 1.253t + 12.34$

Quartic: $S = 0.010739t^4 - 0.26789t^3 + 2.2749t^2 - 5.879t + 18.74$

(c) Linear

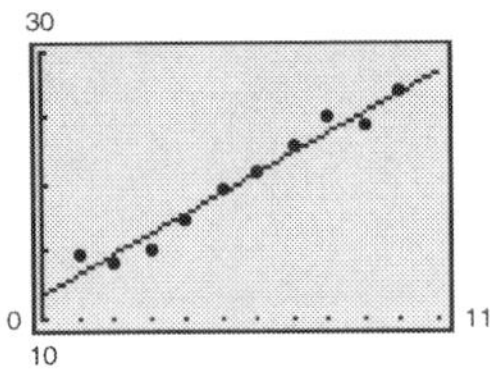

Quadratic

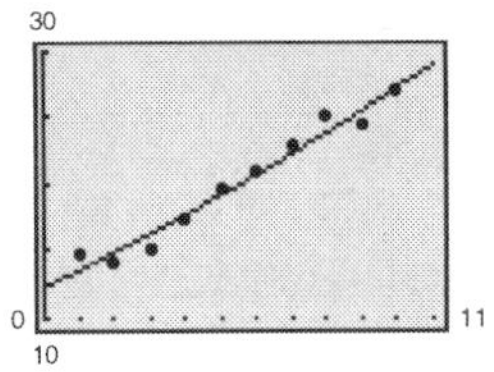

Quartic

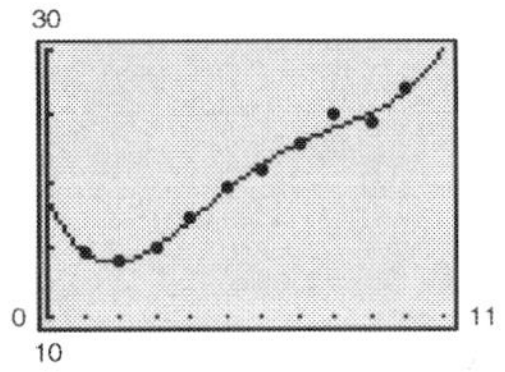

Each model could be considered a good fit for the data.

(d) Linear: 2015 $(t \approx 15.29)$

Quadratic: 2014 $(t \approx 14.21)$

Quartic: 2011 $(t \approx 11.92)$

Answers will vary. Sample answer: All three models give predictions that reflect the increasing data. The quartic model increases most rapidly, while the linear and quadratic models increase at slower rates.

6. $(12 + 3i) + (4 - 5i) = 12 + 3i + 4 - 5i$
$= (12 + 4) + (3 - 5)i$
$= 16 - 2i$

7. $(10 - 2i) - (3 + 8i) = 10 - 2i - 3 - 8i$
$= (10 - 3) + (-2 - 8)i$
$= 7 - 10i$

8. $\left(5 + \sqrt{-12}\right)\left(3 - \sqrt{-12}\right) = \left(5 + 2i\sqrt{3}\right)\left(3 - 2i\sqrt{3}\right)$

$$= 15 - 10i\sqrt{3} + 6i\sqrt{3} - 4i^2(3)$$
$$= 15 - 4i\sqrt{3} + 12$$
$$= 27 - 4\sqrt{3}i$$

9. $(4 + 3i)(2 - 5i) = 8 - 20i + 6i - 15i^2$

$$= 8 - 14i + 15$$
$$= 23 - 14i$$

10. $\dfrac{1+i}{1-i} = \dfrac{1+i}{1-i} \cdot \dfrac{1+i}{1+i} = \dfrac{1 + i + i + i^2}{1+1} = \dfrac{2i}{2} = i$

11. $x^2 + 5x + 7 = 0$

$$x = \frac{-5 \pm \sqrt{(5)^2 - 4(1)(7)}}{2(1)}$$
$$= \frac{-5 \pm \sqrt{-3}}{2}$$
$$= \frac{-5 \pm \sqrt{-3}i}{2}$$
$$= -\frac{5}{2} \pm \frac{\sqrt{3}}{2}i$$

12. $2x^2 - 5x + 11 = 0$

$$x = \frac{-(-5) \pm \sqrt{(-5)^2 - 4(2)(11)}}{2(2)}$$
$$= \frac{5 \pm \sqrt{-63}}{4}$$
$$= \frac{5 \pm 3\sqrt{7}i}{4}$$
$$= \frac{5}{4} \pm \frac{3\sqrt{7}}{4}i$$

13. $P(x) = (x - 2)(x - 5)(x - 3i)(x + 3i)$

$$= (x - 2)(x - 5)(x^2 + 9)$$
$$= (x^2 - 7x + 10)(x^2 + 9)$$
$$= x^4 - 7x^3 + 19x^2 - 63x + 90$$

14. $f(x) = x^3 + 2x^2 + 5x + 10$

Because $\sqrt{5}i$ is a zero, so is $-\sqrt{5}i$.

$$\begin{array}{r|rrrr} \sqrt{5}i & 1 & 2 & 5 & 10 \\ & & \sqrt{5}i & -5 + 2\sqrt{5}i & -10 \\ \hline -\sqrt{5}i & 1 & 2 + \sqrt{5}i & 2\sqrt{5}i & 0 \\ & & -\sqrt{5}i & -2\sqrt{5}i & \\ \hline & 1 & 2 & 0 & \end{array}$$

$f(x) = \left(x - \sqrt{5}i\right)\left(x + \sqrt{5}i\right)(x + 2)$

Zeros: $\pm\sqrt{5}i, -2$

15.

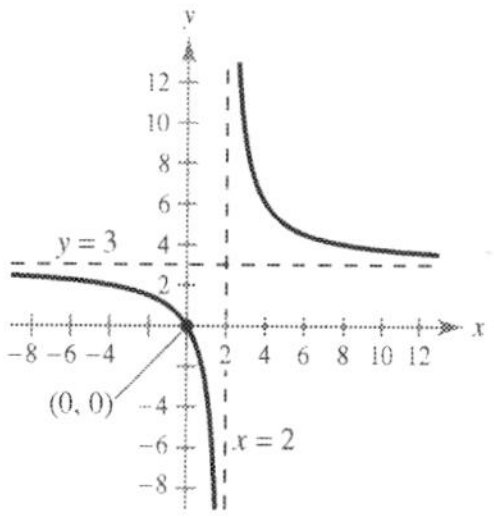

Domain: all real numbers except $x = 2$

CHAPTER 4
Exponential and Logarithmic Functions

CHAPTER 4
Exponential and Logarithmic Functions

Section 4.1 Inverse Functions

Skills Warm Up

1. All real numbers x

2. $x + 1 \ge 0$

$x \ge -1 \Rightarrow [-1, \infty)$

3. $x^2 - 2x \ne 0$

$x(x-2) \ne 0 \Rightarrow x \ne 0, x \ne 2$

All real numbers except $x = 0$ and $x = 2$

4. $3x + 5 \ne 0$

$x \ne -\frac{5}{3}$

All real numbers except $x = -\frac{5}{3}$

5. $2\left(\dfrac{x+5}{2}\right) - 5 = x + 5 - 5 = x$

6. $7 - 10\left(\dfrac{7-x}{10}\right) = 7 - 7 + x = x$

7. $\sqrt[3]{2\left(\dfrac{x^3}{2} - 2\right) + 4} = \sqrt[3]{x^3 - 4 + 4} = \sqrt[3]{x^3} = x$

8. $\sqrt[5]{(x+2)^5} - 2 = x + 2 - 2 = x$

9.
$$y = \frac{2x-6}{3}$$
$$3y = 2x - 6$$
$$3y + 6 = 2x$$
$$\frac{3}{2}y + 3 = x$$

10.
$$y = \sqrt[3]{2x-4}$$
$$y^3 = 2x - 4$$
$$y^3 + 4 = 2x$$
$$\frac{y^3+4}{2} = x$$
$$x = \frac{1}{2}y^3 + 2$$

---

1. $f^{-1} = \{(4, 1), (5, 2), (6, 3), (7, 4)\}$

3. $f^{-1} = \{(1, -1), (2, -2), (3, -3), (4, -4)\}$

5. $f(x) = 2x;\ f^{-1}(x) = \dfrac{x}{2}$

$$f(f^{-1}(x)) = f\left(\frac{x}{2}\right) = 2\left(\frac{x}{2}\right) = x$$
$$f^{-1}(f(x)) = f^{-1}(2x) = \frac{2x}{2} = x$$

7. $f(x) = x - 5;\ f^{-1}(x) = x + 5$

$$f(f^{-1}(x)) = f(x+5) = (x+5) - 5 = x$$
$$f^{-1}(f(x)) = f^{-1}(x-5) = (x-5) + 5 = x$$

9. (a) $f(g(x)) = f\left(\dfrac{x-1}{5}\right) = 5\left(\dfrac{x-1}{5}\right) + 1 = x$

$$g(f(x)) = g(5x+1) = \frac{(5x+1)-1}{5} = x$$

(b)

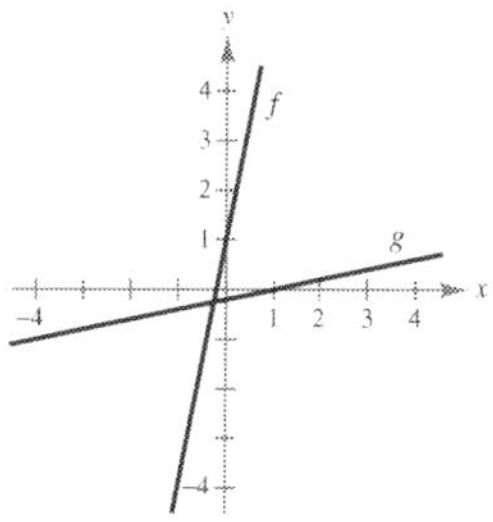

11. (a) $f(g(x)) = f(\sqrt[3]{x}) = (\sqrt[3]{x})^3 = x$

$$g(f(x)) = g(x^3) = \sqrt[3]{x^3} = x$$

(b)

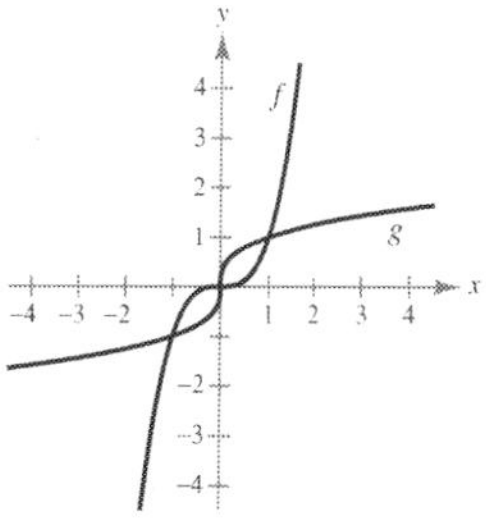

13. (a) $f(g(x)) = f(x^2 + 4),\ x \ge 0$

$= \sqrt{(x^2 + 4) - 4} = x$

$g(f(x)) = g(\sqrt{x - 4})$

$= (\sqrt{x - 4})^2 + 4 = x$

(b)

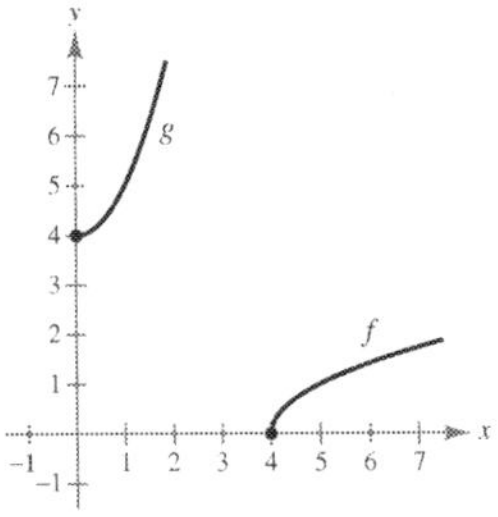

15. (a) $f(g(x)) = f(\sqrt[3]{1 - x}) = 1 - (\sqrt[3]{1 - x})^3 = x$

$g(f(x)) = g(1 - x^3) = \sqrt[3]{1 - (1 - x^3)} = x$

(b)

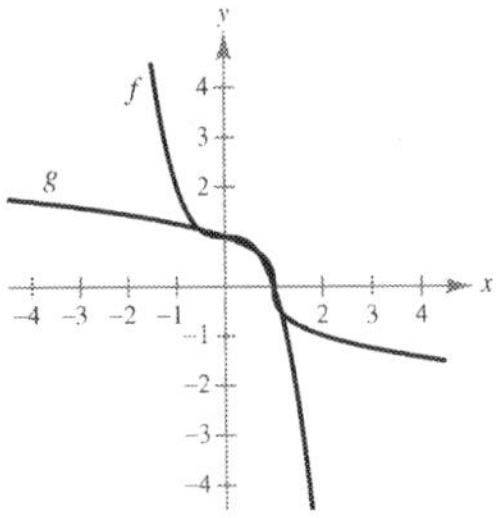

17.

x	0	1	2	3	4
$f^{-1}(x)$	-2	0	1	2	4

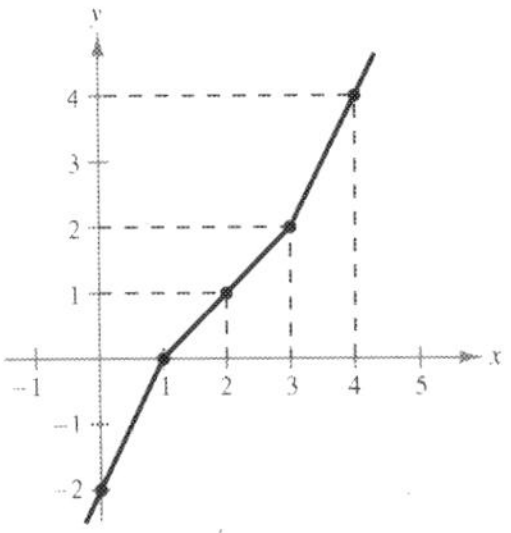

19. $f(x) = x^4$

$y = x^4$

$x = y^4$

$y = \pm\sqrt[4]{x}$

This does not represent y as a function of x. f does not have an inverse.

21. $g(x) = \dfrac{x}{8}$

$y = \dfrac{x}{8}$

$x = \dfrac{y}{8}$

$y = 8x$

This is a function of x, so g has an inverse.

$g^{-1}(x) = 8x$

23. $p(x) = -4$

$y = -4$

Since $y = -4$ for all x, the graph is a horizontal line and fails the Horizontal Line Test. p does not have an inverse.

25. $f(x) = (x + 3)^2,\ x \ge -3 \Rightarrow y \ge 0$

$y = (x + 3)^2,\ x \ge -3,\ y \ge 0$

$x = (y + 3)^2,\ y \ge -3,\ x \ge 0$

$\sqrt{x} = y + 3,\ y \ge -3,\ x \ge 0$

$y = \sqrt{x} - 3,\ x \ge 0,\ y \ge -3$

This is a function of x, so f has an inverse.

$f^{-1}(x) = \sqrt{x} - 3,\ x \ge 0$

27. $h(x) = \dfrac{1}{x}$

$y = \dfrac{1}{x}$

$xy = 1$

$y = \dfrac{1}{x}$

This is a function of x, so h has an inverse.

$h^{-1}(x) = \dfrac{1}{x}$

29. $f(x) = \sqrt{2x + 3} \Rightarrow x \ge -\dfrac{3}{2},\ y \ge 0$

$y = \sqrt{2x + 3},\ x \ge -\dfrac{3}{2},\ y \ge 0$

$x = \sqrt{2y + 3},\ y \ge -\dfrac{3}{2},\ x \ge 0$

$x^2 = 2y + 3,\ x \ge 0,\ y \ge -\dfrac{3}{2}$

$y = \dfrac{x^2 - 3}{2},\ x \ge 0,\ y \ge -\dfrac{3}{2}$

This is a function of x, so f has an inverse.

$f^{-1}(x) = \dfrac{x^2 - 3}{2},\ x \ge 0$

31. $g(x) = x^2 - x^4$

The graph fails the Horizontal Line Test, so g does not have an inverse.

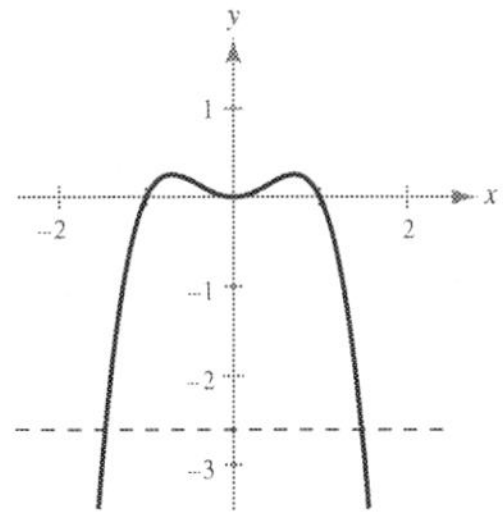

33. $f(x) = 25 - x^2, x \le 0 \Rightarrow y \le 25$

$$y = 25 - x^2, x \le 0, y \le 25$$
$$x = 25 - y^2, y \le 0, x \le 25$$
$$y^2 = 25 - x, x \le 25, y \le 0$$
$$y = -\sqrt{25 - x}, x \le 25, y \le 0$$

This is a function of x, so f has an inverse.

$f^{-1}(x) = -\sqrt{25 - x}, x \le 25$

35. $f(x) = \frac{x}{3}, f^{-1}(x) = 3x$

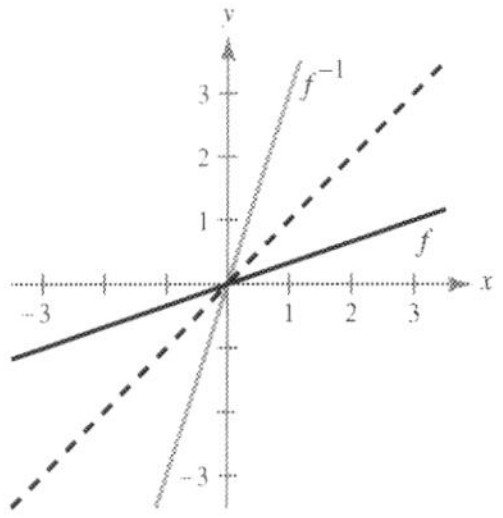

x	-3	0	3
$f(x)$	-1	0	1

x	-1	0	1
$f^{-1}(x)$	-3	0	3

37. $f(x) = 7x + 1, f^{-1}(x) = \frac{x - 1}{7}$

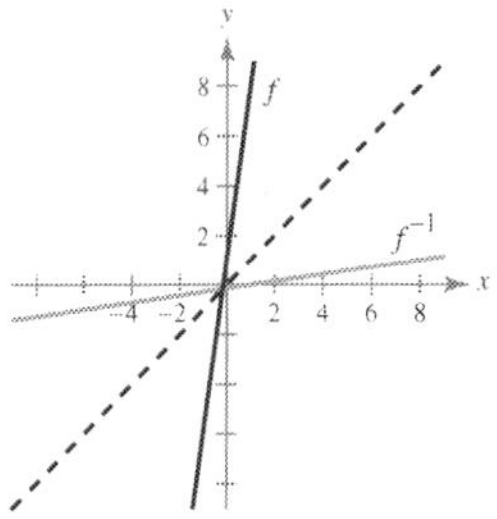

x	-1	0	1
$f(x)$	-6	1	8

x	-6	1	8
$f^{-1}(x)$	-1	0	1

39. $f(x) = x^5 - 2, f^{-1}(x) = \sqrt[5]{x + 2}$

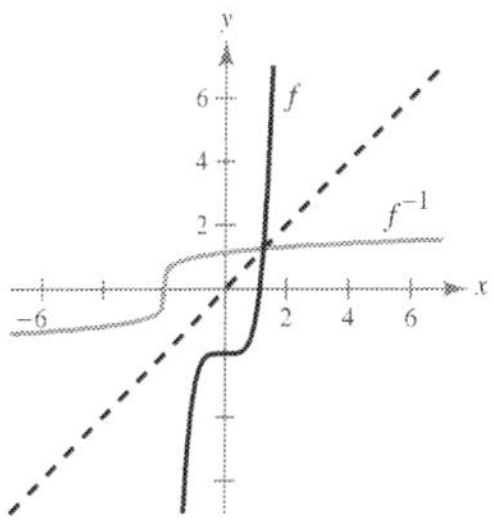

x	-1	0	1
$f(x)$	-3	-2	-1

x	-3	-2	-1
$f^{-1}(x)$	-1	0	1

41. The error is that

$$f^{-1}(x) \ne \frac{1}{\sqrt{2x - 5}}.$$

The notation $f^{-1}(x)$ does not mean take the reciprocal of $f(x)$. $f^{-1}(x)$ is read "f inverse of x." The student could use the method of finding an inverse of a function on page 337.

43.

$$f(x) = 2x - 3$$
$$y = 2x - 3$$
$$x = 2y - 3$$
$$y = \frac{x + 3}{2}$$
$$f^{-1}(x) = \frac{x + 3}{2}$$

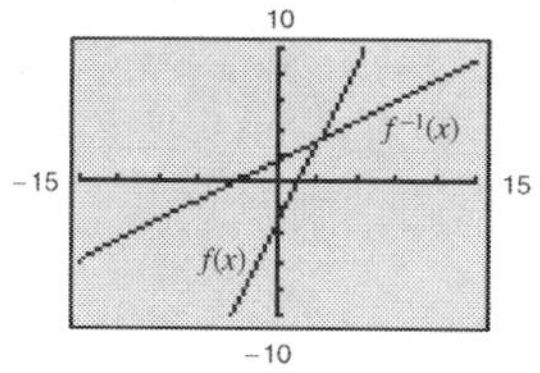

45.

$$f(x) = x^5$$
$$y = x^5$$
$$x = y^5$$
$$y = \sqrt[5]{x}$$
$$f^{-1}(x) = \sqrt[5]{x}$$

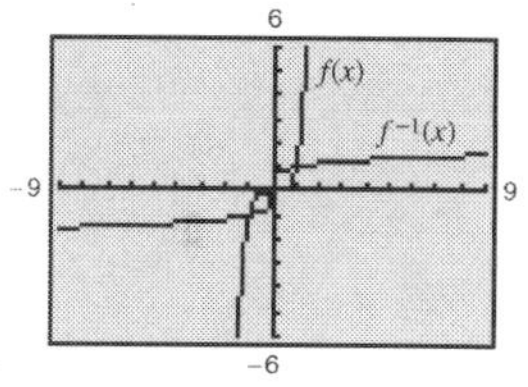

47.

$$f(x) = \sqrt{x}$$
$$y = \sqrt{x}$$
$$x = \sqrt{y}$$
$$y = x^2$$
$$f^{-1}(x) = x^2, x \ge 0$$

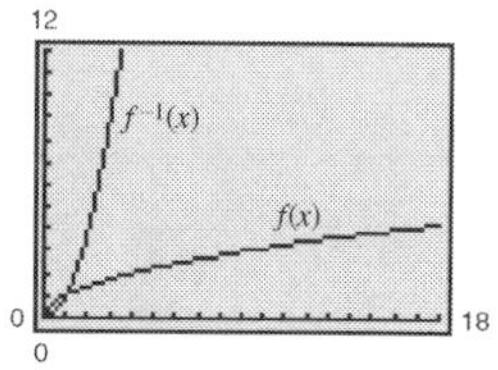

49.

$$y = \sqrt{16 - x^2}, 0 \le x \le 4$$
$$x = \sqrt{16 - y^2}$$
$$x^2 = 16 - y^2$$
$$x^2 - 16 = -y^2$$
$$16 - x^2 = y^2$$
$$\sqrt{16 - x^2} = y, 0 \le x \le 4$$
$$f^{-1}(x) = \sqrt{16 - x^2}, 0 \le x \le 4$$

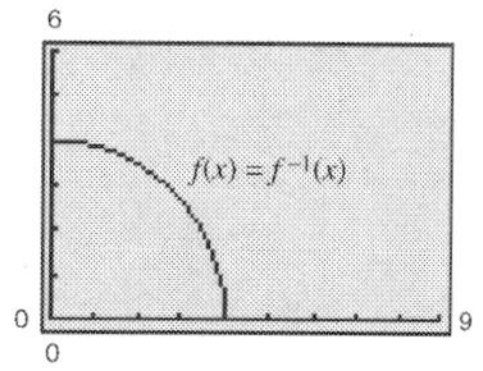

51.
$$y = \sqrt[3]{x+2}$$
$$x = \sqrt[3]{y+2}$$
$$x^3 = y + 2$$
$$y = x^3 - 2$$
$$f^{-1}(x) = x^3 - 2$$

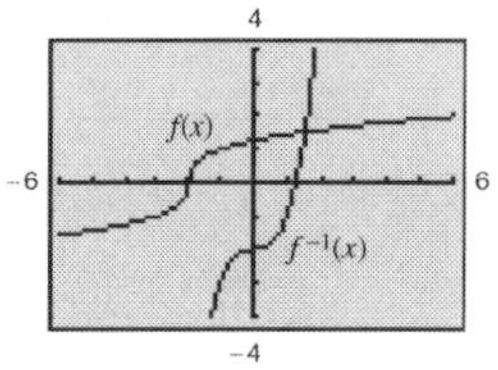

53. Since no horizontal line crosses the graph of f at more than one point, f **has** an inverse.

55. Since some horizontal lines cross the graph of f twice, f does **not** have an inverse.

57. g **has** an inverse function.

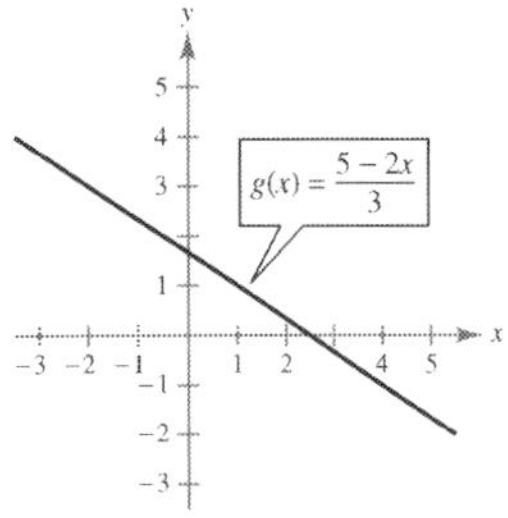

59. h does **not** have an inverse function.

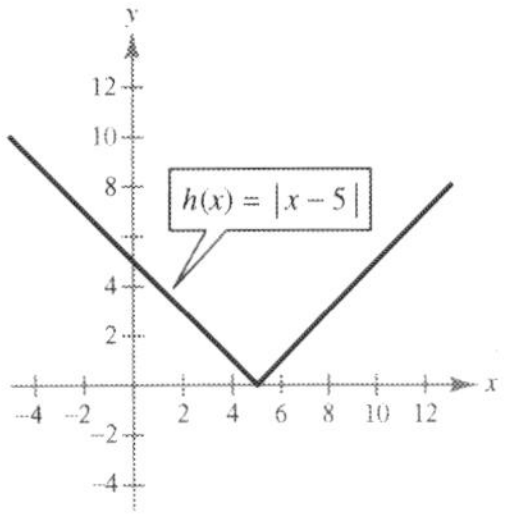

61. f does **not** have an inverse function.

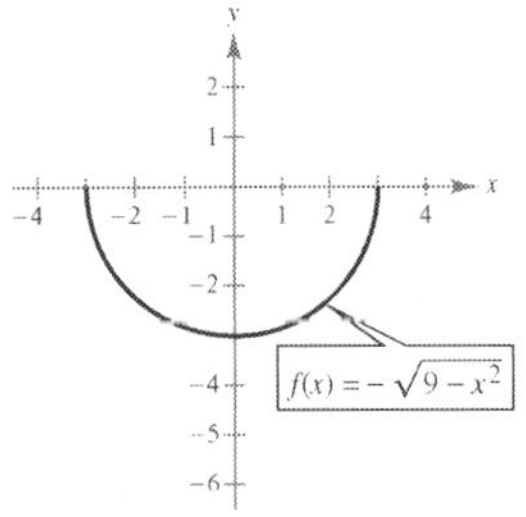

For Exercises 63 and 65:

$$f(x) = x + 4 \qquad g(x) = 2x - 6$$
$$f^{-1}(x) = x - 4 \qquad g^{-1}(x) = \frac{x+6}{2}$$

63.
$$g^{-1} \circ f^{-1} = g^{-1}\left(f^{-1}\right) = g^{-1}(x-4) = \frac{x-4+6}{2} = \frac{x+2}{2}$$

65. $f \circ g = f(2x - 6) = 2x - 6 + 4 = 2x - 2$

Find $(f \circ g)^{-1}$:
$$y = 2x - 2$$
$$x = 2y - 2$$
$$x + 2 = 2y$$
$$\frac{x+2}{2} = y$$

So, $(f \circ g)^{-1} = \frac{x+2}{2}$. **Note:** $(f \circ g)^{-1} = g^{-1} \circ f^{-1}$

For Exercises 67 and 69:

$$f(x) = \tfrac{1}{8}x - 3 \qquad g(x) = x^2,\ x \ge 0$$
$$f^{-1}(x) = 8x + 24 \qquad g^{-1}(x) = \sqrt{x},\ x \ge 0$$

67. $\left(f^{-1} \circ g^{-1}\right)(1) = f^{-1}\left(g^{-1}(1)\right) = f^{-1}(1) = 32$

69.
$$\left(f^{-1} \circ f^{-1}\right)(6) = f^{-1}\left(f^{-1}(6)\right) = f^{-1}(72) = 8(72) + 24 = 600$$

71.
$$y = 47x - 5736$$
$$x = 47y - 5736$$
$$x + 5736 = 47y$$
$$\frac{x + 5736}{47} = y$$
$$P^{-1}(x) = \frac{x + 5736}{47}$$

$P^{-1}(x)$ represents the number of units that must be sold to obtain a profit of x.

Domain of $P(x)$: $[0, \infty)$

Domain of $P^{-1}(x)$: $[-5736, \infty)$

73. (a)

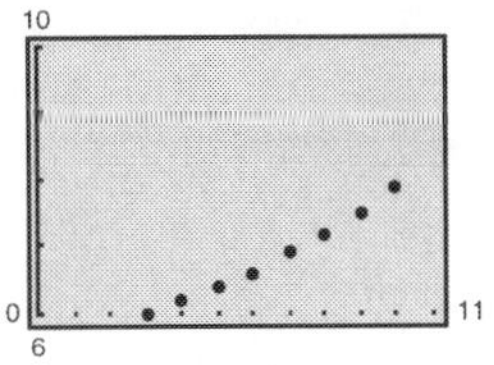

(b) $y = 0.263t + 5.12$

(c)

$$y = 0.263t + 5.12$$

$$t = 0.263y + 5.12 \quad \text{Interchange } y \text{ and } t.$$

$$t - 5.12 = 0.263y \quad \text{Solve for } y.$$

$$\frac{t - 5.12}{0.263} = y$$

So, $y^{-1} = \dfrac{t - 5.12}{0.263}$.

y^{-1} represents the year in which the average price is t dollars.

(d) $y^{-1} = \dfrac{9 - 5.12}{0.263} \approx 14.75$

In 2014, the average admission price to a movie theater will reach $9.

75.

$$f(x) = \sqrt{0.0209x^3 - 0.021}$$

$$y = \sqrt{0.0209x^3 - 0.021}$$

$$x = \sqrt{0.0209y^3 - 0.021}$$

$$x^2 = 0.0209y^3 - 0.021$$

$$\frac{x^2 + 0.021}{0.0209} = y^3$$

$$y = \sqrt[3]{\frac{x^2 + 0.021}{0.0209}}$$

So, $f^{-1}(x) = \sqrt[3]{\dfrac{x^2 + 0.021}{0.0209}}$

$$f^{-1}(0.93) = \sqrt[3]{\frac{(0.93)^2 + 0.021}{0.0209}} \approx 3.49.$$

In 2008, Wal-Mart had earnings per share of $3.49.

Section 4.2 Exponential Functions

Skills Warm Up

1. $5^{2x}(5^{-x}) = 5^x$
2. $3^{-x}(3^{3x}) = 3^{2x}$
3. $\dfrac{4^{5x}}{4^{2x}} = 4^{3x}$
4. $\dfrac{10^{2x}}{10^x} = 10^x$
5. $(4^x)^2 = 4^{2x}$
6. $(4^{2x})^5 = 4^{10x}$
7. $\left(\dfrac{2^x}{3^x}\right)^{-1} = \dfrac{3^x}{2^x} = \left(\dfrac{3}{2}\right)^x$
8. $(4^{6x})^{1/2} = 4^{3x}$
9. $(2^{3x})^{-1/3} = 2^{-x} = \dfrac{1}{2^x}$
10. $\left(\dfrac{3^{4x}}{5^{4x}}\right)^{1/4} = \dfrac{3^x}{5^x} = \left(\dfrac{3}{5}\right)^x$
11. $(16^x)^{1/4} = 2^x$
12. $(27^x)^{1/3} = 3^x$

1. $(2.6)^{1.3} \approx 3.463$

3. $100(1.03)^{-9/5} \approx 94.818$

5. $6^{-\sqrt{2}} \approx 0.079$

7. $f(x) = 2^x$ rises to the right.

Asymptote: $y = 0$

Intercept: $(0, 1)$

Graph (g)

9. $f(x) = -2^x$ falls to the right.

Asymptote: $y = 0$

Intercept: $(0, -1)$

Graph (b)

11. $f(x) = 2^x + 3$ rises to the right.

Asymptote: $y = 3$

Intercept: $(0, 4)$

Graph (d)

13. $f(x) = 2^{x+1}$ rises to the right.

Asymptote: $y = 0$

Intercept: $(0, 2)$

Graph (f)

15. $g(x) = 4^x$

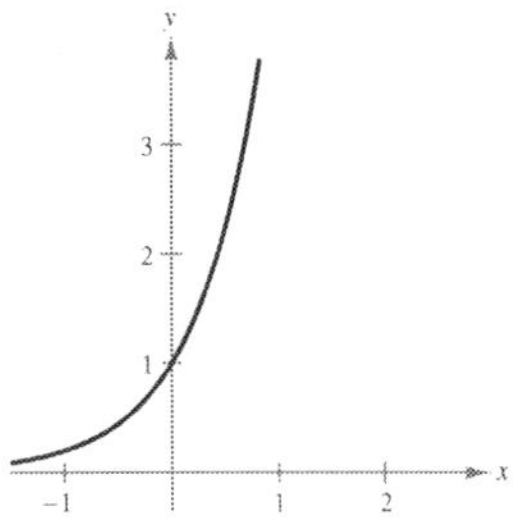

17. $f(x) = 4^{-x}$

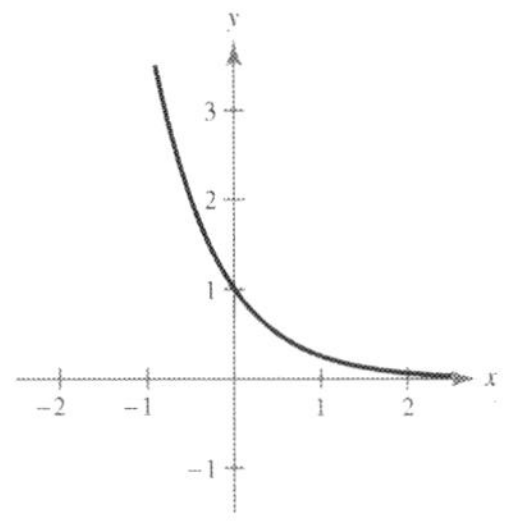

19. $h(x) = 4^{x-3}$

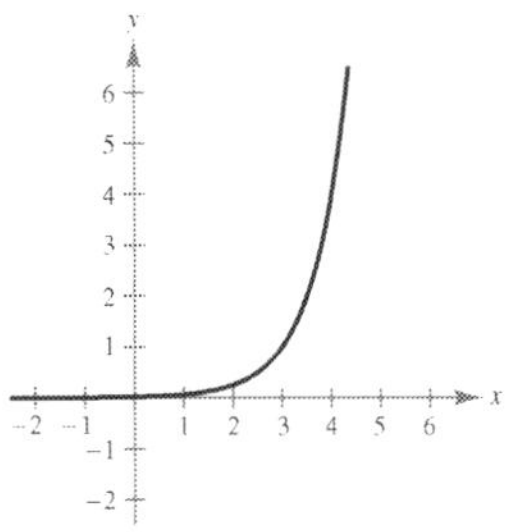

21. $g(x) = 4^{-x} - 2$

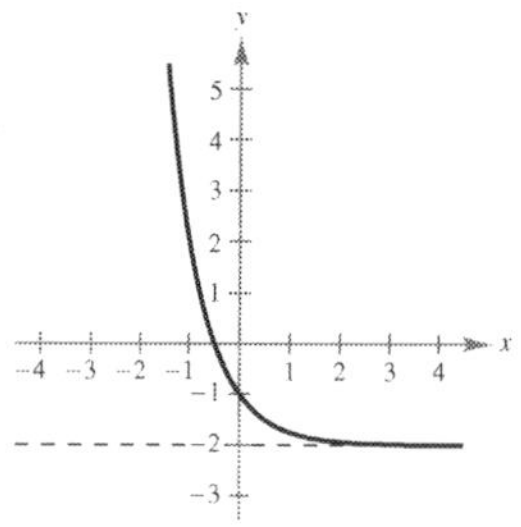

23. $h(x) = -4x$

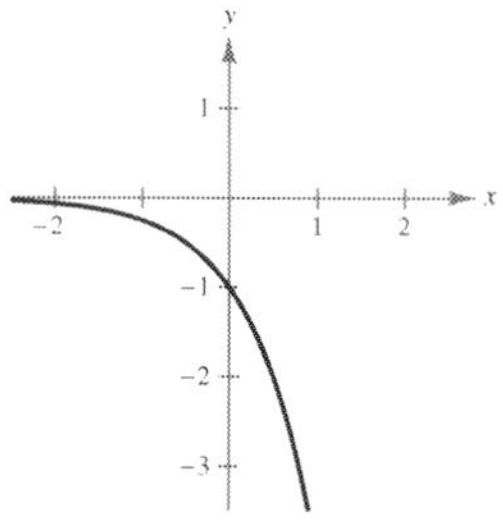

25. $y = 2^{-x^2}$

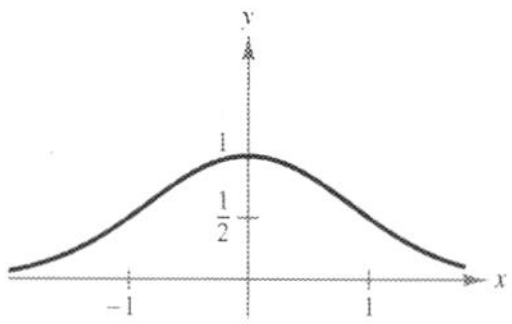

27. $e^4 \approx 54.598$

29. $e^{-2.7} \approx 0.067$

31. $y = e^{-0.1x}$

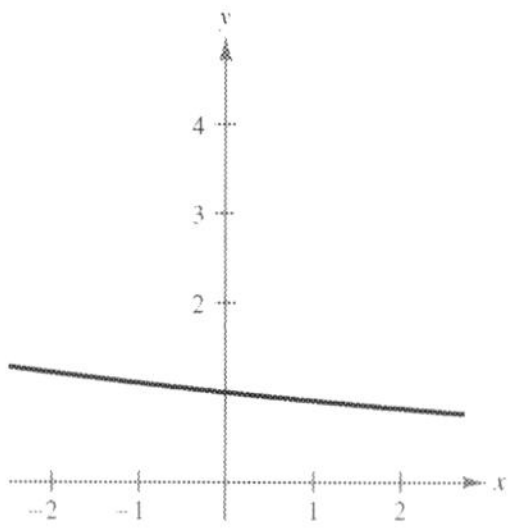

33. $f(x) = 2e^{0.12x}$

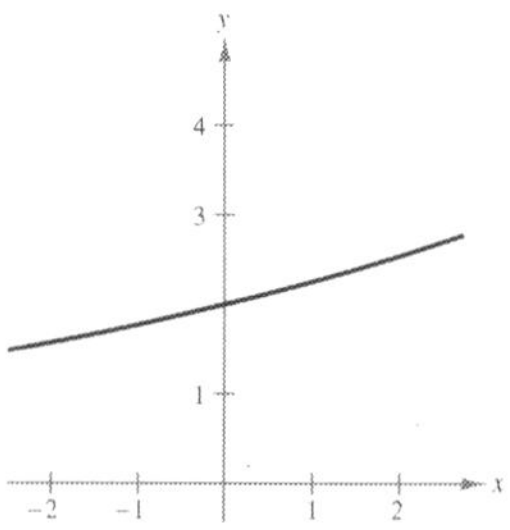

35. $f(x) = e^{x+3}$

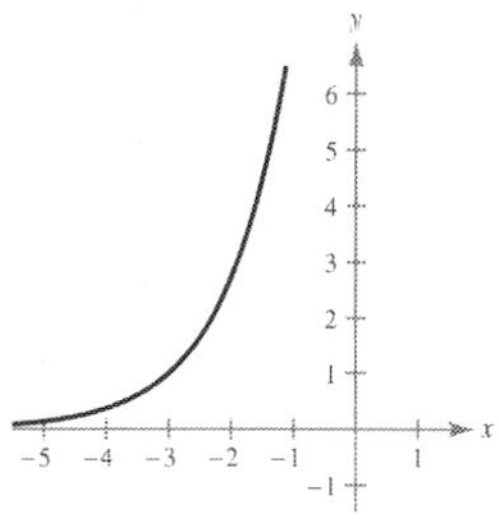

37. $g(x) = 1 + e^{-x}$

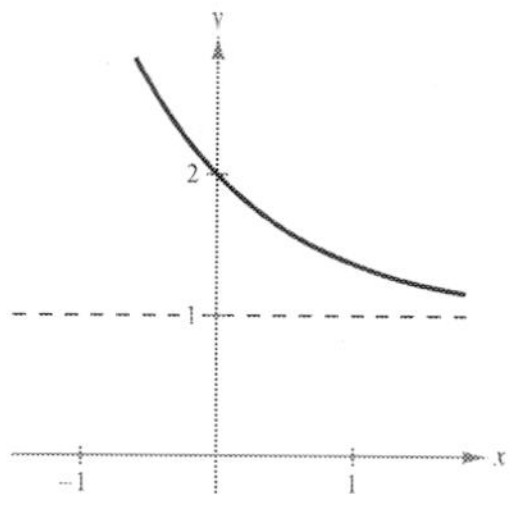

39. Yes. For the natural exponential function, $f(x) = e^x$, and the base is $a = e$.

41. $P = \$5000, r = 8\%, t = 5$ years

Compounded n times per year: $A = 5000\left(1 + \frac{0.08}{n}\right)^{5n}$

Compounded continuously: $A = 5000e^{0.08(5)}$

n	1	2	4	12	365	Continuous compounding
A	\$7346.64	\$7401.22	\$7429.74	\$7449.23	\$7458.80	\$7459.12

43. $P = \$2500, r = 5\%, t = 20$ years

Compounded n times per year: $A = 2500\left(1 + \frac{0.05}{n}\right)^{20n}$

Compounded continuously: $A = 2500e^{(0.05)(20)}$

n	1	2	4	12	365	Continuous compounding
A	\$6633.24	\$6712.66	\$6753.71	\$6781.60	\$6795.24	\$6795.70

45. Let $P = \$1.00$.

Amount in an account at 5% interest compounded quarterly for one year,

$A = 1\left(1 + \frac{0.05}{4}\right)^{4(1)} \approx 1.050945337$ or \$1.05.

Amount in an account at 3% interest compounded continuously for one year,

$A = e^{.03(1)} \approx 1.030454534$ or \$1.03.

The account with 5% interest compounded quarterly earns more money. While the account that compounds interest quarterly calculates the interest less frequently than the continuous interest account, the higher interest rate of 5% overcomes the fewer number of compoundings.

47. $P = 10{,}000\left(1 + \frac{0.06}{4}\right)^{-4(5)} \approx \$7424.70$

49. $P = 20{,}000\left(1 + \frac{0.08}{4}\right)^{-6(4)} \approx \$12{,}434.43$

51. $P = 100{,}000\left(1 + \frac{0.10}{12}\right)^{-12t}$

t	1	10	20	30	40	50
P	\$90,521.24	\$36,940.70	\$13,646.15	\$5040.98	\$1862.17	\$687.90

53. $P = 100{,}000e^{-0.09t}$

t	1	10	20	30	40	50
P	\$91,393.12	\$40,656.97	\$16,529.89	\$6720.55	\$2732.37	\$1110.90

55. $P = \dfrac{100{,}000}{e^{(0.065 \cdot 25)}} \approx \$19{,}691.17$

57. $P = \dfrac{750{,}000}{e^{0.065(25)}} \approx \$147{,}683.76$

59. $P = \dfrac{100{,}000}{e^{0.075(21)}} \approx \$20{,}700.76$

61. $P = \dfrac{750{,}000}{e^{0.075(21)}} \approx \$155{,}255.66$

63. (a) $p = 3000\left(1 - \dfrac{5}{5 + e^{-0.015(75)}}\right) \approx \$182.91$

(b) $p = 3000\left(1 - \dfrac{5}{5 + e^{-0.015(200)}}\right) \approx \$29.58$

(c)

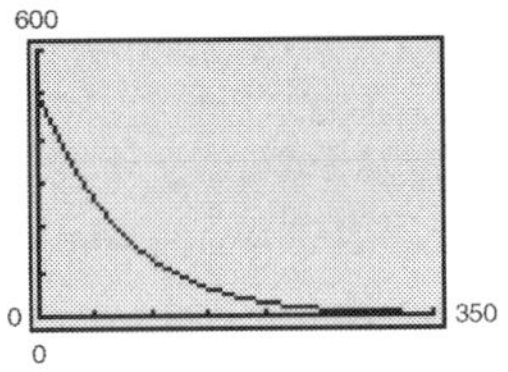

(d) \$117.19

65. (a) $P(0) = 100e^{0.01896(0)} = 100$

(b) $P(5) = 100e^{0.01896(5)} \approx 109.94$

(c) $P(10) = 100e^{0.01896(10)} \approx 120.88$

(d) $P(24) = 100e^{0.01896(24)} \approx 157.62$

67. (a) $P(10) = 4500e^{0.0272(10)} \approx 5907$

(b) $P(12) = 4500e^{0.0272(12)} \approx 6237$

(c) $P(15) = 4500e^{0.0272(15)} \approx 6767$

(d) $P(20) = 4500e^{0.0272(20)} \approx 7753$

69. $S = 100e^{-0.0238t}$

$S = 100e^{-0.0238(50)} \approx 30.42212641$

After 50 years, about 30.4 kilograms of Strontium 90 will remain.

71. (a)

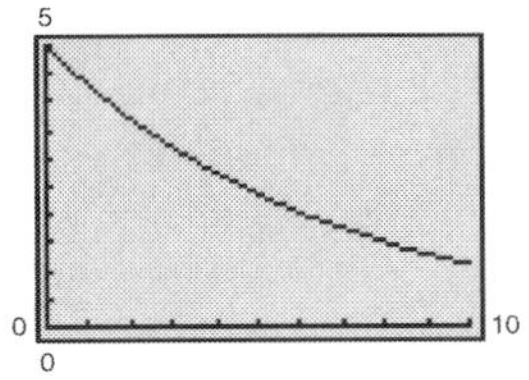

(b) $P = 5e^{-0.1507t}$

$P(10) = 5e^{-0.1507(10)} \approx 1.1079$

After ten months, about 1.11 pounds will remain.

(c)

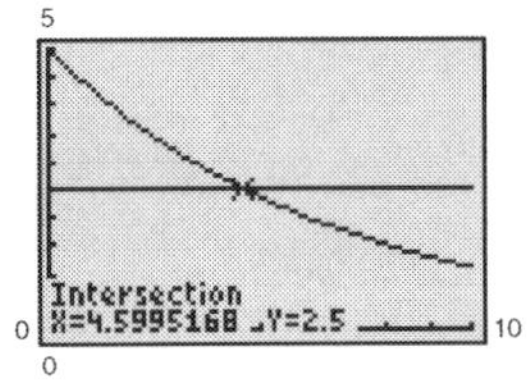

Using an intersect feature or trace feature, the half-life of ^{230}Pu is about 4.6 months.

73. (a) In 2008, there were about 9000 screens.

In 2010, there were about 39,000 screens.

(b) For 2008, $t = 8$: $y = 25.87e^{0.7323(8)} \approx 9059$

For 2010, $t = 10$: $y = 25.87e^{0.7323(10)} \approx 39{,}186$

In 2008 there were 9059 screens and in 2010 there were 39,186 screens.

75. (a) 1980: about 22 years
1990: about 24 years
2000: about 25 years
2010: about 26 years

(b) For 1980, $t = 0$: $A = -45.10 + \dfrac{72.02}{1 + e^{-0.0523(0) - 2.6314}} \approx 22.08$

For 1990, $t = 10$: $A = -45.10 + \dfrac{72.02}{1 + e^{-0.0523(10) - 2.6314}} \approx 23.97$

For 2000, $t = 20$: $A = -45.10 + \dfrac{72.02}{1 + e^{-0.0523(20) - 2.6314}} \approx 25.14$

For 2010, $t = 30$: $A = -45.10 + \dfrac{72.02}{1 + e^{-0.0523(30) - 2.6314}} \approx 25.86$

The median age of an American woman at her first marriage in 1980 was 22 years, in 1990 was 24 years, in 2000 was 25 years, and in 2010 was 26 years.

77. (a)

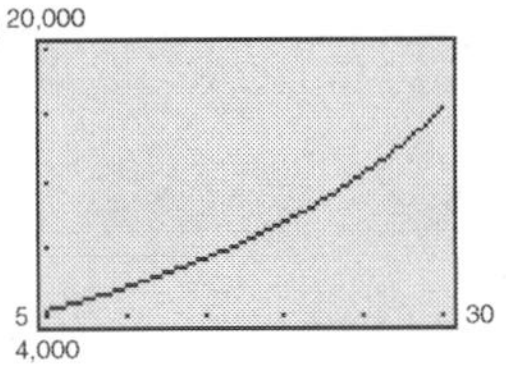

(b) 1990: \$5682.8 billion
2000: \$9624.6 billion
2009: \$15,463.8 billion

79. Answers will vary.

Section 4.3 Logarithmic Functions

Skills Warm Up

1. $2^x = 8$
$2^3 = 8$
$x = 3$

2. $4^x = 1$
$4^0 = 1$
$x = 0$

3. $10^x = 0.1$
$10^{-1} = 0.1$
$x = -1$

4. $e^x = e$
$e^1 = e$
$x = 1$

5. $e^2 \approx 7.389$

6. $e^{-1} \approx 0.368$

7. The graph of g is the graph of f shifted two units to the left.

8. The graph of g is the graph of f reflected about the x-axis.

9. The graph of g is the graph of f shifted downward one unit.

10. The graph of g is the graph of f reflected about the y-axis.

1. c

3. b

5. a

7. $4^4 = 256$
$4 = \log_4 256$

9. $81^{1/4} = 3$
$\frac{1}{4} = \log_{81} 3$

11. $6^{-2} = \frac{1}{36}$

$-2 = \log_6\left(\frac{1}{36}\right)$

13. $\log_4 16 = 2 \Rightarrow 4^2 = 16$

15. $\log_2 \frac{1}{2} = -1 \Rightarrow 2^{-1} = \frac{1}{2}$

17. $\log_5 0.2 = -1 \Rightarrow 5^{-1} = 0.2$

19. $\log_{27} 3 = \frac{1}{3} \Rightarrow 27^{1/3} = 3$

21. $\log_3 9 = \log_3 3^2 = 2$

23. $\log_2 \frac{1}{16} = \log_2 2^{-4} = -4$

25. $\log_8 2 = \log_8 8^{1/3} = \frac{1}{3}$

27. $\log_7 7 = \log_7 7^1 = 1$

29. $\log_{10} 0.0001 = \log_{10} 10^{-4} = -4$

31. $\log_{10} 345 \approx 2.538$

33. $\log_{10}\left(\frac{4}{5}\right) \approx -0.097$

35. $\log_{10} \sqrt{8} \approx 0.452$

37. $\log_5 5 = x$

$x = 1$

39. $\log_2 2^{-1} = x$

$x = -1$

41. $\log_2(x + 1) = \log_2 4$

$x + 1 = 4$

$x = 3$

43. $\log(2x + 1) = \log 15$

$2x + 1 = 15$

$2x = 14$

$x = 7$

45. $f(x) = 7x,\ g(x) = \log_7 x$

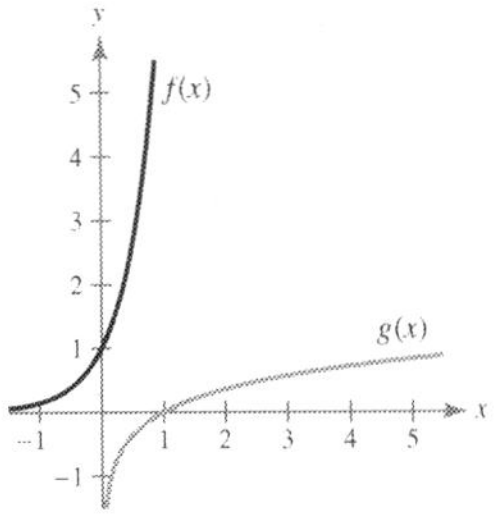

47. $f(x) = 6^x,\ g(x) = \log_6 x$

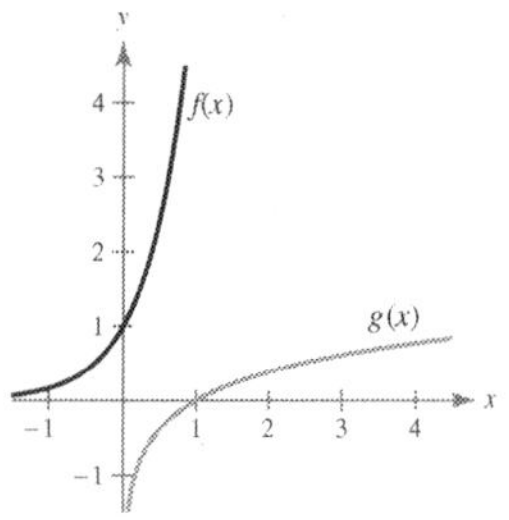

49. $f(x) = \log_2 x$

Domain: $(0, \infty)$

Asymptote: $x = 0$

x-intercept: $(1, 0)$

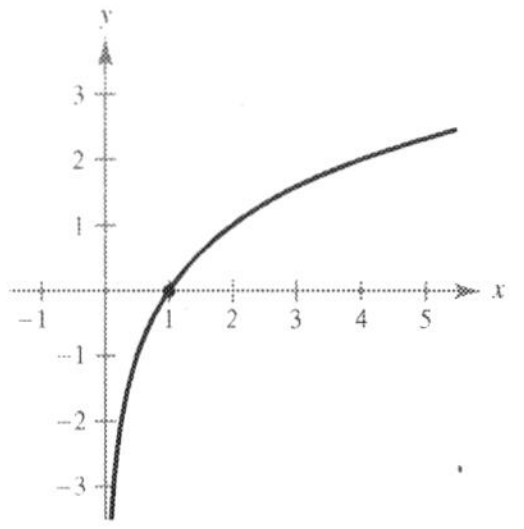

51. $h(x) = \log_2(x + 4)$

Domain: $(-4, \infty)$

Asymptote: $x = -4$

x-intercept: $(-3, 0)$

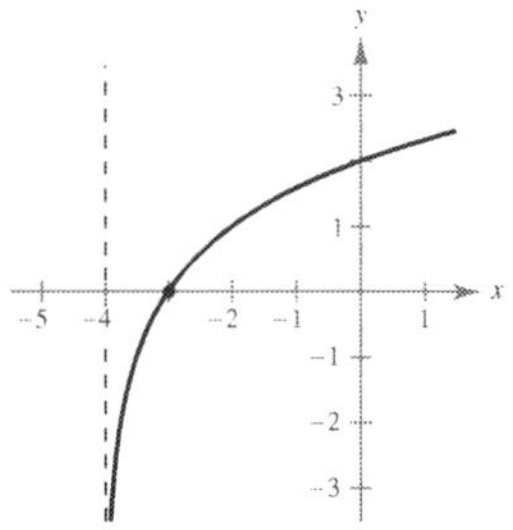

53. $f(x) = -\log_2 x$

Domain: $(0, \infty)$

Asymptote: $x = 0$

x-intercept: $(1, 0)$

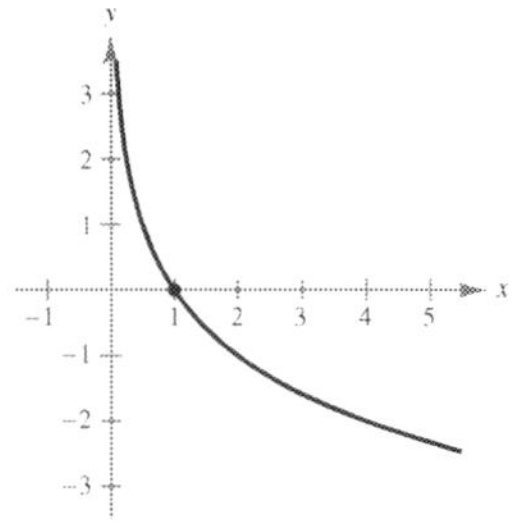

55. $e^1 = e$

$1 = \ln e$

57. $e^x = 4$

$x = \ln 4$

59. $\ln e = 1 \Rightarrow e^1 = e$

61. $\ln 1 = 0 \Rightarrow e^0 = 1$

63. $\ln e = \ln e^1 = 1$

65. $\ln e^{-4} = -4$

67. $\ln 7 \approx 1.946$

69. $\ln 18.42 \approx 2.913$

71. $\ln \sqrt{6} \approx 0.896$

73. $\ln(x + 2) = \ln 6$

$x + 2 = 6$

$x = 4$

75. $\ln(x^2 - 2) = \ln 23$

$x^2 - 2 = 23$

$x^2 = 25$

$x = \pm 5$

77. $f(x) = e^x$, $g(x) = \ln x$

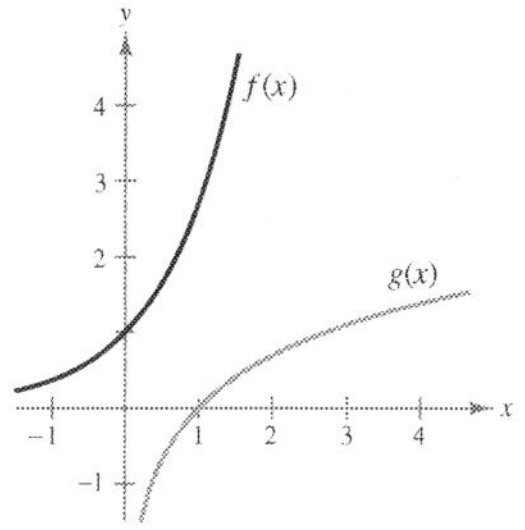

79. $g(x) = \ln(-x)$

Domain: $(-\infty, 0)$

Asymptote: $x = 0$

x-intercept: $(-1, 0)$

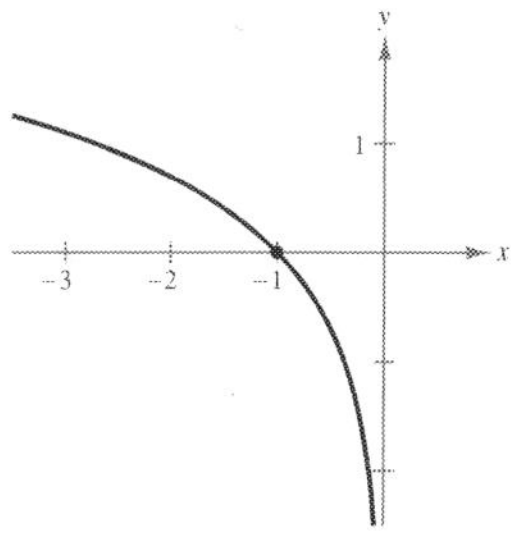

81. $h(x) = \ln(x + 1)$

Domain: $(-1, \infty)$

Asymptote: $x = -1$

x-intercept and y-intercept: $(0, 0)$

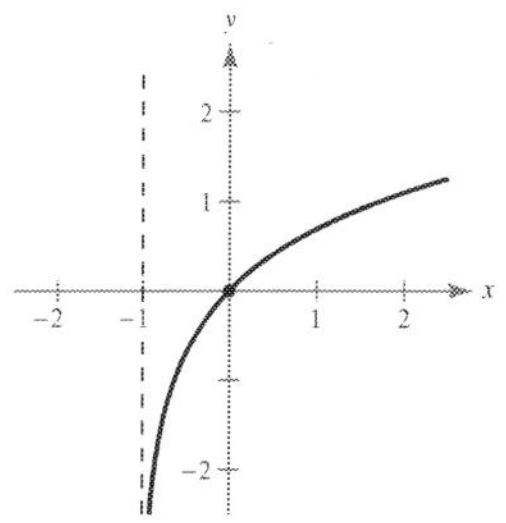

83. $f(x) = \ln x + 2$ rises to the right.

Asymptote: $x = 0$

Point on graph: $(1, 2)$

Graph (d)

85. $f(x) = -3 \ln(x - 2)$ falls to the right.

Asymptote: $x = 2$

Intercept: $(3, 0)$

Graph (a)

87. $f(x) = \log(x + 1)$

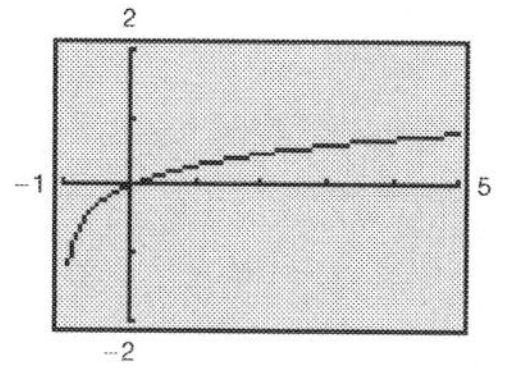

89. $f(x) = \ln(x - 1)$

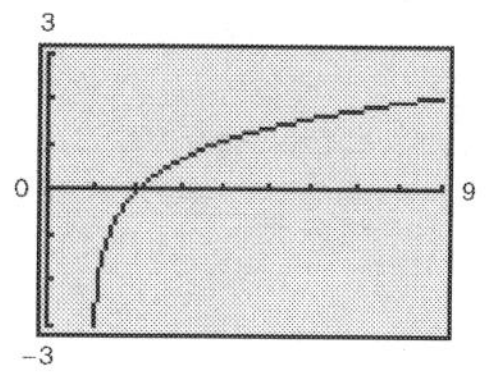

91. $f(x) = \ln x + 1$

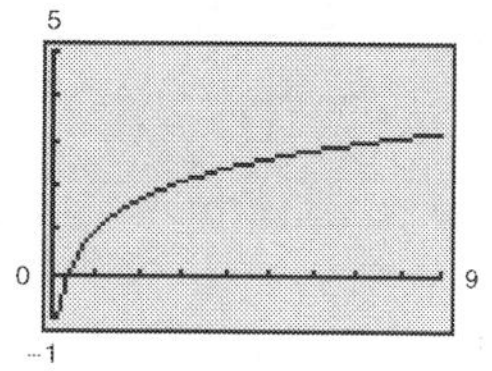

93. $t = \dfrac{8 \ln 3}{\ln 63 - \ln 45} \approx 26.12$ years

95.

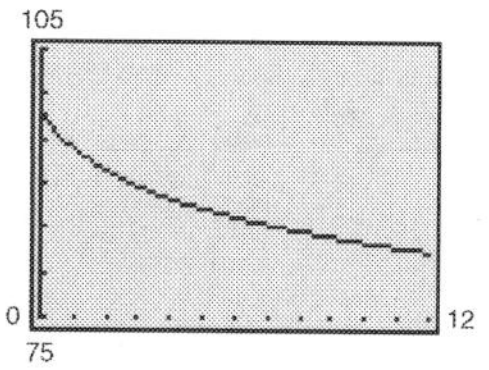

The domain $0 \le t \le 12$ covers the period of 12 months or 1 year. The scores range from the average score on the original exam to the average score after 12 months, so the range is $82.40 \le f(t) \le 98$.

97. $g(t) = 87 - 16\log_{10}(t+1),\ 0 \le t \le 24$

(a) $g(0) = 87 - 16\log_{10}(0+1) = 87 - 16(0) = 87$

(b) $g(6) = 87 - 16\log_{10}(6+1) \approx 87 - 16(0.845) \approx 73.5$

(c) $g(11) = 87 - 16\log_{10}(11+1) \approx 87 - 16(1.079) \approx 69.7$

The average score dropped below 70 after about 11 months.

99. (a) $t(1100.65) = 12.542\ln\left(\dfrac{1100.65}{1100.65 - 1000}\right) \approx 30$ years

$t(1254.68) = 12.542\ln\left(\dfrac{1254.68}{1254.68 - 1000}\right) \approx 20$ years

(b) (payment)(months)(years)

$(1100.65)(12)(30) = \$396{,}234.00$

$(1254.68)(12)(20) = \$301{,}123.20$

(c) Interest = Total payments – 150,000

For \$1100.65 payment:

Interest = \$396,234 – \$150,000 = \$246,234

For \$1254.68 payment:

Interest = \$301,123.20 – \$150,000 = \$151,123.20

(d) The vertical asymptote is $x = 1000$. A payment of \$1000 would not be enough to retire the debt.

101. $t = \dfrac{\ln K}{0.0525}$

(a)

K	1	2	4	6	8	10	12
t	0	13.2	26.4	34.1	39.6	43.9	47.3

(b)

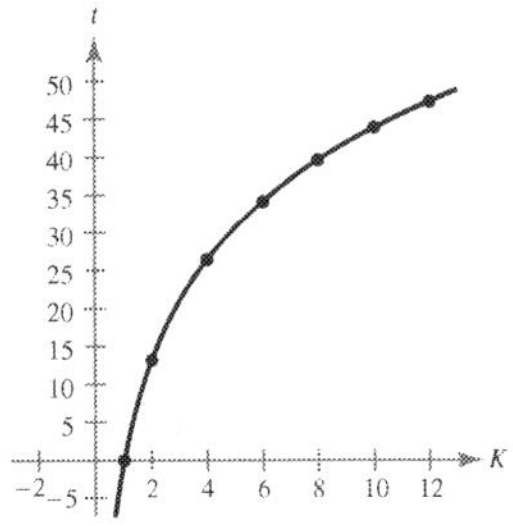

Chapter 4 Quiz Yourself

1. The function $f(x) = 3x^2$ does not have an inverse because it is not one-to-one.

2. The function $f(x) = \sqrt[3]{x+1}$ is one-to-one, and so it has an inverse function.

$y = \sqrt[3]{x+1}$

$x = \sqrt[3]{y+1}$

$x^3 = y + 1$

$x^3 - 1 = y$

So, $f^{-1}(x) = x^3 - 1$.

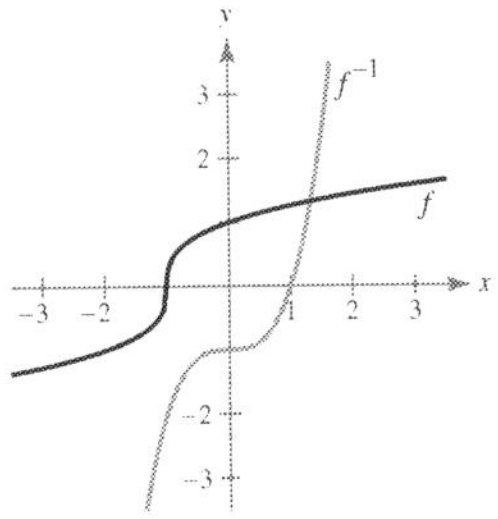

3. $f(x) = \frac{1}{2}x - 3$

(a) $y = \frac{1}{2}x - 3$

$x = \frac{1}{2}y - 3$

$x + 3 = \frac{1}{2}y$

$2(x+3) = y$

So, $f^{-1}(x) = 2(x+3)$.

(b)

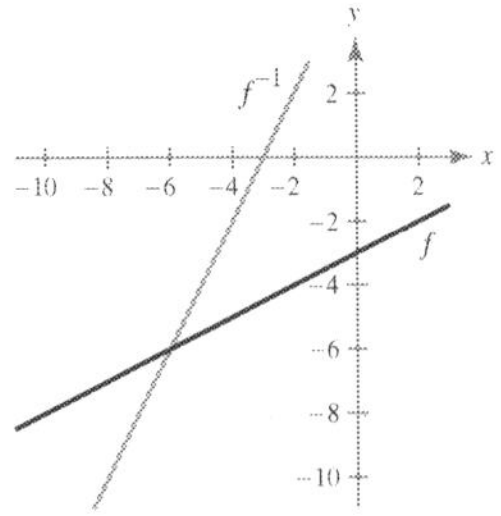

(c) $f^{-1}(f(x)) = 2\left(\frac{1}{2}x - 3 + 3\right) = x$

$f(f^{-1}(x)) = \frac{1}{2}[2(x+3)] - 3 = x + 3 - 3 = x$

4. $f(x) = \sqrt{x + 1}$

(a) $y = \sqrt{x + 1}$

$x = \sqrt{y + 1}$

$x^2 = y + 1$

$x^2 - 1 = y$

So, $f^{-1}(x) = x^2 - 1, x \geq 0.$

(b)

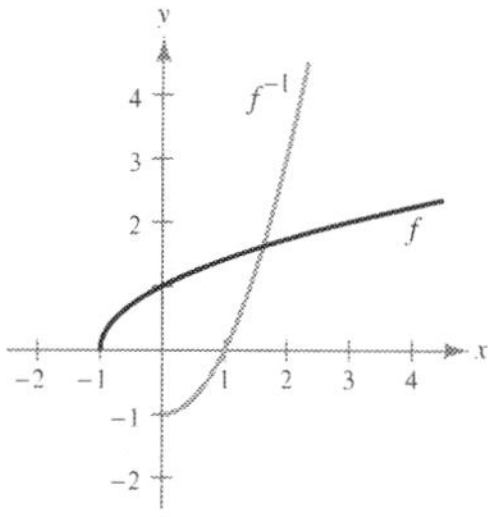

(c) $f^{-1}(f(x)) = \left(\sqrt{x + 1}\right)^2 - 1 = x$

$f(f^{-1}(x)) = \sqrt{x^2 - 1 + 1} = x, x \geq 0.$

5. $g(x) = 3^x - 2$

Asymptote: $y = -2$

Intercepts: $(0, -1), (\log_3 2, 0)$ or $(\approx 0.63, 0)$

x	-2	-1	0	1	2
y	-1.89	-1.67	-1	1	7

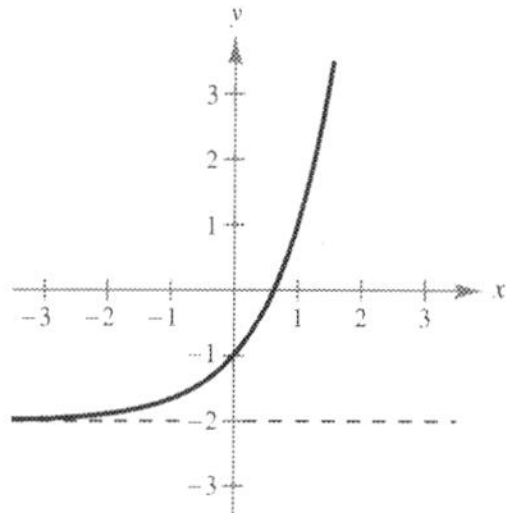

6. $h(x) = 3^{-x}$

Asymptote: $y = 0$

Intercept: $(0, 1)$

x	-2	-1	0	1	2
y	9	3	1	$\frac{1}{3}$	$\frac{1}{9}$

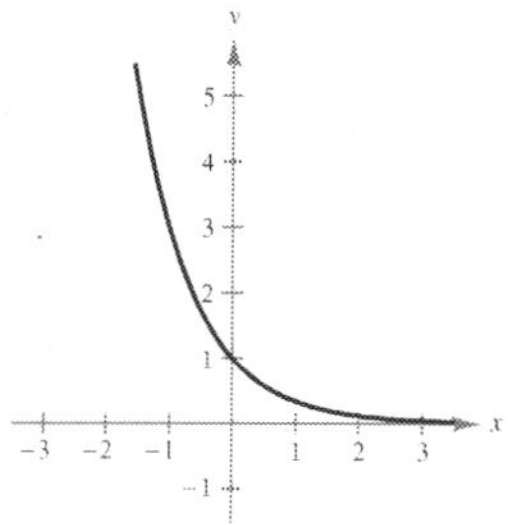

7. $k(x) = \log_3 x \Rightarrow x = 3^y$

Asymptote: $x = 0$

Intercept: $(1, 0)$

x	$\frac{1}{9}$	$\frac{1}{3}$	1	3	9
y	-2	-1	0	1	2

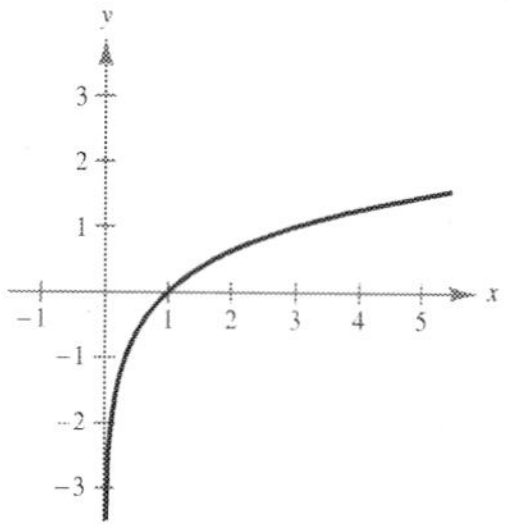

8. $j(x) = \log_3(x - 1) \Rightarrow x - 1 = 3^y$

$x = 3^y + 1$

Asymptote: $x = 1$

Intercept: $(2, 0)$

x	$1\frac{1}{9}$	$1\frac{1}{3}$	2	4	10
y	-2	-1	0	1	2

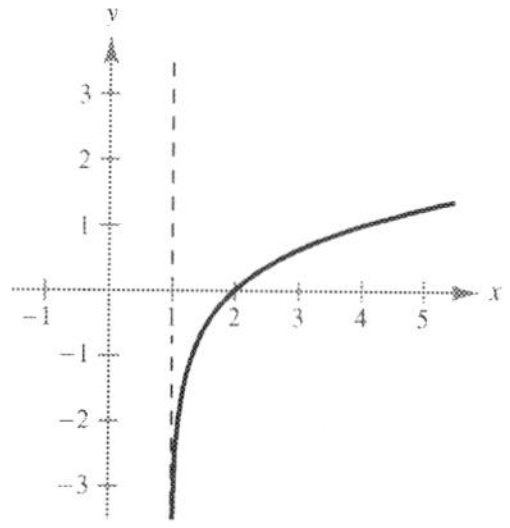

9. (a) $A = 10{,}000\left(1 + \dfrac{0.0515}{12}\right)^{12(8)} \approx \$15{,}085.04$

(b) $A = 10{,}000e^{0.0515(8)} \approx \$15{,}098.34$

10. (a) For 2006, $t = 6$: $C = 7.50(1.04)^6 \approx 9.489$

For 2008, $t = 8$: $C = 7.50(1.04)^8 \approx 10.264$

In 2006, about 9.49 million children participated in the Federal School Breakfast Program and in 2008, about 10.26 million children participated.

(b) For 2014, $t = 14$: $C = 7.5(1.04)^{14} \approx 12.988$

For 2015, $t = 15$: $C = 7.5(1.04)^{15} \approx 13.507$

According to the model, the number of children participating in the Federal School Breakfast Program will be 12.99 million in 2014 and 13.51 million in 2015.

11. (a) $P(0) = 100e^{0.2154(0)} = 100$

(b) $P(6) = 100e^{0.2154(6)} \approx 364.15$

(c) $P(12) = 100e^{0.2154(12)} \approx 1326.06$

12. $P = 4000\left(1 - \dfrac{8}{8 + e^{-0.003(500)}}\right) \approx \$108.54$

13. $\log_{10} 100 = 2$ because $10^2 = 100$.

14. $\ln e^4 = 4$

15. $\log_4 \frac{1}{16} = -2$ because $4^{-2} = \frac{1}{16}$.

16. $\ln 1 = 0$ because $e^0 = 1$.

17. $f(x) = 3^x$, $g(x) = \log_3 x$

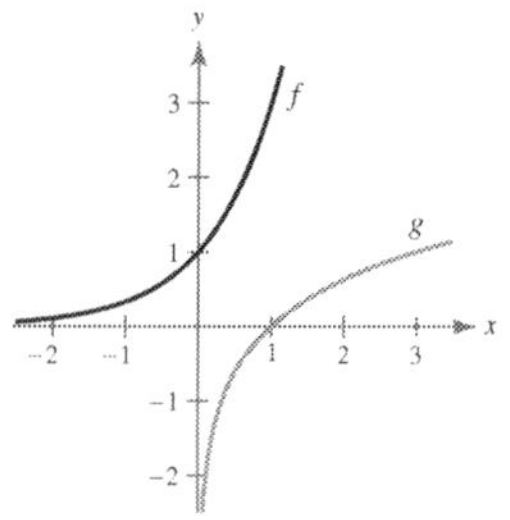

Domain of $f(x)$: $(-\infty, \infty)$

Domain of $g(x)$: $(0, \infty)$

The graphs are reflections of each other about the line $y = x$.

Section 4.4 Properties of Logarithms

Skills Warm Up

1. $\log_7 49 = 2$ because $7^2 = 49$.

2. $\log_2 \frac{1}{32} = -5$ because $2^{-5} = \frac{1}{32}$.

3. $\ln \dfrac{1}{e^2} = \ln e^{-2} = -2$

4. $\log_{10} 0.001 = -3$ because $10^{-3} = 0.001$.

5. $e^2 e^3 = e^{2+3} = e^5$

6. $\dfrac{e^2}{e^3} = e^{2-3} = e^{-1} = \dfrac{1}{e}$

7. $(e^2)^3 = e^{2 \cdot 3} = e^6$

8. $(e^2)^0 = e^{2 \cdot 0} = e^0 = 1$

9. $y = \dfrac{1}{x^2} = x^{-2}$

10. $y = \sqrt{x} = x^{1/2}$

11. $\log_4 64 = 3 \Rightarrow 4^3 = 64$

12. $\log_{16} 4 = \frac{1}{2} \Rightarrow 16^{1/2} = 4$

1. $\log_5 8 = \dfrac{\log_{10} 8}{\log_{10} 5} \approx \dfrac{0.90309}{0.69897} \approx 1.292$

3. $\log_4 30 = \dfrac{\log_{10} 30}{\log_{10} 4} \approx \dfrac{1.47712}{0.60206} \approx 2.453$

5. $\log_5 8 = \dfrac{\ln 8}{\ln 5} \approx \dfrac{2.07944}{1.60944} \approx 1.292$

7. $\log_6 14 = \dfrac{\ln 14}{\ln 6} \approx \dfrac{2.63906}{1.79176} \approx 1.473$

9. $\log_2 6 = \dfrac{\log_{10} 6}{\log_{10} 2} = \dfrac{\ln 6}{\ln 2} \approx 2.585$

11. $\log_{27} 35 = \dfrac{\log_{10} 35}{\log_{10} 27} = \dfrac{\ln 35}{\ln 27} \approx 1.079$

13. $\log_{15} 1250 = \dfrac{\log_{10} 1250}{\log_{10} 15} = \dfrac{\ln 1250}{\ln 15} \approx 2.633$

15. $\log_5 \dfrac{1}{3} = \dfrac{\log_{10} \frac{1}{3}}{\log_{10} 5} = \dfrac{\ln \frac{1}{3}}{\ln 5} \approx -0.683$

17. $\log_{1/4} 10 = \dfrac{\log_{10} 10}{\log_{10} \frac{1}{4}} = \dfrac{\ln 10}{\ln \frac{1}{4}} \approx -1.661$

19. $\ln 10 = \ln(2 \cdot 5) = \ln 2 + \ln 5$

21. $\ln \frac{5}{2} = \ln 5 - \ln 2$

23. $\ln 100 = \ln(4.25)$

$= \ln(2^2 \cdot 5^2)$

$= 2 \ln 2 + 2 \ln 5$

$= 2(\ln 2 + \ln 5)$

25. $-\log_{10} \frac{11}{8} = -\log_{10}\left(\frac{8}{11}\right)^{-1} = -(-1)\log_{10} \frac{8}{11} = \log_{10} \frac{8}{11}$

27. $-2 \ln \frac{e}{7} = -2 \ln\left(\frac{7}{e}\right)^{-1}$

$= -2(-1) \ln \frac{7}{e}$

$= 2 \ln \frac{7}{e}$

$= 2 \ln 7 - 2 \ln e$

$= 2 \ln 7 - 2$

29. $\log_b 10 = \log_b(2 \cdot 5) = \log_b 2 + \log_b 5$

$\approx 0.3562 + 0.8271 \approx 1.1833$

31. $\log_b \frac{2}{3} = \log_b 2 - \log_b 3 \approx 0.3562 - 0.5646$

≈ -0.2084

33. $\log_b 8 = \log_b 2^3 = 3 \log_b 2 \approx 3(0.3562) \approx 1.0686$

35. $\log_b \sqrt{2} = \frac{1}{2} \log_b 2 \approx \frac{1}{2}(0.3562) \approx 0.1781$

37. $\log_b 40 = \log_b(2^3 \cdot 5) = 3 \log_b 2 + \log_b 5$

$\approx 3(0.3562) + 0.8271 \approx 1.8957$

39. $\log_b(2b)^{-2} = -2[\log_b 2 + \log_b b] \approx -2[0.3562 + 1]$

≈ -2.7124

41. $\log_b \sqrt[3]{4b} = \frac{1}{3}(\log_b 4 + \log_b b) = \frac{1}{3}(2 \log_b 2 + 1)$

$\approx \frac{1}{3}(2(0.3562) + 1) \approx 0.5708$

43. $\log_9 \frac{1}{18} = \log_9\left(\frac{1}{2} \cdot \frac{1}{9}\right) = \log_9 \frac{1}{2} + \log_9 \frac{1}{9}$

$= \log_9 \frac{1}{2} - 1 = -\log_9 2 - 1$

45. $\log_7 \sqrt{70} = \frac{1}{2} \log_7(7 \cdot 10) = \frac{1}{2}(\log_7 7 + \log_7 10)$

$= \frac{1}{2} + \frac{1}{2} \log_7 10$

47. $\log_5 \frac{1}{250} = \log_5\left(\frac{1}{2} \cdot \frac{1}{125}\right) = \log_5 \frac{1}{2} + \log_5 \frac{1}{125}$

$= \log_5 2^{-1} + \log_5 5^{-3} = -\log_5 2 - 3$

49. $\ln(5e^6) = \ln 5 + \ln e^6 = \ln 5 + 6$

51. The change-of-base formula was applied incorrectly. The formula is $\log_a x = \dfrac{\ln x}{\ln a}$. So, $\log_2 5 = \dfrac{\ln 5}{\ln 2}$.

53. The Quotient Property was used incorrectly. The Quotient Property is $\ln \dfrac{u}{v} = \ln u - \ln v$.

So, $\ln 8 - \ln 2 = \ln \dfrac{8}{2} = \ln 4$.

55. $\log_2(4^3 \cdot 3^5) = \log_2 4^3 + \log_2 3^5$

$= 3 \log_2 4 + 5 \log_2 3$

$= 6 + 5 \log_2 3$

57. $\log_3 4n = \log_3 4 + \log_3 n$

59. $\log_5 \dfrac{x}{25} = \log_5 x - \log_5 25 = \log_5 x - 2$

61. $\log_2 x^4 = 4 \log_2 x$

63. $\ln \sqrt{z} = \ln z^{1/2} = \frac{1}{2} \ln z$

65. $\ln xyz = \ln x + \ln y + \ln z$

67. $\ln \sqrt{a-1} = \ln(a-1)^{1/2} = \frac{1}{2} \ln(a-1),\ a > 1$

69. $\ln \dfrac{(z-1)^2}{z} = \ln(z-1)^2 - \ln z = 2 \ln(z-1) - \ln z$

71. $\ln \dfrac{yz}{\sqrt[3]{z+3}} = \ln yz - \ln \sqrt[3]{z+3}$

$= \ln y + \ln z - \ln(z+3)^{1/3}$

$= \ln y + \ln z - \dfrac{1}{3} \ln(z+3)$

73. $\ln 5 \sqrt[3]{\dfrac{x}{y}} = \ln 5\left(\dfrac{x}{y}\right)^{1/3}$

$= \ln 5 + \dfrac{1}{3} \ln\left(\dfrac{x}{y}\right)$

$= \ln 5 + \dfrac{1}{3}(\ln x - \ln y)$

75. $\ln \sqrt[4]{x^3(x^2+3)} = \frac{1}{4}(\ln x^3 + \ln(x^2+3))$

$= \frac{1}{4}(3 \ln x + \ln(x^2+3))$

or $\frac{3}{4} \ln x + \frac{1}{4} \ln(x^2+3)$

77. $\log_3 x + \log_3 5 = \log_3 5x$

79. $\log_4 8 - \log_4 x = \log_4 \dfrac{8}{x}$

81. $2 \log_{10}(x+4) = \log_{10}(x+4)^2$

83. $-\ln x - 3 \ln 6 = \ln \dfrac{1}{6^3 x} = \ln \dfrac{1}{216x}$

85. $\dfrac{1}{3} \ln 5x - \ln(x+1) = \ln \sqrt[3]{5x} - \ln(x+1) = \ln\left(\dfrac{\sqrt[3]{5x}}{x+1}\right)$

87. $\log_8(x-2) - \log_8(x+2) = \log_8 \dfrac{x-2}{x+2}$

89. $2 \ln 4 + 2 \ln 3 - \frac{1}{2}\ln(x^2 + 1) = \ln 4^2 + \ln 3^2 - \frac{1}{2}\ln(x^2 + 1)$

$$= \ln 4^2 \cdot 3^2 - \frac{1}{2}\ln(x^2 + 1)$$

$$= \ln 144 - \ln(x^2 + 1)^{1/2}$$

$$= \ln \frac{144}{\sqrt{x^2 + 1}}$$

91. $2 \ln x - [\ln(x + 2) + \ln(x - 2)] = \ln x^2 - \ln(x + 2)(x - 2) = \ln \frac{x^2}{(x + 2)(x - 2)}$

93. 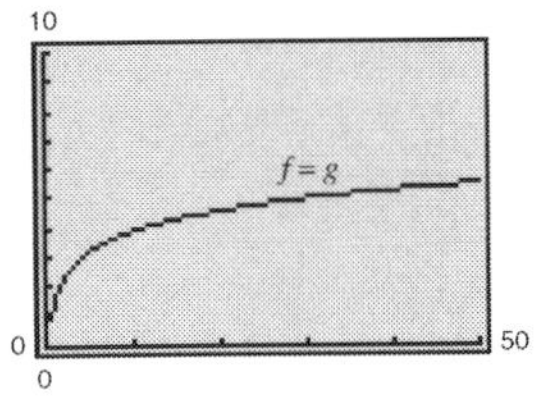

$\ln 5x = \ln 5 + \ln x$

They have the same graph.

Property: $\log_a(uv) = \log_a u + \log_a v$

95. 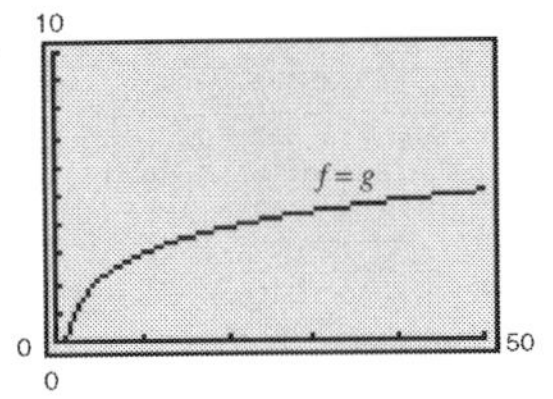

The two graphs are the same. The graphs demonstrate the Power Property for logarithms.

97. $\log_4 \sqrt[3]{4} = \log_4 4^{1/3} = \frac{1}{3}$

99. $\ln \frac{1}{\sqrt{e}} = \ln e^{-1/2} = -\frac{1}{2}$

101. $\log_5 \frac{1}{125} = \log_5 5^{-3} = -3$

103.

x	1	2	3	4	5	6
y	1	1.189	1.316	1.414	1.495	1.565
$\ln x$	0	0.693	1.099	1.386	1.609	1.792
$\ln y$	0	0.173	0.275	0.346	0.402	0.448

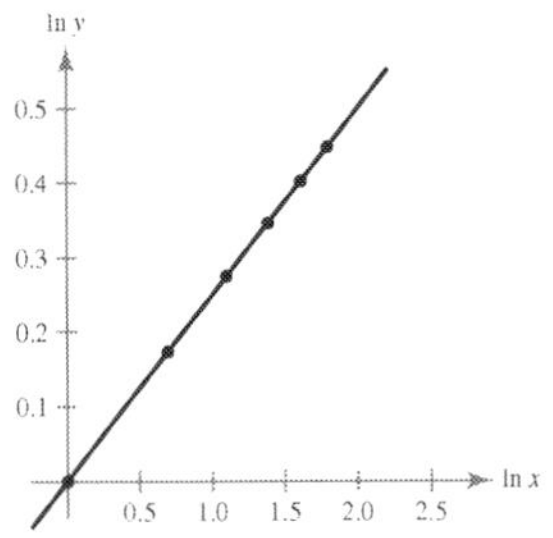

The slope of the line is $\frac{1}{4}$.

So,

$$\ln y = \frac{1}{4}\ln x$$

$$\ln y = \ln \sqrt[4]{x}$$

$$y = \sqrt[4]{x}.$$

105.

Length, x	1	2	3	4	5	6
Diameter, y	0.070	0.111	0.146	0.176	0.204	0.231
$\ln x$	0	0.693	1.099	1.386	1.609	1.792
$\ln y$	−2.659	−2.198	−1.924	−1.737	−1.590	−1.465

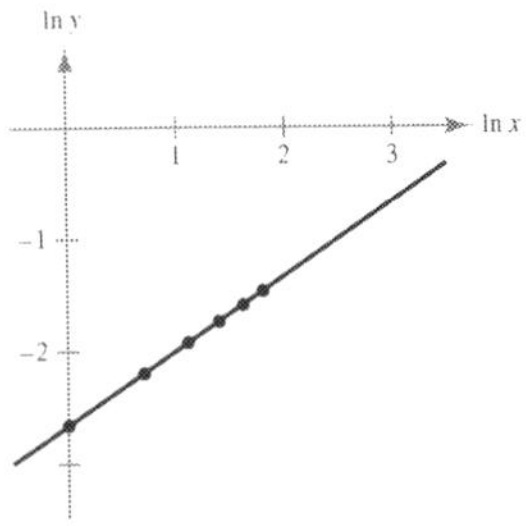

The slope of the line is $\frac{2}{3}$.

So,

$$\ln y = \frac{2}{3}\ln x + \ln 0.07$$

$$\ln y - \ln x^{2/3} = \ln 0.07$$

$$\ln \frac{y}{x^{2/3}} = \ln 0.07$$

$$\frac{y}{x^{2/3}} = 0.07$$

$$y = 0.07x^{2/3}.$$

107. $L = 10 \log_{10} \dfrac{I}{I_0}$

L_1 = Office loudness

L_2 = Broadcast studio loudness

$$L_1 - L_2 = 10 \log_{10} \frac{1.26 \times 10^{-7}}{10^{-12}} - 10 \log_{10} \frac{3.16 \times 10^{-10}}{10^{-12}} \approx 26$$

The office is about 26 decibels louder.

109. (a) $\log_a(uv) = \log_a u + \log_a v$ is related to the exponent property $a^m a^n = a^{m+n}$.

(b) $\log_a \dfrac{u}{v} = \log_a u - \log_a v$ is related to the exponent property $\dfrac{a^m}{a^n} = a^{m-n}$.

(c) $\log_a u^n = n \log_a u$ is related to the exponent property $\left(a^m\right)^n = a^{mn}$.

111. Prove: $\log_a \dfrac{u}{v} = \log_a u - \log_a v$

Let $\log_a u = x$ and let $\log_a v = y$.

$a^x = u$ and $a^y = v$	Rewrite in exponential form.
$\dfrac{u}{v} = \dfrac{a^x}{a^y} = a^{x-y}$	Divide and substitute for u and v.
$\log_a a^{x-y} = x - y$	Rewrite in logarithmic form.
$\log_a \dfrac{u}{v} = \log_a u - \log_a v$	Substitute for x and y.

Section 4.5 Solving Exponential and Logarithmic Equations

Skills Warm Up

1. $x \ln 2 = \ln 3$

$$x = \frac{\ln 3}{\ln 2}$$

2. $(x - 1) \ln 4 = 2$

$$x - 1 = \frac{2}{\ln 4}$$

$$x = \frac{2}{\ln 4} + 1$$

3. $2xe^2 = e^3$

$$x = \frac{e^3}{2e^2} = \frac{e}{2}$$

4. $4xe^{-1} = 8$

$$x = \frac{8e}{4} = 2e$$

5. $x^2 - 4x + 5 = 0$

$$x = \frac{4 \pm \sqrt{16 - 4(5)}}{2} = \frac{4 \pm \sqrt{-4}}{2} = 2 \pm i$$

6. $2x^2 - 3x + 1 = 0$

$$(2x - 1)(x - 1) = 0$$

$$x = \tfrac{1}{2} \text{ or } x = 1$$

7. $\log_{10} 10^x = x \log_{10} 10 = x$

8. $\log_{10} 10^{2x} = 2x \log_{10} 10 = 2x$

9. $\ln e^{2x} = 2x \ln e = 2x$

10. $\ln e^{-x^2} = -x^2 \ln e = -x^2$

1. $$\begin{aligned} 5^x &= 25 \\ 5^x &= 5^2 \\ x &= 2 \end{aligned}$$

3. $$\begin{aligned} 7^x &= \tfrac{1}{49} \\ 7^x &= 7^{-2} \\ x &= -2 \end{aligned}$$

5. $$\begin{aligned} 4^{x-1} &= 64 \\ 4^{x-1} &= 4^3 \\ x - 1 &= 3 \\ x &= 4 \end{aligned}$$

7. $$\begin{aligned} \log_4 x &= 3 \\ x &= 4^3 \\ x &= 64 \end{aligned}$$

9. $$\begin{aligned} \log_5 5x &= \log_5 2 \\ 5x &= 2 \\ x &= \tfrac{2}{5} \end{aligned}$$

11. $\ln e^{x^2} = x^2$

13. $\log_{10} 10^x + 1 = x + 1$

15. $\log_5 5^{x^3} - 7 = x^3 - 7$

17. $8 + e^{\ln x^3} = 8 + x^3$

19. $10^{\log 10(x+5)} = x + 5$

21. $2^{\log 2\, x^2} = x^2$

23. $$\begin{aligned} 3e^x &= 9 \\ e^x &= 3 \\ \ln e^x &= \ln 3 \\ x &= \ln 3 \approx 1.099 \end{aligned}$$

25. $$\begin{aligned} 2(3^x) &= 16 \\ 3^x &= 8 \\ \log_3 3^x &= \log_3 8 \\ x &= \log_3 8 \\ x &= \frac{\ln 8}{\ln 3} \approx 1.893 \end{aligned}$$

27. $$\begin{aligned} e^x - 9 &= 19 \\ e^x &= 28 \\ \ln e^x &= \ln 28 \\ x &= \ln 28 \approx 3.332 \end{aligned}$$

29. $$\begin{aligned} 3^{2x} &= 80 \\ \log_3 3^{2x} &= \log_3 80 \\ 2x &= \log_3 80 \\ x &= \frac{1}{2}\log_3 80 = \frac{1}{2}\,\frac{\ln 80}{\ln 3} \approx 1.994 \end{aligned}$$

31. $$\begin{aligned} 5^{-t/2} &= 0.20 \\ \log_5 5^{-t/2} &= \log_5 0.20 \\ -\frac{t}{2} &= \log_5 \frac{1}{5} \\ -\frac{t}{2} &= -1 \\ t &= 2 \end{aligned}$$

33. $$\begin{aligned} 3^{x-1} &= 28 \\ \log_3 3^{x-1} &= \log_3 28 \\ x - 1 &= \log_3 28 \\ x &= \log_3 28 + 1 \\ &= \frac{\ln 28}{\ln 3} + 1 \approx 4.033 \end{aligned}$$

35. $$\begin{aligned} 2^{3-x} &= 565 \\ \log_2 2^{3-x} &= \log_2 565 \\ 3 - x &= \log_2 565 \\ -x &= \log_2 565 - 3 \\ x &= 3 - \log_2 565 \\ x &= 3 - \frac{\ln 565}{\ln 2} \approx -6.142 \end{aligned}$$

37. $$\begin{aligned} 8(10^{3x}) &= 12 \\ 10^{3x} &= \tfrac{3}{2} \\ \log_{10} 10^{3x} &= \log_{10} \tfrac{3}{2} \\ x &= \frac{\log_{10} \frac{3}{2}}{3} \approx 0.059 \end{aligned}$$

39. $$\begin{aligned} 3(5^{x-1}) &= 21 \\ 5^{x-1} &= 7 \\ \log_5 5^{x-1} &= \log_5 7 \\ x &= \log_5 7 + 1 = \frac{\ln 7}{\ln 5} + 1 \approx 2.209 \end{aligned}$$

41. $$\begin{aligned} e^{3x} &= 12 \\ 3x &= \ln 12 \\ x &= \frac{\ln 12}{3} \approx 0.828 \end{aligned}$$

43. $500e^{-x} = 300$

$e^{-x} = \frac{3}{5}$

$-x = \ln \frac{3}{5}$

$x = -\ln \frac{3}{5} = \ln \frac{5}{3} \approx 0.511$

45. $7 - 2e^x = 6$

$-2e^x = -1$

$e^x = \frac{1}{2}$

$x = \ln \frac{1}{2} \approx -0.693$

47. $6(2^{3x-1}) - 7 = 9$

$2^{3x-1} = \frac{8}{3}$

$\log_2 2^{3x-1} = \log_2 \frac{8}{3}$

$3x - 1 = \log_2 \frac{8}{3}$

$x = \frac{\log_2 \frac{8}{3} + 1}{3}$

$x = \frac{1}{3}\left(\frac{\ln \frac{8}{3}}{\ln 2} + 1\right) \approx 0.805$

49. $e^{2x} - 8e^x + 12 = 0$

$(e^x - 2)(e^x - 6) = 0$

$e^x - 2 = 0$ $\quad$ $e^x - 6 = 0$

$e^x = 2$ $\quad$ $e^x = 6$

$x = \ln 2$ $\quad$ $x = \ln 6$

≈ 0.693 $\quad$ ≈ 1.792

51. $e^{2x} - 3e^x - 4 = 0$

$(e^x - 4)(e^x + 1) = 0$

$e^x - 4 = 0$ $\quad$ $e^x + 1 = 0$

$e^x = 4$ $\quad$ $e^x = -1$

$x = \ln 4$ $\quad$ No solution

≈ 1.386

53. $\frac{500}{100 - e^{x/2}} = 20$

$20(100 - e^{x/2}) = 500$

$100 - e^{x/2} = 25$

$e^{x/2} = 75$

$\frac{x}{2} = \ln 75$

$x = 2 \ln 75 \approx 8.635$

55. $\frac{3000}{2 + e^{2x}} = 2$

$2(2 + e^{2x}) = 3000$

$2 + e^{2x} = 1500$

$e^{2x} = 1498$

$2x = \ln 1498$

$x = \frac{1}{2} \ln 1498 \approx 3.656$

57. $\left(1 + \frac{0.065}{365}\right)^{365t} = 4$

$\ln\left(1 + \frac{0.065}{365}\right)^{365t} = \ln 4$

$365t \ln\left(1 + \frac{0.065}{365}\right) = \ln 4$

$t = \frac{\ln 4}{365 \ln\left(1 + \frac{0.065}{365}\right)} \approx 21.330$

59. $\left(1 + \frac{0.10}{12}\right)^{12t} = 2$

$\ln\left(1 + \frac{0.10}{12}\right)^{12t} = \ln 2$

$12t \ln\left(1 + \frac{0.10}{12}\right) = \ln 2$

$t = \frac{\ln 2}{12 \ln\left(1 + \frac{0.10}{12}\right)} \approx 6.960$

61. $\log_{10} x = 4$

$10^{\log_{10} x} = 10^4$

$x = 10{,}000$

63. $\ln x = -3$

$e^{\ln x} = e^{-3}$

$x = e^{-3} \approx 0.050$

65. $\ln 2x = 2.4$

$e^{\ln 2x} = e^{2.4}$

$2x = e^{2.4}$

$x = \frac{1}{2}e^{2.4} \approx 5.512$

67. $\log_{10} 2x = 7$

$10^{\log_{10} 2x} = 10^7$

$2x = 10^7$

$x = \frac{10^7}{2} = 5{,}000{,}000$

69. $4\log_3(x+1) = 12$

$\log_3(x+1) = 3$

$3^{\log_3(x+1)} = 3^3$

$x + 1 = 27$

$x = 26$

71. $3\ln 5x = 10$

$\ln 5x = \frac{10}{3}$

$e^{\ln 5x} = e^{10/3}$

$5x = e^{10/3}$

$x = \frac{1}{5}e^{10/3} \approx 5.606$

73. $\ln\sqrt{x+2} = 1$

$\sqrt{x+2} = e^1$

$x + 2 = e^2$

$x = e^2 - 2 \approx 5.389$

75. $7 + 3\ln x = 5$

$3\ln x = -2$

$x = e^{-2/3} \approx 0.513$

77. $\ln x - \ln(x+1) = 2$

$\ln\left(\dfrac{x}{x+1}\right) = 2$

$\dfrac{x}{x+1} = e^2$

$e^2(x+1) = x$

$e^2x + e^2 = x$

$x(e^2 - 1) = -e^2$

$x = \dfrac{-e^2}{e^2-1} \approx -1.157$

No solution; $\ln\left(\dfrac{-e^2}{e^2-1}\right)$ does not exist.

79. $\ln(x+5) = \ln(x-1) - \ln(x+1)$

$\ln(x+5) = \ln\dfrac{x-1}{x+1}$

$x + 5 = \dfrac{x-1}{x+1}$

$(x+5)(x+1) = x - 1$

$x^2 + 6x + 5 = x - 1$

$x^2 + 5x + 6 = 0$

$(x+2)(x+3) = 0$

$x = -2$ or $x = -3$

No solution

81. $\log_2(2x-3) = \log_2(x+4)$

$2x - 3 = x + 4$

$x = 7$

83. $\log_{10}(x+4) - \log_{10} x = \log_{10}(x+2)$

$\log_{10}\left(\dfrac{x+4}{x}\right) = \log_{10}(x+2)$

$\dfrac{x+4}{x} = x + 2$

$x^2 + 2x = x + 4$

$x^2 + x - 4 = 0$

$x = \dfrac{-1 \pm \sqrt{1^2 - 4(1)(-4)}}{2(1)}$

$x = \dfrac{-1 \pm \sqrt{17}}{2}$

$x = \dfrac{-1 + \sqrt{17}}{2} \approx 1.562$

$x = \dfrac{-1 - \sqrt{17}}{2} \approx -2.562$

Using the positive value for x, you have $x = 1.562$.

85. $\log_4 x - \log_4(x-1) = \dfrac{1}{2}$

$\log_4\left(\dfrac{x}{x-1}\right) = \dfrac{1}{2}$

$\dfrac{x}{x-1} = 4^{1/2}$

$\dfrac{x}{x-1} = 2$

$2x - 2 = x$

$x = 2$

87. $\ln y = \ln(2x+1) + \ln 1$

$\ln y = \ln(2x+1)(1)$

$\ln y = \ln(2x+1)$

$y = 2x + 1$

89. $\log_{10} y = 2\log_{10}(x-1) - \log_{10}(x+2)$

$\log_{10} y = \log_{10}(x-1)^2 - \log_{10}(x+2)$

$\log_{10} y = \log_{10}\dfrac{(x-1)^2}{(x+2)}$

$y = \dfrac{(x-1)^2}{(x+2)}$

91. $2^x - 7 = 0$

$$2^x = 7$$

$$x = \log_2 7 = \frac{\ln 7}{\ln 2} \approx 2.807$$

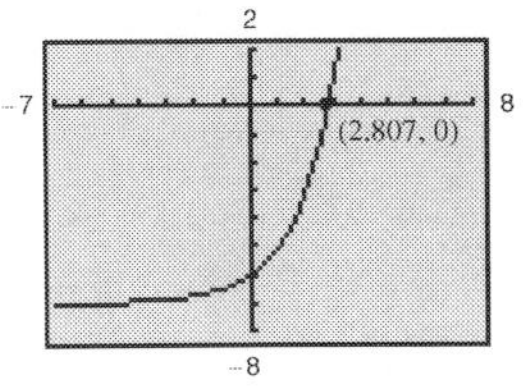

93. $3 - \ln x = 0$

$$3 = \ln x$$

$$x = e^3 \approx 20.086$$

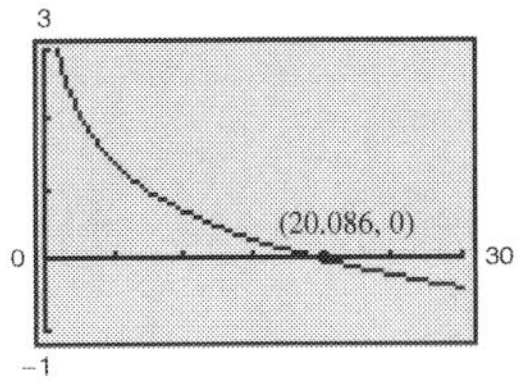

95.
$$\begin{aligned} Pe^{rt} &= A \\ 1000e^{0.0725t} &= 2000 \\ e^{0.0725t} &= 2 \\ \ln e^{0.0725t} &= \ln 2 \\ 0.0725t &= \ln 2 \\ t &= \frac{1}{0.0725}\ln 2 \\ t &\approx 9.56 \end{aligned}$$

The $1000 will double in about 9.56 years.

97.
$$\begin{aligned} Pe^{rt} &= A \\ 1000e^{0.0825t} &= 3000 \\ e^{0.0825t} &= 3 \\ \ln e^{0.0825t} &= \ln 3 \\ 0.0825t &= \ln 3 \\ t &= \frac{1}{0.0825}\ln 3 \\ t &\approx 13.32 \end{aligned}$$

The $1000 will triple in about 13.32 years.

99.
$$\begin{aligned} 15 \cdot 10^{0.02x} &= 60 \\ 10^{0.02x} &= 4 \\ \log_{10} 10^{0.02x} &= \log_{10} 4 \\ 0.02x &= \log_{10} 4 \\ x &= \frac{\log_{10} 4}{0.02} \approx 30.10 \end{aligned}$$

The community was completed about 30 months ago.

101. $p = 1000\left(1 - \dfrac{5}{5 + e^{-0.001x}}\right)$

(a)
$$\begin{aligned} 139.50 &= 1000\left(1 - \frac{5}{5 + e^{-0.001x}}\right) \\ \frac{279}{2000} &= 1 - \frac{5}{5 + e^{-0.001x}} \\ \frac{5}{5 + e^{-0.001x}} &= \frac{1721}{2000} \\ 1721\left(5 + e^{-0.001x}\right) &= 10{,}000 \\ 5 + e^{-0.001x} &= \frac{10{,}000}{1721} \\ e^{-0.001x} &= \frac{1395}{1721} \\ \ln e^{-0.001x} &= \ln \frac{1395}{1721} \\ -0.001x &= \ln \frac{1395}{1721} \\ x &= \frac{\ln \frac{1395}{1721}}{-0.001} \approx 210.011 \end{aligned}$$

When the price is $139.50, the demand is for about 210 coin sets.

(b)
$$\begin{aligned} 99.99 &= 1000\left(1 - \frac{5}{5 + e^{-0.001x}}\right) \\ \frac{9999}{100{,}000} &= 1 - \frac{5}{5 + e^{-0.001x}} \\ \frac{5}{5 + e^{-0.001x}} &= \frac{90{,}001}{100{,}000} \\ 90{,}001\left(5 + e^{-0.001x}\right) &= 500{,}000 \\ 5 + e^{-0.001x} &= \frac{500{,}000}{90{,}001} \\ e^{-0.001x} &= \frac{49{,}995}{90{,}001} \\ \ln e^{-0.001x} &= \ln \frac{49{,}995}{90{,}001} \\ -0.001x &= \ln \frac{49{,}995}{90{,}001} \\ x &= \frac{\ln \frac{49{,}995}{90{,}001}}{-0.001} \approx 587.898 \end{aligned}$$

When the price is $99.99, the demand is for about 588 coin sets.

(c)

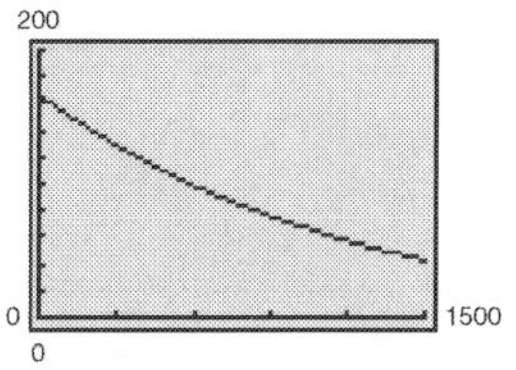

103. $V = 6.7^{-48.1/t},\ 0 \le t$

(a) $1.3 = 6.7e^{-48.1/t}$

$$t = -\frac{-48.1}{\ln(1.3/6.7)} \approx 29.3 \text{ years}$$

(b) $2 = 6.7e^{-48.1/t}$

$$t = \frac{-48.1}{\ln(2/6.7)} \approx 39.8 \text{ years}$$

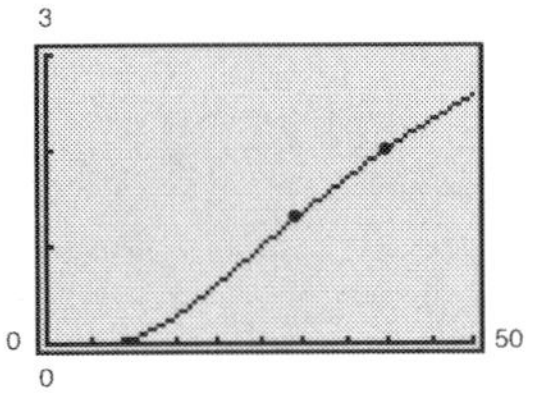

105. (a)

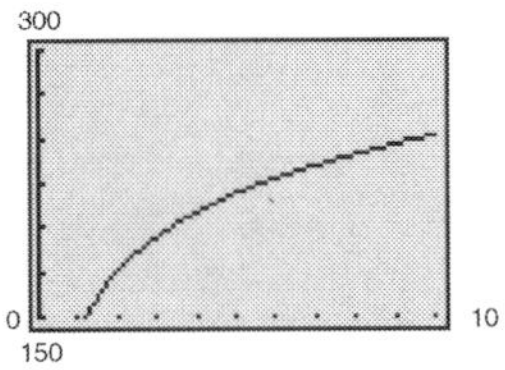

(b) Prescriptions exceeded 200 million in 2003.

(c) $143.09 + 47.2 \ln t = 200$

$$47.2 \ln t = 56.91$$

$$\ln t = \frac{56.91}{47.2}$$

$$e^{\ln t} = e^{56.91/47.2}$$

$$t = e^{56.91/47.2} \approx 3.339$$

Prescriptions exceeded 200 million in 2003.

107. (a)

x	0.2	0.4	0.6	0.8	1.0
y	162.58	78.46	52.498	40.524	33.928

(b)

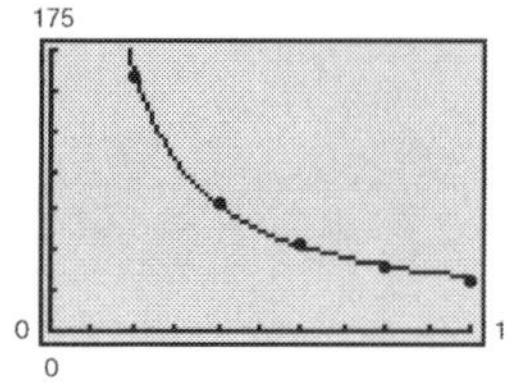

The model is a good fit for the data.

(c) $y = -3.00 + 11.88 \ln x + \dfrac{36.94}{x}$

Using a *trace* feature, you can conclude that when $x = 1.2$ meters, the impact is less than 30 g's.

(d) No. To reduce the g's to fewer than 23 requires a crumple zone of more than 2.27 meters, a length that exceeds the front width of most cars.

109. Yes, a logarithmic equation can have more than one extraneous solution. See Exercise 83.

111. $\log_a uv = \log_a u + \log_a v$

True by the Product Rule in Section 4.4.

113. $\log_a(u - v) = \log_a u - \log_a v$

False

$1.95 \approx \log(100 - 10) \ne \log 100 - \log 10 = 1$

115. Yes. Time to double: $t = \dfrac{\ln 2}{r}$

Time to quadruple: $t = \dfrac{\ln 4}{r} = 2\left(\dfrac{\ln 2}{r}\right)$

Section 4.6 Exponential and Logarithmic Models

Skills Warm Up

1. $y = e^{0.1x}$

x	−2	−1	0	1	2
y	0.819	0.905	1	1.105	1.22

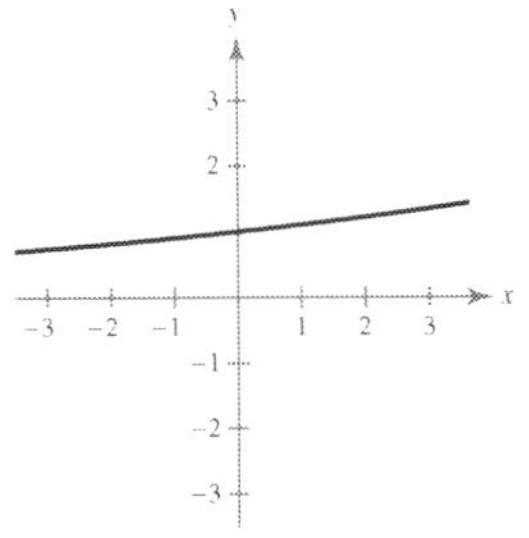

2. $y = e^{-0.25x}$

x	−2	−1	0	1	2
y	1.649	1.284	1	0.779	0.607

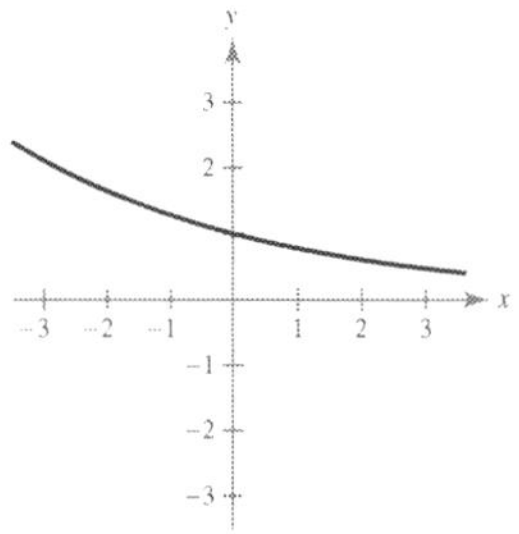

Skills Warm Up —*continued*—

3. $y = e^{-x^2/5}$

x	-2	-1	0	1	2
y	0.449	0.819	1	0.819	0.449

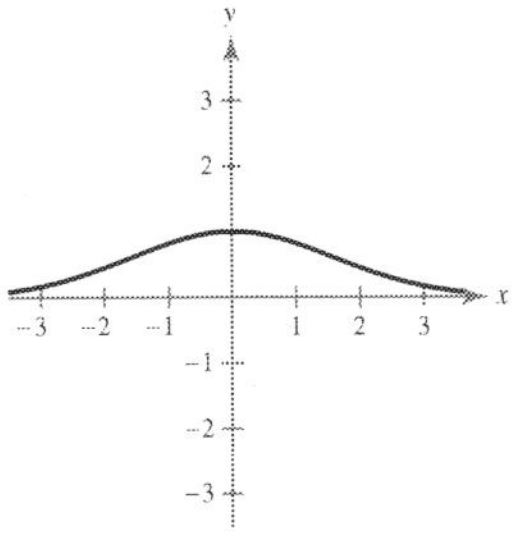

4. $y = \dfrac{2}{1 + e^{-x}}$

x	-2	-1	0	1	2
y	0.238	0.538	1	1.462	1.762

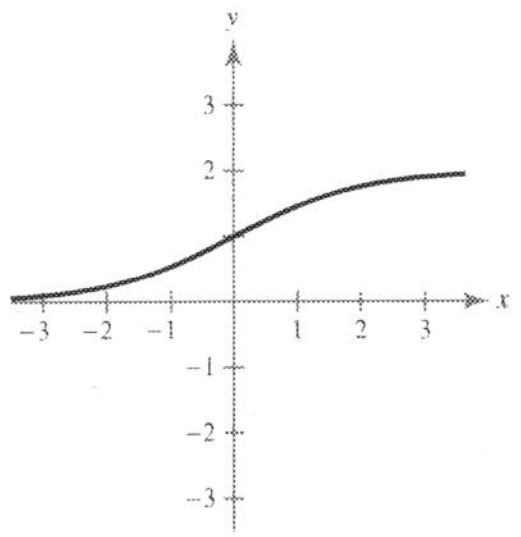

5. $y = \log_{10} 2x$

x	0.25	0.5	1	2
y	-0.301	0	0.301	0.602

x	3	4	5
y	0.778	0.903	1

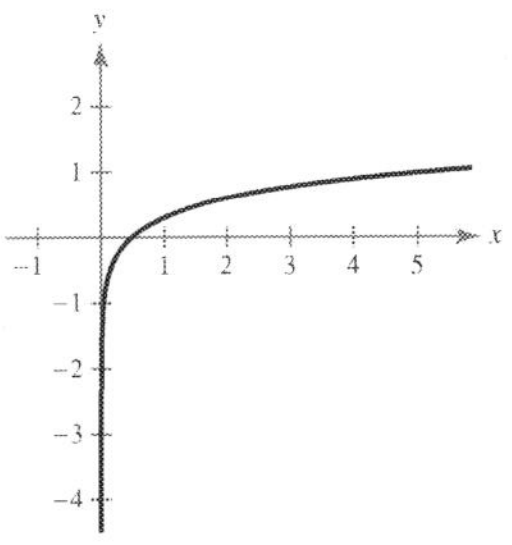

6. $y = \ln 4x$

x	$\frac{1}{4}$	$\frac{1}{2}$	1	2	3	4
y	0	0.693	1.386	2.079	2.485	2.773

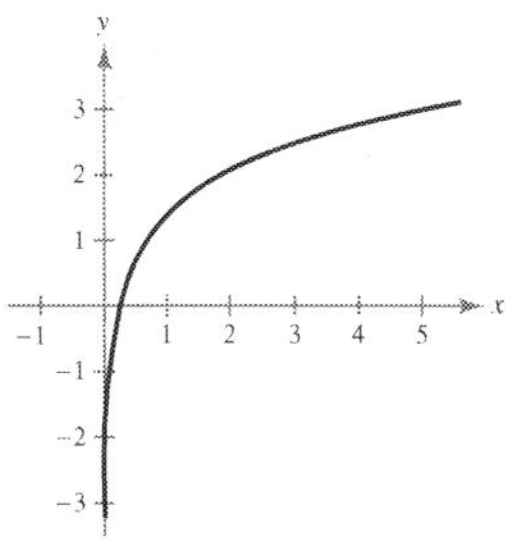

7. $3e^{2x} = 7$

$$e^{2x} = \frac{7}{3}$$

$$2x = \ln \frac{7}{3}$$

$$x = \frac{\ln \frac{7}{3}}{2} \approx 0.424$$

8. $4 \ln 5x = 14$

$$\ln 5x = \frac{7}{2}$$

$$5x = e^{7/2}$$

$$x = \frac{e^{7/2}}{5} \approx 6.623$$

9. $2e^{-0.2x} = 0.002$

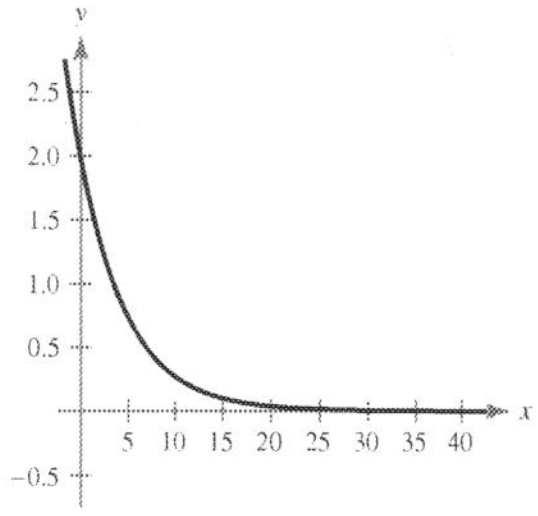

The x-intercept is at $x \approx 34.539$.

10. $6 \ln 2x = 12$

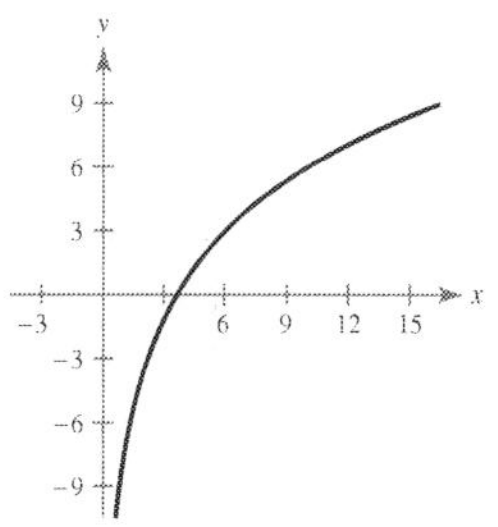

The x-intercept is at $x \approx 3.695$.

1. c

3. b

5. d

7. $y = 3e^{0.5t}$

$k = 0.5 > 0$

Exponential growth

9. $y = 20e^{-1.5t}$

$k = -1.5 < 0$

Exponential decay

11. $P = 120{,}000e^{0.016t}$

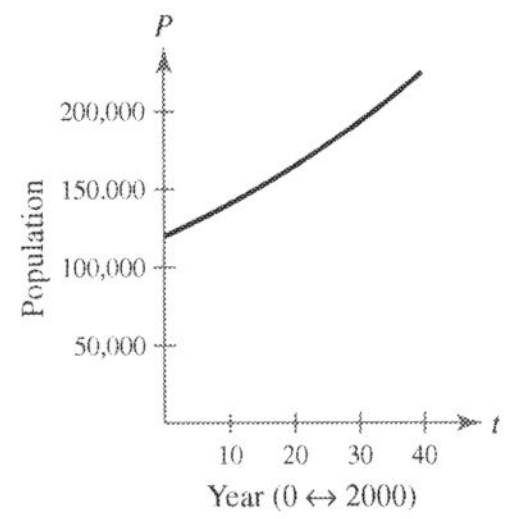

$$180{,}000 = 120{,}000e^{0.016t}$$

$$e^{0.016t} = \frac{3}{2}$$

$$0.016t = \ln\frac{3}{2}$$

$$t = \frac{\ln\frac{3}{2}}{0.016} \approx 25.3$$

The city will reach a population of about 180,000 in 2025.

13. $y = Ce^{kt}$

$1 = Ce^{k(0)} \Rightarrow C = 1$

$10 = e^{k(4)}$

$$k = \frac{\ln 10}{4} \approx 0.5756$$

$y = e^{0.5756t}$

15. $1 = Ce^{k(0)} \Rightarrow C = 1$

$$\frac{1}{4} = e^{k(4)}$$

$$k = \frac{\ln 0.25}{4} \approx -0.3466$$

$y = e^{-0.3466t}$

17. Because $A = 5000e^{0.07t}$, the time to double is given by $10{,}000 = 5000e^{0.07t}$ and you have

$$t = \frac{\ln 2}{0.07} \approx 9.90 \text{ years.}$$

Amount after 10 years: $A = 5000e^{0.7} \approx \$10{,}068.76$

19. Because $A = 500e^{rt}$ and $A = 1000$ when $t = 10$, you have the following.

$1000 = 500e^{10r}$

$$r = \frac{\ln 2}{10} \approx 0.0693 = 6.93\%$$

Amount after 10 years: $A = 500e^{0.693} \approx \$999.85$

21. Because $A = 1000e^{rt}$ and $A = 2281.88$ when $t = 10$, you have the following.

$2281.88 = 1000e^{10r}$

$$r = \frac{\ln(2281.88/1000)}{10} \approx 0.0825 = 8.25\%$$

The time to double is given by

$2000 = 1000e^{0.0825t}$

$$t = \frac{\ln 2}{0.0825} \approx 8.40 \text{ years.}$$

23. Because $A = Pe^{0.11t}$ and $A = 19{,}205$ when $t = 10$, you have the following.

$19{,}205 = Pe^{0.11(10)}$

$$P = \frac{19{,}205}{e^{1.1}} \approx \$6392.79$$

The time to double is given by

$$t = \frac{\ln 2}{0.11} \approx 6.30 \text{ years.}$$

25. Because $A = 5000e^{rt}$ and $A = 11{,}127.70$ when $t = 10$, you have the following.

$11{,}127.70 = 5000e^{r(10)}$

$$r = \frac{\ln(11{,}127.70/5000)}{10} \approx 0.08 = 8\%$$

The time to double is given by $t = \dfrac{\ln 2}{0.08} \approx 8.66$ years.

27. $\frac{1}{2}C = Ce^{k(1599)}$

$$k = \frac{\ln 0.5}{1599}$$

Given $C = 4$ grams, after 1000 years, you have

$y = 4e^{[(\ln 0.5)/1599](1000)} \approx 2.59$ grams.

29. $\frac{1}{2}C = Ce^{k(5715)}$

$$k = \frac{\ln 0.5}{5715}$$

Given $y = 3.5$ grams, after 1000 years, you have

$3.5 = Ce^{[(\ln 0.5)/5715](1000)}$

$C \approx 3.95$ grams.

31. $\frac{1}{2}C = Ce^{k(24,100)}$

$k = \frac{\ln 0.5}{24,100}$

Given $y = 1.6$ grams, after 1000 years, you have

$1.6 = Ce^{[(\ln 0.5)/24,100](1000)}$

$C \approx 1.65$ grams.

33. For 1980, $t = -20$:

$25,000e^{-20k} = 15,000$

$e^{-20k} = \frac{3}{5}$

$\ln e^{-20k} = \ln \frac{3}{5}$

$-20k = \ln \frac{3}{5}$

$k = \frac{\ln \frac{3}{5}}{-20} \approx 0.0255412812$

For 2015, $t = 15$:

$P = 25,000e^{0.0255412812(15)} \approx 36,671.32$

The population of the city will be about 36,671 people in 2015.

35. $N = 100e^{kt}$

When $t = 6$:

$140 = 100e^{6k}$

$e^{6k} = \frac{7}{5}$

$6k = \ln \frac{7}{5}$

$k = \frac{\ln \frac{7}{5}}{6} \approx 0.056$

So, $N = 100e^{0.056t}$.

When the bacteria doubles, $N = 200$.

$200 = 100e^{0.056t}$

$e^{0.056t} = 2$

$0.056t = \ln 2$

$t = \frac{\ln 2}{0.056} \approx 12.38$

The bacteria population doubles in size after about 12.38 hours.

When the bacteria triples, $N = 300$.

$300 = 100e^{0.056t}$

$e^{0.056t} = 3$

$0.056t = \ln 3$

$t = \frac{\ln 3}{0.056} \approx 19.62$

The bacteria population triples after about 19.62 hours.

37. $R = \frac{1}{10^{12}}e^{-t/8223}$

$\frac{1}{8^{14}} = \frac{1}{10^{12}}e^{-t/8223}$

$8^{14}e^{-t/8223} = 10^{12}$

$e^{-t/8223} = \frac{10^{12}}{8^{14}}$

$\frac{-t}{8223} = \ln \frac{10^{12}}{8^{14}}$

$t = -8223 \ln\left(\frac{10^{12}}{8^{14}}\right) \approx 12,179$

The wood is about 12,179 years old.

39. $0.5 = e^{30k}$

$k = \frac{\ln 0.5}{30} \approx -0.0231$

For $t = 100$ and letting the original amount be one unit, $A = e^{(\ln 0.5/30)(100)} = 0.0992$.

Percent remaining: $\frac{0.0992}{1}(100) = 9.92\%$

41. $N = 40(1 - e^{kt})$

(a) When $t = 20$ and $N = 25$:

$25 = 40(1 - e^{20k})$

$\frac{5}{8} = 1 - e^{20k}$

$e^{20k} = \frac{3}{8}$

$20k = \ln \frac{3}{8}$

$k = \frac{\ln \frac{3}{8}}{20} \approx -0.049$

The worker's learning curve is represented by $N = 40(1 - e^{-0.049t})$.

(b) $35 = 40(1 - e^{-0.049t})$

$\frac{7}{8} = 1 - e^{-0.049t}$

$e^{-0.049t} = \frac{1}{8}$

$-0.049t = \ln \frac{1}{8}$

$t = \frac{\ln \frac{1}{8}}{-0.049} \approx 42.44$

The worker should produce 35 units per day after about 42.44 days.

43. The maximum occurs at $x = 100$, so the average IQ score at the college is 100.

45. $P = \dfrac{10{,}000}{1 + 19e^{-t/5}}, \ t \geq 0$

(a)

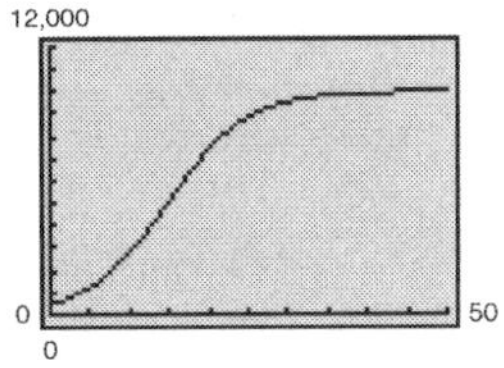

(b) $P(5) = \dfrac{10{,}000}{1 + 19e^{-1}} \approx 1252$ fish

(c)
$$2000 = \frac{10{,}000}{1 + 19e^{-t/5}}$$
$$1 + 19e^{-t/5} = 5$$
$$e^{-t/5} = \frac{4}{19}$$
$$t = -5 \ln \frac{4}{19} \approx 7.8 \text{ months}$$

47. No, because the function is always increasing.

49. (a)

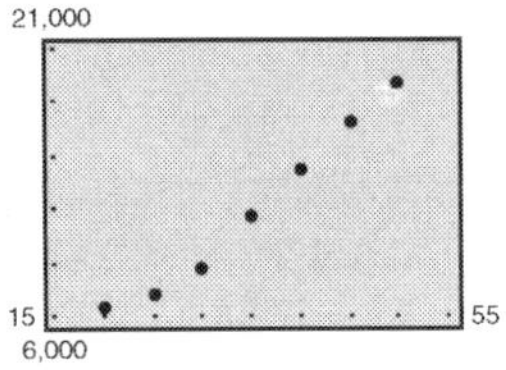

(b) $P = 2922.32(1.0391)^t$;

$P = 2922.32\left(e^{\ln 1.0391}\right)^t \approx 2922.32e^{0.0384t}$

(c)

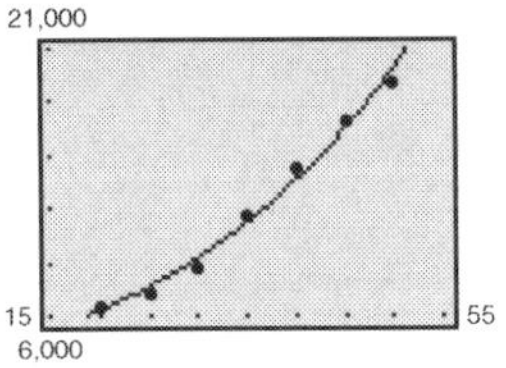

(d) For 2022, use $t = 22$:

$P = 2922.32e^{0.0384(22)} \approx 6801.7356$

For 2042, use $t = 42$:

$P = 2922.32e^{0.0384(42)} \approx 14{,}660.8081$

In 2022, the population will be about 6,801,736 and in 2042, it will be about 14,660,808.

51. (a)

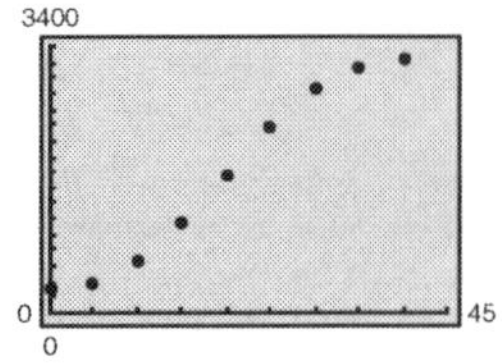

(b) The logistic growth model fits the data.

The data points seem to be leveling off in the later years as though approaching an asymptote.

(c) $P = \dfrac{3455}{1 + 16.51e^{-0.1415t}}$

(d)

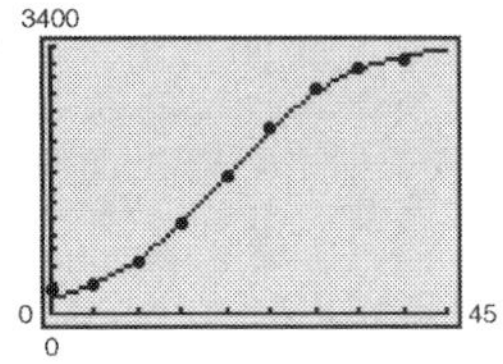

The model is a good fit.

(e) $P = \dfrac{3455}{1 + 16.51e^{-0.1415(50)}} \approx 3407.41$

In year 50 the population will be about 3407 people.

53. (a) $R = \log_{10} I$

$R = \log_{10} 80{,}500{,}000 \approx 7.91$

(b) $R = \log_{10} 48{,}275{,}000 \approx 7.68$

55. $L = 10 \log_{10} \dfrac{I}{I_0} = 10\left[\log_{10} I - \log_{10} I_0\right]$

$= 10\left[\log_{10} I - \log_{10} 10^{-12}\right]$

$= 10\left[\log_{10} I + 12\right] = 120 + 10 \log_{10} I$

(a) $L\left(10^{-10}\right) = 120 + 10 \log_{10} 10^{-10}$

$= 120 - 100 = 20$ decibels

(b) $L\left(10^{-5}\right) = 120 + 10 \log_{10} 10^{-5}$

$= 120 - 50 = 70$ decibels

57.
$$\text{pH} = -\log_{10}\left[H^+\right]$$
$$5.8 = -\log_{10}\left[H^+\right]$$
$$\log_{10}\left[H^+\right] = -5.8$$
$$H^+ = 10^{-5.8} \approx 1.585 \times 10^{-6}$$

59. $\text{pH} = -\log_{10}\left[H^+\right]$ so $\left[H^+\right] = 10^{-\text{pH}}$.

For a grape: $\left[H^+\right] = 10^{-3.5} = 3.163 \times 10^{-4}$ mole of hydrogen per liter

For baking soda: $\left[H^+\right] = 10^{-8} = 1.00 \times 10^{-8}$ mole of hydrogen per liter

$$\frac{\text{concentration}}{\text{concentration}} = \frac{\text{grape}}{\text{baking soda}} = \frac{3.163 \times 10^{-4}}{1.00 \times 10^{-8}} \approx 31{,}623$$

The hydrogen ion concentration of a grape is about 31,623 times the hydrogen ion concentration of baking soda.

61. $t = -10 \ln \dfrac{T - 70}{98.6 - 70}$

$t = -10 \ln \dfrac{85.7 - 70}{98.6 - 70}$

≈ 5.997 or 6 hours

9:00 A.M. − 6 hours = 3:00 A.M.

63. (a)

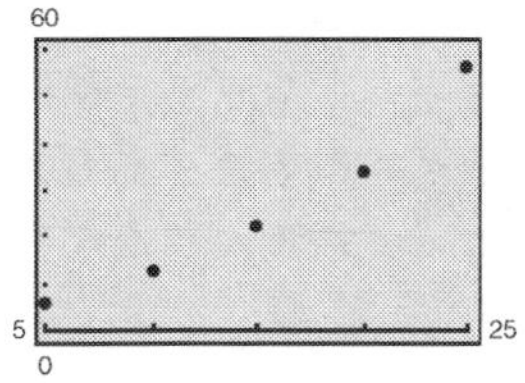

The data fits an exponential model.

(b) Exponential model: $n = 3.9405e^{0.1086t}$

(c)

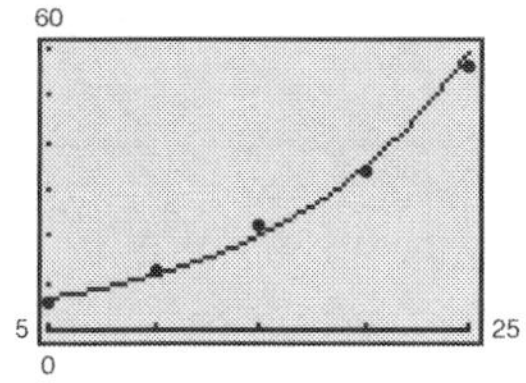

A logistic growth model would be more appropriate for this data because after the initial rapid growth in productivity, the worker's production rate will eventually level off.

Review Exercises for Chapter 4

1. $f(x) = 3x + 5$

$f(g(x)) = 3\left(\dfrac{x - 5}{3}\right) + 5$

$= x - 5 + 5$

$= x$

$g(x) = \dfrac{x - 5}{3}$

$g(f(x)) = \dfrac{(3x + 5) - 5}{3}$

$= \dfrac{3x}{3}$

$= x$

3. $f(x) = \dfrac{1}{x}$

$x = \dfrac{1}{y}$

$xy = 1$

$y = \dfrac{1}{x}$

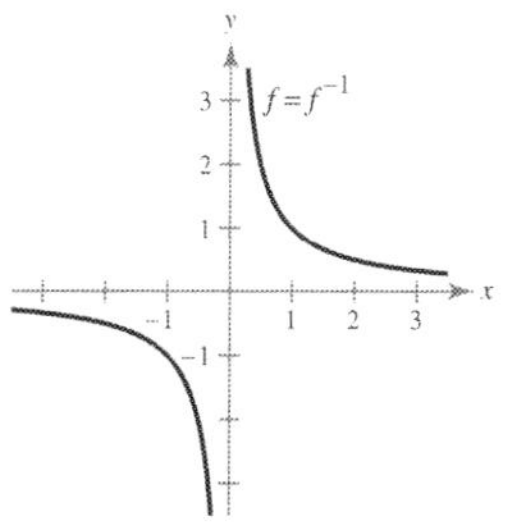

5. $f(x) = x^2, x \geq 0$

(a) $x = y^2$

$\sqrt{x} = \sqrt{y^2}$

$\sqrt{x} = y$

$f^{-1}(x) = \sqrt{x}, x \geq 0$

(b)

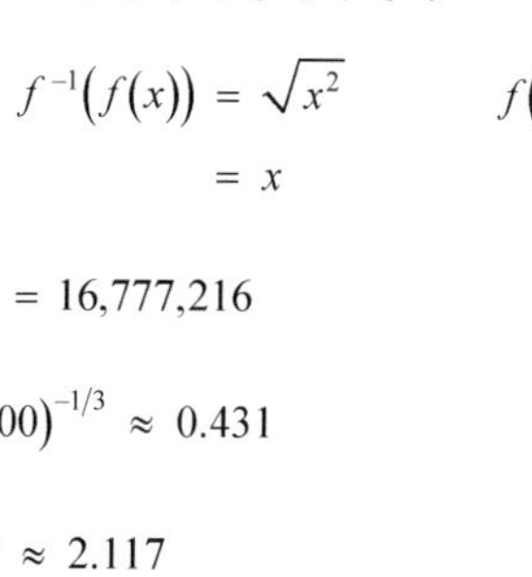

(c) $f^{-1}(f(x)) = \sqrt{x^2}$
$= x$

$f(f^{-1}(x)) = \left(\sqrt{x}\right)^2$
$= x$

7. $4^{12} = 16{,}777{,}216$

9. $2(100)^{-1/3} \approx 0.431$

11. $e^{3/4} \approx 2.117$

13. $f(x) = -3^x$

Intercept: $(0, -1)$

Horizontal asymptote: x-axis

Decreasing on $(-\infty, \infty)$

Matches graph (d)

15. $f(x) = -3^{-x}$

Intercept: $(0, -1)$

Horizontal asymptote: x-axis

Increasing on $(-\infty, \infty)$

Matches graph (a)

17. $f(x) = 4^x$

x	-1	$-\frac{1}{2}$	0	$\frac{1}{2}$	1
y	$\frac{1}{4}$	$\frac{1}{2}$	1	2	4

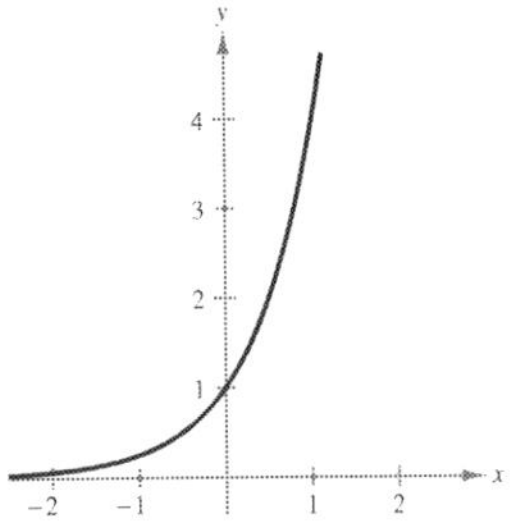

19. $f(x) = \left(\frac{1}{2}\right)^x$

x	-2	-1	0	1	2
y	4	2	1	$\frac{1}{2}$	$\frac{1}{4}$

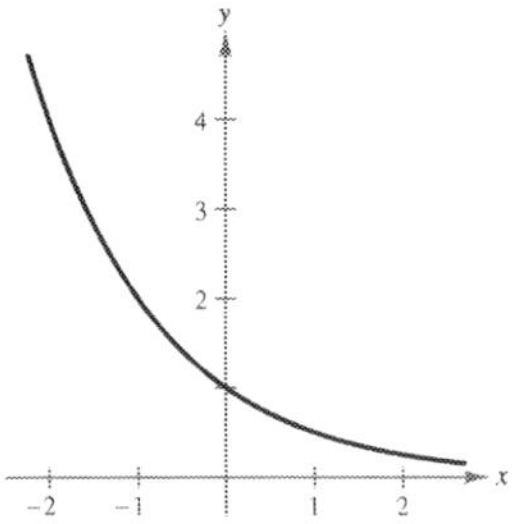

21. $f(x) = 3e^{0.2x}$

x	-2	-1	0	1	2
y	2.011	2.456	3	3.664	4.475

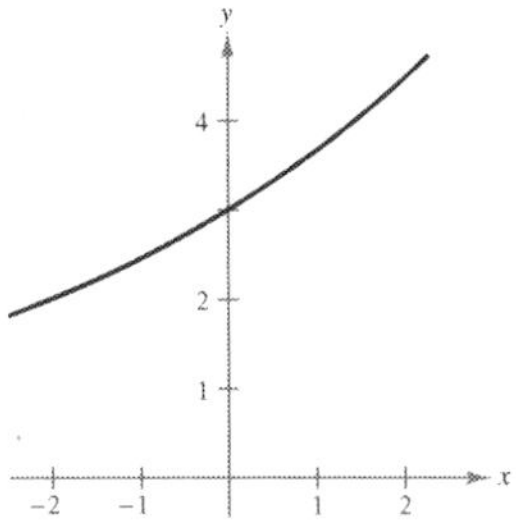

23. $f(x) = 3^{-x^2}$

x	-2	-1	0	1	2
y	$\frac{1}{81}$	$\frac{1}{3}$	1	$\frac{1}{3}$	$\frac{1}{81}$

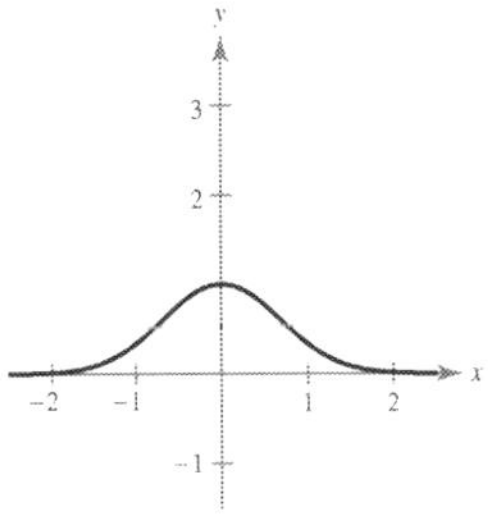

25. $P = 5000,\ r = 8.5,\ t = 12$

$$A = P\left(1 + \frac{r}{n}\right)^{nt} = 5000\left(1 + \frac{0.085}{n}\right)^{12n}$$

n	1	2	4	12	365	Continuous
A	\$13,308.43	\$13,576.74	\$13,718.59	\$13,816.21	\$13,864.32	\$13,865.97

27. $A = Pe^{rt}$ so $P = \dfrac{A}{e^{rt}} = \dfrac{200{,}000}{e^{0.0775t}}$.

t	1	10	20
P	\$185,085.40	\$92,140.76	\$42,449.59

t	30	40	50
P	\$19,556.69	\$9009.84	\$4150.87

29. $P = \$6000,\ r = 0.0575,\ t = 6$

$A = Pe^{rt} = 6000e^{0.0575(6)} \approx 8471.94$

The fund will have about \$8471.94 after about 6 years.

31. $4^3 = 64 \Rightarrow \log_4 64 = 3$

33. $e^2 = 7.3890... \Rightarrow \ln 7.390... = 2$

35. $\log_3 81 = 4 \Rightarrow 3^4 = 81$

37. $\ln 1 = 0 \Rightarrow e^0 = 1$

39. $\log_2 32 = \log_2 2^5 = 5$

41. $\ln e^7 = 7$

43. $\ln e^{-1/2} = -\frac{1}{2}$

45. $f(x) = \log_2 x$

Intercept: $(1, 0)$

Vertical asymptote: y-axis

Increasing on $(0, \infty)$

Matches graph (d)

47. $f(x) = -\log_2 x$

Intercept: $(1, 0)$

Vertical asymptote: y-axis

Decreasing on $(0, \infty)$

Matches graph (c)

49. $f(x) = 10^x,\ g(x) = \log_{10} x$

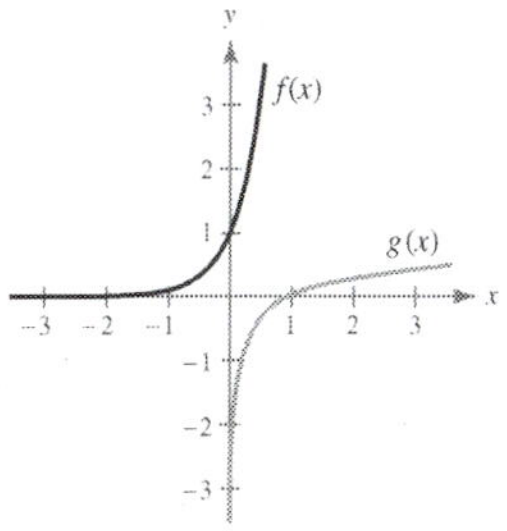

51. $f(x) = \log_2(x - 3)$

Domain: $(3, \infty)$

Vertical asymptote: $x = 3$

x-intercept: $(4, 0)$

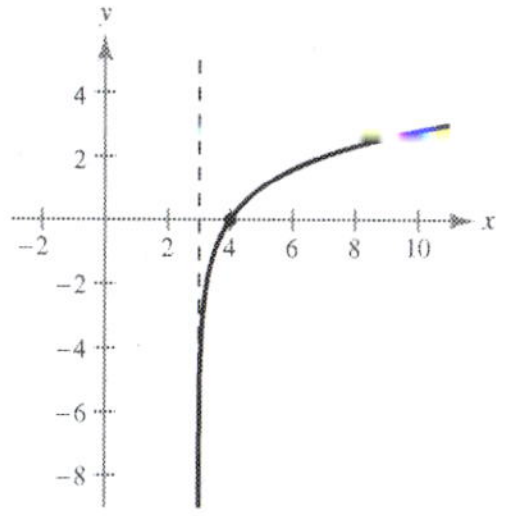

53. $g(x) = 2 \ln x$

Domain: $(0, \infty)$

Vertical asymptote: $x = 0$

x-intercept: $(1, 0)$

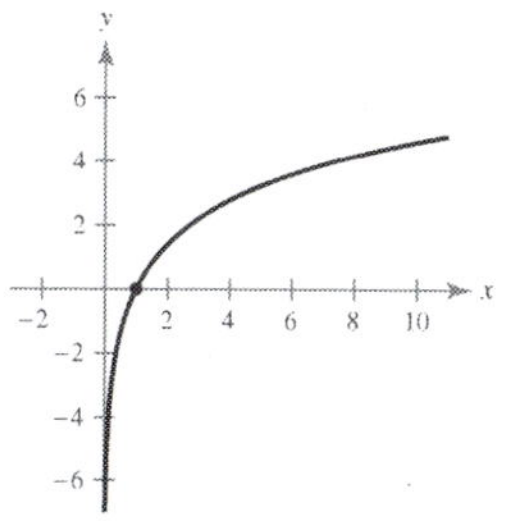

55. $f(t) = 82 - 16 \log_{10}(t + 1),\ 0 \le t \le 6$

When $t = 0$: $f(0) = 82 - 16 \log_{10} 1 \approx 82$

When $t = 6$: $f(6) = 82 - 16 \log_{10}(6 + 1) \approx 68.48$

Average scores decreased from 82 to approximately 68.

57. (a) $h = 116 \log_{10}(a + 40) - 176$

$h = 116 \log_{10}(55 + 40) - 176 \approx 53.4$

The shoulder height of the elk is about 53.4 inches.

(b)

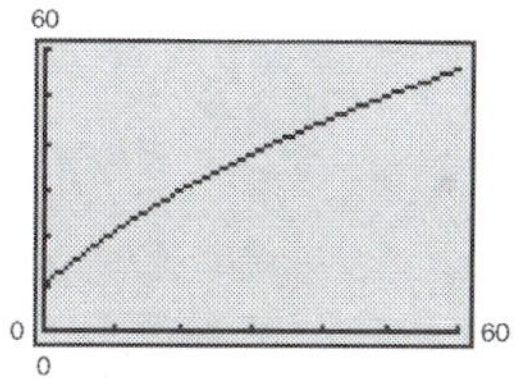

59. $\log_3 10 = \dfrac{\log_{10} 10}{\log_{10} 3} = \dfrac{\ln 10}{\ln 3} \approx 2.096$

61. $\log_{12} 200 = \dfrac{\log_{10} 200}{\log_{10} 12} = \dfrac{\ln 200}{\ln 12} \approx 2.132$

63. $\log_b 6 = \log_b(2 \cdot 3) = \log_b 2 + \log_b 3$

$\approx 0.3562 + 0.5646 \approx 0.9208$

65. $\log_b \sqrt{3} = \frac{1}{2} \log_b 3 = \frac{1}{2}(0.5646) = 0.2823$

67. $\log_7 49 = \log_7 7^2 = 2$

69. $\ln e^{3.2} = 3.2$

71. $\log_{10} \dfrac{x}{y} = \log_{10} x - \log_{10} y$

73. $\ln\left(x\sqrt{x-3}\right) = \ln x + \ln\sqrt{x-3}$

$= \ln x + \frac{1}{2}\ln(x-3)$

75. $\log_5(y-3)^4 = 4\log_5(y-3)$

77. $\log_4 2 + \log_4 3 = \log_4(2 \cdot 3) = \log_4 6$

79. $\frac{1}{2}\ln x = \ln x^{1/2} = \ln\sqrt{x}$

81. $\ln x - \ln(x-3) - \ln(x+1) = \ln x - \left[\ln(x-3) + \ln(x+1)\right] = \ln x - \ln\left[(x-3)(x+1)\right] = \ln\left[\dfrac{x}{(x-3)(x+1)}\right]$

83. The process to find the model:

1. Take the natural logarithm of x and y.

$\ln x$	0	0.693	1.099	1.386	1.609	1.792
$\ln y$	0	0.924	1.465	1.848	2.146	2.389

2. Plot these points in a scatter plot using a graphing utility. The data points will be linear.
3. Find a linear model using the *regression* feature.

The model is $\ln y = \frac{4}{3}\ln x$.

85. $e^x = 8$

$x = \ln 8 \approx 2.079$

87. $3^{4x-1} - 4 = 23$

$3^{4x-1} = 27$

$3^{4x-1} = 3^3$

$4x - 1 = 3$

$4x = 4$

$x = 1$

89. $e^{2x} - 3e^x - 4 = 0$

$(e^x - 4)(e^x + 1) = 0$

$e^x - 4 = 0$ $e^x + 1 = 0$

$e^x = 4$ $e^x = -1$

$x = \ln 4 \approx 1.386$ Not a solution

91. $\ln 3x = 8.2$

$3x = e^{8.2}$

$x = \dfrac{e^{8.2}}{3} \approx 1213.650$

93. $-2 + \ln 5x = 0$

$\ln 5x = 2$

$5x = e^2$

$x = \dfrac{e^2}{5} \approx 1.478$

95. $\ln x - \ln 3 = 2$

$\ln\dfrac{x}{3} = 2$

$\dfrac{x}{3} = e^2$

$x = 3e^2 \approx 22.167$

97. $\log_2\sqrt[3]{x+1} = 1$

$\sqrt[3]{x+1} = 2$

$x + 1 = 8$

$x = 7$

99. $p = 6000\left(1 - \dfrac{5}{5 + e^{-0.004x}}\right)$

(a) $500 = 6000\left(1 - \dfrac{5}{5 + e^{-0.004x}}\right)$

$\dfrac{1}{12} = 1 - \dfrac{5}{5 + e^{-0.004x}}$

$\dfrac{5}{5 + e^{-0.004x}} = \dfrac{11}{12}$

$11\left(5 + e^{-0.004x}\right) = 60$

$e^{-0.004x} = \dfrac{5}{11}$

$x = \dfrac{\ln\frac{5}{11}}{-0.004} \approx 197.11$

The demand will be about 197 desks.

(b) $400 = 6000\left(1 - \dfrac{5}{5 + e^{-0.004x}}\right)$

$\dfrac{5}{5 + e^{-0.004x}} = \dfrac{14}{15}$

$e^{-0.004x} = \dfrac{5}{14}$

$x = \dfrac{\ln\frac{5}{14}}{-0.004} \approx 257.4$

The demand will be about 257 desks.

101. $A = 12e^{\left[\ln(1/2)/5715\right]t}$

Amount after 1000 years: $A(1000) \approx 10.63$ grams

103. (a)

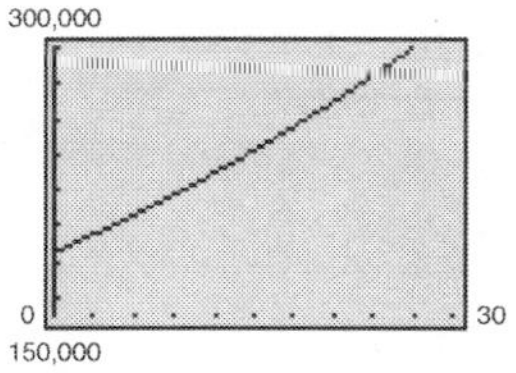

(b) $185{,}000e^{0.018t} = 270{,}000$

$$e^{0.018t} = \frac{54}{37}$$

$$\ln e^{0.018t} = \ln \frac{54}{37}$$

$$0.018t = \ln \frac{54}{37}$$

$$t = \frac{\ln \frac{54}{37}}{0.018} \approx 21.004$$

The population will reach 270,000 people in 2021.

105. $N = 250e^{kt}$

When $t = 6$ and $N = 380$:

$$380 = 250e^{6k}$$

$$e^{6k} = \frac{38}{25}$$

$$6k = \ln \frac{38}{25}$$

$$k = \frac{\ln \frac{38}{25}}{6} \approx 0.070$$

So, $N = 250e^{0.07t}$.

Double:

$$500 = 250e^{0.07t}$$

$$e^{0.07t} = 2$$

$$0.07t = \ln 2$$

$$t = \frac{\ln 2}{0.07} \approx 9.902$$

The bacteria population will double in about 9.902 hours.

Triple:

$$750 = 250e^{0.07t}$$

$$e^{0.07t} = 3$$

$$0.07t = \ln 3$$

$$t = \frac{\ln 3}{0.07} \approx 15.694$$

The bacteria population will triple in about 15.694 hours.

107. $N = 50\left(1 - e^{kt}\right)$

(a) $31 = 50\left(1 - e^{20k}\right)$

$$0.38 = e^{20k}$$

$$k = \frac{\ln 0.38}{20} \approx -0.04838$$

$$N = 50\left(1 - e^{-0.04838t}\right)$$

(b) $45 = 50\left(1 - e^{-0.04838}\right)$

$$\ln 0.1 = \left[-0.04838\right]t$$

$$t \approx 48 \text{ days}$$

109. $P = \dfrac{400}{1 + 3e^{-kt}}, \ t \ge 0$

(a) $135 = \dfrac{400}{1 + 3e^{-2k}}$

$$1 + 3e^{-2k} = \frac{80}{27}$$

$$3e^{-2k} = \frac{53}{27}$$

$$e^{-2k} = \frac{53}{81}$$

$$-2k = \ln \frac{53}{81}$$

$$k = \frac{\ln \frac{53}{81}}{-2} \approx 0.212$$

So, $P = \dfrac{400}{1 + 3e^{-0.212t}}$.

(b) When $t = 5$: $P = \dfrac{400}{1 + 3e^{-0.212(5)}} \approx 196.139$

After 5 years, the deer population will be about 196.

When $t = 10$: $P = \dfrac{400}{1 + 3e^{-0.212(10)}} \approx 294.097$

After 10 years, the deer population will be about 294.

When $t = 20$: $P = \dfrac{400}{1 + 3e^{-0.212(20)}} \approx 383.427$

After 20 years, the deer population will be about 383.

111. $t = -3.95 \ln \dfrac{T - 40}{0 - 40}$

6:00 P.M. $\Rightarrow t = 8$

$$8 = -3.95 \ln \frac{T - 40}{0 - 40}$$

$$\frac{8}{-3.95} = \ln \frac{T - 40}{0 - 40}$$

$$-40e^{8/-3.95} = T - 40$$

$$T = 40 - 40e^{8/-3.95} \approx 34.72^\circ \text{ or } 35^\circ\text{F}$$

The steaks will be thawed by 6:00 P.M.

Chapter 4 Test Yourself

1. Because it is one-to-one, $f(x) = 5x - 7$ has an inverse function.

$$y = 5x - 7$$
$$x = 5y - 7$$
$$x + 7 = 5y$$
$$\frac{x+7}{5} = y$$

So, $f^{-1}(x) = \dfrac{x+7}{5}$.

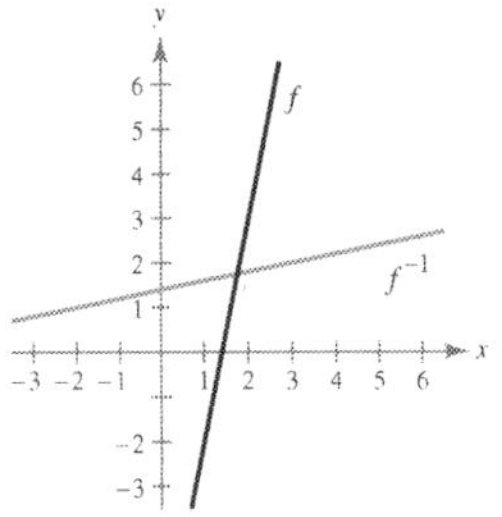

2. Because $f(x) = x^2 + 1$ is not a one-to-one function, it does not have an inverse function.

3. $y = 2^x$

x	-2	-1	0	1	2
y	$\frac{1}{4}$	$\frac{1}{2}$	1	2	4

Horizontal asymptote: x-axis

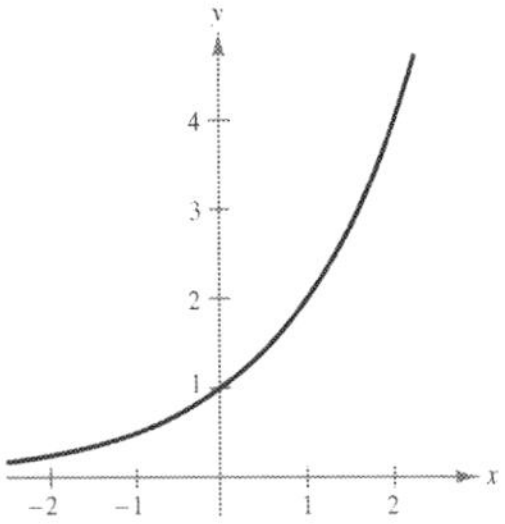

4. $y = e^{-2x}$

x	-2	-1	0	1	2
y	54.60	7.39	1	0.14	0.02

Horizontal asymptote: x-axis

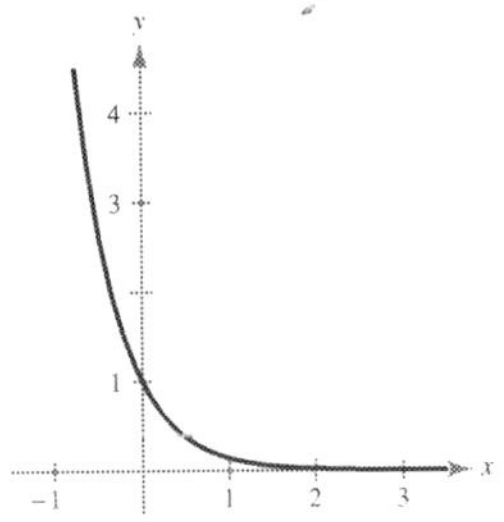

5. $y = \ln x$

x	$\frac{1}{4}$	$\frac{1}{2}$	1	2	3
y	-1.39	-0.69	0	0.69	1.1

Vertical asymptote: y–axis

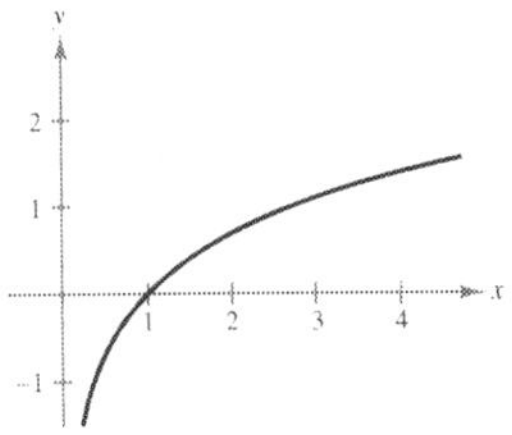

6. $y = \log_3(x - 1)$

x	$\frac{10}{9}$	$\frac{4}{3}$	2	4	10
y	-2	-1	0	1	2

Vertical asymptote: $x = 1$

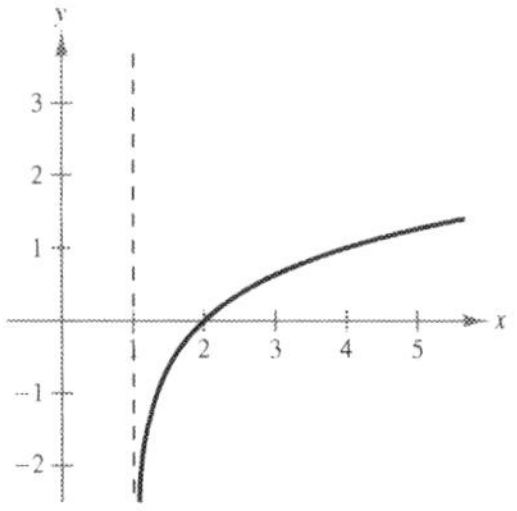

7. $f(0) = 87 - 15\log_{10}(0 + 1) = 87$

$f(2) = 87 - 15\log_{10}(2 + 1) \approx 79.8$

$f(4) = 87 - 15\log_{10}(4 + 1) \approx 76.5$

8. $f(t) = 87 - 15\log_{10}(t + 1)$

After 6 months: $f(6) \approx 74.3$

After 12 months: $f(12) \approx 70.3$

After 18 months: $f(18) \approx 67.8$

Human memory diminishes over time.

9. $\ln \dfrac{x^2y^3}{z} = \ln(x^2y^3) - \ln z = \ln x^2 + \ln y^3 - \ln z$
$= 2\ln x + 3\ln y - \ln z$

10. $\log_{10}(3xyz^2) = \log_{10} 3 + \log_{10} x + \log_{10} y + \log_{10} z^2$
$= \log_{10} 3 + \log_{10} x + \log_{10} y + 2\log_{10} z$

11. $\log_2\left(x\sqrt[3]{x-2}\right) = \log_2 x + \log_2 \sqrt[3]{x-2}$
$= \log_2 x + \frac{1}{3}\log_2(x - 2)$

12. $\log_8 \sqrt[5]{x^2 + 1} = \frac{1}{5}\log_8(x^2 + 1)$

13. $2\ln x + 3\ln y - \ln z = \ln \dfrac{x^2y^3}{z}$

14. $\frac{2}{3}(\log_{10} x + \log_{10} y) = \frac{2}{3}\log_{10}(xy) = \log_{10}(xy)^{2/3}$
$= \log_{10}\sqrt[3]{x^2y^2}$

15.
$$2^{4x} = 21$$
$$\log_2 2^{4x} = \log_2 21$$
$$4x = \log_2 21$$
$$x = \tfrac{1}{4}\log_2 21 \approx 1.098$$

16. $e^{2x} - 8e^x + 12 = 0$

$$\left(e^x - 2\right)\left(e^x - 6\right) = 0$$

$$e^x = 2 \quad \text{or} \quad e^x = 6$$

$$x = \ln 2 \qquad x = \ln 6$$

$$\approx 0.693 \qquad \approx 1.792$$

17. $\log_2(x + 1) - 7 = 0$

$$\log_2(x + 1) = 7$$

$$x + 1 = 2^7$$

$$x = 2^7 - 1$$

$$x = 127$$

18. $\ln\sqrt{x + 2} = 3$

$$\sqrt{x + 2} = e^3$$

$$x + 2 = e^6$$

$$x = e^6 - 2 \approx 401.429$$

19. $P = \$40{,}000$, $r = 0.0675$

$$A = Pe^{rt} = 40{,}000e^{0.0675t}$$

$$120{,}000 = 40{,}000e^{0.0675t}$$

$$e^{0.0675t} = 3$$

$$0.0675t = \ln 3$$

$$t = \ln\frac{3}{0.0675} \approx 16.276$$

The balance in the account will be greater than \$120,000 after about 16.28 years.

20. $P = 85{,}000e^{0.025t}$

$$125{,}000 = 85{,}000e^{0.025t}$$

$$e^{0.025t} = \frac{25}{17}$$

$$0.025t = \ln\frac{25}{17}$$

$$t = \frac{\ln\frac{25}{17}}{0.025} \approx 15.426$$

The city will have a population of 125,000 when $t \approx 15.426$, which corresponds to the year 2015.

21. $N = 100e^{kt}$

$$t = 0,\ N = 100$$

$$t = 8,\ N = 175$$

$$175 = 100e^{8k}$$

$$e^{8k} = 1.75$$

$$8k = \ln 1.75$$

$$k = \frac{\ln 1.75}{8} \approx 0.070$$

So, $N = 100e^{0.070t}$.

Double: $200 = 100e^{0.070t}$

$$e^{0.070t} = 2$$

$$0.070t = \ln 2$$

$$t = \frac{\ln 2}{0.070} \approx 9.902$$

The bacteria population will double in about 9.9 hours.

22. $A = 10e^{kt}$

$$5 = 10e^{5715k}$$

$$\ln\left(\frac{1}{2}\right) = 5715k$$

$$k = \frac{\ln(1/2)}{5715}$$

$$A = 10e^{[\ln(1/2)/(5715)]t}$$

$$A(10{,}000) = 10e^{[\ln(1/2)/5715](10{,}000)} \approx 2.97 \text{ grams}$$

$$A(20{,}000) = 10e^{[\ln(1/2)/5715](20{,}000)} \approx 0.88 \text{ gram}$$

23. Answers will vary. Sample answer: The growth of the bear population will slow down as the population approaches the carrying capacity of the island. So, the population will grow logistically.

CHAPTER 5
Systems of Equations and Inequalities

CHAPTER 5
Systems of Equations and Inequalities

Section 5.1 Solving Systems Using Substitution

Skills Warm Up

1. $y = -\frac{1}{3}x + 6$

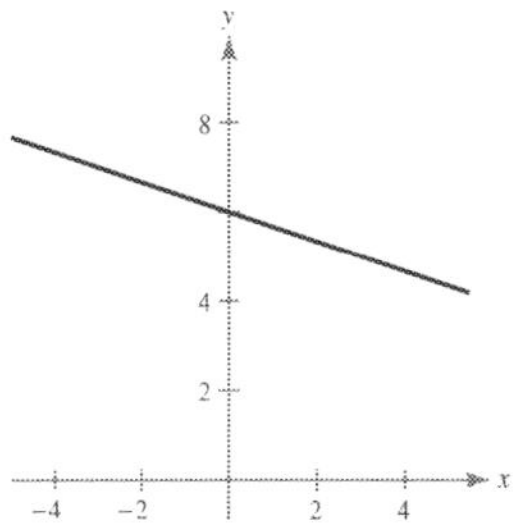

2. $y = 2(x - 3)$

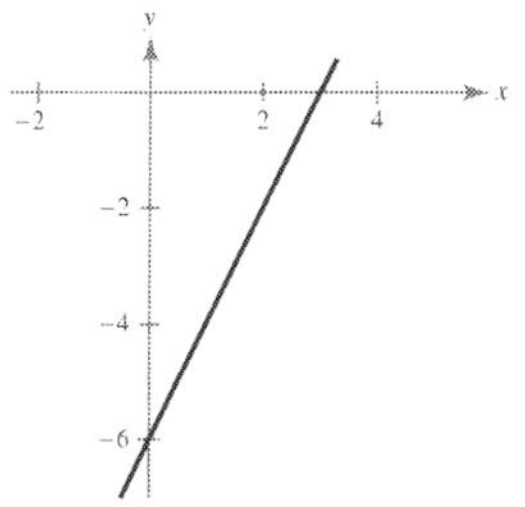

3. $x^2 + y^2 = 4$

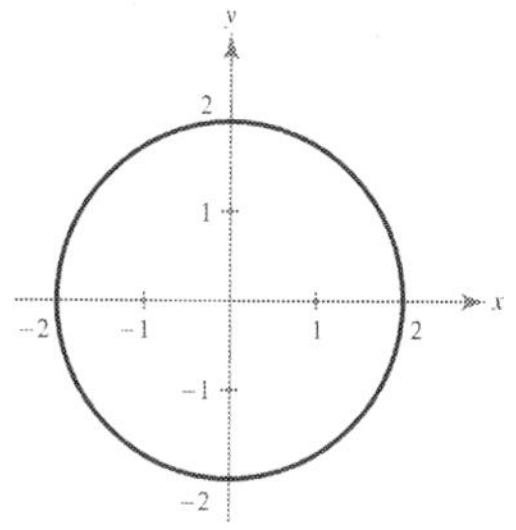

4. $y = 5 - (x - 3)^2$

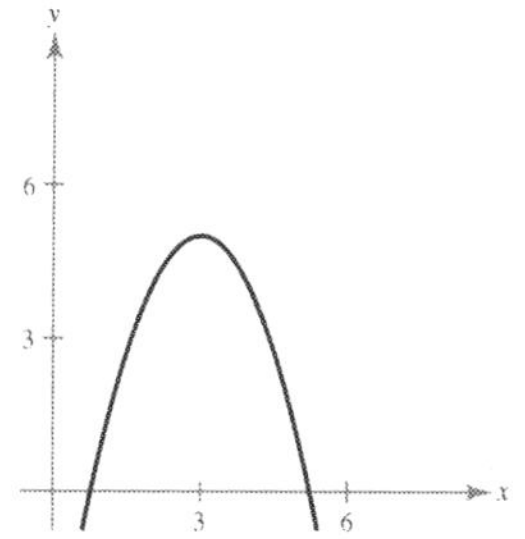

5. $(3x + 2y) - 2(x + y) = 3x + 2y - 2x - 2y = x$

6. $(-10u + 3v) + 5(2u - 8v) = -10u + 3v + 10u - 40v = -37v$

7. $x^2 + (x - 3)^2 + 6x = x^2 + x^2 - 6x + 9 + 6x = 2x^2 + 9$

8. $$\begin{aligned} y^2 - (y + 1)^2 + 2y &= y^2 - (y^2 + 2y + 1) + 2y \\ &= y^2 - y^2 - 2y - 1 + 2y \\ &= -1 \end{aligned}$$

9. $$\begin{aligned} 3x + (x - 5) &= 15 + 4 \\ 4x &= 24 \\ x &= 6 \end{aligned}$$

10. $$\begin{aligned} y^2 + (y - 2)^2 &= 2 \\ y^2 + y^2 - 4y + 4 &= 2 \\ 2y^2 - 4y + 2 &= 0 \\ 2(y^2 - 2y + 1) &= 0 \\ 2(y - 1)^2 &= 0 \\ y - 1 &= 0 \\ y &= 1 \end{aligned}$$

1. $\begin{cases} x + 4y = -3 \\ 5x - y = 6 \end{cases}$

(a) $(-1) + 4(-1) \neq -3$
$5(-1) - (-1) \neq 6$
$(-1, -1)$ is *not* a solution.

(b) $1 + 4(-1) = -3$
$5(1) - (-1) = 6$
$(1, -1)$ *is* a solution.

3. $\begin{cases} 2x + 5y = -5 \\ 2x - y^2 = 1 \end{cases}$

(a) $2(5) + 5(-3) = -5$
$2(5) - (-3)^2 = 1$
$(5, -3)$ *is* a solution.

(b) $2(0) + 5(-1) = -5$
$2(0) - (-1)^2 \neq 1$
$(0, -1)$ is *not* a solution.

5. $\begin{cases} y = -2e^x \\ 3x - y = 2 \end{cases}$

(a) $0 \neq -2e^{-2}$
$3(-2) - 0 \neq 2$
$(-2, 0)$ is *not* a solution.

(b) $2 \neq -2e^{-1}$
$3(-1) - 2 \neq 2$
$(-1, 2)$ is *not* a solution.

7. $\begin{cases} 2x - y = -3 & \text{Equation 1} \\ -3x - 4y = -1 & \text{Equation 2} \end{cases}$

Solve for y in Equation 1: $y = 2x + 3$

Substitute for y in Equation 2: $-3x - 4(2x + 8) = -1$

Solve for x: $-11x = 11 \Rightarrow x = -1$

Back-substitute $x = -1$: $y = 2x + 3 = 2(-1) + 3 = 1$

Solution: $(-1, 1)$

9. $\begin{cases} 2x - y + 2 = 0 & \text{Equation 1} \\ 4x + y - 5 = 0 & \text{Equation 2} \end{cases}$

Solve for y in Equation 1: $y = 2x + 2$

Substitute for y in Equation 2 and solve for x:
$4x + (2x + 2) - 5 = 0 \Rightarrow x = \frac{1}{2}$

Back-substitute $x = \frac{1}{2}$: $y = 2x + 2 = 2\left(\frac{1}{2}\right) + 2 = 3$

Solution: $\left(\frac{1}{2}, 3\right)$

11. $\begin{cases} x - y = 7 & \text{Equation 1} \\ 2x + y = 23 & \text{Equation 2} \end{cases}$

Solve for y in Equation 1: $y = x - 7$

Substitute for y in Equation 2: $2x + x - 7 = 23$

Solve for x: $3x = 30 \Rightarrow x = 10$

Back-substitute $x = 10$: $y = 10 - 7 = 3$

Solution: $(10, 3)$

13. $\begin{cases} 0.3x - 0.4y - 0.33 = 0 & \text{Equation 1} \\ 0.1x + 0.2y - 0.21 = 0 & \text{Equation 2} \end{cases}$

Solve for y in Equation 1: $y = 0.75x - 0.825$

Substitute for y in Equation 2:
$0.1x + 0.2(0.75x - 0.825) - 0.21 = 0$
$\Rightarrow 0.25x = 0.375 \Rightarrow x = 1.5$

Back-substitute $x = 1.5$: $y = 0.75(1.5) - 0.825 = 0.3$

Solution: $(1.5, 0.3)$

15. $\begin{cases} \frac{1}{5}x + \frac{1}{2}y = 8 & \text{Equation 1} \\ x + y = 20 & \text{Equation 2} \end{cases}$

Solve for x in Equation 2: $x = 20 - y$

Substitute for x in Equation 1: $\frac{1}{5}(20 - y) + \frac{1}{2}y = 8$

Solve for y: $4 + \frac{3}{10}y = 8 \Rightarrow y = \frac{40}{3}$

Back-substitute: $y = \frac{40}{3}$: $x = 20 - y = 20 - \frac{40}{3} = \frac{20}{3}$

Solution: $\left(\frac{20}{3}, \frac{40}{3}\right)$

17. $\begin{cases} 6x + 5y = -3 & \text{Equation 1} \\ -x - \frac{5}{6}y = -7 & \text{Equation 2} \end{cases}$

Solve for x in Equation 2: $x = -\frac{5}{6}y + 7$

Substitute for x in Equation 1:
$6\left(-\frac{5}{6} + 7\right) + 5y = -3 \Rightarrow 42 \neq -3$

No points of intersection; no solution

19. Amount in 3% fund + Amount in 5% fund = Total investment

$$x + y = 12{,}000$$

Interest for 3% fund + Interest for 5% fund = Total interest

$$0.03x + 0.05y = 500$$

$\begin{cases} x + y = 12{,}000 & \text{Equation 1} \\ 0.03x + 0.05y = 500 & \text{Equation 2} \end{cases}$

Multiply Equation 2 by 100: $100(0.03x + 0.05y) = 100(500)$

$$3x + 5y = 50{,}000$$

Solve Equation 1 for x: $x = 12{,}000 - y$

Substitute for x in revised Equation 2:

$$\begin{aligned} 3x + 5y &= 50{,}000 \\ 3(12{,}000 - y) + 5y &= 50{,}000 \\ 36{,}000 - 3y + 5y &= 50{,}000 \\ 36{,}000 + 2y &= 50{,}000 \\ 2y &= 14{,}000 \\ y &= 7000 \end{aligned}$$

Back-substitute $y = 7000$:

$$\begin{aligned} x &= 12{,}000 - y \\ x &= 12{,}000 - 7000 \\ x &= 5000 \end{aligned}$$

Solution: $(5000, 7000)$

So, \$5000 is invested at 3% and \$7000 is invested at 5%.

21. Amount in 2.8% fund + Amount in 3.8% fund = Total investment

$$x + y = 12{,}000$$

Interest for 2.8% fund + Interest for 3.8% = Total interest

$$0.028x + 0.038y = 396$$

$$\begin{cases} x + y = 12{,}000 & \text{Equation 1} \\ 0.028x + 0.038y = 396 & \text{Equation 2} \end{cases}$$

Multiply Equation 2 by 1000: $1000(0.028x + 0.038y) = 1000(396)$

$$28x + 38y = 396{,}000$$

Solve Equation 1 for x: $x = 12{,}000 - y$

Substitute for x in revised Equation 2:

$$\begin{aligned} 28x + 38y &= 396{,}000 \\ 28(12{,}000 - y) + 38y &= 396{,}000 \\ 336{,}000 - 28y + 38y &= 396{,}000 \\ 336{,}000 + 10y &= 396{,}000 \\ 10y &= 60{,}000 \\ y &= 6000 \end{aligned}$$

Back-substitute $y = 6000$:

$$\begin{aligned} x &= 12{,}000 - y \\ x &= 12{,}000 - 6000 \\ x &= 6000 \end{aligned}$$

Solution: $(6000, 6000)$

So, \$6000 was invested at 2.8% and \$6000 was invested at 3.8%.

23. $\begin{cases} y = 2x & \text{Equation 1} \\ y = x^2 - 1 & \text{Equation 2} \end{cases}$

Substitute for y in Equation 2: $2x = x^2 - 1$

Solve for x: $x^2 - 2x - 1 = 0 \Rightarrow x = 1 \pm \sqrt{2}$

Back-substitute $x = 1 \pm \sqrt{2}$ in Equation 1:

$y = 2x = 2 \pm 2\sqrt{2}$

Solution: $\left(1 \pm \sqrt{2},\ 2 \pm 2\sqrt{2}\right)$

25. $\begin{cases} 3x - 7y + 6 = 0 & \text{Equation 1} \\ x^2 - y^2 = 4 & \text{Equation 2} \end{cases}$

Solve for y in Equation 1: $y = \dfrac{3x + 6}{7}$

Solve for y in Equation 2: $x^2 - \left(\dfrac{3x + 6}{7}\right)^2 = 4$

Solve for x: $x^2 - \left(\dfrac{9x^2 + 36x + 36}{49}\right) = 4$

$$\begin{aligned} 49x^2 - \left(9x^2 + 36x + 36\right) &= 196 \\ 40x^2 - 36x - 232 &= 0 \end{aligned}$$

$$10x^2 - 9x - 58 = 0 \Rightarrow x = \frac{9 \pm \sqrt{81 + 40(58)}}{20} \Rightarrow x = \frac{29}{10}, -2$$

Back-substitute $x = \dfrac{29}{10}$: $y = \dfrac{3x + 6}{7} = \dfrac{3\left(\frac{29}{10}\right) + 6}{7} = \dfrac{21}{10}$

Back-substitute $x = -2$: $y = \dfrac{3x + 6}{7} = 0$

Solutions: $\left(\frac{29}{10}, \frac{21}{10}\right), (-2, 0)$

27. $\begin{cases} x - 2y = 4 & \text{Equation 1} \\ x^2 - y = 0 & \text{Equation 2} \end{cases}$

Solve for y in Equation 2: $y = x^2$

Substitute for y in Equation 1:

$x - 2x^2 = 4 \Rightarrow x = \dfrac{1 \pm \sqrt{-31}}{4}$

Because the discriminant is negative, there is no real solution.

29. $\begin{cases} y = x^4 - 2x^2 + 1 & \text{Equation 1} \\ y = 1 - x^2 & \text{Equation 2} \end{cases}$

Substitute for y in Equation 1: $1 - x^2 = x^4 - 2x^2 + 1$

Solve for x:

$x^4 - x^2 = 0 \Rightarrow x^2(x^2 - 1) = 0 \Rightarrow x = 0, \pm 1$

Back-substitute $x = 0$: $1 - x^2 = 1$

Back-substitute $x = 1$: $1 - x^2 = 1 - 1^2 = 0$

Back-substitute $x = -1$: $1 - x^2 = 1 - (-1)^2 = 0$

Solutions: $(0, 1), (\pm 1, 0)$

31. $\begin{cases} xy - 2 = 0 & \text{Equation 1} \\ y = \sqrt{x - 1} & \text{Equation 2} \end{cases}$

Substitute for y in Equation 1: $x\sqrt{x - 1} - 2 = 0$

Solve for x:

$x\sqrt{x - 1} - 2 = 0 \Rightarrow x^2(x - 1) = 4$

$x^3 - x^2 - 4 = 0 \Rightarrow (x - 2)(x^2 + x + 2)$

$= 0 \Rightarrow x = 2$

$(x^2 + x + 2 = 0$ has no real solution.$)$

Back-substitute $x = 2$ in Equation 2:

$y = \sqrt{x - 1} = \sqrt{2 - 1} = 1$

Solution: $(2, 1)$

33. $\begin{cases} x + y = -1 & \text{Equation 1} \\ -2x + y = -7 & \text{Equation 2} \end{cases}$

Solve for y in Equation 1: $y = -x - 1$

Substitute for y in Equation 2: $-2x + (-x - 1) = -7$

Solve for x: $-3x = -6 \Rightarrow x = 2$

Back-substitute $x = 2$: $y = -2 - 1 = -3$

Solution: $(2, -3)$

35. $\begin{cases} x - y = -3 & \text{Equation 1} \\ x^2 - y = -1 & \text{Equation 2} \end{cases}$

Solve for y in Equation 1: $y = x + 3$

Substitute for y in Equation 2: $x^2 - (x + 3) = -1$

Solve for x:

$x^2 - x - 3 = -1 \Rightarrow (x - 2)(x + 1) = 0 \Rightarrow x = 2, -1$

Back-substitute $x = 2$: $y = x + 3 = 2 + 3 = 5$

Back-substitute $x = -1$: $y = x + 3 = -1 + 3 = 2$

Solutions: $(-1, 2), (2, 5)$

37. $\begin{cases} x - y = 0 & \text{Equation 1} \\ x^3 - 5x + y = 0 & \text{Equation 2} \end{cases}$

Solve for y in Equation 1: $y = x$

Substitute for y in Equation 2: $x^3 - 5x + x = 0$

Solve for x:

$x^3 - 4x = 0 \Rightarrow x(x^2 - 4) = 0 \Rightarrow x = 0, \pm 2$

Back-substitute $x = 2$: $y = x = 2$

Back-substitute $x = -2$: $y = x = -2$

Back-substitute $x = 0$: $y = x = 0$

Solutions: $(-2, -2), (0, 0), (2, 2)$

39. $\begin{cases} -x + 2y = 2 \Rightarrow y = \dfrac{x + 2}{2} \\ 3x + y = 15 \Rightarrow y = -3x + 15 \end{cases}$

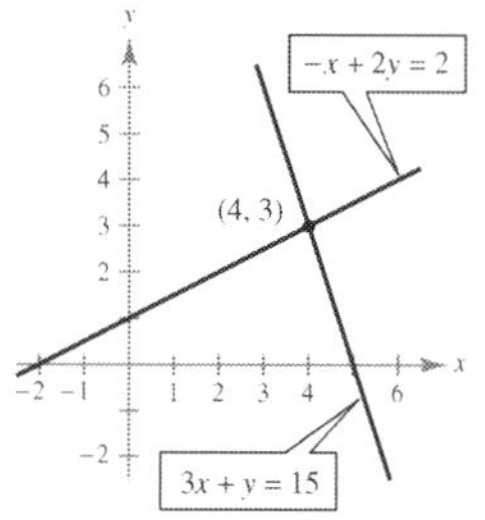

Point of intersection: $(4, 3)$

41. $\begin{cases} x - 3y = -2 \\ 5x + 3y = 17 \end{cases}$

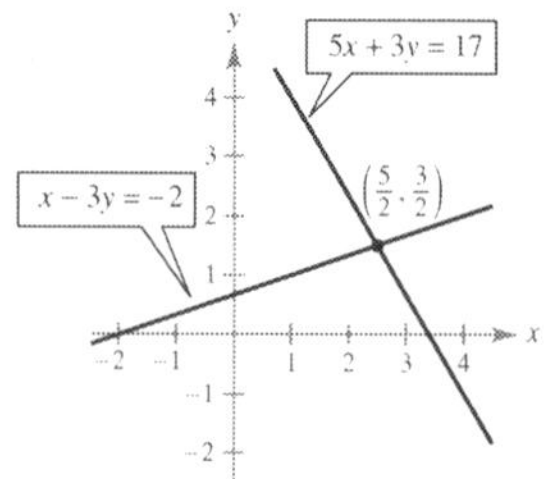

Point of intersection: $\left(\frac{5}{2}, \frac{3}{2}\right)$

43. $\begin{cases} x + y = 4 \\ x^2 + y^2 - 4x = 0 \end{cases}$

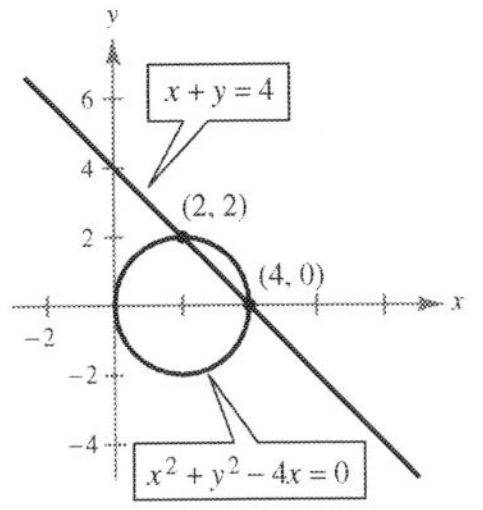

Points of intersection: $(2, 2), (4, 0)$

45. $\begin{cases} x - y + 3 = 0 \\ x^2 - 4x + 7 = y \end{cases}$

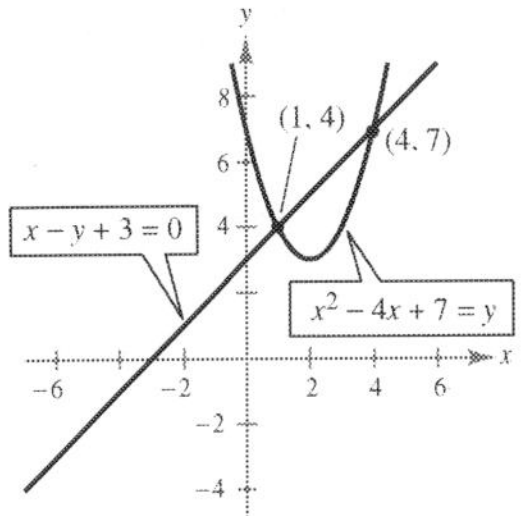

Points of intersection: $(1, 4), (4, 7)$

47. $\begin{cases} 7x + 8y = 24 \\ x - 8y = 8 \end{cases}$

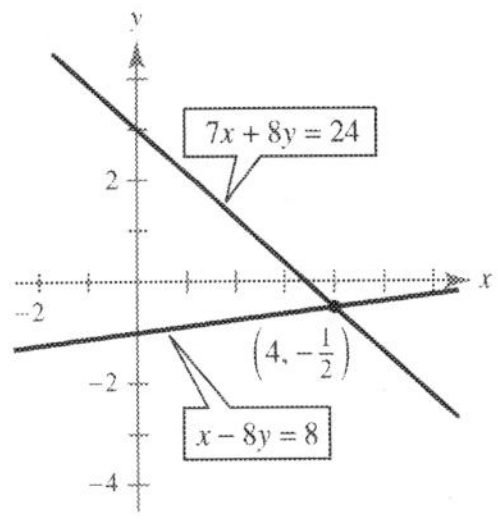

Point of intersection: $\left(4, -\frac{1}{2}\right)$

49. $\begin{cases} 3x - 2y = 0 \Rightarrow y = \dfrac{3}{2}x \\ x^2 - y^2 = 4 \Rightarrow \dfrac{x^2}{4} - \dfrac{y^2}{4} = 1 \end{cases}$

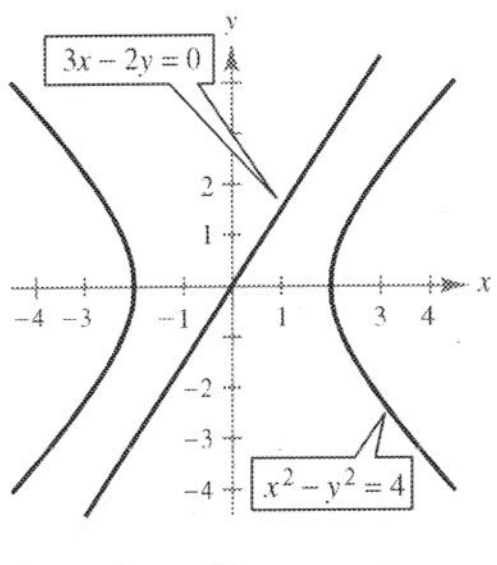

No point of intersection $\Rightarrow$ No solution

51. $\begin{cases} y = -5x + 1 \\ y = x + 3 \end{cases}$

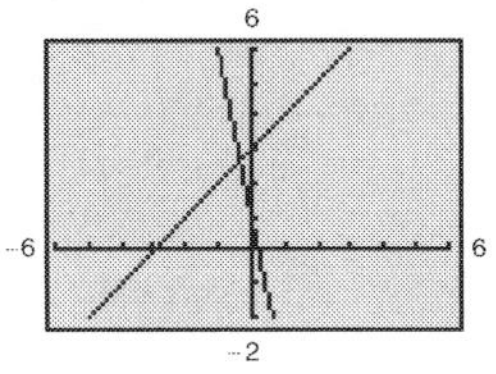

One solution

53. $\begin{cases} y = x^2 + 2x - 1 \\ y = 2x + 5 \end{cases}$

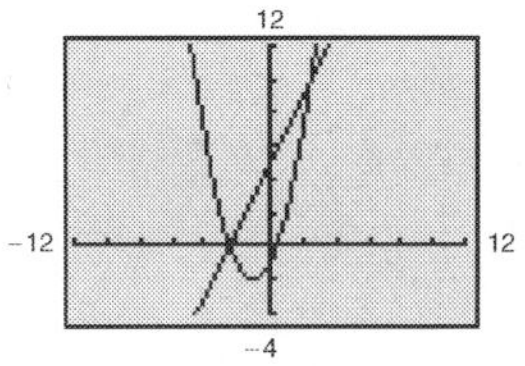

Two solutions

55. $\begin{cases} y = x^2 + 3x + 7 \\ y = -x^2 - 3x + 1 \end{cases}$

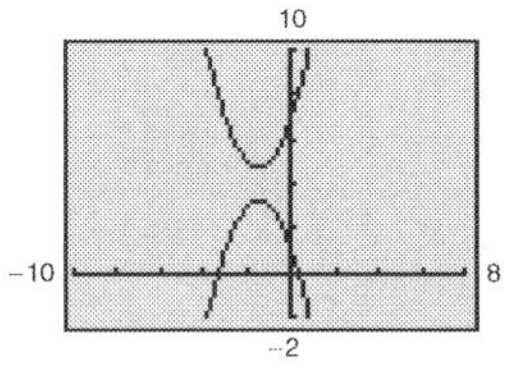

No solution

57. $\begin{cases} y = x^2 + 3x - 1 & \text{Equation 1} \\ y = -x^2 - 2x + 2 & \text{Equation 2} \end{cases}$

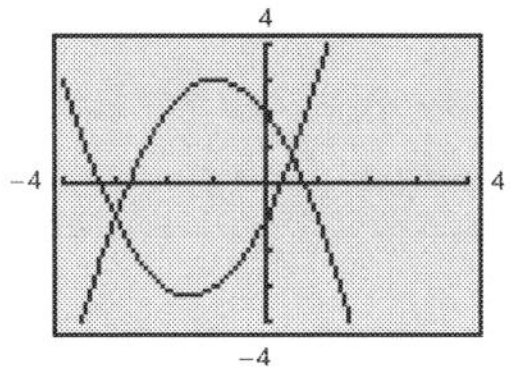

Points of intersection: $(-3, -1), \left(\frac{1}{2}, \frac{3}{4}\right)$

Substitute for y in Equation 2:

$x^2 + 3x - 1 = -x^2 - 2x + 2 \Rightarrow 2x^2 + 5x - 3 = 0,$

$x = -3, \frac{1}{2}$

Back-substitute $x = -3$ in Equation 1:

$y = x^2 + 3x - 1 = (-3)^2 + 3(-3) - 1 = -1$

Back-substitute $x = \frac{1}{2}$ in Equation 1:

$y = x^2 + 3x - 1 = \left(\frac{1}{2}\right)^2 + 3\left(\frac{1}{2}\right) - 1 = \frac{3}{4}$

Solutions: $(-3, -1), \left(\frac{1}{2}, \frac{3}{4}\right)$

59. $\begin{cases} x - y + 3 = 0 & \text{Equation 1} \\ x^2 - 4x + 7 = y \Rightarrow (x-2)^2 + 3 = y & \text{Equation 2} \end{cases}$

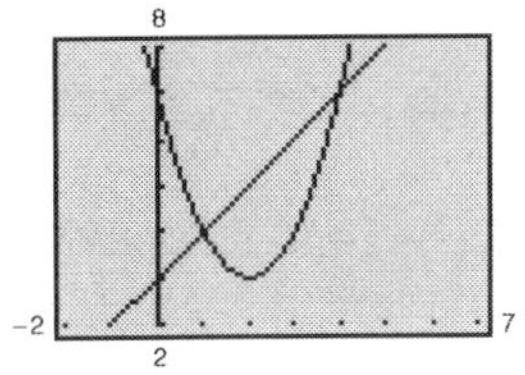

Points of intersection: $(1, 4), (4, 7)$

Solve for y in Equation 1: $y = x + 3$

Substitute for y in Equation 2:

$x^2 - 4x + 7 = x + 3 \Rightarrow x^2 - 5x + 4 = 0 \Rightarrow x = 1, 4$

Back-substitute $x = 1$ in Equation 1:

$y = x + 3 = 1 + 3 = 4$

Back-substitute $x = 4$ in Equation 1:

$y = x + 3 = 4 + 3 = 7$

Solutions: $(1, 4), (4, 7)$

61. $\begin{cases} y = e^x & \text{Equation 1} \\ x - y + 1 = 0 & \text{Equation 2} \end{cases}$

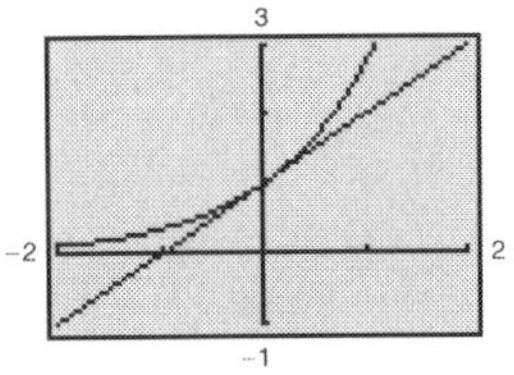

Point of intersection: $(0, 1)$

Check by substitution: $y = e^x$; $1 = e^0$

$x - y + 1 = 0; 0 - 1 + 1 = 0$

Solution: $(0, 1)$

63. Answers will vary. Sample answer:

$\begin{cases} 4x - y = 0 & \text{Equation 1} \\ x^2 - y = -3 & \text{Equation 2} \end{cases}$

Solve Equation 1 for y: $y = 4x$

Substitute for y in Equation 2:

$$x^2 - y = -3$$
$$x^2 - 4x = -3$$
$$x^2 - 4x + 3 = 0$$
$$(x-1)(x-3) = 0$$
$$x = 1 \text{ or } x = 3$$

If $x = 1, y = 4x = 4(1) = 4$

If $x = 3, y = 4x = 4(3) = 12$

Solutions: $(1, 4), (3, 12)$

This method was chosen because it was easy to solve Equation 1 for y.

65. Answers will vary. Sample answer:

$\begin{cases} x^2 + y^2 = 25 & \text{Equation 1} \\ \frac{1}{2}x + y = \frac{5}{2} & \text{Equation 2} \end{cases}$

Solve Equation 2 for y: $y = -\frac{1}{2}x + \frac{5}{2}$

Substitute for y in Equation 1:

$$x^2 + y^2 = 25$$
$$x^2 + \left(-\tfrac{1}{2}x + \tfrac{5}{2}\right)^2 = 25$$
$$x^2 + \tfrac{1}{4}x^2 - \tfrac{5}{2}x + \tfrac{25}{4} = 25$$
$$4x^2 + x^2 - 10x + 25 = 100$$
$$5x^2 - 10x - 75 = 0$$
$$5(x^2 - 2x - 15) = 0$$
$$x^2 - 2x - 15 = 0$$
$$(x-5)(x+3) = 0$$

$x - 5 = 0 \quad x + 3 = 0$

$x = 5 \quad\quad x = -3$

For $x = 5$: $y = -\frac{1}{2}x + \frac{5}{2} = -\frac{1}{2}(5) + \frac{5}{2} = 0$

For $x = -3$:

$y = -\frac{1}{2}x + \frac{5}{2} = -\frac{1}{2}(-3) + \frac{5}{2} = \frac{3}{2} + \frac{5}{2} = 4$

Solutions: $(-3, 4), (5, 0)$

This method was chosen because it was easy to solve Equation 2 for y.

67. Answers will vary. Sample answers:

(a) $\begin{cases} x + 2y = 0 \\ 2x - 4y = 3 \end{cases}$

(b) $\begin{cases} 2x^2 - y = -1 \\ x + y = 4 \end{cases}$

(c) $\begin{cases} 3x + y = 8 \\ 3x + y = -2 \end{cases}$

69. $C = 8650x + 250{,}000,\ R = 9950x$

$$R = C$$
$$9950x = 8650x + 250{,}000$$
$$1300x = 250{,}000$$
$$x \approx 192 \text{ units}$$

71. $C = 2.65x + 350{,}000,\ R = 4.15x$

$$R = C$$
$$4.15x = 2.65x + 350{,}000$$
$$1.50x = 350{,}000$$
$$x \approx 233{,}333 \text{ units}$$

73. Let x = the number of software packages .

$C = 11.70x + 76{,}500,\ R = 49.95x$

$R = C$

$49.95x = 11.70x + 76{,}500$

$38.25x = 76{,}500$

$x = 2000$

The company must sell 2000 software packages to break even.

75. $\begin{cases} P = 30.2t + 4420 & \text{Equation 1 (Alabama)} \\ P = 73.6t + 4332 & \text{Equation 2 (Colorado)} \end{cases}$

$73.6t + 4332 = 30.2t + 4420$

$43.4t + 4332 = 4420$

$43.4t = 88$

$t \approx 2.03$

The population of Colorado exceeded the population of Alabama in 2002.

77.

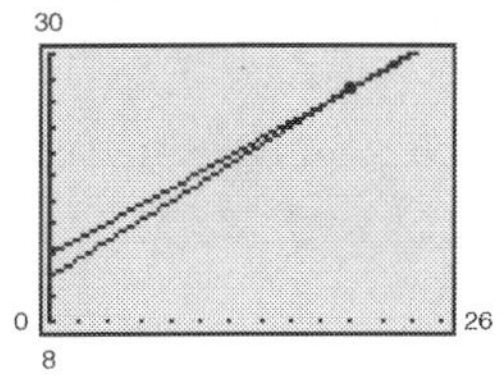

The curves intersect when $a = 20$.

So, the BMI for males exceeds the BMI for females at age 20.

79. Let x = total annual sales.

$\begin{cases} S = 48{,}000 + 0.025x & \text{Company A} \\ S = 44{,}000 + 0.065x & \text{Company B} \end{cases}$

$44{,}000 + 0.065x = 48{,}000 + 0.025x$

$44{,}000 + 0.04x = 48{,}000$

$0.04x = 4000$

$x = 100{,}000$

The two salaries will be the same when sales are \$100,000.

81.

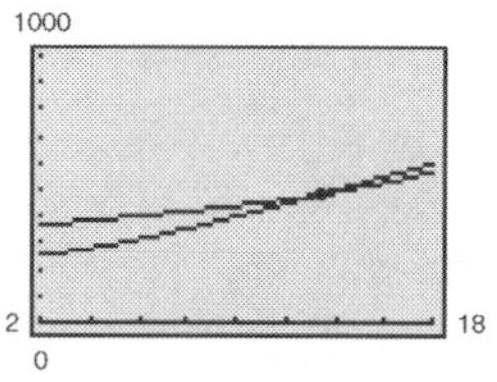

According to the graphs, sales of general merchandise stores exceeded sales of food and beverage stores in 2003 $(t \approx 13.37)$. Both models eventually decrease and become negative. So, it is unlikely that these models will continue to be accurate.

Section 5.2 Solving Systems Using Elimination

Skills Warm Up

1. $2x + y = 4$

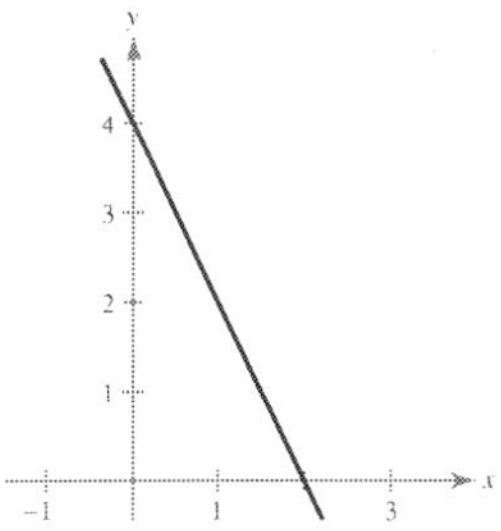

2. $5x - 2y = 3$

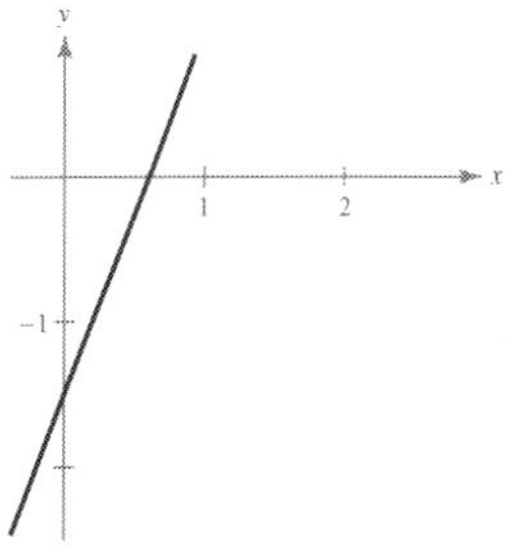

3. $(-1, 3), (4, 8)$

$m = \dfrac{8 - 3}{4 - (-1)} = \dfrac{5}{5} = 1$

$y - 3 = 1(x - (-1))$

$y = x + 4$

4. $(2, 6), (5, 1)$

$m = \dfrac{1 - 6}{5 - 2} = -\dfrac{5}{3}$

$y - 1 = -\dfrac{5}{3}(x - 5)$

$y = -\dfrac{5}{3}x + \dfrac{28}{3}$

5. $3x + 6y = 4$

$y = -\frac{1}{2}x + \frac{2}{3}$

Slope: $m = -\frac{1}{2}$

Skills Warm Up —continued—

6. $7x - 4y = 10$

$y = \frac{7}{4}x - \frac{5}{2}$

Slope: $m = \frac{7}{4}$

7. $2x - 3y = -10$

$y = \frac{2}{3}x + \frac{10}{3}$

$m_1 = \frac{2}{3}$

$3x + 2y = 11$

$y = -\frac{3}{2}x + \frac{11}{2}$

$m_2 = -\frac{3}{2}$

Because $m_1 = -\dfrac{1}{m_2}$, the lines are perpendicular.

8. $4x - 12y = 5$

$y = \frac{1}{3}x - \frac{5}{12}$

$m_1 = \frac{1}{3}$

$-2x + 6y = 3$

$y = \frac{1}{3}x + \frac{1}{2}$

$m_2 = \frac{1}{3}$

Because $m_1 = m_2$, the lines are parallel.

9. $5x + y = 2$

$y = -5x + 2$

$m_1 = -5$

$3x + 2y = 1$

$y = -\frac{3}{2}x + \frac{1}{2}$

$m_2 = -\frac{3}{2}$

Because $m_1 \neq m_2$ and $m_1 \neq -\dfrac{1}{m_2}$, the lines are neither parallel nor perpendicular.

10. $x - 3y = 2$

$y = \frac{1}{3}x - \frac{2}{3}$

$m_1 = \frac{1}{3}$

$6x + 2y = 4$

$y = -3x + 2$

$m_2 = -3$

Because $m_1 = -\dfrac{1}{m_2}$, the lines are perpendicular.

1. $\begin{cases} 3x - 2y = 2 & \text{Equation 1} \\ x + 2y = 6 & \text{Equation 2} \end{cases}$

Add to eliminate y: $4x = 8 \Rightarrow x = 2$

Substitute $x = 2$ in Equation 2: $2 + 2y = 6 \Rightarrow y = 2$

Solution: $(2, 2)$

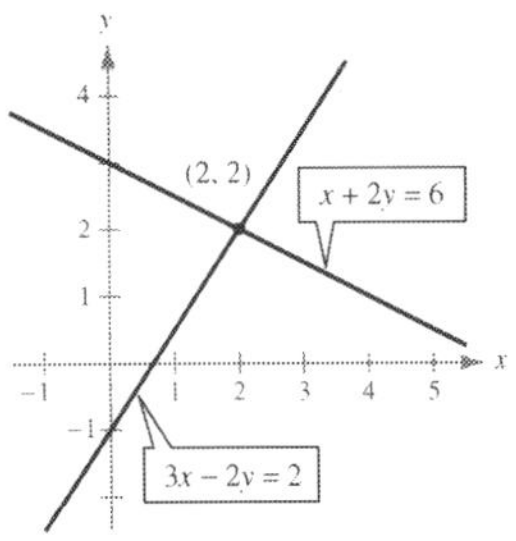

3. $\begin{cases} x - 4y = 2 & \text{Equation 1} \\ 2x + y = 4 & \text{Equation 2} \end{cases}$

Multiply Equation 2 by 4: $8x + 4y = 16$

Add this to Equation 1 to eliminate y:

$9x = 18 \Rightarrow x = 2$

Substitute $x = 2$ in Equation 2:

$2(2) + y = 4 \Rightarrow y = 0$

Solution: $(2, 0)$

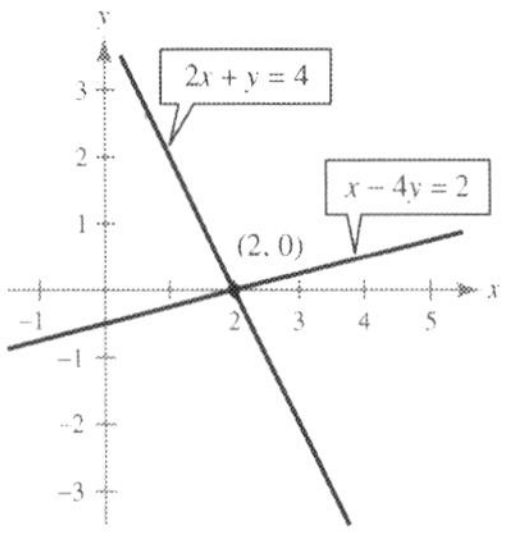

5. $\begin{cases} x - y = 1 & \text{Equation 1} \\ -2x + 2y = 5 & \text{Equation 2} \end{cases}$

Multiply Equation 1 by 2: $2x - 2y = 2$

Add this to Equation 2: $0 = 7$

There are no solutions.

The system is inconsistent.

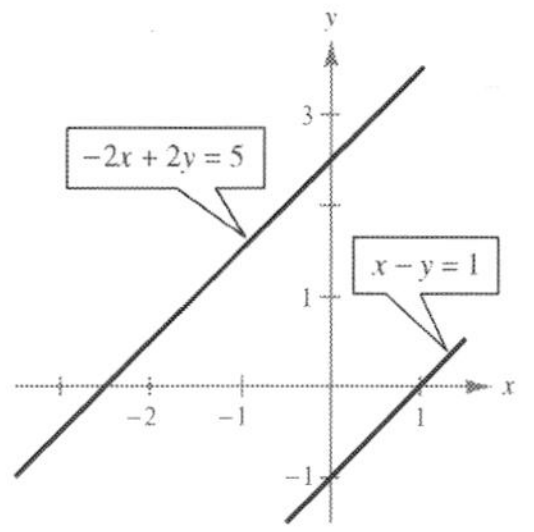

7. $\begin{cases} 3x - 2y = 6 & \text{Equation 1} \\ -6x + 4y = -12 & \text{Equation 2} \end{cases}$

Multiply Equation 1 by 2 and add to Equation 2: $0 = 0$

The equations are dependent.

The solution set consists of all points (x, y) lying on the line

$3x - 2y = 6$

$y = \frac{3}{2}x - 3.$

Let $x = 2a$, then $y = \frac{3}{2}(2a) - 3 = 3a - 3.$

Solution: $(2a, 3a - 3)$ for any real number a.

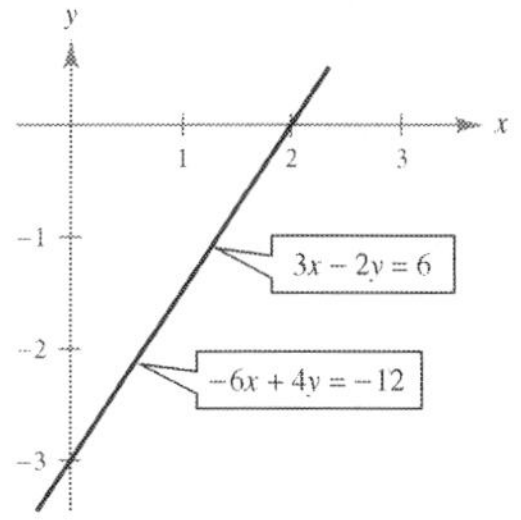

9. $\begin{cases} x + 2y = 3 & \text{Equation 1} \\ x - 2y = 1 & \text{Equation 2} \end{cases}$

Add to eliminate y:

$2x = 4 \Rightarrow x = 2$

Substitute $x = 2$ in Equation 2:

$2 - 2y = 1 \Rightarrow y = \frac{1}{2}$

Solution: $\left(2, \frac{1}{2}\right)$; consistent

11. $\begin{cases} 4x - 3y = 11 & \text{Equation 1} \\ -6x + 3y = 3 & \text{Equation 2} \end{cases}$

Add to eliminate y: $-2x = 14 \Rightarrow x = -7$

Substitute $x = -7$ in Equation 1:

$4(-7) - 3y = 11 \Rightarrow y = -13$

Solution: $(-7, -13)$; consistent

13. $\begin{cases} 3x - y = 17 & \text{Equation 1} \\ 5x + 5y = -5 & \text{Equation 2} \end{cases}$

Multiply Equation 1 by 5: $15x - 5y = 85$

Add this to Equation 2 to eliminate y:

$20x = 80 \Rightarrow x = 4$

Substitute $x = 4$ in Equation 1:

$3(4) - y = 17 \Rightarrow y = -5$

Solution: $(4, -5)$; consistent

15. $\begin{cases} 3x + 2y = 10 & \text{Equation 1} \\ 2x + 5y = 3 & \text{Equation 2} \end{cases}$

Multiply Equation 1 by 2 and Equation 2 by (-3):

$6x + 4y = 20$

$-6x - 15y = -9$

Add to eliminate x:

$-11y = 11 \Rightarrow y = -1$

Substitute $y = -1$ in Equation 1:

$3x - 2 = 10 \Rightarrow x = 4$

Solution: $(4, -1)$; consistent

17. $\begin{cases} 2u + v = 120 & \text{Equation 1} \\ u + 2v = 120 & \text{Equation 2} \end{cases}$

Multiply Equation 2 by (-2):

$-2u - 4v = -240$

Add this to Equation 1 to eliminate u:

$-3v = -120 \Rightarrow v = 40$

Substitute $v = 40$ in Equation 2:

$u + 80 = 120 \Rightarrow u = 40$

Solution: $(40, 40)$; consistent

19. $\begin{cases} 4b + 3m = 3 & \text{Equation 1} \\ 3b + 11m = 13 & \text{Equation 2} \end{cases}$

Multiply Equation 1 by 3 and Equation 2 by (-4):

$12b + 9m = 9$

$-12b - 44m = -52$

Add to eliminate b:

$-35m = -43 \Rightarrow m = \frac{43}{35}$

Substitute $m = \frac{43}{35}$ in Equation 1:

$4b + 3\left(\frac{43}{35}\right) = 3 \Rightarrow b = -\frac{6}{35}$

Solution: $\left(-\frac{6}{35}, \frac{43}{35}\right)$; consistent

21. $\begin{cases} \dfrac{x}{4} + \dfrac{y}{6} = 1 & \text{Equation 1} \\ x - y = 3 & \text{Equation 2} \end{cases}$

Multiply Equation 1 by 6: $\frac{3}{2}x + y = 6$

Add this to Equation 2 to eliminate y:

$\frac{5}{2}x = 9 \Rightarrow x = \frac{18}{5}$

Substitute $x = \frac{18}{5}$ in Equation 2: $\frac{18}{5} - y = 3 \Rightarrow y = \frac{3}{5}$

Solution: $\left(\frac{18}{5}, \frac{3}{5}\right)$; consistent

23. $\begin{cases} \dfrac{x+3}{4} + \dfrac{y-1}{3} = 1 & \text{Equation 1} \\ x - y = 3 & \text{Equation 2} \end{cases}$

Multiply Equation 1 by 12 and Equation 2 by 4:

$3x + 4y = 7$

$4x - 4y = 12$

Add to eliminate y:

$7x = 19 \Rightarrow x = \frac{19}{7}$

Substitute $x = \frac{19}{7}$ in Equation 2:

$\frac{19}{7} - y = 3 \Rightarrow y = -\frac{2}{7}$

Solution: $\left(\frac{19}{7}, -\frac{2}{7}\right)$; consistent

25.
$$\begin{aligned} x + y &= 13 \\ x - y &= 3 \\ \hline 2x &= 16 \\ x &= 8 \\ x + y &= 13 \\ 8 + y &= 13 \\ y &= 5 \end{aligned}$$

Solution: $(8, 5)$

27.
$$\begin{aligned} 2r + s &= 8 \\ r - s &= 7 \\ \hline 3r &= 15 \\ r &= 5 \\ r - s &= 7 \\ 5 - s &= 7 \\ s &= -2 \end{aligned}$$

Solution: $(5, -2)$

29. $\begin{cases} 6r - 5s = 3 & \text{Equation 1} \\ -1.2r + s = 0.5 & \text{Equation 2} \end{cases}$

Multiply Equation 2 by 5:

$-6r + 5s = 2.5$

Add this to Equation 1:

$0 = 5.5$

Inconsistent; no solution

31. $\begin{cases} 2.5x - 3y = 1.5 & \text{Equation 1} \\ 10x - 12y = 6 & \text{Equation 2} \end{cases}$

Multiply Equation 1 by (-4): $-10x + 12y = -6$

Add this to Equation 2 to eliminate x: $0 = 0$ (dependent)

The solution set consists of all points lying on the line $10x - 12y = 6$.

Let $x = a$, then $y = \frac{5}{6}a - \frac{1}{2}$.

Solution: $\left(a, \frac{5}{6}a - \frac{1}{2}\right)$, where a is any real number;

Consistent

33. $\begin{cases} 0.05x - 0.03y = 0.21 & \text{Equation 1} \\ 0.07x + 0.02y = 0.16 & \text{Equation 2} \end{cases}$

Multiply Equation 1 by 200 and Equation 2 by 300:

$10x - 6y = 42$

$21x + 6y = 48$

Add to eliminate y: $31x = 90 \Rightarrow x = \frac{90}{31}$

Substitute $x = \frac{90}{31}$ in Equation 2:

$0.07\left(\frac{90}{31}\right) + 0.02y = 0.16 \Rightarrow y = -\frac{67}{31}$

Solution: $\left(\frac{90}{31}, -\frac{67}{31}\right)$; consistent

35. $\begin{cases} 8x + 12y = 8 & \text{Equation 1} \\ 16x + 30y = 22 & \text{Equation 2} \end{cases}$

The method of elimination can be used because it is easy to eliminate x.

Multiply Equation 1 by -2:

$$-2(8x + 12y) = -2(8)$$
$$-16x - 24y = -16$$

Add this to Equation 2:

$$\begin{aligned} -16x - 24y &= -16 \\ 16x + 30y &= 22 \\ \hline 6y &= 6 \\ y &= 1 \end{aligned}$$

Back-substitute $y = 1$ into Equation 1:

$$\begin{aligned} 8x + 12y &= 8 \\ 8x + 12(1) &= 8 \\ 8x &= -4 \\ x &= -\tfrac{1}{2} \end{aligned}$$

Solution: $\left(-\frac{1}{2}, 1\right)$

37. $\begin{cases} -3x + 10y = -84 & \text{Equation 1} \\ y = x \quad 7 & \text{Equation 2} \end{cases}$

Because y is written in terms of x in Equation 2, the method of substitution can be used easily.

Substitute $x - 7$ for y in Equation 1:

$$\begin{aligned} -3x + 10(x - 7) &= -84 \\ -3x + 10x - 70 &= -84 \\ 7x - 70 &= -84 \\ 7x &= -14 \\ x &= -2 \end{aligned}$$

Back-substitute $x = -1$ into Equation 2:

$$\begin{aligned} y &= x - 7 \\ y &= -2 - 7 \\ y &= -9 \end{aligned}$$

Solution: $(-2, -9)$

39. $\begin{cases} y = e^x & \text{Equation 1} \\ 2x - 3y + 3 = 0 & \text{Equation 2} \end{cases}$

Because neither x nor y can be eliminated without resulting in a complicated equation, the graphing method can be used.

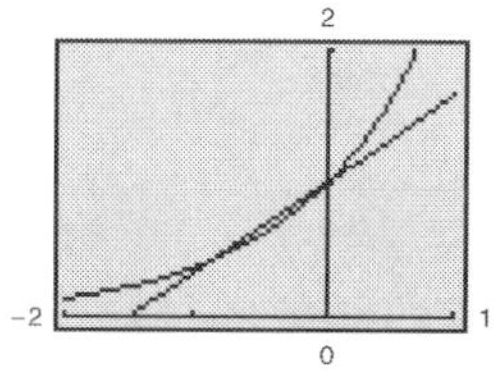

The graphs of $y = e^x$ and $2x - 3y + 3 = 0$ intersect at $(0, 1)$ and about $(-0.874, 0.417)$.

41. $\begin{cases} 4x + 3y = 140 & \text{Equation 1} \\ 3x + 4y = 140 & \text{Equation 2} \end{cases}$

In this system, either x or y can be eliminated. Eliminate y by multiplying Equation 1 by 4 and Equation 2 by -3:

$$4(4x + 3y) = 4(140)$$
$$16x + 12y = 560$$
$$-3(3x + 4y) = -3(140)$$
$$-9x - 12y = -420$$

Add the revised equations:

$$\begin{aligned} 16x + 12y &= 560 && \text{Revised Equation 1} \\ -9x - 12y &= -420 && \text{Revised Equation 2} \\ \hline 7x &= 140 \\ x &= 20 \end{aligned}$$

Back-substitute $x = 20$ into Equation 1:

$$\begin{aligned} 4x + 3y &= 140 \\ 4(20) + 3y &= 140 \\ 80 + 3y &= 140 \\ 3y &= 60 \\ y &= 20 \end{aligned}$$

Solution: $(20, 20)$

43. Answers will vary. Sample answers:

(a) $\begin{cases} 3x + y = 18 \\ x + y = 6 \end{cases}$

(b) $\begin{cases} -2x + 8y = 38 \\ 4x + y = -8 \end{cases}$

(c) $\begin{cases} 8x - y = 1 \\ -4x + 4y = -25 \end{cases}$

45. $\begin{cases} 200y - x = 200 & \text{Equation 1} \\ 199y - x = -198 & \text{Equation 2} \end{cases}$

Subtract to eliminate x: $y = 398$

Substitute $y = 398$ in Equation 1:

$200(398) - x = 200 \Rightarrow x = 79{,}400$

Solution: (79,400, 398)

Because the slopes of the two lines are not equal, the lines are *not* parallel. To verify the intersection graphically, the graph needs to be zoomed out considerably to show the point of intersection.

47. Let x = the ground speed and y = the wind speed.

$\begin{cases} 3.6(x - y) = 1800 & \text{Equation 1} \\ 3(x + y) = 1800 & \text{Equation 2} \end{cases}$

$$\begin{aligned} x - y &= 500 \\ x + y &= 600 \\ \hline 2x &= 1100 \\ x &= 550 \\ 550 + y &= 600 \\ y &= 50 \end{aligned}$$

Solution: (550, 50)

The ground speed is 550 mph and the wind speed is 50 mph.

49. Let x = the number of gallons at 20%

y = the number of gallons at 50%

$\begin{cases} x + y = 10 & \text{Equation 1} \\ 0.2x + 0.5y = 0.3(10) & \text{Equation 2} \end{cases}$

$$\begin{aligned} -2x - 2y &= -20 \\ 2x + 5y &= 30 \\ \hline 3y &= 10 \\ y &= \tfrac{10}{3} \\ x + \tfrac{10}{3} &= 10 \\ &= \tfrac{20}{3} \end{aligned}$$

Solution: $\left(\frac{20}{3}, \frac{10}{3}\right)$

The amounts required are $x = 6\frac{2}{3}$ gallons at 20% and $y = 3\frac{1}{3}$ gallons at 50%.

51. Amount in 4.75% bond + Amount in 5.5% bond = Total investment

$x + y = 25{,}000$

Interest for 4.75% bond + Interest for 5.5% bond = Total interest

$0.0475x + 0.055y = 1315$

$\begin{cases} x + y = 25{,}000 & \text{Equation 1} \\ 0.0475x + 0.055y = 1315 & \text{Equation 2} \end{cases}$

Multiply Equation 2 by 10,000: $10{,}000(0.0475x + 0.055y) = 10{,}000(1315)$

$475x + 550y = 13{,}150{,}000$

Solve Equation 1 for x: $x = 25{,}000 - y$

Substitute $x = 25{,}000 - y$ into Revised Equation 2:

$$\begin{aligned} 475x + 550y &= 13{,}150{,}000 \\ 475(25{,}000 - y) + 550y &= 13{,}150{,}000 \\ 11{,}875{,}000 - 475y + 550y &= 13{,}150{,}000 \\ 11{,}875{,}000 + 75y &= 13{,}150{,}000 \\ 75y &= 1{,}275{,}000 \\ y &= 17{,}000 \end{aligned}$$

Back-substitute $y = 17{,}000$ into Revised Equation 1:

$$\begin{aligned} x &= 25{,}000 - y \\ x &= 25{,}000 - 17{,}000 \\ x &= 8000 \end{aligned}$$

Solution: $(8000, 17{,}000)$

So, \$8000 is invested in the 4.75% bond and \$17,000 is invested in the 5.5% bond.

53. Demand = Supply

$$\begin{aligned} 36 - 0.0001x &= 22 + 0.00001x \\ 34 &= 0.00011x \\ x &\approx 309{,}091 \text{ units} \\ p &\approx \$25.09 \end{aligned}$$

55. Demand = Supply

$$\begin{aligned} 140 - 0.00002x &= 80 + 0.00001x \\ 60 &= 0.00003x \\ x &= 2{,}000{,}000 \text{ units} \\ p &= \$100.00 \end{aligned}$$

57. (a) Fast-food

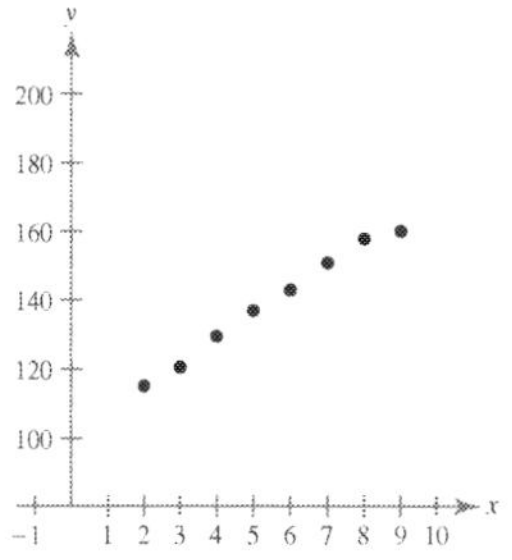

$y = 6.79x + 101.8$

Full-service

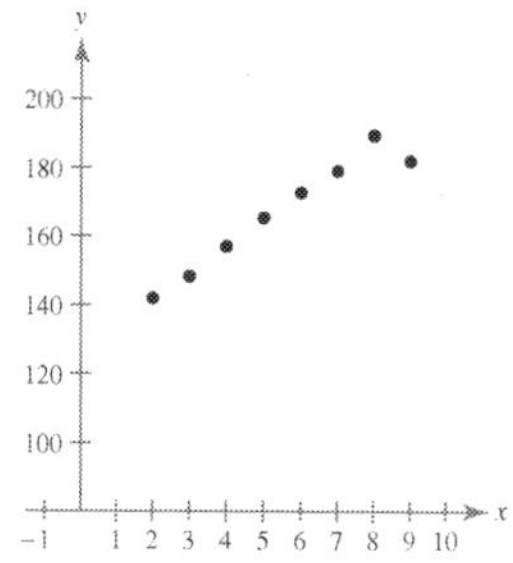

$y = 6.67x + 130.3$

(b)
$$\begin{aligned} 6.79x + 101.8 &= 6.67x + 130.3 \\ 0.12x + 101.8 &= 130.3 \\ 0.12x &= 28.5 \\ x &= 237.5 \end{aligned}$$

In the near future, fast-food sales will not equal full-service sales. According to the models, fast-food sales will equal full-service sales in the year 2237. So, these models are not appropriate to use for predicting when fast-food sales will equal full-service sales.

59. $\begin{cases} p + 0.53x = 1542 & \text{Demand} \\ p - 0.37x = 300 & \text{Supply} \end{cases}$

$$\begin{aligned} p + 0.53x &= 1542 &\Rightarrow\quad p + 0.53x &= 1542 \\ p - 0.37x &= 300 &\xrightarrow{\times(-1)}\quad -p + 0.37x &= -300 \\ & & 0.90x &= 1242 \\ & & x &= 1380 \end{aligned}$$

$$\begin{aligned} p + 0.53x &= 1542 \\ p + 0.53(1380) &= 1542 \\ p &= 810.6 \end{aligned}$$

The number of units demanded will equal the quantity supplied when 1380 units are sold at a price of \$810.60.

61.
$$\begin{aligned} 5b + 10a = 20.2 &\Rightarrow -10b - 20a = -40.4 \\ 10b + 30a = 50.1 &\Rightarrow \underline{\ \ 10b + 30a = 50.1\ \ } \\ & \qquad\qquad 10a = 9.7 \\ & \qquad\qquad a = 0.97 \\ & \qquad\qquad b = 2.1 \end{aligned}$$

Least squares regression line: $y = 0.97x + 2.1$

63.
$$\begin{aligned} 7b + 21a = 35.1 &\Rightarrow -21b - 63a = -105.3 \\ 21b + 91a = 114.2 &\Rightarrow \underline{\ \ 21b + 91a = 114.2\ \ } \\ & \qquad\qquad 28a = 8.9 \\ & \qquad\qquad a = \tfrac{89}{280} \approx 0.318 \\ & \qquad\qquad b = \tfrac{1137}{280} \approx 4.061 \end{aligned}$$

Least squares regression line:

$y = \frac{1}{280}(89x + 1{,}137) \approx 0.318x + 4.061$

65. (a) $\begin{cases} 5b + 10a = 1943.26 & \text{Equation 1} \\ 10b + 30a = 3906.83 & \text{Equation 2} \end{cases}$

Eliminate b by multiplying Equation 1 by -2 and adding the result to Equation 2:

$$-2(5b + 10a) = -2(1943.26)$$
$$-10b - 20a = -3886.52$$

$$\begin{aligned} -10b - 20a &= -3886.52 \\ 10b + 30a &= 3906.83 \\ \hline 10a &= 20.31 \\ a &= 2.031 \end{aligned}$$

Back-substitute $a = 2.031$ into Equation 1:

$$\begin{aligned} 5b + 10a &= 1943.26 \\ 5b + 10(2.031) &= 1943.26 \\ 5b + 20.31 &= 1943.26 \\ 5b &= 1922.95 \\ b &= 384.59 \end{aligned}$$

So, the model is $y = 2.031t + 384.59$.

(b) 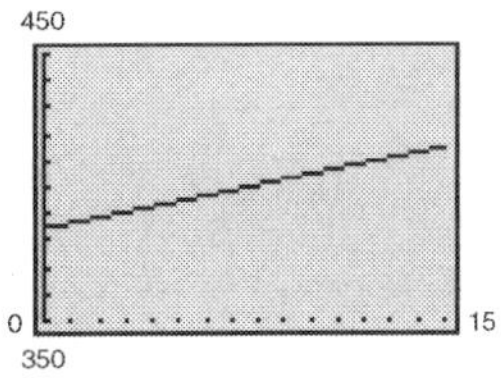

The greatest monthly carbon dioxide concentration in 2016 will be about 404.9 parts per million.

(c) The regression model using a graphing utility is $y = 2.031t + 384.59$, which is the same as the one found in part (a).

67. Answers will vary.

5.3 Linear Systems in Three or More Variables

Skills Warm Up

1. $\begin{cases} x + y = 25 \\ y = 15 \end{cases}$

Back-substitute $y = 15$:

$$\begin{aligned} x + y &= 25 \\ x + 15 &= 25 \\ x &= 10 \end{aligned}$$

Solution: $(10, 15)$

2. $\begin{cases} 2x - 3y = 4 \\ 6x = -18 \end{cases}$

Solve for x: $6x = -18 \Rightarrow x = -3$

Back-substitute $x = -3$:

$$\begin{aligned} 2x - 3y &= 4 \\ 2(-3) - 3y &= 4 \\ -6 - 3y &= 4 \\ -3y &= 10 \\ y &= -\tfrac{10}{3} \end{aligned}$$

Solution: $\left(-3, -\frac{10}{3}\right)$

3. $\begin{cases} x + y = 32 \\ x - y = 24 \end{cases}$ $\xrightarrow{\times(-1)}$

$$\begin{aligned} x + y &= 32 \\ -x + y &= -24 \\ \hline 2y &= 8 \\ y &= 4 \end{aligned}$$

Back-substitute $y = 4$: $x + 4 = 32 \Rightarrow x = 28$

Solution: (28, 4)

4. $\begin{cases} 2r - s = 5 \\ r + 2s = 10 \end{cases}$ $\xrightarrow{\times 2}$

$$\begin{aligned} 4r - 2s &= 10 \\ r + 2s &= 10 \\ \hline 5r &= 20 \\ r &= 4 \end{aligned}$$

Back-substitute $r = 4$:

$4 + 2s = 10 \Rightarrow s = 3$

Solution: (4, 3)

5. $5x - 3y + 4z = 2,\ (-1, -2, 1)$

$$5(-1) - 3(-2) + 4(1) \stackrel{?}{=} 2$$
$$5 \neq 2$$

Not a solution

Skills Warm Up —*continued*—

6. $x - 2y + 12z = 9,\ (6, 3, 2)$

$6 - 2(3) + 12(2) \stackrel{?}{=} 9$

$24 \neq 9$

Not a solution

7. $2x - 5y + 3z = -9,\ (a - 2,\ a + 1,\ a)$

$2(a - 2) - 5(a + 1) + 3a \stackrel{?}{=} -9$

$2a - 4 - 5a - 5 + 3a \stackrel{?}{=} -9$

$-9 = -9$ ✓

Solution

8. $-5x + y + z = 21,\ (a - 4,\ 4a + 1,\ a)$

$-5(a - 4) + 4a + 1 + a \stackrel{?}{=} 21$

$-5a + 20 + 4a + 1 + a \stackrel{?}{=} 21$

$21 = 21$ ✓

Solution

9. $x + 2y - 3z = 4,\ y = 1 - a,\ z = a$

$x + 2(1 - a) - 3a = 4$

$x = 2 + 5a$

10. $x - 3y + 5z = 4,\ y = 2a + 3,\ z = a$

$x - 3(2a + 3) + 5a = 4$

$x = 13 + a$

1. C;

$$\begin{cases} x + 2y + 5z = -11 & \text{Interchange Eq.1 and Eq.3} \\ 3x - 4y + z = -1 \\ -2x + 3y - 2z = 5 \end{cases}$$

$$\begin{cases} x + 2y + 5z = -11 \\ -10y - 14z = 32 & -3\text{Eq.1} + \text{Eq.2} \\ 7y + 8z = -17 & 2\text{Eq.1} + \text{Eq.3} \end{cases}$$

$$\begin{cases} x + 2y + 5z = -11 \\ y + \frac{7}{5}z = -\frac{16}{5} & -\frac{1}{10}\text{Eq.2} \\ 7y + 8z = -17 \end{cases}$$

$$\begin{cases} x + 2y + 5z = -11 \\ y + \frac{7}{5}z = -\frac{16}{5} \\ -\frac{9}{5}z = \frac{27}{5} & -7\text{Eq.2} + \text{Eq.3} \end{cases}$$

$-\frac{9}{5}z = \frac{27}{5} \Rightarrow z = -3$

$y + \frac{7}{5}(-3) = -\frac{16}{5} \Rightarrow y = 1$

$x + 2(1) + 5(-3) = -11 \Rightarrow x = 2$

Solution: $(2, 1, -3)$

3. B;

$$\begin{cases} x + 6y - 10z = 38 & \text{Interchange Eq.1 and Eq.2} \\ 2x + 5y - 7z = 36 \\ x - 4y + 8z = -18 \end{cases}$$

$$\begin{cases} x + 6y - 10z = 38 \\ -7y + 13z = -40 & -2\text{Eq.1} + \text{Eq.2} \\ -10y + 18z = -56 & -\text{Eq.1} + \text{Eq.3} \end{cases}$$

$$\begin{cases} x + 6y - 10z = 38 \\ y - \frac{13}{7}z = \frac{40}{7} & -\frac{1}{7}\text{Eq.2} \\ -10y + 18z = -56 \end{cases}$$

$$\begin{cases} x + 6y - 10z = 38 \\ y - \frac{13}{7}z = \frac{40}{7} \\ -\frac{4}{7}z = \frac{8}{7} & 10\text{Eq. }2 + \text{Eq. }3 \end{cases}$$

$-\frac{4}{7}z = \frac{8}{7} \Rightarrow z = -2$

$y - \frac{13}{7}(-2) = \frac{40}{7} \Rightarrow y = 2$

$x + 6(2) - 10(-2) = 38 \Rightarrow x = 6$

Solution: $(6, 2, -2)$

5. Yes, the system is in row–echelon form because it has a stair-step pattern with leading coefficients of 1.

7. No, the system is not in row-echelon form because the leading coefficient of the second equation is not 1.

9. $\begin{cases} x - y + z = 4 \\ y + 2z = -6 \\ z = -2 \end{cases}$

Using back-substitution in Equation 2

$y + 2(-2) = -6 \Rightarrow y = -2$

Using back-substitution in Equation 1

$x - (-2) + (-2) = 4 \Rightarrow x = 4$

Solution: $(4, -2, -2)$

11. $\begin{cases} x + y + z = -3 \\ 4x + y - 3z = 11 \\ 2x - 3y + 2z = 9 \end{cases}$ Interchange the equations.

$\begin{cases} x + y + z = -3 \\ -3y - 7z = 23 & -4\text{Eq.}1 + \text{Eq.}2 \\ -5y = 15 & -2\text{Eq.}1 + \text{Eq.}3 \end{cases}$

$-5y = 15 \Rightarrow y = -3$

$-3(3) - 7z = 23 \Rightarrow z = -2$

$x - 3 - 2 = -3 \Rightarrow x = 2$

Solution: $(2, -3, -2)$

13. $\begin{cases} 3x + 2z = 13 \\ x + 2y + z = -5 \\ -3y - z = 10 \end{cases}$

$$\begin{array}{rl} -6y - z = 28 & -3\text{Eq.}2 + \text{Eq.}1 \\ 3y + z = -10 & -\text{Eq.}3 \\ \hline -3y = 18 & \\ y = -6 & \end{array}$$

$-3(-6) - z = 10 \Rightarrow z = 8$

$3x + 2(8) = 13 \Rightarrow x = -1$

Solution: $(-1, -6, 8)$

15. $\begin{cases} x + y - 2z = 3 \\ 3x - 2y + 4z = 1 \\ 2x - 3y + 6z = 8 \end{cases}$ Interchange the equations.

$\begin{cases} x + y - 2z = 3 \\ -5y + 10z = -8 & -3\text{Eq.}1 + \text{Eq.}2 \\ -5y + 10z = 2 & -2\text{Eq.}1 + \text{Eq.}3 \end{cases}$

$\begin{cases} x + y - 2z = 3 \\ -5y + 10z = -8 \\ 0 = 10 \rightarrow\leftarrow & -\text{Eq.}2 + \text{Eq.}3 \end{cases}$

No solution.

The system is inconsistent.

17. $\begin{cases} 3x + 3y + 5z = 1 \\ 3x + 5y + 9z = 0 \\ 5x + 9y + 17z = 0 \end{cases}$

$\begin{cases} 6x + 6y + 10z = 2 & 2\text{Eq.}1 \\ 3x + 5y + 9z = 0 \\ 5x + 9y + 17z = 0 \end{cases}$

$\begin{cases} x - 3y - 7z = 2 & -\text{Eq.}3 + \text{Eq.}1 \\ 3x + 5y + 9z = 0 \\ 5x + 9y + 17z = 0 \end{cases}$

$\begin{cases} x - 3y - 7z = 2 \\ 14y + 30z = -6 & -3\text{Eq.}1 + \text{Eq.}2 \\ 24y + 52z = -10 & -5\text{Eq.}1 + \text{Eq.}3 \end{cases}$

$\begin{cases} x - 3y - 7z = 2 \\ 84y + 180z = -36 & 6\text{Eq.}2 \\ 84y + 182z = -35 & 3.5\text{Eq.}3 \end{cases}$

$\begin{cases} x - 3y - 7z = 2 \\ 84y + 180z = -36 \\ 2z = 1 & -\text{Eq.}2 + \text{Eq.}3 \end{cases}$

$2z = 1 \Rightarrow z = \frac{1}{2}$

$84y + 180\left(\frac{1}{2}\right) = -36 \Rightarrow y = -\frac{3}{2}$

$x - 3\left(-\frac{3}{2}\right) - 7\left(\frac{1}{2}\right) = 2 \Rightarrow x = 1$

Solution: $\left(1, -\frac{3}{2}, \frac{1}{2}\right)$

19. $\begin{cases} x + 2y - 7z = -4 \\ 2x + y + z = 13 \\ 3x + 9y - 36z = -33 \end{cases}$

$\begin{cases} x + 2y - 7z = -4 \\ -3y + 15z = 21 & -2\text{Eq.}1 + \text{Eq.}2 \\ 3y - 15z = -21 & -3\text{Eq.}1 + \text{Eq.}3 \end{cases}$

$\begin{cases} x + 2y - 7z = -4 \\ -3y + 15z = 21 \\ 0 = 0 & \text{Eq.}2 + \text{Eq.}3 \end{cases}$

$\begin{cases} x + 2y - 7z = -4 \\ y - 5z = -7 & \frac{1}{3}\text{Eq.}2 \end{cases}$

$\begin{cases} x + 3z = 10 & -2\text{Eq.}2 + \text{Eq.}1 \\ y - 5z = -7 \end{cases}$

Let $z = a$. Then

$y = 5a - 7$

$x = -3a + 10$

Solution: $(-3a + 10, 5a - 7, a)$

21. $\begin{cases} x + 4z = 13 \\ 4x - 2y + z = 7 \\ 2x - 2y - 7z = -19 \end{cases}$

$\begin{cases} x + 4z = 13 & \\ -2y - 15z = -45 & -4\text{Eq.1} + \text{Eq.2} \\ -2y - 15z = -45 & -2\text{Eq.1} + \text{Eq.3} \end{cases}$

$\begin{cases} x + 4z = 13 & \\ -2y - 15z = -45 & \\ 0 = 0 & -\text{Eq.2} + \text{Eq.3} \end{cases}$

Let $z = a$. Then

$y = -\frac{15}{2}a + \frac{45}{2}$

$x = -4a + 13$

Solution: $\left(-4a + 13, -\frac{15}{2}a + \frac{45}{2}, a\right)$

23. $\begin{cases} x + 4z = 1 \\ x + y + 10z = 10 \\ 2x - y + 2z = -5 \end{cases}$

$\begin{cases} x + 4z = 1 & \\ y + 6z = 9 & -\text{Eq.1} + \text{Eq.2} \\ -y + 6z = -7 & -2\text{Eq.1} + \text{Eq.3} \end{cases}$

$\begin{cases} x + 4z = 1 & \\ y + 6z = 9 & \\ 0 = 2 \rightarrow\leftarrow & \text{Eq.2} + \text{Eq.3} \end{cases}$

No solution.

The system is inconsistent.

25. $\begin{cases} 4x + 3y + 5z = 10 \\ 5x + 2y + 10z = 13 \\ 3x + y - 2z = -9 \end{cases}$

$\begin{cases} 4x + 3y + 5z = 10 & \\ -7y + 15z = 2 & -5\text{Eq.1} + 4\text{Eq.2} \\ -5y - 23z = -66 & -3\text{Eq.1} + 4\text{Eq.2} \end{cases}$

$\begin{cases} 4x + 3y + 5z = 10 & \\ -7y + 15z = 2 & \\ -236z = -472 & -5\text{Eq.2} + 7\text{Eq.3} \end{cases}$

$-236z = -472 \Rightarrow z = 2$

$-7y + 15z = 2 \Rightarrow y = 4$

$4x + 3y + 5z = 10 \Rightarrow x = -3$

Solution: $(-3, 4, 2)$

27. $\begin{cases} x - 2y + z = 7 \\ 2x + 3y - z = 1 \\ 3x + y + 2z = 12 \end{cases}$ Interchange the equations.

$\begin{cases} x - 2y + z = 7 & \\ 7y - 3z = -13 & -2\text{Eq.1} + \text{Eq.2} \\ 7y - z = -9 & -3\text{Eq.1} + \text{Eq.3} \end{cases}$

$\begin{cases} x - 2y + z = 7 & \\ y - \frac{3}{7}z = -\frac{13}{7} & \frac{1}{7}\text{Eq.2} \\ 2z = 4 & -\text{Eq.2} + \text{Eq.3} \end{cases}$

$2z = 4 \Rightarrow z = 2$

$y - \frac{3}{7}(2) = -\frac{13}{7} \Rightarrow y = -1$

$x - 2(-1) + 2 = 7 \Rightarrow x = 3$

Solution: $(3, -1, 2)$

29. $\begin{cases} 12x + 5y + z = 0 \\ 12x + 4y - z = 0 \end{cases}$

$\begin{cases} 12x + 5y + z = 0 & \\ -y - 2z = 0 & -\text{Eq.1} + \text{Eq.2} \end{cases}$

$\begin{cases} 12x - 9z = 0 & 5\text{Eq.2} + \text{Eq.1} \\ -y - 2z = 0 & \end{cases}$

Let $z = a$. Then

$y = -2a$

$x = \frac{3}{4}a.$

Solution: $\left(\frac{3}{4}a, -2a, a\right)$

31. $\begin{cases} x - 3y + 2z = 18 \\ 5x - 13y + 12z = 80 \end{cases}$

$\begin{cases} x - 3y + 2z = 18 & \\ 2y + 2z = -10 & -5\text{Eq.1} + \text{Eq.2} \end{cases}$

$\begin{cases} x - 3y + 2z = 18 & \\ y + z = -5 & \frac{1}{2}\text{Eq.2} \end{cases}$

$\begin{cases} x + 5z = 3 & 3\text{Eq.2} + \text{Eq.1} \\ y + z = -5 & \end{cases}$

Let $z = a$. Then

$y = -a - 5$

$x = -5a + 3$

Solution: $(-5a + 3, -a - 5, a)$

33. $\begin{cases} -x + y - 2z = 3 \\ 3x - 3y + 6z = 7 \\ 2x + 3y - 4z = 8 \end{cases}$ Interchange the equations.

$\begin{cases} -x + y - 2z = 3 & \\ 0 = 16 \rightarrow\leftarrow & 3\text{Eq.}1 + \text{Eq.}2 \\ 5y = 14 & 2\text{Eq.}1 + \text{Eq.}3 \end{cases}$

No solution. The system is inconsistent.

35. $\begin{cases} x + 3w = 4 \\ 2y - z - w = 0 \\ 3y - 2w = 1 \\ 2x - y + 4z = 5 \end{cases}$

$\begin{cases} x + 3w = 4 & \\ 2y - z - w = 0 & \\ 3y - 2w = 1 & \\ -y + 4z - 6w = -3 & -2\text{Eq.}1 + \text{Eq.}4 \end{cases}$

$\begin{cases} x + 3w = 4 & \\ y - 4z + 6w = 3 & -\text{Eq.}4 \text{ and interchange} \\ 2y - z - w = 0 & \text{the equations.} \\ 3y - 2w = 1 & \end{cases}$

$\begin{cases} x + 3w = 4 & \\ y - 4z + 6w = 3 & \\ 7z - 13w = -6 & -2\text{Eq.}2 + \text{Eq.}3 \\ 12z - 20w = -8 & -3\text{Eq.}2 + \text{Eq.}4 \end{cases}$

$\begin{cases} x + 3w = 4 & \\ y - 4z + 6w = 3 & \\ z - 3w = -2 & -\tfrac{1}{2}\text{Eq.}4 + \text{Eq.}3 \\ 12z - 20w = -8 & \end{cases}$

$\begin{cases} x + 3w = 4 & \\ y - 4z + 6w = 3 & \\ z - 3w = -2 & \\ 16w = 16 & -12\text{Eq.}3 + \text{Eq.}4 \end{cases}$

$16w = 16 \Rightarrow w = 1$

$z - 3(1) = -2 \Rightarrow z = 1$

$y - 4(1) + 6(1) = 3 \Rightarrow y = 1$

$x + 3(1) = 4 \Rightarrow x = 1$

Solution: (1, 1, 1, 1)

37. Answers will vary.

$\begin{cases} ax + by + cz = d \\ ey + fz = g \\ hz = i \end{cases}$

When $x = 3, y = -1$, and $z = 2$:

$\begin{cases} 3a - b + 2c = d \\ -e + 2f = g \\ 2h = i \end{cases}$

Find two sets of values for $a, b, c, \ldots, i$ that satisfy the above equations:

(1) Let $a = 2$, $b = -1$, and $c = -4$.
Then $d = 3(2) - (-1) + 2(-4) = -1$.
Let $h = 1$. Then $i = 2(1) = 2$.
System 1:

$\begin{cases} 2x - y - 4z = -1 \\ 5y + 6z = 7 \\ z = 2 \end{cases}$

(2) Let $a = 6$, $b = 9$, and $c = -1$.
Then $d = 3(6) - 9 + 2(-1) = 7$.
Let $e = \frac{5}{2}$ and $f = \frac{1}{4}$.
Then $g = -\left(\frac{5}{2}\right) + 2\left(\frac{1}{4}\right) = -2$
Let $h = -3$. Then $i = 2(-3) = -6$.
System 2:

$\begin{cases} 6x + 9y - z = 7 \\ \frac{5}{2}y + \frac{1}{4}z = -2 \\ -3z = -6 \end{cases}$

39. Answers will vary.

$$\begin{cases} ax + by + cz = d \\ ey + fz = g \\ hz = i \end{cases}$$

When $x = 1$, $y = -5$, and $z = -3$:

$$\begin{cases} a - 5b - 3c = d \\ -5e - 3f = g \\ -3h = i \end{cases}$$

Find two sets of values for $a, b, c, \ldots, i$ that satisfy the above equations:

(1) Let $a = 8$, $b = 1$, and $c = 2$. Then $d = 8 - 5(1) - 3(2) = -3$.

Let $e = 7$ and $f = -8$. Then $g = -5(7) - 3(-8) = -11$.

Let $h = 1$. Then $i = -3(1) = -3$.

System 1:

$$\begin{cases} 8x + y + 2z = -3 \\ 7y - 8z = -11 \\ z = -3 \end{cases}$$

(2) Let $a = 1$, $b = -2$, and $c = 2$. Then $d = 1 - 5(-2) - 3(2) = 5$.

Let $e = 1$ and $f = 1$. Then $g = -5(1) - 3(1) = -8$.

Let $h = -2$. Then $i = -3(-2) = 6$.

System 2:

$$\begin{cases} x - 2y + 2z = 5 \\ y + z = -8 \\ -2z = 6 \end{cases}$$

41. $\left(a, a + 2, \frac{1}{3}a\right)$

Answers will vary. Sample answer:

$a = 1$: $\left(1, 3, \frac{1}{3}\right)$

$a = 2$: $\left(2, 4, \frac{2}{3}\right)$

$a = 0$: $(0, 2, 0)$

43. $\left(\frac{1}{2}a, 4a, 7\right)$

Answers will vary.

Letting $a = 2$, the ordered triple is $(1, 8, 7)$.

Letting $a = 4$, the ordered triple is $(2, 16, 7)$.

Letting $a = 0$, the ordered triple is $(0, 0, 7)$.

45. $y = ax^2 + bx + c$

Passing through $(0, -4)$, $(1, 1)$, and $(2, 10)$

$(0, -4)$: $-4 = c$ Equation 1

$(1, 1)$: $1 = a + b + c$ Equation 2

$(2, 10)$: $10 = 4a + 2b + c$ Equation 3

Solution: $a = 2$, $b = 3$, $c = -4$

The equation of the parabola is:

$y = 2x^2 + 3x - 4.$

47. $y = ax^2 + bx + c$

Passing through $(-1, -5)$, $(0, 1)$, and $(1, -1)$

$(-1, -5)$: $-5 = a - b + c$ Equation 1

$(0, 1)$: $1 = c$ Equation 2

$(1, -1)$: $-1 = a + b + c$ Equation 3

Solution: $a = -4$, $b = 2$, $c = 1$

The equation of the parabola is:

$y = -4x^2 + 2x + 1.$

49. $x^2 + y^2 + Dx + Ey + F = 0$

Passing through $(0, 0)$, $(2, -2)$, and $(4, 0)$

$(0, 0)$: $F = 0$ Equation 1

$(2, -2)$: $8 + 2D - 2E + F = 0$ Equation 2

$(4, 0)$: $16 + 4D + F = 0$ Equation 3

Solution: $D = -4$, $E = 0$, $F = 0$

The equation is $x^2 + y^2 - 4x = 0$.

51. $x^2 + y^2 + Dx + Ey + F = 0$

Passing through $(0, -3)$, $(3, 0)$, and $(3, -6)$

$(0, -3)$: $9 - 3E + F = 0$

$(3, 0)$: $9 + 3D + F = 0$

$(3, -6)$: $45 + 3D - 6E + F = 0$

Solution: $D = -6$, $E = 6$, $F = 9$

The equation is $x^2 + y^2 - 6x + 6y + 9 = 0$.

53. Let x = amount borrowed at 3%

y = amount borrowed at 4%

z = amount borrowed at 6%

$$\begin{cases} x + y + z = 1{,}500{,}000 \\ 0.03x + 0.04y + 0.06z = 53{,}000 \\ y - z = 0 \end{cases}$$

Substitute $z = y$ into Equations 1 and 2:

$$\begin{cases} x + 2y = 1{,}500{,}000 & \xrightarrow{\times 10} & 10x + 20y = 15{,}000{,}000 \\ 0.03x + 0.1y = 53{,}000 & \xrightarrow{\times(-200)} & -6x - 20y = -10{,}600{,}000 \end{cases}$$

$$4x = 4{,}400{,}000$$

$$x = 1{,}100{,}000$$

Back-substitute $x = 1{,}100{,}000$:

$$\begin{aligned} x + 2y &= 1{,}500{,}000 \\ 1{,}100{,}000 + 2y &= 1{,}500{,}000 \\ 2y &= 400{,}000 \\ y &= 200{,}000 \\ z &= 200{,}000 \end{aligned}$$

Solution: x = \$1,100,000 at 3%

y = \$200,000 at 4%

z = \$200,000 at 6%

55. Let x = units of \$5 candles

y = units of \$10 candles

z = units of \$15 candles

$$\begin{cases} x + y + z = 50{,}000 \\ 5x + 10y + 15z = 550{,}000 \\ z = \frac{1}{2}y \end{cases}$$

$$\begin{cases} x + y + z = 50{,}000 \\ 5y + 10z = 300{,}000 & -5\text{Eq.1} + \text{Eq.2} \\ -\frac{1}{2}y + z = 0 \end{cases}$$

$$\begin{cases} x + y + z = 50{,}000 \\ y + 2z = 60{,}000 & \frac{1}{5}\text{Eq.2} \\ 2z = 30{,}000 & \frac{1}{10}\text{Eq.2} + \text{Eq.3} \end{cases}$$

Back-substitute $z = 15{,}000$:

$y + 2(15{,}000) = 60{,}000 \Rightarrow y = 30{,}000$

Back-substitute $y = 30{,}000$ and $z = 15{,}000$:

$x + 30{,}000 + 15{,}000 = 50{,}000 \Rightarrow x = 5000$

Solution: x = 5000 candles at \$5

y = 30,000 candles at \$10

z = 15,000 candles at \$15

57. Let x = gallons of spray X
y = gallons of spray Y
z = gallons of spray Z

$$\begin{aligned}\text{Chemical A: } &\tfrac{1}{6}x + \tfrac{1}{3}z = 5 &\xrightarrow{\times 6}\quad & x + 2z = 30\\ \text{Chemical B: } &\tfrac{1}{3}x + \tfrac{1}{3}z = 8 &\xrightarrow{\times 3}\quad & x + z = 24\\ \text{Chemical C: } &\tfrac{1}{2}x + y + \tfrac{1}{3}z = 12 &\xrightarrow{\times 6}\quad & 3x + 6y + 2z = 72\end{aligned}$$

Back-substitute $x = 24 - z$:

$(24 - z) + 2z = 30 \Rightarrow z = 6$

Back-substitute $z = 6$:

$x + 6 = 24 \Rightarrow x = 18$

Back-substitute $x = 18$ and $z = 6$:

$3(18) + 6y + 2(6) = 72 \Rightarrow y = 1$

Solution: x = 18 gallons of spray X
y = 1 gallon of spray Y
z = 6 gallons of spray Z

59. Let C = amount in certificates of deposit.
Let M = amount in municipal bonds.
Let B = amount in blue-chip stocks.
Let G = amount in growth or speculative stocks.

$$\begin{cases} C + M + B + G = 500{,}000 \\ 0.025C + 0.10M + 0.12B + 0.18G = 0.10(500{,}000) \\ B + G = \tfrac{1}{4}(500{,}000)\end{cases}$$

This system has infinitely many solutions.

Let $G = a$. Then $B = 125{,}000 - a$
$M = 341{,}666.67 - 0.8a$
$C = 33{,}333.33 + 0.8a$

Solution:

$(33{,}333.33 + 0.8a, 341{,}666.67 - 0.8a, 125{,}000 - a, a)$

where $0 \le a \le 125{,}000$

61.
$$\begin{aligned} 5c \quad + 10a &= 15.5\\ 10b \quad &= 6.3\\ 10c \quad + 34a &= 32.1\end{aligned}$$

Solution: $a = \frac{11}{140}$, $b = \frac{63}{100}$, $c = \frac{103}{35}$

Least squares regression parabola:

$y = \frac{11}{140}x^2 + \frac{63}{100}x + \frac{103}{35}$
$\approx 0.079x^2 + 0.63x + 2.9$

63.
$$\begin{cases} 6c + 3b + 19a = 23.9\\ 3c + 9b + 27a = -7.2\\ 19c + 27b + 115a = 48.8\end{cases}$$

Solution: $c = \frac{1779}{350}$, $b = -\frac{621}{700}$, $a = -\frac{29}{140}$

Least square regression parabola:

$y = -\frac{29}{140}x^2 - \frac{621}{700}x + \frac{1779}{350}$
$\approx -0.207x^2 - 0.89x + 5.1$

65. (a)
$$\begin{cases} 5c \quad + 10a = 20.9 & \text{Eq.1}\\ 10b \quad = 4.0 & \text{Eq.2}\\ 10c \quad + 34a = 43.4 & \text{Eq.3}\end{cases}$$

$$\begin{cases} c \quad + 2a = 4.18 & \tfrac{1}{5}\text{Eq.1}\\ b \quad = 0.4 & \tfrac{1}{10}\text{Eq.1}\\ 14a = 1.6 & -2\text{Eq.1} + \text{Eq.3.}\end{cases}$$

$14a = 1.6 \Rightarrow a = \frac{4}{35} \approx 0.114$
$b = 0.4$
$c = -2a + 4.18 = -2\left(\frac{4}{35}\right) + 4.18$
≈ 4.0

The least squares parabola is
$y = 0.114x^2 + 0.4x + 4.0.$

(b) The model is the same:
$y = 0.114x^2 + 0.4x + 4.0$

67. (a) $\begin{cases} 5c + 5b + 15a = 119 \\ 5c + 15b + 35a = 124 \\ 15c + 35b + 99a = 382 \end{cases}$

$$\begin{cases} c + b + 3a = 23.8 & \frac{1}{5}\text{Eq.1} \\ 10b + 20a = 5 & -\text{Eq.1} + \text{Eq.2} \\ 20b + 54a = 25 & -3\text{Eq.1} + \text{Eq.3} \end{cases}$$

$$\begin{cases} c + b + 3a = 23.8 & \\ b + 2a = 0.5 & \frac{1}{10}\text{Eq.2} \\ 14a = 15 & -2\text{Eq.2} + \text{Eq.3} \end{cases}$$

$14a = 15 \Rightarrow a = \frac{15}{14} \approx 1.071$

$b + 2\left(\frac{15}{14}\right) = 0.5 \Rightarrow b = -\frac{23}{14} \approx -1.64$

$c + \left(-\frac{23}{14}\right) + 3\left(\frac{15}{14}\right) = 23.8 \Rightarrow c = \frac{778}{35} \approx 22.2$

The least squares parabola is $y = 1.07x^2 - 1.6x + 22$.

(b) When $x = 4$:

$y = 1.07(4)^2 - 1.6(4) + 22 = 32.72$

In 2010, 32.72% of U.S. households were involved in vegetable gardening.

69. (a) $\begin{cases} 5c + 250b + 13{,}500a = 1140 & \text{Eq.1} \\ 250c + 13{,}500b + 775{,}000a = 66{,}950 & \text{Eq.2} \\ 13{,}500c + 775{,}000b + 46{,}590{,}000a = 4{,}090{,}500 & \text{Eq.3} \end{cases}$

$$\begin{matrix} 1000b + 100{,}000a = 9{,}950(-50\text{Eq.1} + \text{Eq.2}) \\ 1000b + 101{,}400a = 10{,}125(-2700\text{Eq.1} + \text{Eq.3}) \end{matrix} \Rightarrow \begin{matrix} a = 0.125 \\ b = -2.55 \end{matrix}$$

Using Eq.1 $5c + 250(-2.55) + 13{,}500(0.125) = 1140 \Rightarrow c = 18$

Model: $y = 0.125x^2 - 2.55x + 18$

(b) Using a regression feature

Model: $y = 0.125x^2 - 2.55x + 18$

(c) When $x = 75$: $y = 0.125(75)^2 - 2.55(75) + 18 = 529.875$

The stopping distance of the automobile traveling 75 miles per hour is about 530 feet.

71. $\begin{cases} 9a + 3b + c = 3 & \text{Eq.1} \\ 16a + 4b + c = 6 & \text{Eq.2} \\ 25a + 5b + c = 10 & \text{Eq.3} \end{cases}$

$\begin{cases} 9a + 3b + c = 3 & \\ 7a + b = 3 & \text{Eq.2} - \text{Eq.1} \\ 16a + 2b = 7 & \text{Eq.3} - \text{Eq.1} \end{cases}$

$\begin{cases} 9a + 3b + c = 3 & \\ 7a + b = 3 & \\ 2a = 1 & -2\text{Eq.2} + \text{Eq.3} \end{cases}$

$\begin{cases} 9a + 3b + c = 3 & \\ 7a + b = 3 & \\ a = \frac{1}{2} & \frac{1}{2}\text{Eq. 3} \end{cases}$

$a = \frac{1}{2}$

$7\left(\frac{1}{2}\right) + b = 3 \Rightarrow b = -\frac{1}{2}$

$9\left(\frac{1}{2}\right) + 3\left(-\frac{1}{2}\right) + c = 3 \Rightarrow c = 0$

The quadratic model is $y = \frac{1}{2}x^2 - \frac{1}{2}x$.

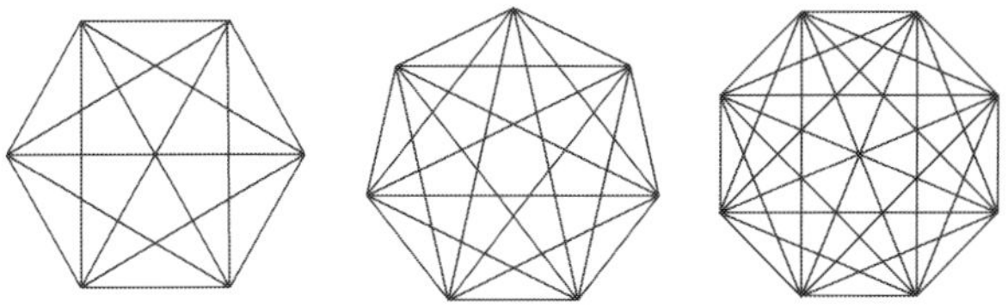

$x = 6, y = 15$ $x = 7, y = 21$ $x = 8, y = 28$

When $x = 6: y = \frac{1}{2}(6)^2 - \frac{1}{2}(6) = 15$

When $x = 7: y = \frac{1}{2}(7)^2 - \frac{1}{2}(7) = 21$

When $x = 8: y = \frac{1}{2}(8)^2 - \frac{1}{2}(8) = 28$

Yes. The relationship between x and y is represented by the equation for the regular polygons with 6, 7, and 8 sides.

73. Yes, it is possible for a square linear system to have infinitely many solutions because the graphs of the equations may intersect along a line or a plane.

75. $\begin{cases} x + y = 2 \\ y + z = 2 \\ x + z = 2 \\ ax + by + cz = 0 \end{cases}$

Answers will vary. Sample answer:

(a) $a = 4, b = -2, c = -2$

(b) $a = 3, b = 2, c = 1$

(c) Not possible

77. Answers will vary.

Chapter 5 Quiz Yourself

1. $\begin{cases} 3x + y = 11 \\ x - 2y = -8 \end{cases}$ $\Rightarrow$ $3x + y = 11$; $\times(-3)$ $\Rightarrow$ $-3x + 6y = 24$

$7y = 35$

$y = 5$

Back-substitute $y = 5$: $3x + 5 = 11 \Rightarrow x = 2$

Solution: $(2, 5)$

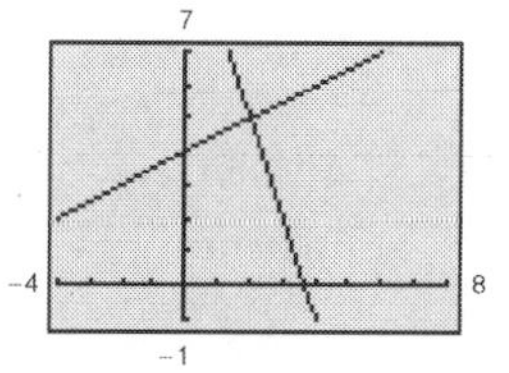

2. $\begin{cases} 4x + 8y = 8 \\ x + 2y = 6 \end{cases}$ $\Longrightarrow$ $\begin{aligned} 4x + 8y &= 8 \\ -4x - 8y &= -24 \\ \hline 0 &= -16 \end{aligned}$ (second equation $\times(-4)$)

No solution

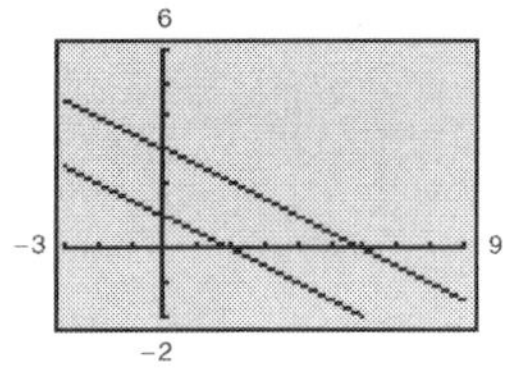

3. $\begin{cases} x + y = 4 & \text{Equation 1} \\ y = 2\sqrt{x} + 1 & \text{Equation 2} \end{cases}$

$$\begin{aligned} -x + 4 &= 2\sqrt{x} + 1 \\ -x + 3 &= 2\sqrt{x} \\ x^2 - 6x + 9 &= 4x \\ x^2 - 10x + 9 &= 0 \\ (x - 1)(x - 9) &= 0 \\ x &= 1 \end{aligned}$$

Solution: $(1, 3)$

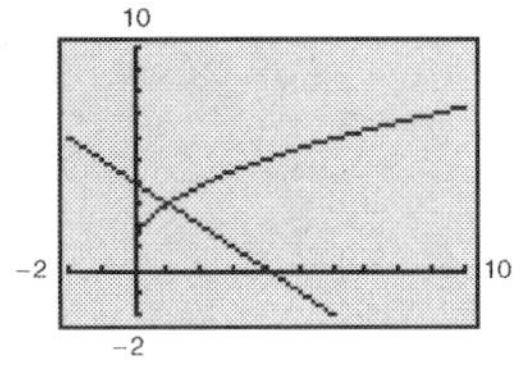

4. $\begin{cases} x^2 + y^2 = 9 & \text{Equation 1} \\ y = 2x + 1 & \text{Equation 2} \end{cases}$

Substitute $y = 2x + 1$ for y in Equation 1

$$\begin{aligned} x^2 + (2x + 1)^2 &= 9 \\ 5x^2 + 4x - 8 &= 0 \\ x &= \frac{-4 \pm \sqrt{4^2 - 4(5)(-8)}}{10} \\ &= -\frac{2}{5} \pm \frac{2\sqrt{11}}{5} \text{ or} \\ x &\approx -1.727,\ 0.927 \end{aligned}$$

Solution: $\left(-\frac{2}{5} \pm \frac{2\sqrt{11}}{5},\ \frac{1}{5} \pm \frac{4\sqrt{11}}{5}\right)$ or

$(-1.727, -2.454), (0.927, 2.854)$

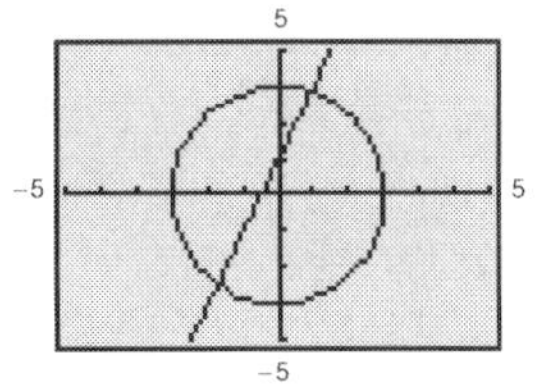

5. $C = 10.50x + 9000,\ R = 16.50x$

$$\begin{aligned} R &= C \\ 16.50x &= 10.50x + 9000 \\ 6x &= 9000 \\ x &= 1500 \text{ units} \end{aligned}$$

6. $C = 3.79x + 400{,}000,\ R = 4.59x$

$$\begin{aligned} R &= C \\ 4.59x &= 3.79x + 400{,}000 \\ 0.80x &= 400{,}000 \\ x &= 500{,}000 \text{ units} \end{aligned}$$

7. $2.5x - y = 6 \Rightarrow 10x - 4y = 24$

$$\begin{aligned} 3x + 4y &= 2 \\ \hline 13x &= 26 \\ x &= 2 \end{aligned}$$

$$\begin{aligned} 3(2) + 4y &= 2 \\ 4y &= -4 \\ y &= -1 \end{aligned}$$

Solution: $(2, -1)$

8. $\frac{1}{2}x + \frac{1}{3}y = 1 \Rightarrow 3x + 2y = 6$

$$\begin{aligned} x - 2y &= -2 \\ \hline 4x &= 4 \\ x &= 1 \end{aligned}$$

$$\begin{aligned} 1 - 2y &= -2 \\ -2y &= -3 \\ y &= \tfrac{3}{2} \end{aligned}$$

Solution: $\left(1, \frac{3}{2}\right)$

9.
$$\begin{aligned} \text{Supply} &= \text{Demand} \\ 20 + 0.004x &= 50 - 0.002x \\ 0.006x &= 30 \\ x &= 5000 \end{aligned}$$

$p = 50 - 0.002(5000) = 40$

Point of equilibrium: $(5000, 40)$

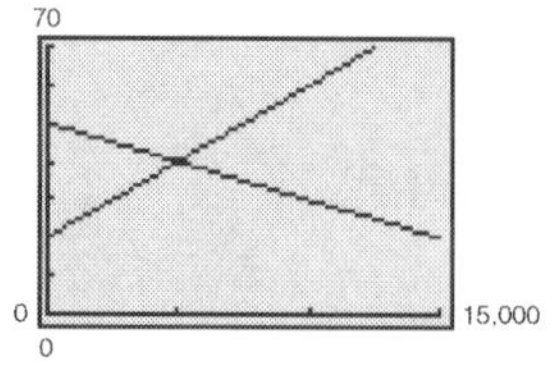

10. $\begin{cases} 5b + 10a = 221.4 & \text{Eq.1} \\ 10b + 30a = 451.2 & \text{Eq.2} \end{cases}$

$\begin{cases} b + 2a = 44.28 & \frac{1}{5}\text{Eq.1} \\ 10a = 8.4 & \text{Eq.2} - 2\text{Eq.1} \end{cases}$

$\begin{cases} b + 2a = 44.28 \\ a = 0.84 & \frac{1}{10}\text{Eq.2} \end{cases}$

$a = 0.84$

$b + 2(0.84) = 44.28 \Rightarrow b = 42.6$

The least squares regression line is $y = 0.84t + 42.6$.

11. $\begin{cases} x + 3z = 10 & \text{Interchange the equations.} \\ 2x + 3y - z = -7 \\ 2y + z = -1 \end{cases}$

$\begin{cases} x + 3z = 10 \\ 3y - 7z = -27 & -2\text{Eq.1} + \text{Eq.2} \\ 2y + z = -1 \end{cases}$

$\begin{cases} x + 3z = 10 \\ y - 8z = -26 & -1\text{Eq.3} + \text{Eq.2} \\ 17z = 51 & -2\text{Eq.2} + \text{Eq.3} \end{cases}$

$z = 3$ Solve and use back-substitution.

$y - 8(3) = -26 \Rightarrow y = -2$

$x + 3(3) = 10 \Rightarrow x = 1$

Solution: $(1, -2, 3)$

12. $\begin{cases} x + y - 2z = 12 \\ 2x - y - z = 6 \\ y - z = 6 \end{cases}$

$\begin{cases} x + y - 2z = 12 \\ -3y + 3z = 18 & -2\text{Eq.1} + \text{Eq.2} \\ y - z = 6 \end{cases}$

$\begin{cases} x + y - 2z = 12 \\ -3y + 3z = -18 \\ 0 = 0 & \text{Eq.2} + 3\text{Eq.3} \end{cases}$

Let $z = a$. Then

$y = a + 6$

$x = a + 6$

Solution: $(a + 6, a + 6, a)$; a is any real number.

13. $\begin{cases} 3x + 2y + z = 17 \\ -x + y + z = 4 \\ x - y - z = 3 \end{cases}$

$\begin{cases} 3x + 2y + z = 17 \\ 5y + 4z = 29 & \text{Eq.1} + 3\text{Eq.2} \\ -5y - 4z = -26 & -\text{Eq.1} + 3\text{Eq.3} \end{cases}$

$\begin{cases} 3x + 2y + z = 17 \\ 5y + 4z = 29 \\ 0 = 3 & \text{Eq.2} + \text{Eq.3} \end{cases}$

Inconsistent system

No solution

14. (a) $\begin{cases} 5c + 10a = 350.56 & \text{Eq.1} \\ 10b = 32.04 & \text{Eq.2} \\ 10c + 34a = 703.08 & \text{Eq.3} \end{cases}$

$\begin{cases} c + 2a = 70.112 & \frac{1}{5}\text{Eq.1} \\ b = 3.204 & \frac{1}{10}\text{Eq.2} \\ 14a = 1.96 & \text{Eq.3} - 2\text{Eq.1} \end{cases}$

$\begin{cases} c + 2a = 70.112 \\ b = 3.204 \\ a = 0.1400 & \frac{1}{14}\text{Eq.3} \end{cases}$

$a = 0.1400$

$b = 3.204$

$c + 2(0.14) = 70.112 \Rightarrow c = 69.832$

The least squares regression parabola is $y = 0.1400x^2 + 3.204x + 69.83$.

(b) When $x = 3$:

$y = 0.1400(3)^2 + 3.204(3) + 69.83 = 80.702$

When $x = 4$:

$y = 0.1400(4)^2 + 3.204(4) + 69.83 = 84.886$

In 2010, the average retail price of prescription drugs was \$80.70, and in 2011, it was \$84.89.

Section 5.4 Systems of Inequalities

Skills Warm Up

1. $x + y = 3$
   Line

2. $y = x^2 - 4$
   Parabola

3. $x^2 + y^2 = 9$
   Circle

4. $y = -x^2 + 1$
   Parabola

5. $4x - y = 8$
   Line

6. $y^2 = 16 - x^2$
   Circle

7. $\begin{cases} x + 2y = 3 \\ 4x - 7y = -3 \end{cases}$ $\xrightarrow{\times(-4)}$

$$\begin{aligned} -4x - 8y &= -12 \\ 4x - 7y &= -3 \\ \hline -15y &= -15 \\ y &= 1 \end{aligned}$$

$x + 2(1) = 3 \Rightarrow x = 1$

Solution: (1, 1)

8. $\begin{cases} 2x - 3y = 4 \\ x + 5y = 2 \end{cases}$ $\xrightarrow{\times(-2)}$ (second equation)

$$\begin{aligned} 2x - 3y &= 4 \\ -2x - 10y &= -4 \\ \hline -13y &= 0 \\ y &= 0 \end{aligned}$$

$x + 5(0) = 2 \Rightarrow x = 2$

Solution: (2, 0)

9. $\begin{cases} x^2 + y = 5 \Rightarrow y = -x^2 + 5 \\ 2x - 4y = 0 \end{cases}$

$$\begin{aligned} 2x - 4(-x^2 + 5) &= 0 \\ 4x^2 + 2x - 20 &= 0 \\ 2(2x^2 + x - 10) &= 0 \\ 2(2x + 5)(x - 2) &= 0 \\ x &= -\tfrac{5}{2}, 2 \end{aligned}$$

$x = -\frac{5}{2}$: $y = -\left(-\frac{5}{2}\right)^2 + 5 = -\frac{5}{4}$

$x = 2$: $y = -(2)^2 + 5 = 1$

Solution: $\left(-\frac{5}{2}, -\frac{5}{4}\right)$ and (2, 1)

10. $\begin{cases} x^2 + y^2 = 13 \\ x + y = 5 \Rightarrow y = 5 - x \end{cases}$

$$\begin{aligned} x^2 + (5 - x)^2 &= 13 \\ x^2 + 25 - 10x + x^2 &= 13 \\ 2x^2 - 10x + 12 &= 0 \\ 2\left(x^2 - 5x + 6\right) &= 0 \\ 2(x - 2)(x - 3) &= 0 \\ x &= 2, 3 \end{aligned}$$

$x = 2$: $y = 5 - 2 = 3$

$x = 3$: $y = 5 - 3 = 2$

Solution: (2, 3) and (3, 2)

1. d; The graph of $2x + 3y \le 6$ is the half-plane below the line $2x + 3y = 6$ together with the line.

3. a; The graph of $x^2 - y \le 2$ is the interior of the parabola $y = x^2 - 2$.

5. f; The graph of $y > x^4 - 5x^2 + 4$ is the interior of $y = x^4 - 5x^2 + 4$.

7. $x \ge 2$

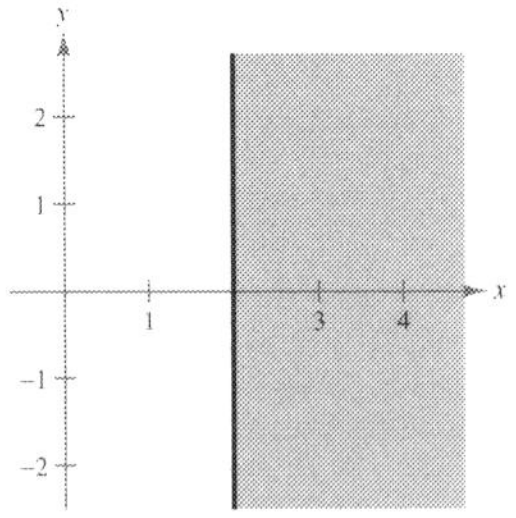

9. $y + 2x > 0$

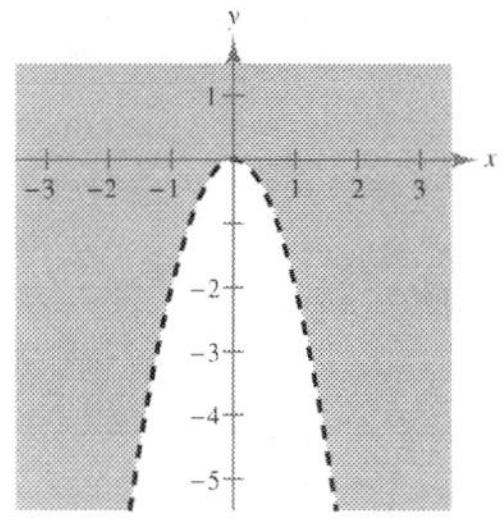

11. $y > -1$

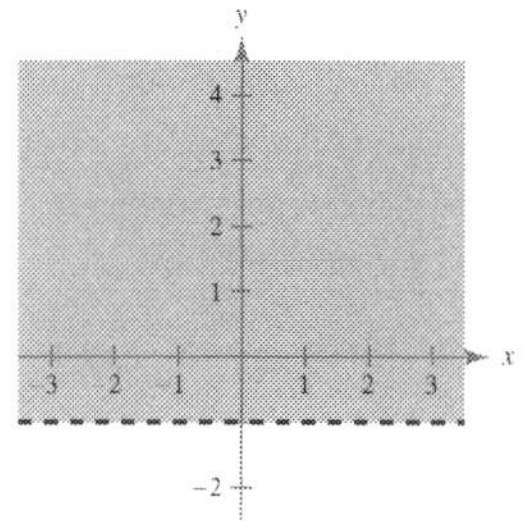

13. $y < 2 - x$

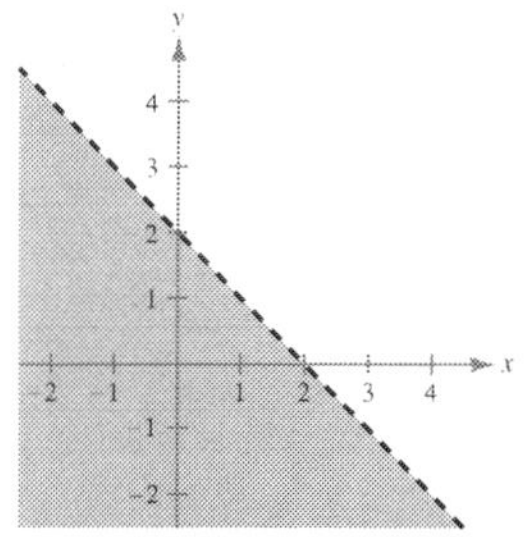

15. $2y - x \geq 4$

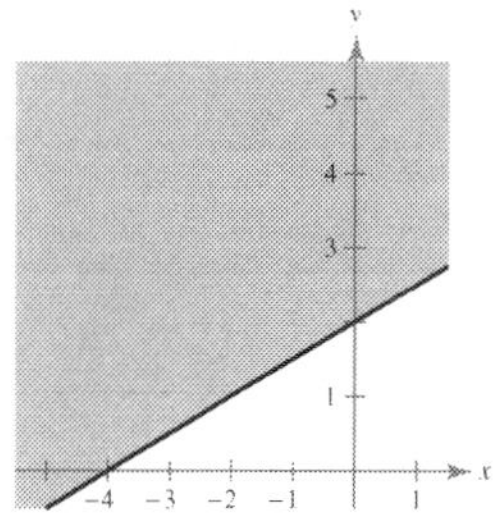

17. $y < \ln x$

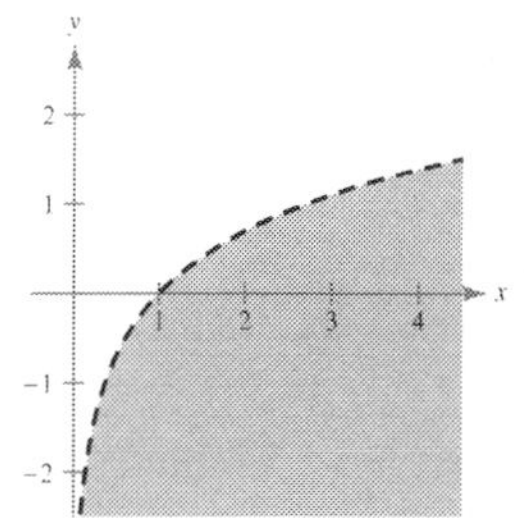

19. $x^2 + y^2 \leq 4$

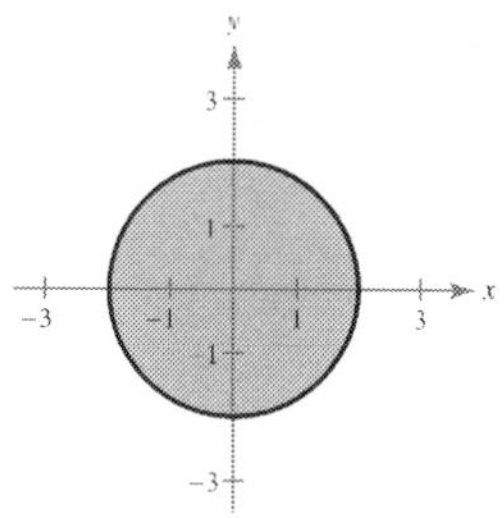

21. $y - (x + 2)^3 \leq 0$

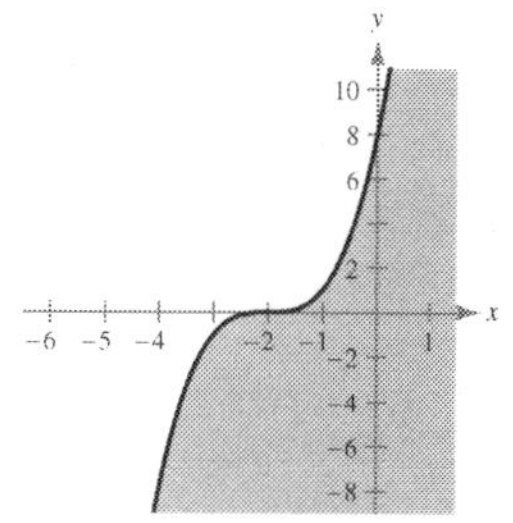

23. $\begin{cases} x + y \leq 2 \\ -x + y \leq 2 \\ y \geq 0 \end{cases}$

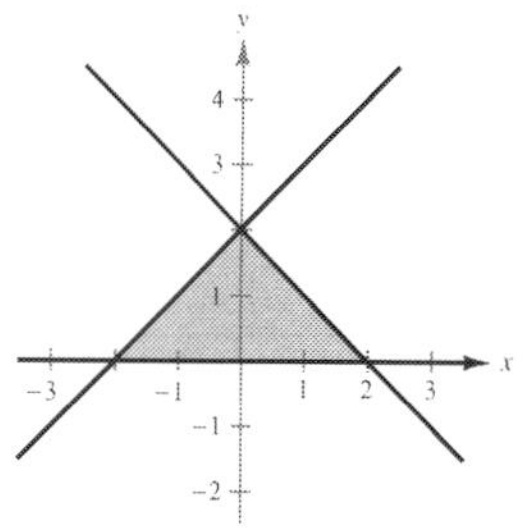

25. $\begin{cases} x + y \leq 5 \\ x \geq 2 \\ y \geq 0 \end{cases}$

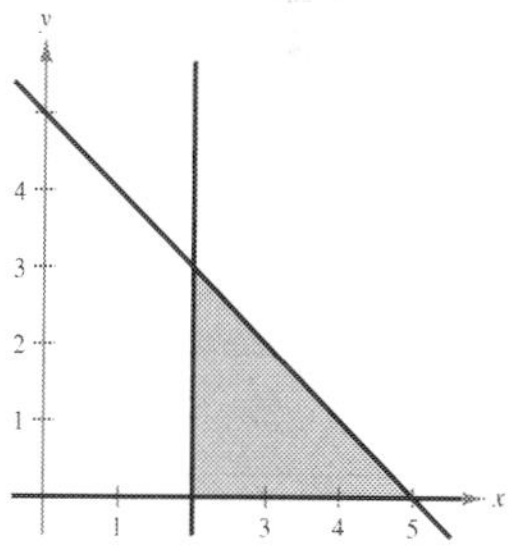

27. $\begin{cases} -3x + 2y < 6 \\ x + 4y < -2 \\ 2x + y < 3 \end{cases}$

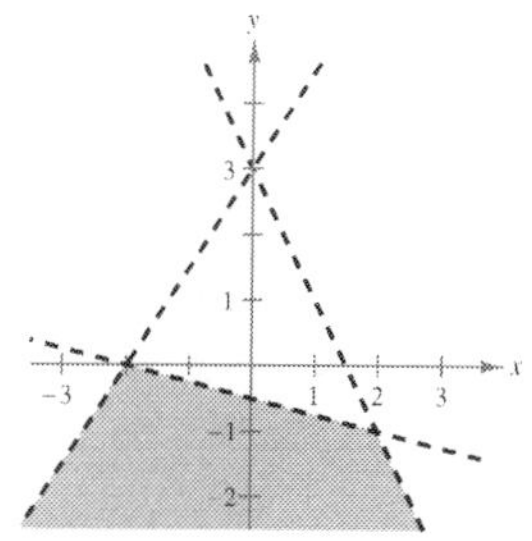

29. $\begin{cases} x^2 + y \le 6 \\ x \ge -1 \\ y \ge 0 \end{cases}$

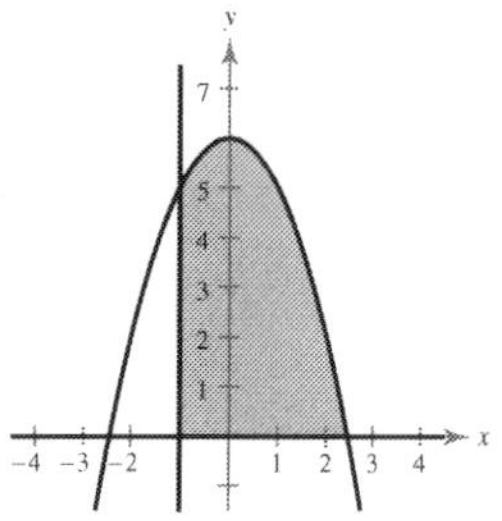

31. $2x + y > 2$

$6x + 3y < 2$

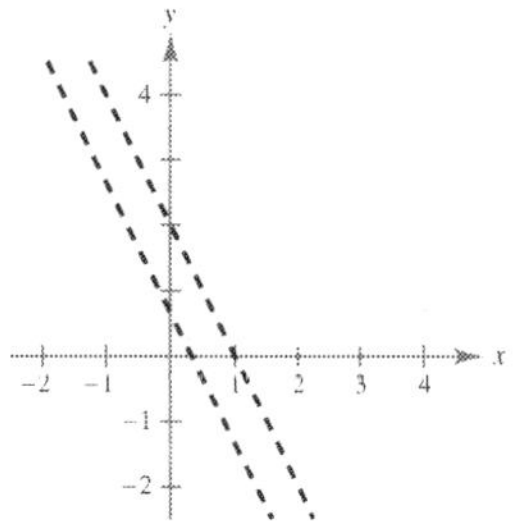

33. $y \ge -3$

$y \le 1 - x^2$

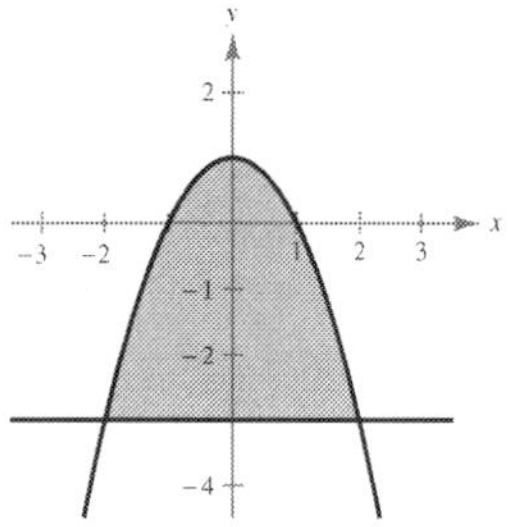

35. $x^2 + y^2 \le 16$

$x^2 + y^2 < 1$

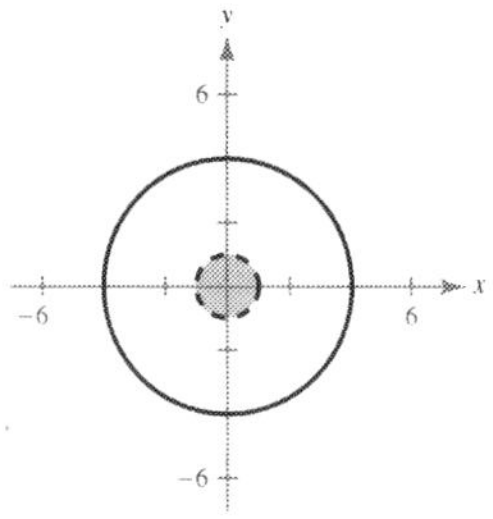

37. $x > y^2$

$x < y + 2$

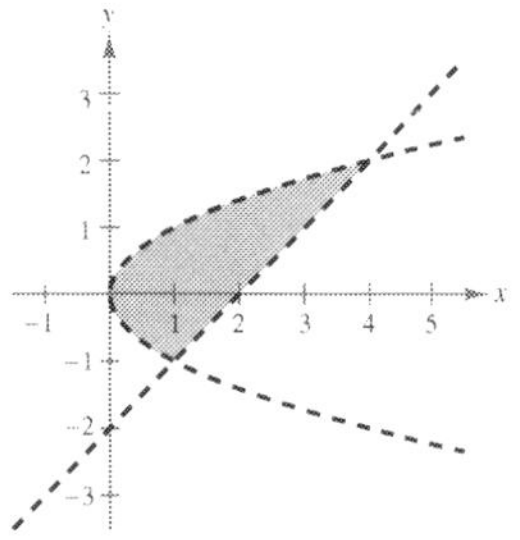

39. $y \le \sqrt{3x} + 1$

$y \ge x + 1$

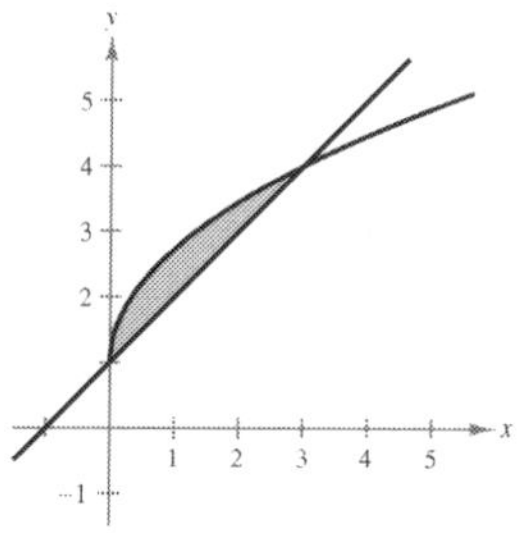

41. $y < x^3 - 2x + 1$

$y > -2x$

$x \le 1$

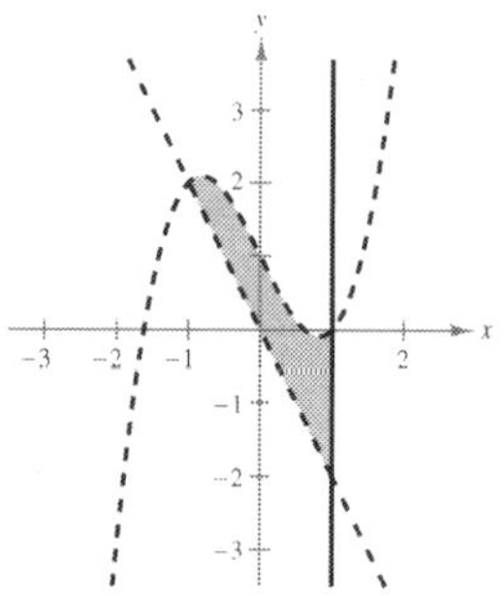

43. $y \le e^x$

$y \ge \ln x$

$x \ge \frac{1}{2}$

$x \le 2$

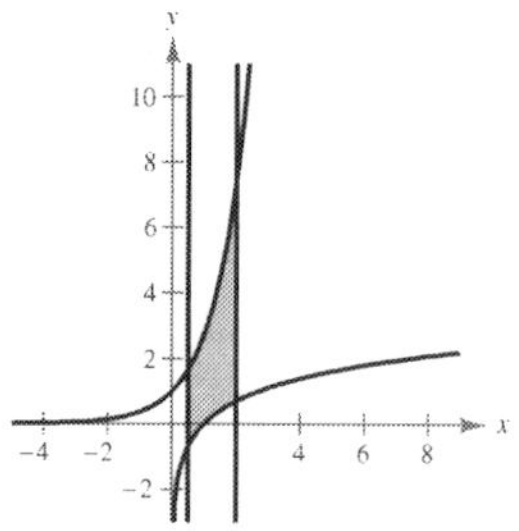

45.

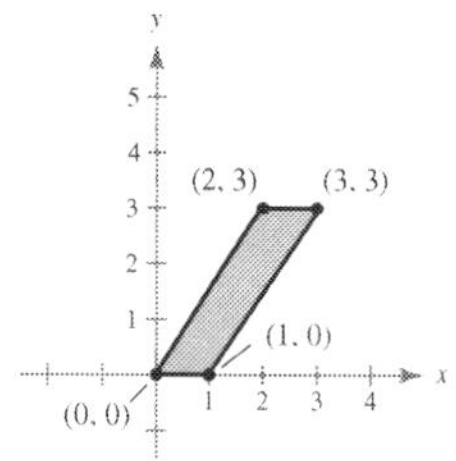

$$m = \frac{3 - 0}{2 - 0} = \frac{3}{2}$$

$$y - 0 = \frac{3}{2}(x - 0) \qquad y - 0 = \frac{3}{2}(x - 1)$$

$$y = \frac{3}{2}x \qquad y = \frac{3}{2}x - \frac{3}{2}$$

$$\begin{cases} y \le \frac{3}{2}x \\ y \ge \frac{3}{2}x - \frac{3}{2} \\ y \le 3 \\ y \ge 0 \end{cases}$$

47.

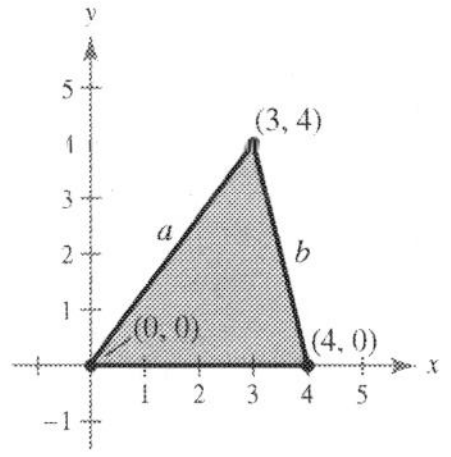

Line a:

$$m = \frac{4-0}{3-0} = \frac{4}{3}$$

$$y - 0 = \frac{4}{3}(x - 0)$$

$$y = \frac{4}{3}x$$

Line b:

$$m = \frac{4-0}{3-4} = -4$$

$$y - 0 = -4(x - 4)$$

$$y = -4x + 16$$

$$\begin{cases} y \le \frac{4}{3}x \\ y \le -4x + 16 \\ y \ge 0 \end{cases}$$

49.

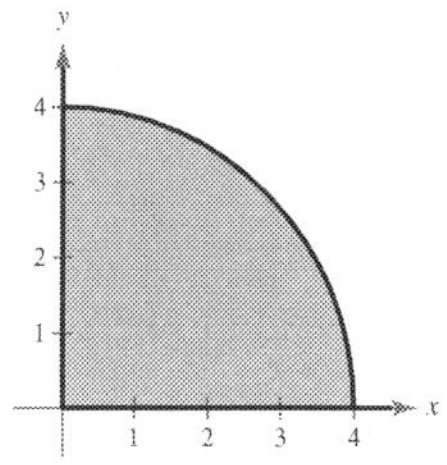

$$\begin{cases} x^2 + y^2 \le 16 \\ x \ge 0 \\ y \ge 0 \end{cases}$$

51. (a) $\begin{cases} 2x + \frac{3}{2}y \le 18 \\ \frac{3}{2}x + \frac{3}{4}y \le 12 \\ x \ge 0 \\ y \ge 0 \end{cases}$

(b)

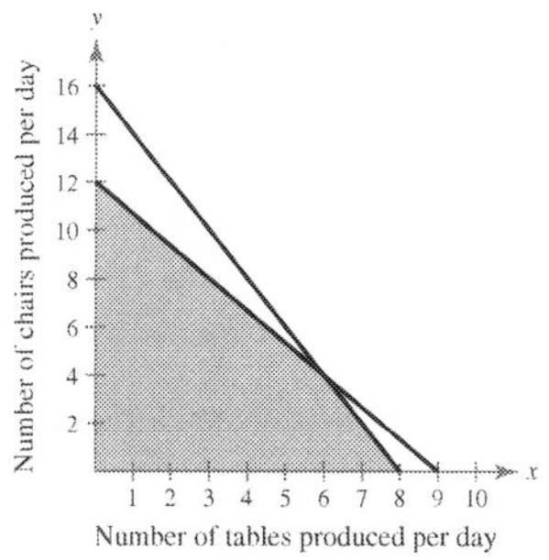

53. Demand = Supply

$56 - 0.0001x = 22 + 0.00001x$

From Exercise 57 in Section 5.2, $x \approx 309{,}091$ units and $p \approx \$25.09$.

Consumer Surplus $= \frac{1}{2}(309{,}091)(30.91)$
$\approx \$4{,}777{,}001.41$

Producer Surplus $= \frac{1}{2}(309{,}091)(3.09)$
$\approx \$477{,}545.60$

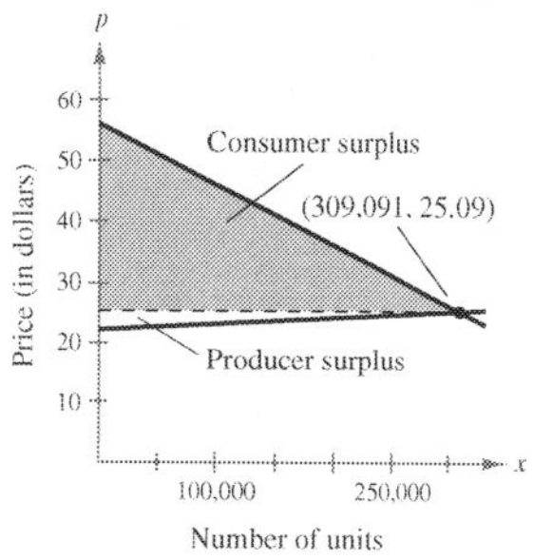

55. Demand = Supply

$140 - 0.00002x = 80 + 0.00001x$

From Exercise 59 in Section 5.2, $x = 2{,}000{,}000$ units and $p = \$100$.

Consumer Surplus $= \frac{1}{2}(2{,}000{,}000)(40) = \$40{,}000{,}000$

Producer Surplus $= \frac{1}{2}(2{,}000{,}000)(20) = \$20{,}000{,}000$

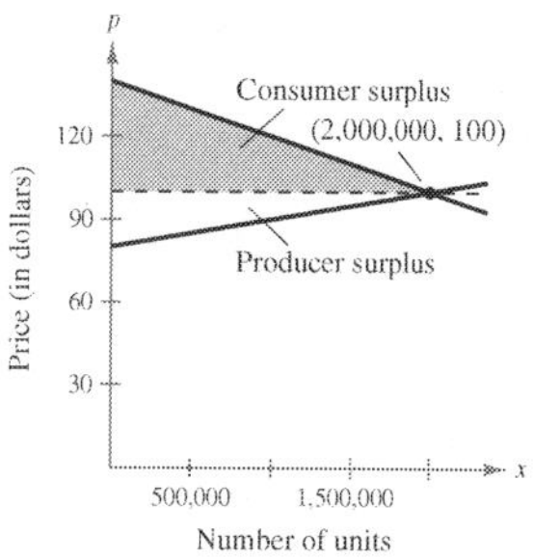

57. Let m_d represent the slope of the demand equation, and let m_s represent the slope of the supply equation. The consumer surplus and producer surplus are equal when $m_d = -m_s$.

59. (a) Let x = one account

y = another account

$$\begin{cases} x + y \le 30{,}000 \\ x \ge 7000 \\ y \ge 7000 \\ x - 2y \ge 0 \end{cases}$$

(b)

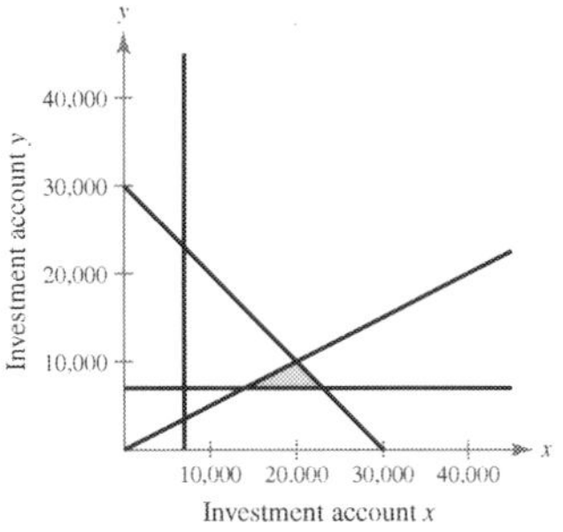

61. (a) x = number of ounces of food X

y = number of ounces of food Y

$$\begin{cases} 20x + 15y \ge 400 & \text{Calcium} \\ 10x + 20y \ge 250 & \text{Iron} \\ 15x + 20y \ge 220 & \text{Vitamin B} \\ x \ge 0 \\ y \ge 0 \end{cases}$$

(b)

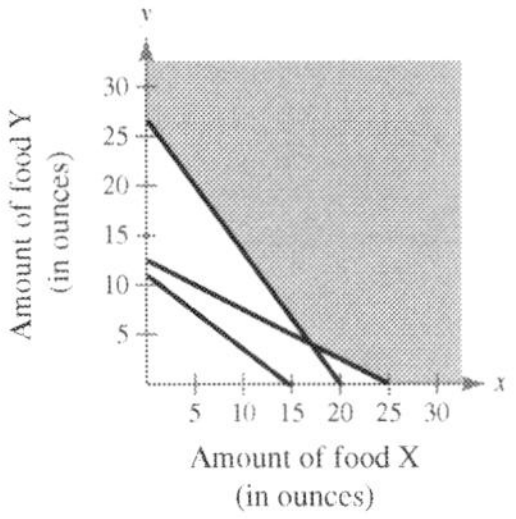

63. (a)

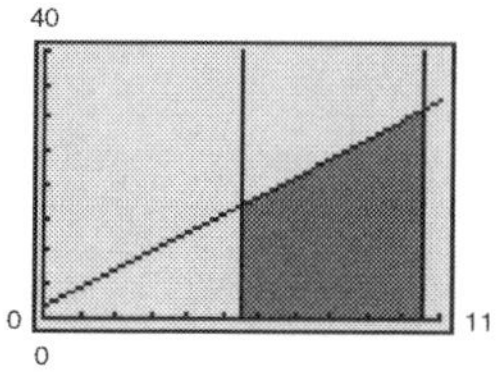

(b) Area of a trapezoid $= \dfrac{h}{2}(a + b)$

$h = 10.5 - 5.5 = 5$

$a = 2.8(5.5) + 1.8 = 17.2$

$b = 2.8(10.5) + 1.8 = 31.2$

$A = \dfrac{h}{2}(a + b) = \dfrac{5}{2}(17.2 + 31.2) = 121$

The total number of nesting pairs is 121.

65. The graph is a half-line on the real number line; on the rectangular coordinate system, the graph is a half-plane.

67. Answers will vary. Sample answer:

$$\begin{cases} x \ge 1 \\ x \le 4 \\ y \ge 1 \\ x \le 4 \end{cases}$$

69. x = radius of smaller circle

y = radius of larger circle

(a) Constraints on circles:

$$\begin{cases} \pi y^2 - \pi x^2 \ge 10 \\ y > x \\ x > 0 \end{cases}$$

(b)

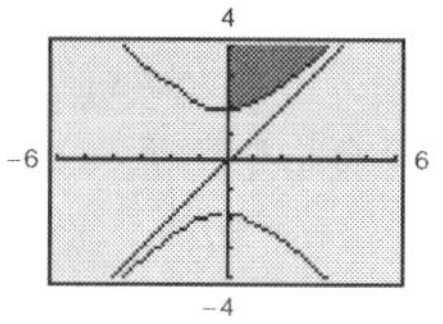

(c) The line is an asymptote to the boundary. The larger the circles, the closer the radii can be and the constraint still be satisfied.

Section 5.5 Linear Programming

Skills Warm Up

1. $y + x = 3$

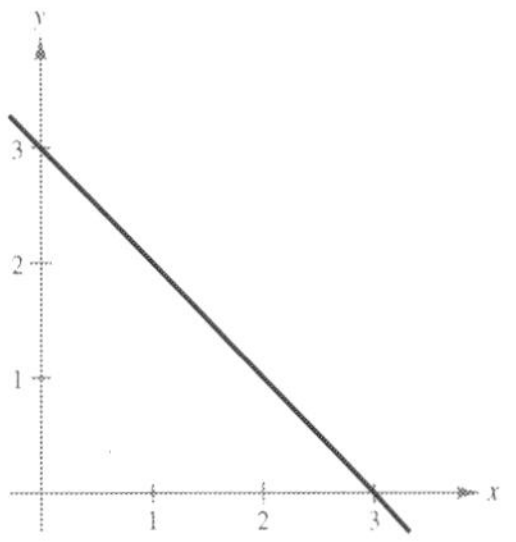

2. $y - x = 12$

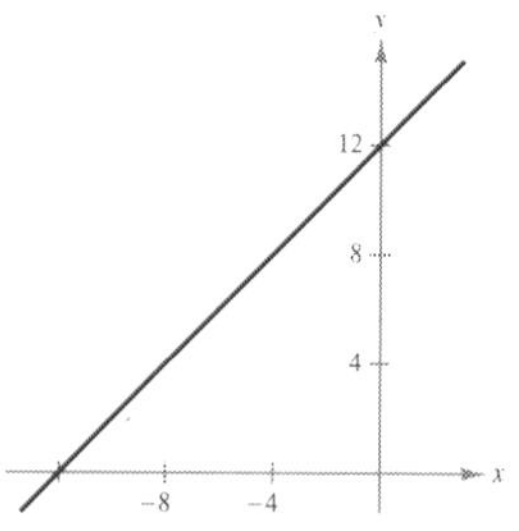

Skills Warm Up —*continued*—

3. $x = 0$

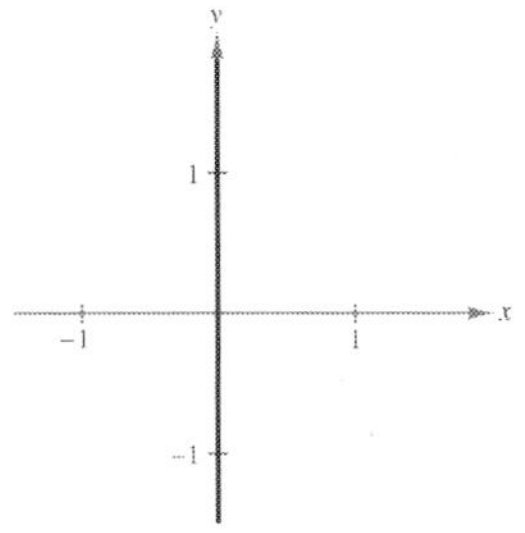

4. $y = 4$

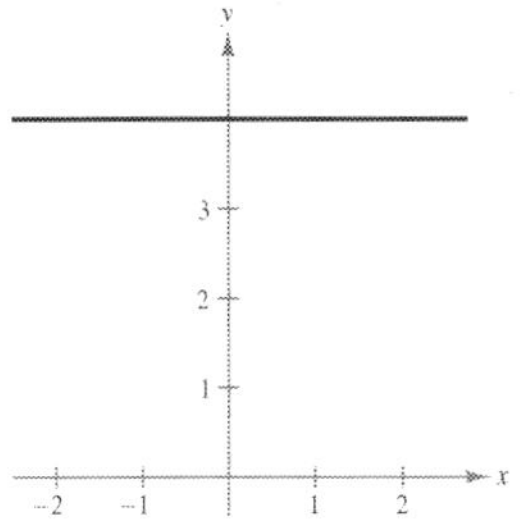

5. $\begin{cases} x + y = 4 \\ x \quad\;\; = 0 \end{cases}$

$0 + y = 4 \Rightarrow y = 4$

Point of intersection: (0, 4)

6. $\begin{cases} x + 2y = 12 \\ \quad\;\; y = 0 \end{cases}$

$x + 2(0) = 12 \Rightarrow x = 12$

Point of intersection: (12, 0)

7. $\begin{cases} x + y = 4 \Rightarrow y = 4 - x \\ 2x + 3y = 9 \end{cases}$

$2x + 3(4 - x) = 9 \Rightarrow x = 3$

$y = 4 - 3 = 1$

Point of intersection: (3,1)

8. $\begin{cases} x + 2y = 12 \Rightarrow x = 12 - 2y \\ 2x + y = 9 \end{cases}$

$2(12 - 2y) + y = 9 \Rightarrow y = 5$

$x = 12 - 2(5) = 2$

Point of intersection: (2, 5)

9. $2x + 3y \geq 18$

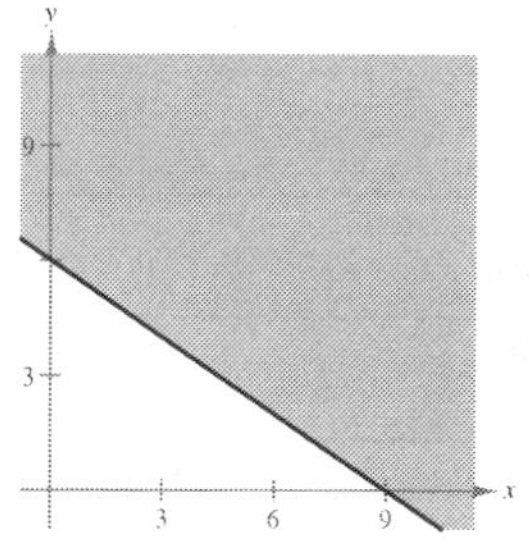

10. $4x + 3y \geq 12$

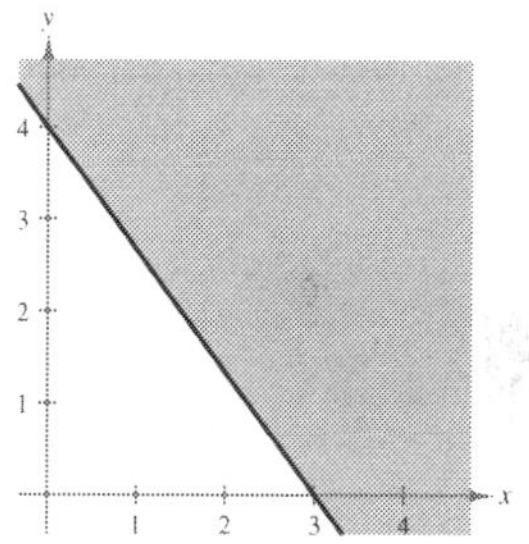

1. $z = 3x + 4y$

At $(0, 0)$: $z = 3(0) + 4(0) = 0$

At $(0, 5)$: $z = 3(0) + 4(5) = 20$

At $(5, 0)$: $z = 3(5) + 4(0) = 15$

The minimum value is 0 at $(0, 0)$.

The maximum value is 20 at $(0, 5)$.

3. $z = 9x + 6y$

At $(0, 0)$: $z = 9(0) + 6(0) = 0$

At $(0, 5)$: $z = 9(0) + 6(5) = 30$

At $(5, 0)$: $z = 9(5) + 6(0) = 45$

The minimum value is 0 at $(0, 0)$.

The maximum value is 45 at $(5, 0)$.

5. $z = 6x + 2y$

At $(0, 0)$: $z = 6(0) + 3(0) = 0$

At $(4, 0)$: $z = 6(4) + 3(0) = 24$

At $(3, 4)$: $z = 6(3) + 3(4) = 30$

At $(0, 5)$: $z = 6(0) + 3(5) = 15$

The minimum value is 0 at $(0, 0)$.

The maximum value is 30 at $(3, 4)$.

7. $z = 5x + 0.5y$

At $(0, 5)$: $z = 5(0) + \frac{5}{2} = \frac{5}{2}$

At $(4, 0)$: $z = 5(4) + \frac{0}{2} = 20$

At $(3, 4)$: $z = 5(3) + \frac{4}{2} = 17$

At $(0, 0)$: $z = 5(0) + \frac{0}{2} = 0$

The minimum value is 0 at (0, 0).

The maximum value is 20 at (4, 0).

9. $z = 6x + 10y$

At $(0, 3)$: $z = 6(0) + 10(3) = 30$

At $(5, 0)$: $z = 6(5) + 10(0) = 30$

At $(0, 0)$: $z = 6(0) + 10(0) = 0$

The minimum value is 0 at (0, 0).

The maximum value is 30 at (0, 3), (5, 0), or at any point on the line segment connecting these points.

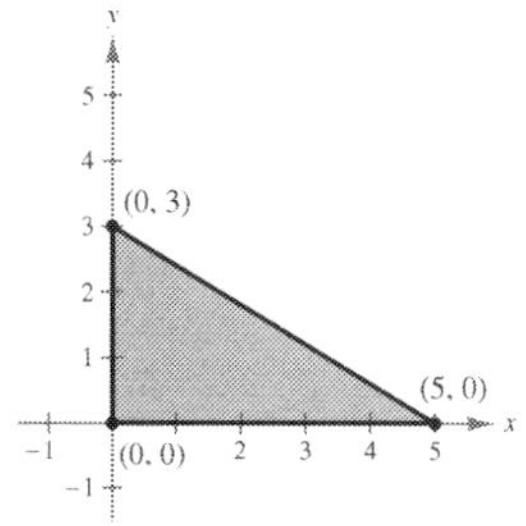

11. $z = 8x + 4y$

At $(0, 0)$: $z = 8(0) + 4(0) = 0$

At $(0, 3)$: $z = 8(0) + 4(3) = 12$

At $(5, 0)$: $z = 8(5) + 4(0) = 40$

The minimum value is 0 at $(0, 0)$.

The maximum value is 40 at $(5, 0)$.

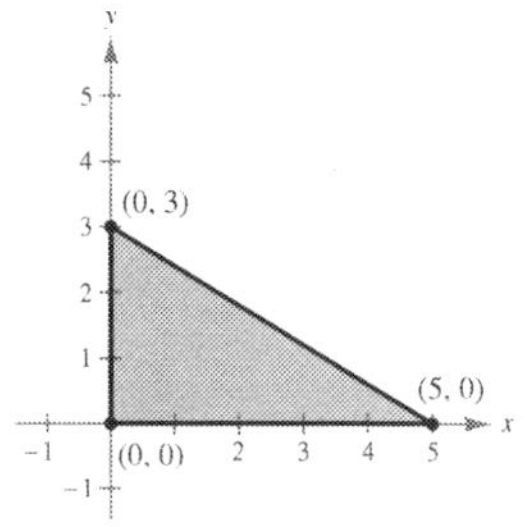

13. $z = 4x + 5y$

At $(10, 0)$: $z = 4(10) + 5(0) = 40$

At $(5, 3)$: $z = 4(5) + 5(3) = 35$

At $(0, 8)$: $z = 4(0) + 5(8) = 40$

The minimum value is 35 at (5, 3).

C is unbounded. Therefore, there is no maximum.

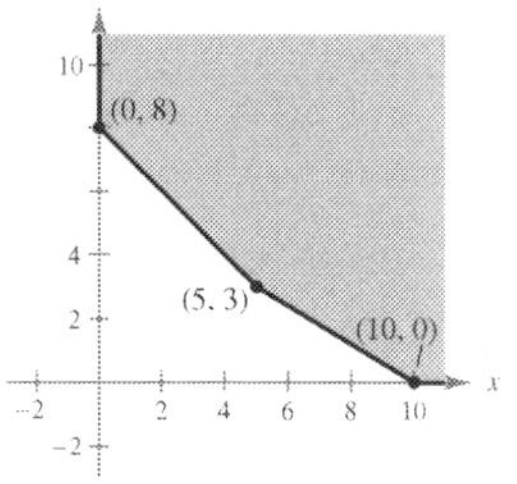

15. $z = 2x + 7y$

At $(10, 0)$: $z = 2(10) + 7(0) = 20$

At $(5, 3)$: $z = 2(5) + 7(3) = 31$

At $(0, 8)$: $z = 2(0) + 7(8) = 56$

The minimum value is 20 at (10, 0).

The region is unbounded. Therefore, there is no maximum.

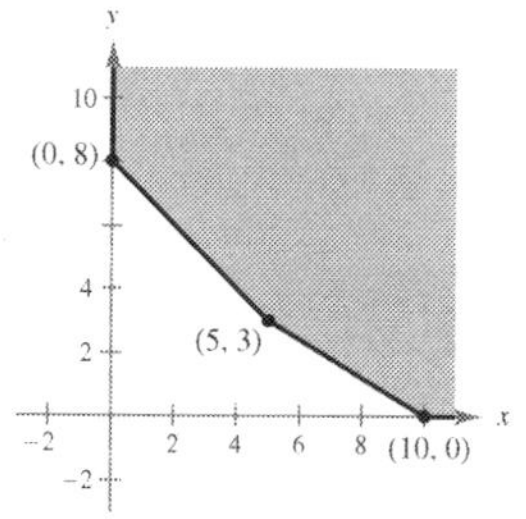

17. $z = 3x + 6y$

At $(0, 0)$: $z = 3(0) + 6(0) = 0$

At $(0, 20)$: $z = 3(0) + 6(20) = 120$

At $(10, 15)$: $z = 3(10) + 6(15) = 120$

At $(25, 5)$: $z = 3(25) + 6(5) = 105$

At $(30, 0)$: $z = 3(30) + 6(0) = 90$

The minimum value is 0 at (0, 0).

The maximum value is 120 at any point on the line segment connecting $(0, 20)$ and $(10, 15)$.

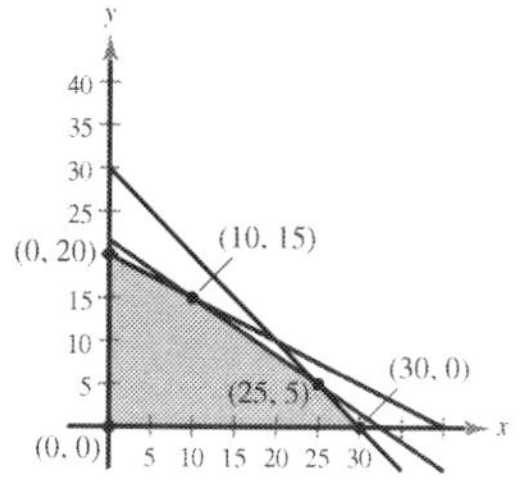

19. $z = x + y$

At $(0, 0)$: $z = 0 + 0 = 0$

At $(0, 20)$: $z = 0 + 20 = 20$

At $(10, 15)$: $z = 10 + 15 = 25$

At $(25, 5)$: $z = 25 + 5 = 30$

At $(30, 0)$: $z = 30 + 0 = 30$

The minimum value is 0 at $(0, 0)$.

The maximum value is 30 at any point on the line segment connecting the points $(25, 5)$ and $(30, 0)$.

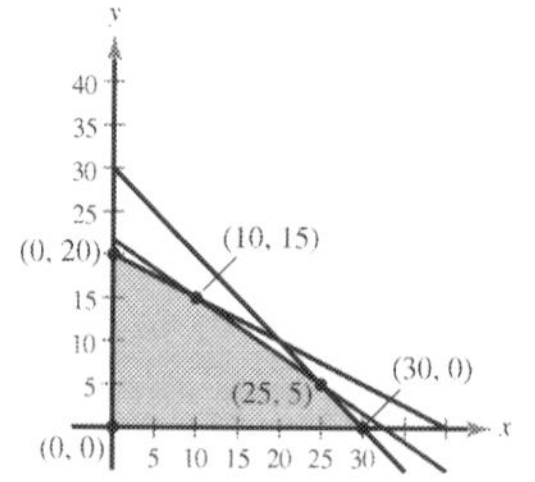

21. $z = 2x + y$

At $(0, 10)$: $z = 2(0) + (10) = 10$

At $(3, 6)$: $z = 2(3) + (6) = 12$

At $(5, 0)$: $z = 2(5) + (0) = 10$

At $(0, 0)$: $z = 2(0) + (0) = 0$

The maximum value is 12 at (3, 6).

23. $z = 4x + 3y$

At $(0, 10)$: $z = 4(0) + 3(10) = 30$

At $(3, 6)$: $z = 4(3) + 3(6) = 30$

At $(5, 0)$: $z = 4(5) + 3(0) = 20$

At $(0, 0)$: $z = 4(0) + 3(0) = 0$

The maximum value is 30 along the line segment connecting (0, 10) and (3, 6).

25. $z = 6x + 7y$

At $(0, 0)$: $z = 6(0) + 7(0) = 0$

At $(0, 5)$: $z = 6(0) + 7(5) = 35$

At $(5, 4)$: $z = 6(5) + 7(4) = 58$

At $(7, 2)$: $z = 6(7) + 7(2) = 56$

At $(8, 0)$: $z = 6(8) + 7(0) = 48$

The maximum value is 58 at $(5, 4)$.

27. $z = 15x + y$

At $(0, 0)$: $z = 15(0) + 0 = 0$

At $(0, 5)$: $z = 15(0) + 5 = 5$

At $(5, 4)$: $z = 15(5) + 4 = 79$

At $(7, 2)$: $z = 15(7) + 2 = 107$

At $(8, 0)$: $z = 15(8) + 0 = 120$

The maximum value is 120 at $(8, 0)$.

29. Answers will vary.

Sample answer: $z = 2x + 15y$

31. Answers will vary.

Sample answer: $z = -x$

33. Answers will vary.

Sample answer: $z = x + 7y$

35. Answers will vary.

Sample answer: $z = 6x + 5y$

37. x = number of acres for crop A

y = number of acres for crop B

Constraints:
$$x + y \le 150$$
$$x + 2y \le 240$$
$$\tfrac{3}{10}x + \tfrac{1}{10}y \le 30$$
$$x \ge 0$$
$$y \ge 0$$

Objective function: $P = 185x + 245y$

Vertices: $(0, 120), (60, 90), (75, 75), (100, 0), (0, 0)$

At $(0, 120)$: $P = 185(0) + 245(120) = 29{,}400$

At $(60, 90)$: $P = 185(60) + 245(90) = 33{,}150$

At $(75, 75)$: $P = 185(75) + 245(75) = 32{,}250$

At $(100, 0)$: $P = 185(100) + 245(0) = 18{,}500$

At $(0, 0)$: $P = 185(0) + 245(0) = 0$

A maximum profit of \$33,150 occurs when 60 acres are used for crop A and 90 acres are used for crop B.

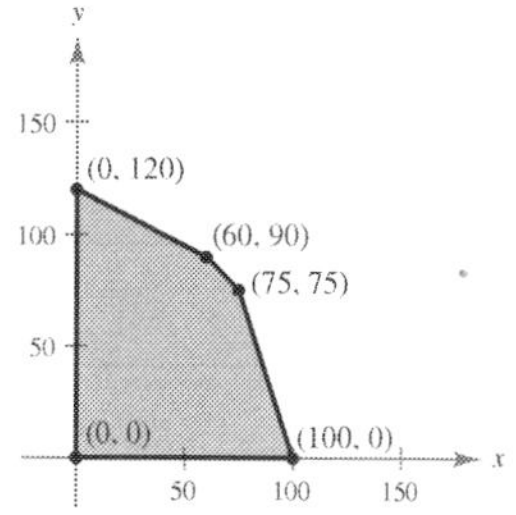

39. x = number of bags of brand X

y = number of bags of brand Y

Constraints:
$$2x + y \ge 12$$
$$2x + 9y \ge 36$$
$$2x + 3y \ge 24$$
$$x \ge 0$$
$$y \ge 0$$

Objective function: $C = 30x + 25y$

Vertices: $(0, 12), (3, 6), (9, 2), (18, 0)$

At $(0, 12)$: $C = 30(0) + 25(12) = 300$

At $(3, 6)$: $C = 30(3) + 25(6) = 240$

At $(9, 2)$: $C = 30(9) + 25(2) = 320$

At $(18, 0)$: $C = 30(18) + 25(0) = 540$

To minimize cost, use three bags of brand X and six bags of brand Y for an optimal cost of \$240.

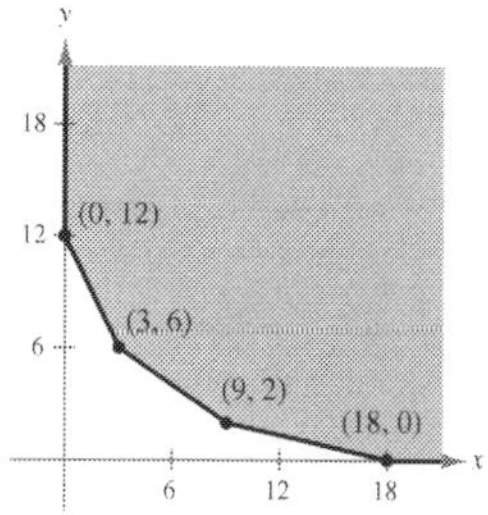

41. x = number of model A

y = number of model B

Constraints:
$$\begin{aligned} 2x + 2.5y &\le 4000 \\ 4x + y &\le 4800 \\ x + 0.75y &\le 1500 \\ x &\ge 0 \\ y &\ge 0 \end{aligned}$$

Objective function: $P = 50x + 75y$

Vertices: $(0, 0), (0, 1600), (750, 1000), (1050, 600), (1200, 0)$

At $(0, 0)$: $P = 50(0) + 75(0) = 0$

At $(0, 1600)$: $P = 50(0) + 75(1600) = 120{,}000$

At $(750, 1000)$: $P = 50(750) + 75(1000) = 112{,}000$

At $(1050, 600)$: $P = 50(1050) + 75(600) = 97{,}500$

At $(1200, 0)$: $P = 50(1200) + 75(0) = 60{,}000$

The optimal profit of $120,000 occurs when no units of model A and 1600 units of model B are produced.

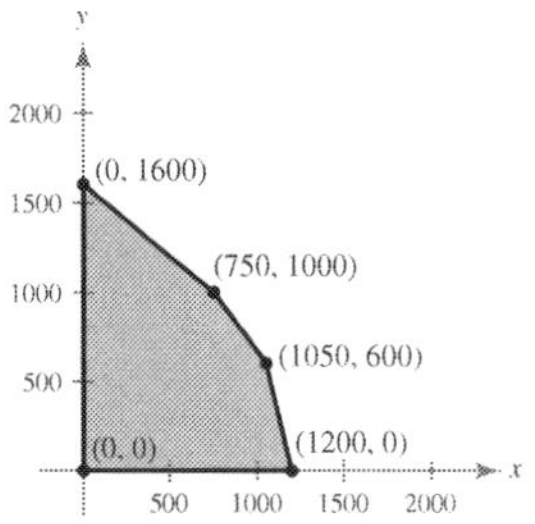

43. Let x = number of audits.

Let y = number of tax returns.

Constraints:
$$\begin{aligned} 175x + 12.5y &\le 900 \\ 10x + 2.5y &\le 155 \\ x &\ge 0 \\ y &\ge 0 \end{aligned}$$

Objective function: $R = 2500x + 350y$

Vertices: $(0, 0), (0, 62), (5, 42), (12, 0)$

At $(0, 0)$: $R = 2500(0) + 350(0) = 0$

At $(0, 62)$: $R = 2500(0) + 350(62) = 21{,}700$

At $(5, 42)$: $R = 2500(5) + 350(42) = 27{,}200$

At $(12, 0)$: $R = 2500(12) + 350(0) = 30{,}000$

The optimal revenue of $30,000 will occur when 12 audits and no tax returns are processed.

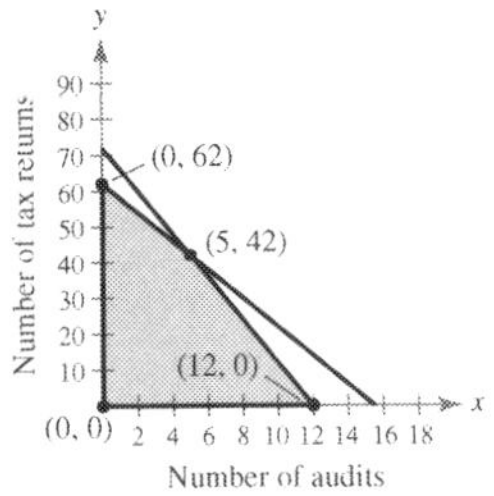

45. Let x = number of TV ads

y = number of newspaper ads

Constraints:
$$\begin{aligned} 400{,}000x + 80{,}000y &\le 4{,}800{,}000 \\ 400{,}000x &\le 3{,}840{,}000 \\ x &\ge 0 \\ y &\ge 0 \end{aligned}$$

Objective function: $z = 20x + 5y$ (in millions)

Vertices: $(0, 0), (0, 60), (9.6, 12), (9.6, 0)$

At $(0, 0)$: $z = 20(0) + 5(0) = 0$

At $(0, 60)$: $z = 20(0) + 5(60) = 300$

At $(9.6, 12)$: $z = 20(9.6) + 5(12) = 252$

At $(9.6, 0)$: $z = 20(9.6) + 5(0) = 192$

The maximum value is 300 at $(0, 60)$.

No advertising dollars should be spent on television ads. The $4,800,000 for advertising should be spent on newspaper ads.

The optimal total audience is 300 million people.

47. Let x = amount in type A investment.

Let y = amount in type B investment.

Constraints: $x \geq \frac{1}{4}(250{,}000)$

$y \geq \frac{1}{4}(250{,}000)$

$x + y \leq 250{,}000$

Objective function: $I = 0.07x + 0.12y$

Vertices: $(62{,}500,\ 62{,}500)$, $(62{,}500,\ 187{,}500)$, $(187{,}500,\ 62{,}500)$

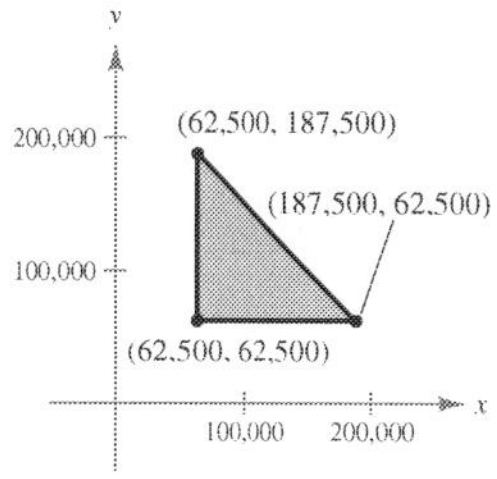

At $(62{,}500,\ 62{,}500)$: $I = 0.07(62{,}500) + 0.12(62{,}500) = 11{,}875$

At $(62{,}500,\ 187{,}500)$: $I = 0.07(62{,}500) + 0.12(187{,}500) = 26{,}875$

At $(187{,}500,\ 62{,}500)$: $I = 0.07(187{,}500) + 0.12(62{,}500) = 20{,}625$

To obtain a maximum return, \$62,500 should be invested in type A and \$187,500 in type B. The optimal return will be \$26,875.

49. Let x = number of model A

Let y = number of model B

Constraints: $3x + 2.75y \leq 3000$

$2.5x + 1.00y \leq 2400$

$0.6x + 1.25y \leq 1200$

$x \geq 0$

$y \geq 0$

Objective function: $P = 100x + 85y$

Vertices: $(0,\ 0)$, $(960,\ 0)$, $(0,\ 960)$, $\left(\frac{28{,}800}{31},\ \frac{2400}{31}\right)$, $\left(\frac{1500}{7},\ \frac{6000}{7}\right)$

At $(0,\ 0)$: $P = \$0$

At $(960,\ 0)$: $P = \$96{,}000$

At $(0,\ 960)$: $P = \$81{,}600$

At $\left(\frac{28{,}800}{31},\ \frac{2400}{31}\right)$: $P \approx \$99{,}483.87$

At $\left(\frac{1500}{7},\ \frac{6000}{7}\right)$: $P \approx \$94{,}285.71$

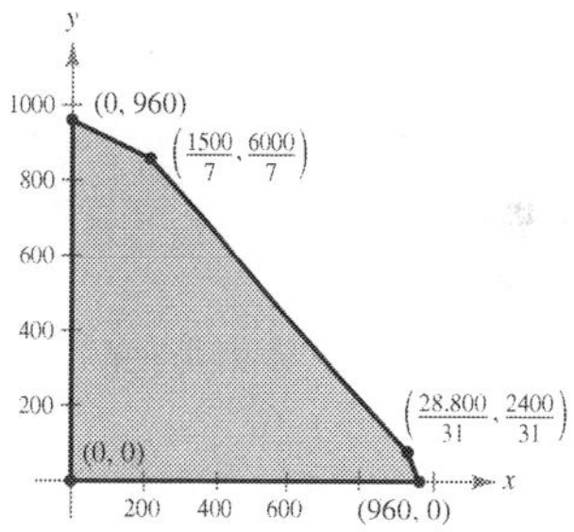

Profit is maximum when $\frac{28{,}800}{31} \approx 929$ units of model A and $\frac{2400}{31} \approx 77$ units of model B are produced. The optimal profit is \$99,445.

51. At $(0,\ 0)$: $z = 0$

At $(2,\ 0)$: $z = 5$

At $\left(\frac{20}{19},\ \frac{45}{19}\right)$: $z = \frac{95}{19} = 5$

At $(0,\ 3)$: $z = 3$

z is maximum at any point on the line segment connecting the vertices $(2,\ 0)$ and $\left(\frac{20}{19},\ \frac{45}{19}\right)$.

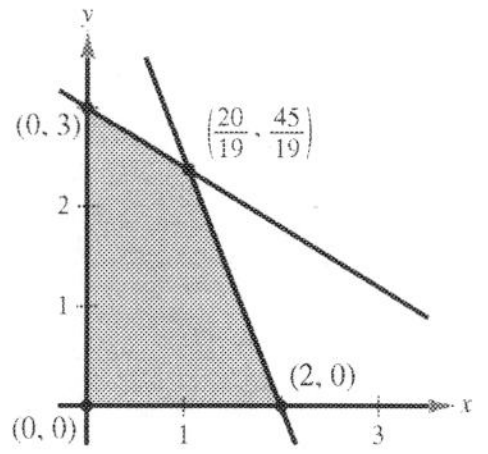

53. At $(0,\ 0)$: $z = -0 + 2(0) = 0$

At $(0,\ 7)$: $z = -0 + 2(7) = 14$

At $(7,\ 0)$: $z = -7 + 2(0) = -7$

The constraint $x \leq 10$ is extraneous.

The maximum value of z occurs at $(0,\ 7)$.

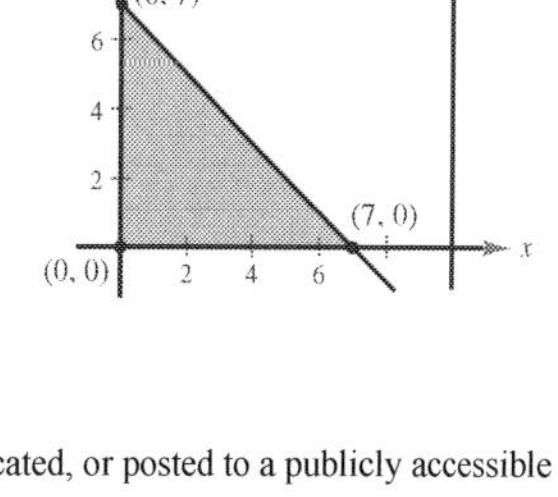

55. The constraint $2x + y \le 4$ is extraneous. The maximum value of z occurs at (0, 1).

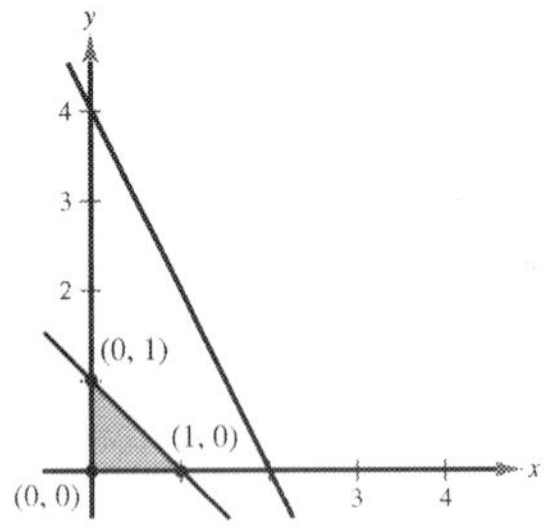

57. No, you cannot conclude that it also has a minimum value at (0, 0). You would have to evaluate the objective function at the points (0, 0) and (20, 0) in order to determine whether the values are equal.

59. Yes, if the objective function has a maximum value at more than one vertex, you can assume that there are an infinite number of points that will produce the maximum value. Any point on the line segment between the vertices will yield a maximum value.

Review Exercises for Chapter 5

1. $$\begin{cases} x + 3y = 10 \Rightarrow x = 10 - 3y \\ 4x - 5y = -28 \Rightarrow 4(10 - 3y) - 5y = -28 \end{cases}$$

$$-17y = -68$$
$$y = 4$$
$$x = 10 - 3(4) = -2$$

Solution: (−2, 4)

3. $$\begin{cases} \frac{1}{2}x + \frac{3}{5}y = -2 \Rightarrow 5x + 6y = -20 \\ 2x + y = 6 \Rightarrow y = 6 - 2x \end{cases}$$

$$5x + 6(6 - 2x) = -20$$
$$-7x = 56$$
$$x = 8$$
$$y = 6 - 2(8) = -10$$

Solution: (8, −10)

5. $$\begin{cases} x^2 + y^2 = 100 \\ x + 2y = 20 \Rightarrow x = 20 - 2y \end{cases}$$

$$(20 - 2y)^2 + y^2 = 100$$
$$5y^2 - 80y + 300 = 0$$
$$5(y - 6)(y - 10) = 0$$
$$y = 6 \text{ or } y = 10$$
$$x = 8 \text{ or } x = 0$$

Solutions: (8, 6), (0, 10)

7. $$\begin{cases} y = x^2 - 3x + 11 \\ y = -x^2 + 2x + 8 \end{cases}$$

Solutions: (1.5, 8.75), (1, 9)

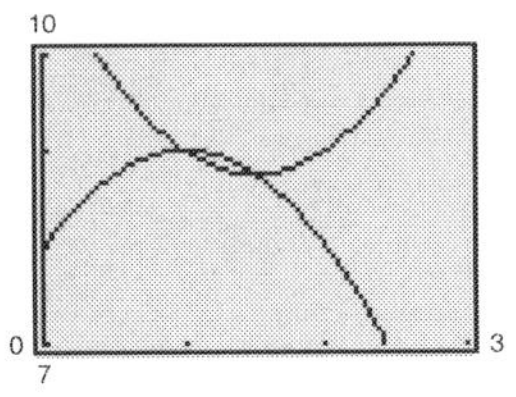

9.
$$C = 6.43x + 5000$$
$$R = 12.68x$$
$$R = C$$
$$12.68x = 6.43x + 5000$$
$$6.25x = 5000$$
$$x = 800$$

You must sell 800 plants to break even.

11. $$\begin{cases} y = 950x + 10{,}000 & \text{Station A} \\ y = -875x + 18{,}000 & \text{Station B} \end{cases}$$

$$950x + 10{,}000 = -875x + 18{,}000$$
$$1825x = 8000$$
$$x \approx 4.384$$

The goal will be achieved during the fourth month of the new contract.

13. $$\begin{cases} 2x - 3y = 21 \\ 3x + y = 4 \end{cases}$$ ⟶ $2x - 3y = 21$; ×3 ⟶ $9x + 3y = 12$

$$11x = 33$$
$$x = 3$$
$$3(3) + y = 4 \Rightarrow y = -5$$

Solution: (3, −5)

15. $$\begin{cases} 4x - 3y = 10 \\ 8x - 6y = 20 \end{cases}$$ ×(−2) ⟶ $-8x + 6y = -20$; ⟶ $8x - 6y = 20$

$$0 = 0$$

Infinitely many solutions. All points on the line $4x - 3y = 10$ are solutions.

Let $y = 4a$. Then $x = \frac{3}{4}(4a) + \frac{5}{2} = 3a + \frac{5}{2}$.

Solution: $\left(3a + \frac{5}{2}, 4a\right)$, a is any real number.

17. $\begin{cases} 1.5x + 2.5y = 8.5 \Rightarrow & 6x + 10y = 34 \\ 6x + 10y = 24 \Rightarrow & -6x - 10y = -24 \end{cases}$

$$0 = 10$$

No solution. The system is inconsistent.

19. $\begin{cases} \dfrac{x-2}{3} + \dfrac{y+3}{4} = 5 \Rightarrow 4x + 3y = 59 \\ 2x - y = 7 \Rightarrow 6x - 3y = 21 \end{cases}$

$$10x = 80$$

$$x = 8$$

$$2(8) - y = 7 \Rightarrow y = 9$$

Solution: (8, 9)

21. $\begin{cases} 2x + y = -1 \Rightarrow y = -2x - 1 \\ 3x - 2y = -5 \end{cases}$

$$3x - 2(-2x - 1) = -5$$

$$7x = -7$$

$$x = -1$$

$$y = -2(-1) - 1 = 1$$

The graph intersects at the point (−1, 1).

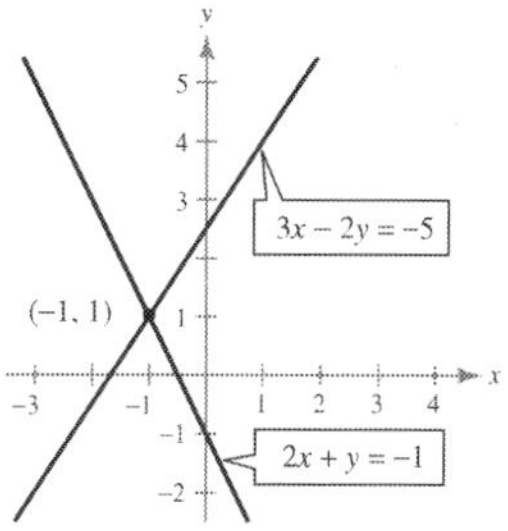

23. Let x = amount of 10% solution.

Let y = amount of 50% solution.

(a) $\begin{cases} x + y = 12 \Rightarrow y = 12 - x \\ 0.10x + 0.50y = 0.25(12) \end{cases}$

$$0.10x + 0.50(12 - x) = 3$$

$$-0.40x = -3$$

$$x = 7.5$$

$$y = 12 - 7.5 = 4.5$$

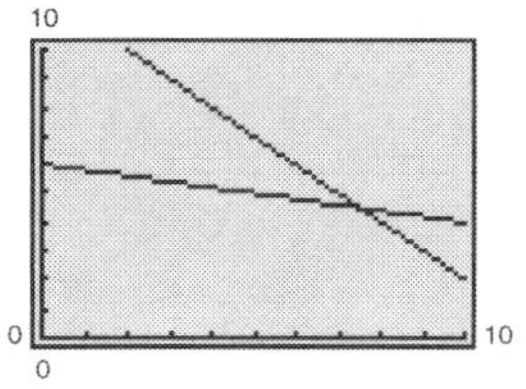

(b) Solution: 7.5 gallons of 10% solution, 4.5 gallons of 50% solution.

25. $\begin{cases} 5b + 10a = 17.8 \Rightarrow & -10b - 20a = -35.6 \\ 10b + 30a = 45.7 \Rightarrow & 10b + 30a = 45.7 \end{cases}$

$$10a = 10.1$$

$$a = 1.01,\ b = 1.54$$

Least squares regression line: $y = 1.01x + 1.54$

27. Supply = Demand

$$22 + 0.00001x = 37 - 0.0002x$$

$$0.00021x = 15$$

$$x \approx 71{,}429 \text{ units}$$

$$p \approx \$22.71$$

29. $\begin{cases} x - 3y + 2z = 1 \\ y - 2z = 1 \\ z = 2 \end{cases}$

Back-substitute $z = 2$:

$$y - 2z = 1$$

$$y - 2(2) = 1$$

$$y - 4 = 1$$

$$y = 5$$

Back-substitute $y = 5$ and $z = 2$:

$$x - 3y + 2z = 1$$

$$x - 3(5) + 2(2) = 1$$

$$x - 15 + 4 = 1$$

$$x - 11 = 1$$

$$x = 12$$

Solution: $(12, 5, 2)$

31. $\begin{cases} 2x + y + z = 6 \\ x - 4y - z = 3 \\ x + y + z = 4 \end{cases}$

$\begin{cases} 2x + y + z = 6 & \\ 9y + 3z = 0 & \text{Eq.1} + (-2)\text{Eq.2} \\ -y - z = -2 & \text{Eq.1} + (-2)\text{Eq.3} \end{cases}$

$\begin{cases} 2x + y + z = 6 & \\ 9y + 3z = 0 & \\ -6z = -18 & \text{Eq.2} + 9\text{Eq.3} \end{cases}$

$$-6z = -18 \Rightarrow z = 3$$

$$9y + 3z = 0 \Rightarrow y = -1$$

$$2x + y + z = 6 \Rightarrow x = 2$$

Solution: (2, −1, 3)

33. $\begin{cases} x + y + z = 10 \Rightarrow 3x + 3y + 3z = 30 \\ -2x + 3y + 4z = 22 \Rightarrow 2x - 3y - 4z = -22 \end{cases}$

$$5x - z = 8 \Rightarrow x = \frac{8 + z}{5}$$

Let $z = a$. Then $x = \dfrac{8 + a}{5}$ and $\dfrac{8 + a}{5} + y + a = 10 \Rightarrow y = \dfrac{42 - 6a}{5}$.

Solution: $\left(\dfrac{8 + a}{5}, \dfrac{42 - 6a}{5}, a\right)$

35. $\begin{cases} x - 3y + z = 2 \\ 2x + 6y - z = 1 \\ 3x + 3y = 12 \end{cases}$ Interchange the equations and multiply one equation by 2 to clear fractions.

$$\begin{cases} x - 3y + z = 2 & \\ 12y - 3z = -3 & -2\text{Eq.}1 + \text{Eq.}2 \\ 12y - 3z = 6 & -3\text{Eq.}1 + \text{Eq.}3 \end{cases}$$

$$\begin{cases} x - 3y + z = 2 & \\ 12y - 3z = -3 & \\ 0 = 9 & -\text{Eq.}2 + \text{Eq.}3 \end{cases}$$

Inconsistent system

There is no solution.

37. $y = ax^2 + bx + c$

At $(0, -6)$: $-6 = c$

At $(1, -3)$: $-3 = a + b + c \Rightarrow a + b = 3 \Rightarrow -2a - 2b = -6$

At $(2, 4)$: $4 = 4a + 2b + c \Rightarrow 4a + 2b = 10 \Rightarrow 4a + 2b = 10$

$$2a = 4$$

$$a = 2$$

$$b = 1$$

Equation of parabola: $y = 2x^2 + x - 6$

39. $x^2 + y^2 + Dx + Ey + F = 0$

At $(2, 2)$: $8 + 2D + 2E + F = 0 \Rightarrow 2D + 2E + F = -8$ Equation 1

At $(5, -1)$: $25 + 5D - E + F = 0 \Rightarrow 5D - E + F = -26$ Equation 2

At $(-1, -1)$: $2 - D - E + F = 0 \Rightarrow D + E - F = 2$ Equation 3

Equation 1 + Equation 3: $3D + 3E = -6 \Rightarrow D + E = -2$

Equation 2 + Equation 3: $6D = -24 \Rightarrow D = -4 \Rightarrow E = 2$

Equation 3: $(-4) + 2 - F = 2 \Rightarrow F = -4$

Equation of circle: $x^2 + y^2 - 4x + 2y - 4 = 0$

41.
$$\begin{aligned} c + \quad m + \quad b + \quad g &= 500{,}000 \\ 0.05c + 0.08m + 0.1b + 0.15g &= 47{,}250 \\ b + \quad g &= 200{,}000 \\ c - \quad 2m &= 0 \end{aligned}$$

$c = \$200{,}000$

$m = \$100{,}000$

$b = \$15{,}000$

$g = \$185{,}000$

You should invest \$200,000 in certificates of deposit, \$100,000 in municipal bonds, \$15,000 in blue-chip stocks, and \$185,000 in growth stocks.

43.
$$\begin{cases} 5c + 10a = 9.1 \Rightarrow -10c - 20a = -18.2 \\ 10b = 8.0 \\ 10c + 34a = 19.8 \Rightarrow \quad 10c + 34a = 19.8 \end{cases}$$

$$14a = 1.6$$
$$a = \frac{4}{35}$$
$$c = \frac{557}{350}$$

From Equation 2: $b = \frac{4}{5}$

Least squares regression parabola:

$y = \frac{4}{35}x^2 + \frac{4}{5}x + \frac{557}{350}$ or $y = 0.114x^2 + 0.8x + 1.59$

45. $x \geq -5$

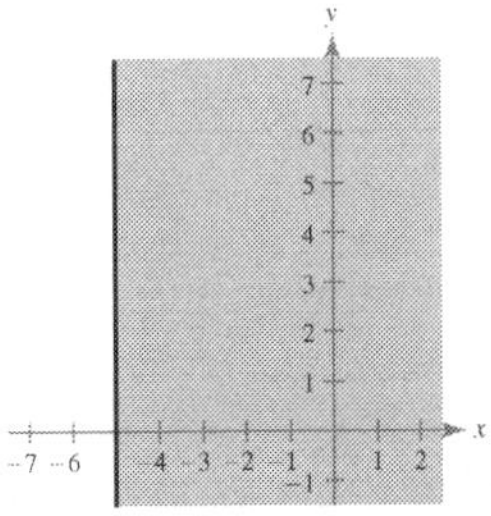

47. $y \leq 5 + \frac{1}{2}x$

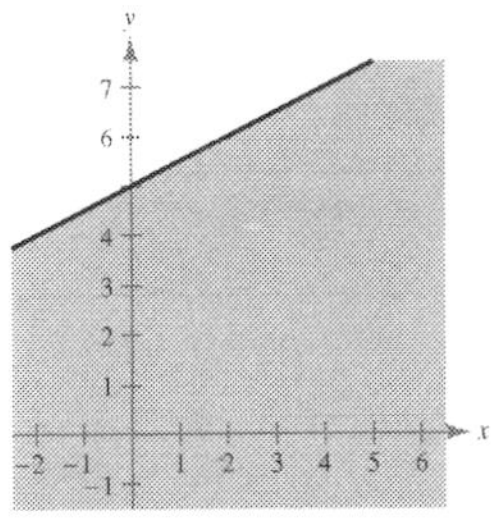

49. $y - 4x^2 > -1$

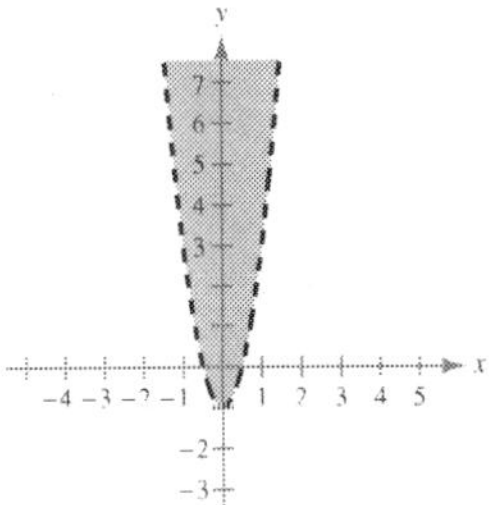

51. $\begin{cases} x + 4y < 8 \\ x > 0 \\ y > 0 \end{cases}$

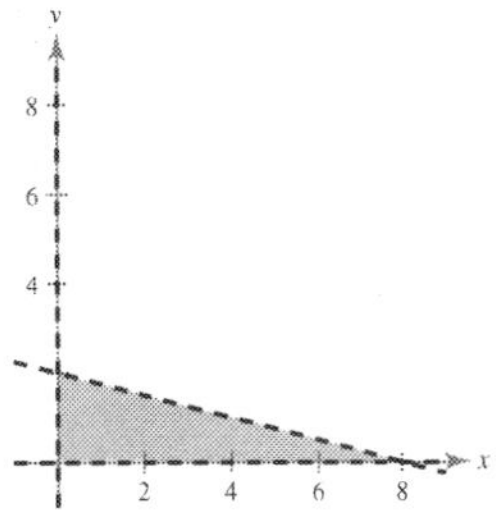

53. $\begin{cases} 3x - y > -4 \\ 2x + y > -1 \\ 7x + y < 4 \end{cases}$

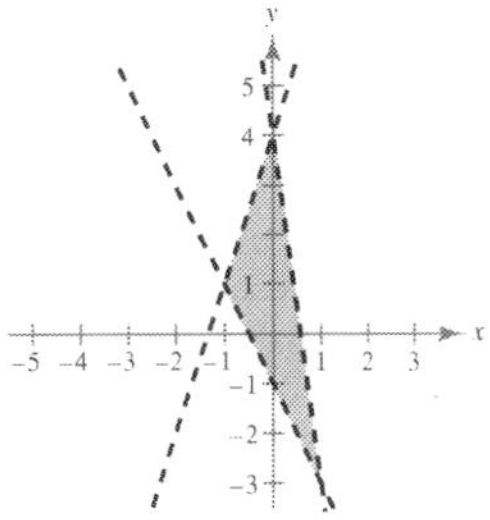

55. $\begin{cases} x^2 + y^2 \leq 9 \\ x^2 - x - 2 \leq y \end{cases}$

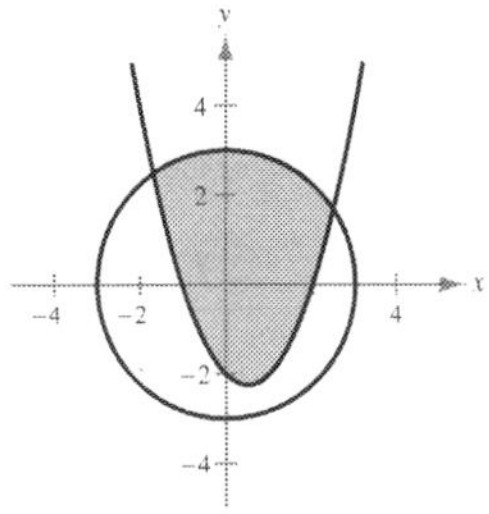

57. $\begin{cases} \ln x < y \\ y > -1 \\ x < 4 \end{cases}$

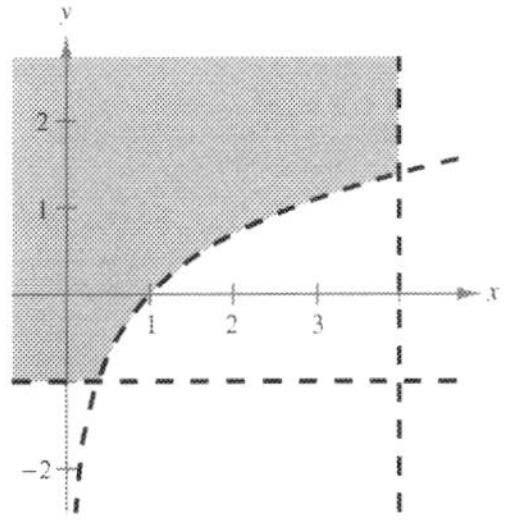

59.

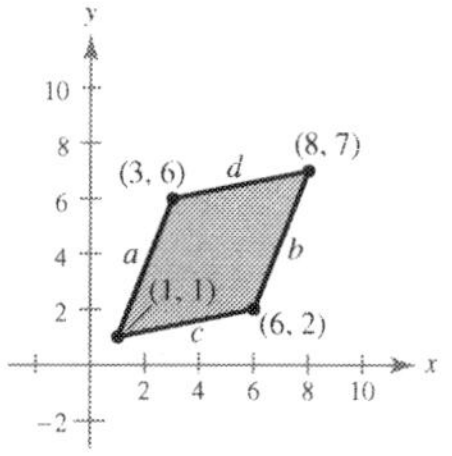

Line a:

$$m = \frac{6 - 1}{3 - 1} = \frac{5}{2}$$

$$y - 1 = \frac{5}{2}(x - 1)$$

$$y = \frac{5}{2}x - \frac{3}{2}$$

$$5x - 2y = 3$$

Line b:

$$m = \frac{5}{2}$$

$$y - 2 = \frac{5}{2}(x - 6)$$

$$y = \frac{5}{2}x - 13$$

$$5x - 2y = 26$$

Line c:

$$m = \frac{2 - 1}{6 - 1} = \frac{1}{5}$$

$$y - 1 = \frac{1}{5}(x - 1)$$

$$y = \frac{1}{5}x - \frac{4}{5}$$

$$x - 5y = -4$$

Line d:

$$m = \frac{1}{5}$$

$$y - 6 = \frac{1}{5}(x - 3)$$

$$y = \frac{1}{5}x + \frac{27}{5}$$

$$x - 5y = -27$$

$$\begin{cases} 5x - 2y \ge 3 \\ 5x - 2y \le 26 \\ x - 5y \le -4 \\ x - 5y \ge -27 \end{cases}$$

61. Point of equilibrium: Demand = Supply

$$160 - 0.0001x = 70 + 0.0002x$$

$$90 = 0.0003x$$

$$x = 300{,}000 \text{ units}$$

$$p = \$130.00$$

Consumer surplus: $\frac{1}{2}(300{,}000)(30) = \$4{,}500{,}000$

Producer surplus: $\frac{1}{2}(300{,}000)(60) = \$9{,}000{,}000$

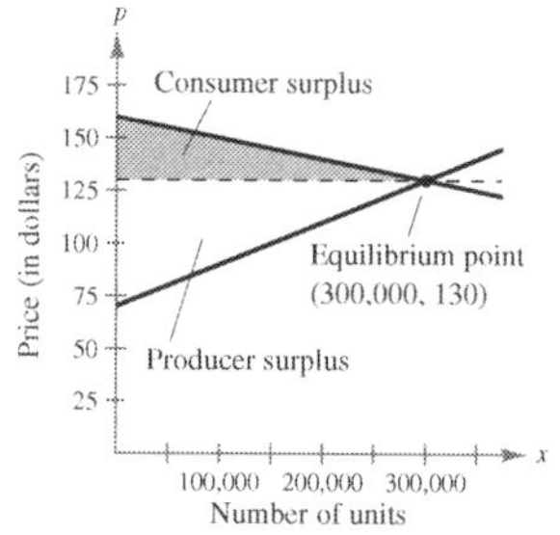

63. Let x = units of model Y

y = units of model Z

$$\begin{cases} x \ge 2y \\ 100x + 150y \le 4000 \\ x \ge 4 \\ y \ge 2 \end{cases}$$

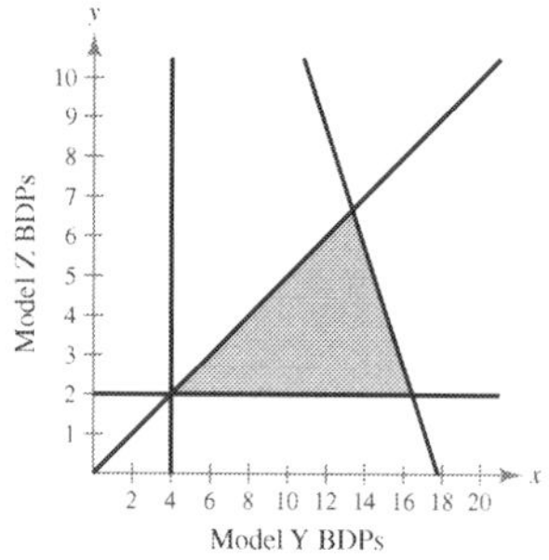

65. Constraints: $x \ge 0,\ y \ge 0$

$x + y \le 8$

Objective function: $z = 5x + 6y$

At $(0, 0)$: $z = 5(0) + 6(0) = 0$

At $(0, 8)$: $z = 5(0) + 6(8) = 48$

At $(8, 0)$: $z = 5(8) + 6(0) = 40$

The minimum value is 0 at (0, 0).

The maximum value is 48 at (0, 8).

67. Constraints: $0 \le x \le 50$

$0 \le y \le 35$

$4x + 5y \le 275$

Objective function: $z = 8x + 3y$

At $(0, 0)$: $z = 8(0) + 10(0) = 0$

At $(0, 35)$: $z = 8(0) + 10(35) = 350$

At $(25, 35)$: $z = 8(25) + 10(35) = 550$

At $(50, 15)$: $z = 8(50) + 10(15) = 550$

At $(50, 0)$: $z = 8(50) + 10(0) = 400$

The minimum value is 0 at (0, 0).

The maximum value is 550 at (25, 35), (50, 15), and the line segment connecting these points.

69. Constraints: $x \ge 0,\ y \ge 0$

$x + 4y \le 16$

$3x + 2 \le 18$

Objective function: $z = 6x + 8y$

At $(0, 0)$: $z = 6(0) + 8(0) = 0$

At $(0, 4)$: $z = 6(0) + 8(4) = 32$

At $(4, 3)$: $z = 6(4) + 8(3) = 48$

At $(6, 0)$: $z = 6(6) + 8(0) = 36$

The minimum value is 0 at (0, 0).

The maximum value is 48 at (4, 3).

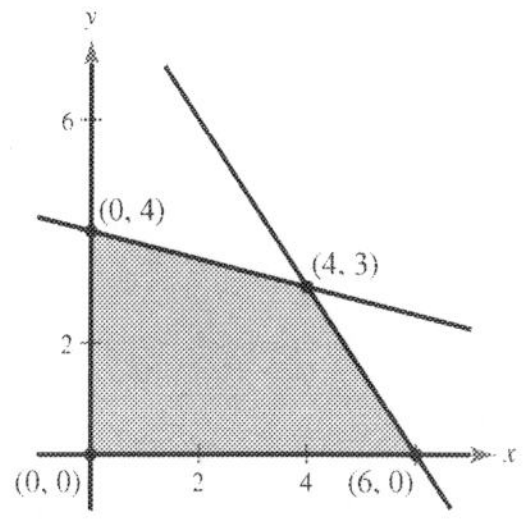

71. Constraints: $0 \le x \le 5,\ 0 \le y \le 7$

$x + y \le 9$

$3x + y < 17$

Objective function: $z = 8x + 3y$

At $(0, 0)$: $z = 8(0) + 3(0) = 0$

At $(0, 7)$: $z = 8(0) + 3(7) = 21$

At $(2, 7)$: $z = 8(2) + 3(7) = 37$

At $(4, 5)$: $z = 8(4) + 3(5) = 47$

At $(5, 2)$: $z = 8(5) + 3(2) = 46$

At $(5, 0)$: $z = 8(5) + 3(0) = 40$

The minimum value is 0 at (0, 0).

The maximum value is 47 at (4, 5).

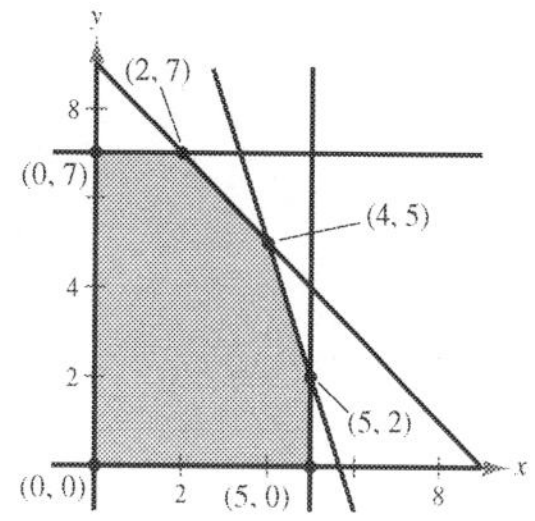

73. Let x = number of audits.

Let y = number of tax returns.

Constraints: $100x + 10y \le 800$

$10x + 2y \le 90$

$x \ge 0,\ y \ge 0$

Objective function: $R = 2500x + 200y$

At $(0, 0)$: $R = 2500(0) + 200(0) = 0$

At $(0, 45)$: $R = 2500(0) + 200(45) = 9000$

At $(7, 10)$: $R = 2500(7) + 200(10) = 19{,}500$

At $(8, 0)$: $R = 2500(8) + 200(0) = 20{,}000$

The revenue will be at its maximum if the firm does 8 audits and 0 tax returns each week. The maximum revenue is \$20,000.

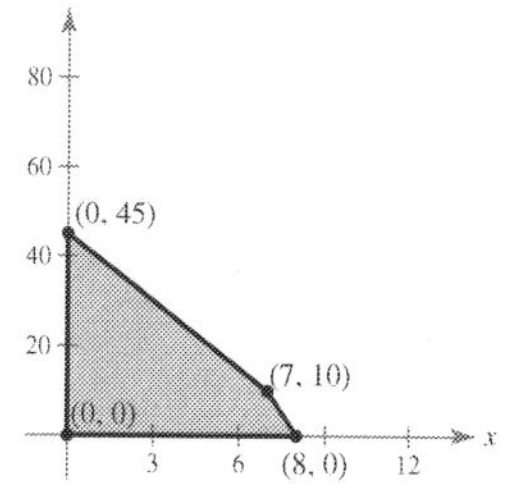

75. Let x = number of basic models.

Let y = number of deluxe models.

Constraints: $2x + 5y \le 3000$

$$x + 2y \le 1300$$
$$x + y \le 1000$$
$$x \ge 0$$
$$y \ge 0$$

Objective function: $P = 100x + 180y$

At $(0, 0)$: $P = 100(0) + 180(0) = 0$

At $(1000, 0)$: $P = 100(1000) + 180(0) = 100{,}000$

At $(700, 300)$: $P = 100(700) + 180(300) = 124{,}000$

At $(500, 400)$: $P = 100(500) + 180(400) = 122{,}000$

At $(0, 600)$: $P = 100(0) + 180(600) = 108{,}000$

The optimal profit is \$124,000 when 700 units of the basic model and 300 units of the deluxe model are sold.

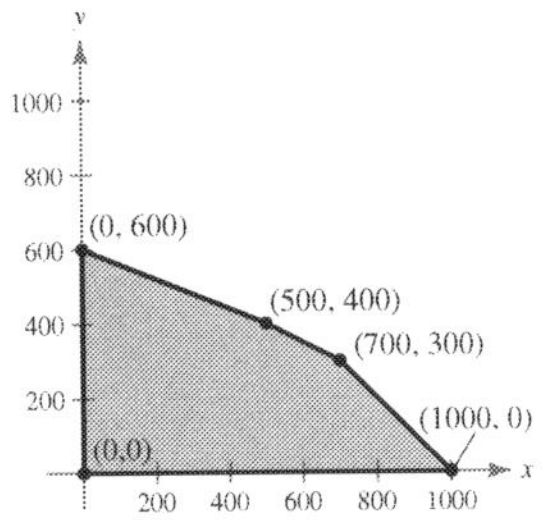

77. Let x = number of \$270 televisions.

Let y = number of \$455 televisions.

Constraints: $x + y \le 100$

$$270x + 455y \le 36{,}250$$
$$x \ge 0;\ y \ge 0$$

Objective function: $P = 30x + 45y$

At $(0, 0)$: $P = 30(0) + 45(0) = 0$

At $(0, 80)$: $P = 30(0) + 45(80) = 3600$

At $(50, 50)$: $P = 30(50) + 45(50) = 3750$

At $(100, 0)$: $P = 30(100) + 45(0) = 3000$

The maximum profit is \$3750 when 50 units of the \$270 television and 50 units of the \$455 television are stocked and sold.

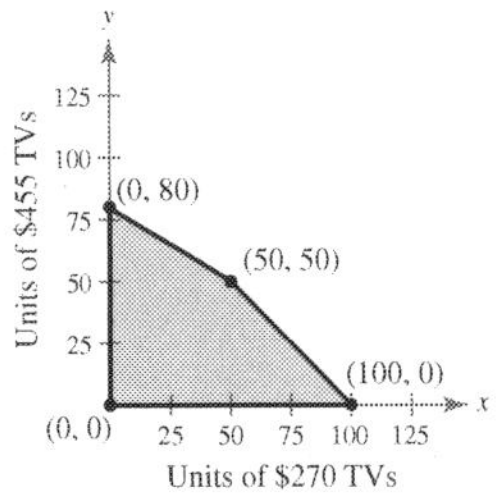

Chapter 5 Test Yourself

1. $5x - 7y = -18 \Rightarrow y = \dfrac{5x + 18}{7}$

$$4x + 3y = 20 \Rightarrow 4x + 3\left(\frac{5x + 18}{7}\right) = 20$$
$$4x + \frac{15}{7}x + \frac{54}{7} = 20$$
$$\frac{43}{7}x = \frac{86}{7}$$
$$x = 2$$
$$y = \frac{5(2) + 18}{7} = 4$$

Solution: $(2, 4)$

2. $x + y = 3 \Rightarrow y = 3 - x$

$$x^2 + y = 9 \Rightarrow x^2 + (3 - x) = 9$$
$$x^2 - x - 6 = 0$$
$$(x - 3)(x + 2) = 0$$

$x = 3$ or $x = -2$

$y = 0$ $\quad y = 5$

Solutions: $(3, 0), (-2, 5)$

3.

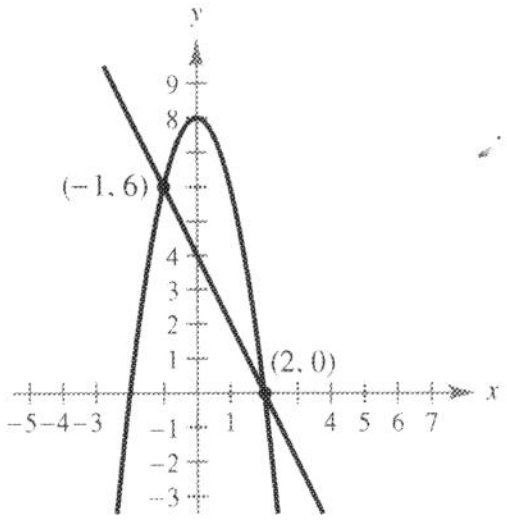

Solutions: $(-1, 6), (2, 0)$

4. $1.5x - 2.25y = 8 \Rightarrow 6x - 9y = 32$

$2.5x + 2y = 5.75 \Rightarrow 10x + 8y = 23$

Solution: $(3.36, -1.32)$

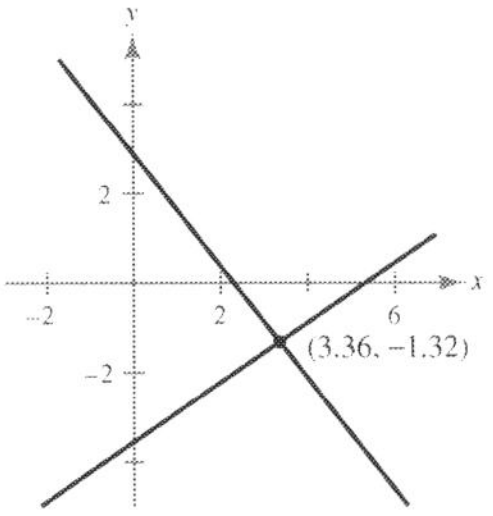

5. $x + 2y + 3z = 9$ Interchange the equations.

$2x - 4y + z = 11$

$3y + 5z = 12$

$x + 2y + 3z = 9$

$-8y - 5z = -7$ $-\text{Eq.}1 + \text{Eq.}2$

$3y + 5z = 12$

$x + 2y + 3z = 9$

$-8y - 5z = -7$

$-5y = 5$ $\text{Eq.}2 + \text{Eq.}3$

$y = -1$

$-8(-1) - 5z = -7 \Rightarrow z = 3$

$x + 2(-1) + 3(3) = 9 \Rightarrow x = 2$

Solution: $(2, -1, 3)$

6. $8x - 2y = 22$ $\text{Eq.}1 + \text{Eq.}2$

$5x - 3y = 19$ $\text{Eq.}1 + \text{Eq.}3$

$24x - 6y = 66$ $3\text{Eq.}1$

$-10x + 6y = -38$ $-2\text{Eq.}2$

$14x = 28$

$x = 2$

$8(2) - 2y = 22 \Rightarrow y = -3$

$3(2) - 2(-3) + z = 16 \Rightarrow z = 4$

Solution: $(2, -3, 4)$

7. Let x = amount invested at 2.3%

y = amount invested at 3.1%

$$\begin{cases} x + y = 80{,}000 & \text{Eq.}1 \\ 0.023x + 0.031y = 2000 & \text{Eq.}2 \end{cases}$$

Multiply Equation 2 by 1000:

$x + y = 80{,}000$ Eq.1

$23x + 31y = 2{,}000{,}000$ Revised Eq.2

Solve Equation 1 for y: $y = 80{,}000 - x$

Substitute $80{,}000 - x$ for y in revised Eq. 2:

$23x + 31(80{,}000 - x) = 2{,}000{,}000$

$23x + 2{,}480{,}000 - 31x = 2{,}000{,}000$

$-8x = -480{,}000$

$x = 60{,}000$

Back-substitute $x = 60{,}000$: $y = 80{,}000 - x$

$y = 80{,}000 - 60{,}000$

$y = 20{,}000$

\$60,000 is invested at 2.3% and \$20,000 is invested at 3.1%.

8. $$\begin{cases} P = 49 - 0.0003x & \text{Demand} \\ P = 33 + 0.00002x & \text{Supply} \end{cases}$$

Demand = Supply

$49 - 0.0003x = 33 + 0.00002x$

$-0.00032x = -16$

$x = 50{,}000$ units

$P = 49 - 0.0003(50{,}000) = \$34.00$

Point of equilibrium: $(50{,}000, 34)$

9. (a) $$\begin{cases} 5c + 10a = 239.6 & \text{Eq.}1 \\ 10b = 49.5 & \text{Eq.}2 \\ 10c + 34a = 487.5 & \text{Eq.}3 \end{cases}$$

$$\begin{cases} c + 2a = 47.92 & \frac{1}{5}\text{Eq.}1 \\ b = 4.95 & \frac{1}{10}\text{Eq.}2 \\ 14a = 8.3 & -2\text{Eq.}1 + \text{Eq.}3 \end{cases}$$

$$\begin{cases} c + 2a = 47.92 \\ b = 4.95 \\ a = \frac{83}{140} \approx 0.593 \end{cases}$$

$a \approx 0.593$

$b = 4.95$

$c + 2\left(\frac{83}{140}\right) = 47.92 \Rightarrow c \approx 46.7$

The least squares regression parabola is $y = 0.593t^2 + 4.95t + 46.7$.

(b) When $t = 3$:

$y = 0.593(3)^2 + 4.95(3) + 46.7 \approx 66.887$

In 2010, about 66.9 million adults baked as a leisure activity.

10. $x \geq 0$

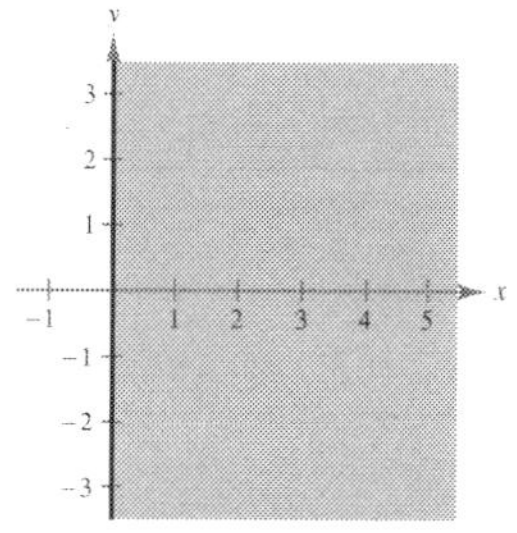

11. $y \geq 0$

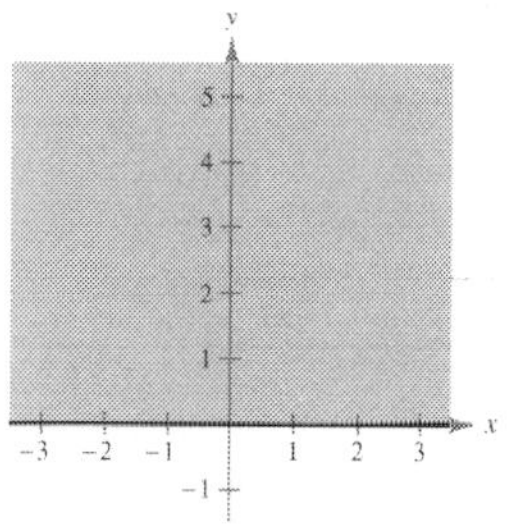

12. $x + 3y \le 12$

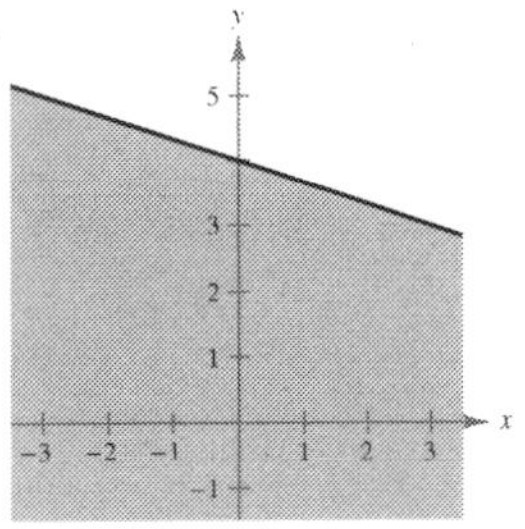

13. $3x + 2y \le 15$

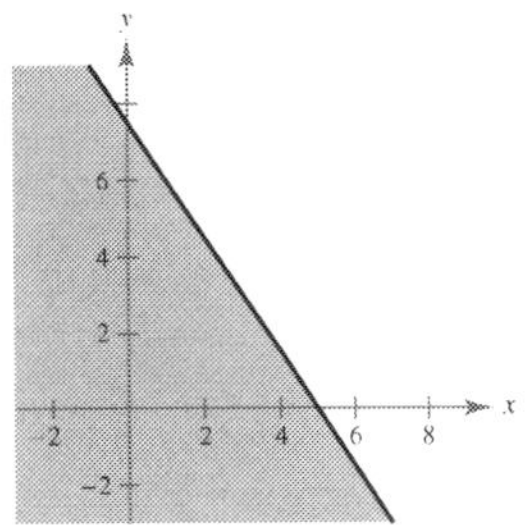

14. $\begin{cases} x \ge 0 \\ y \ge 0 \\ x + 3y \le 12 \\ 3x + 2y \le 15 \end{cases}$

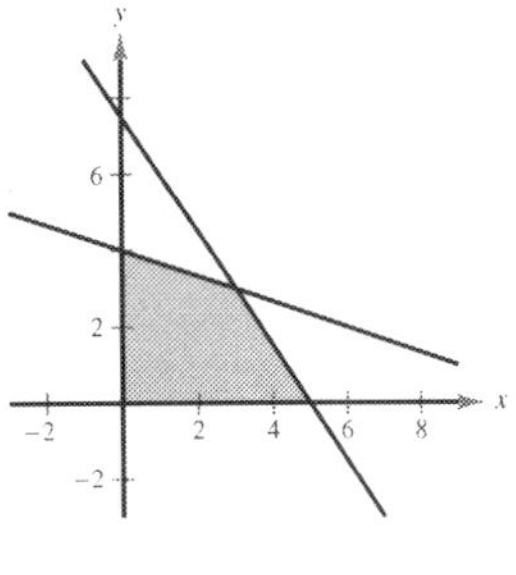

15. At $(0, 0)$: $z = 6(0) + 7(0) = 0$
At $(0, 4)$: $z = 6(0) + 7(4) = 28$
At $(3, 3)$: $z = 6(3) + 7(3) = 39$
At $(5, 0)$: $z = 6(5) + 7(0) = 30$

The minimum value is 0 at (0, 0).

The maximum value is 39 at (3, 3).

16. Objective function: $z = 200x + 275$
At $(0, 0)$: $z = 0$
At $(0, 700)$: $z = 192{,}500$
At $(692, 0)$: $z = 138{,}400$
At $(297, 569)$: $z = 215{,}875$

Constraints:
$3.5x + 8y \le 5600$
$2.5x + 2y \le 2000$
$1.3x + 0.9 \le 900$
$x \ge 0$
$y \ge 0$

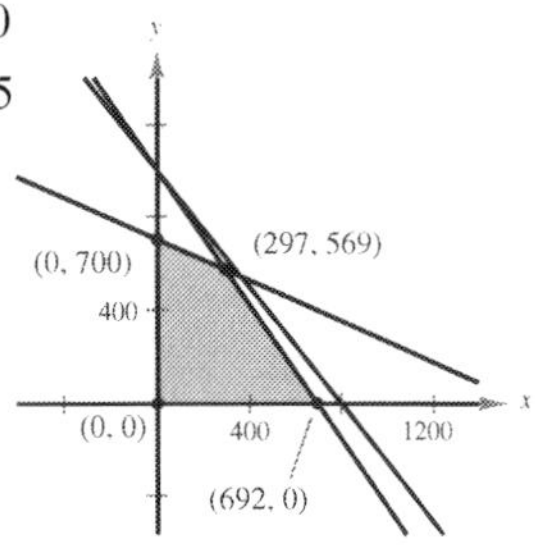

The optimal production level is 297 units of model A and 569 units of model B, because an optimal profit of \$215,875 is generated at that level of production. All other levels of possible production do not yield as great a profit.

CHAPTER 6
Matrices and Determinants

CHAPTER 6
Matrices and Determinants

Section 6.1 Matrices and Linear Systems

Skills Warm Up

1. $2(-1) - 3(5) + 7(2) = -2 - 15 + 14 = -3$

2. $-4(-3) + 6(7) + 8(-3) = 12 + 42 - 24 = 30$

3. $11\left(\frac{1}{2}\right) - 7\left(-\frac{3}{2}\right) - 5(2) = \frac{11}{2} + \frac{21}{2} - 10 = 6$

4. $\frac{2}{3}\left(\frac{1}{2}\right) + \frac{4}{3}\left(-\frac{1}{3}\right) = \frac{1}{3} - \frac{4}{9} = -\frac{1}{9}$

5. $4(1) - 2(3) + 3(-1) = -5$
$4 - 6 - 3 = -5$
$-5 = -5$ ✓

$1 + 3(3) - (-1) = 11$
$1 + 9 + 1 = 11$
$11 = 11$ ✓

$-1 + 2(3) = 5$
$-1 + 6 = 5$
$5 = 5$ ✓

Solution

6. $-1 + 2(3) + (-1) = 4$
$-1 + 6 - 1 = 4$
$4 = 4$ ✓

$2(1) - 3(-1) = 5$
$2 + 3 = 5$ ✓

$3(1) + 5(3) - 2(-1) = 21$
$3 + 15 + 2 = 21$
$20 \neq 21$ ✗

Not a solution

7. $2x - 3(2) = 4$
$2x - 6 = 4$
$2x = 10$
$x = 5$

$(5, 2)$

8. $5x + 4(-3) = 0$
$5x - 12 = 0$
$5x = 12$
$x = \frac{12}{5}$

$\left(\frac{12}{5}, -3\right)$

9. $y - 3(2) = 8$
$y = 14$

$x - 3(14) + 2 = 0$
$x - 42 + 2 = 0$
$x = 40$

$(40, 14, 2)$

10. $y - 4(1) = 0$
$y = 4$

$2x - 5(4) + 3(1) = -2$
$2x - 20 + 3 = -2$
$2x = 15$
$x = \frac{15}{2}$

$\left(\frac{15}{2}, 4, 1\right)$

---

1. 2×3

3. 4×3

5. 4×2

7. 2×4

9. System / Augmented matrix

$$\begin{cases} 4x - 3y = -5 \\ -x + 3y = 12 \end{cases} \qquad \left[\begin{array}{rr:r} 4 & -3 & -5 \\ -1 & 3 & 12 \end{array}\right]$$

The augmented matrix has two rows and three columns, so it is a 2×3 matrix.

11. System

$$\begin{cases} x + 10y - 2z = 2 \\ 5x - 3y + 4z = 0 \\ 2x + y = 6 \end{cases}$$

Augmented matrix

$$\left[\begin{array}{ccc:c} 1 & 10 & -2 & 2 \\ 5 & -3 & 4 & 0 \\ 2 & 1 & 0 & 6 \end{array}\right]$$

The augmented matrix has three rows and four columns, so it is a 3×4 matrix.

13. System

$$\begin{cases} 7x - 5y + z = 13 \\ 19x - 8z = 10 \end{cases}$$

Augmented matrix

$$\left[\begin{array}{ccc:c} 7 & -5 & 1 & 13 \\ 19 & 0 & -8 & 10 \end{array}\right]$$

The augmented matrix has two rows and four columns, so it is a 2×4 matrix.

15. $$\begin{matrix} \\ -5R_1 + R_2 \end{matrix}\begin{bmatrix} 1 & 1 & 1 \\ 0 & -7 & -1 \end{bmatrix}$$

17. $$\begin{matrix} -5R_2 + R_1 \\ \\ \\ \end{matrix}\begin{bmatrix} 1 & 0 & 14 & -11 \\ 0 & 1 & -2 & 2 \\ 0 & 0 & 1 & -7 \end{bmatrix}$$

19. Add 5 times R_2 to R_1.

21. Interchange R_1 and R_2.
Add 4 times R_1 to R_3.

23. $$\begin{aligned} 2x + 4y &= 6 \\ -x + 3y &= -8 \end{aligned}$$

25. $$\begin{aligned} x \qquad + 2z &= -10 \\ 3y - z &= 5 \\ 4x + 2y \qquad &= 3 \end{aligned}$$

27. $$\begin{aligned} x - 5y &= 6 \\ y &= -2 \\ x - 5(-2) &= 6 \\ x &= -4 \end{aligned}$$

Answer: $(-4, -2)$

29. $$\begin{aligned} x + 3y - z &= 15 \\ y + 4z &= -12 \\ z &= -5 \\ y + 4(-5) &= -12 \\ y &= 8 \\ x + 3(8) - (-5) &= 15 \\ x &= -14 \end{aligned}$$

Answer: $(-14, 8, -5)$

31. (a) $$\left[\begin{array}{cc:c} -3 & 4 & 22 \\ 6 & -4 & -28 \end{array}\right]$$

(i) $$\left[\begin{array}{cc:c} 3 & 0 & -6 \\ 6 & -4 & -28 \end{array}\right]$$ Add R_2 to R_1.

(ii) $$\left[\begin{array}{cc:c} 3 & 0 & -6 \\ 0 & -4 & -16 \end{array}\right]$$ Add $-2R_1$ to R_2.

(iii) $$\left[\begin{array}{cc:c} 3 & 0 & -6 \\ 0 & 1 & 4 \end{array}\right]$$ Multiply R_2 by $-\frac{1}{4}$.

(iv) $$\left[\begin{array}{cc:c} 1 & 0 & -2 \\ 0 & 1 & 4 \end{array}\right]$$ Multiply R_1 by $\frac{1}{3}$.

The solution is $(-2, 4)$.

(b) $$\begin{cases} -3x + 4y = 22 \\ 6x - 4y = -28 \end{cases}$$ Eq. 1, Eq. 2

$$\begin{cases} 3x = -6 \\ 6x - 4y = -28 \end{cases}$$ Add Eq. 2 to Eq. 1.

$$\begin{cases} 3x = -6 \\ -4y = -16 \end{cases}$$ Add -2 Eq. 1 to Eq. 2.

$$\begin{cases} x = -2 \\ y = 4 \end{cases}$$ Multiply Eq. 1 by $\frac{1}{3}$. Multiply Eq. 2 by $-\frac{1}{4}$.

The solution is $(-2, 4)$.

(c) Answers will vary. Sample answer: The basic process to solve by the methods in parts (a) and (b) are the same.

33. (i)

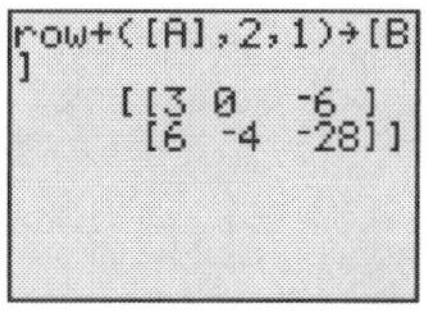

(ii)

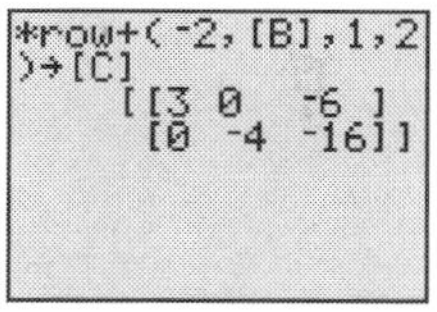

(iii)

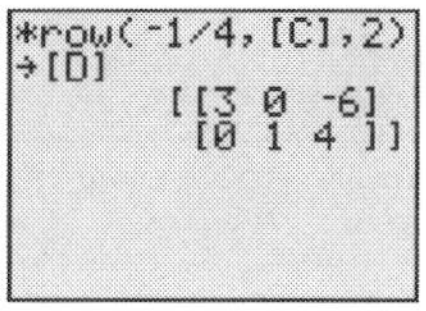

(iv)

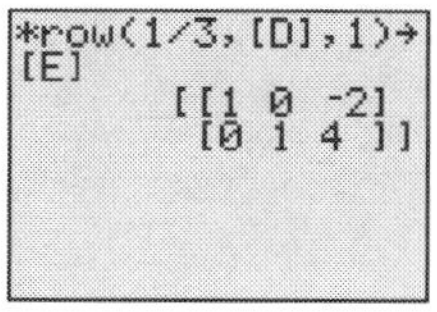

35. Reduced row-echelon form

37. Not in row-echelon form

39. Not in row-echelon form

41. $\begin{bmatrix} 1 & 2 & -1 & 5 \\ 3 & 2 & 1 & 11 \\ 4 & 8 & 1 & 10 \end{bmatrix}$ $\begin{matrix} \\ -3R_1 + R_2 \\ -4R_1 + R_3 \end{matrix} \begin{bmatrix} 1 & 2 & -1 & 5 \\ 0 & -4 & 4 & -4 \\ 0 & 0 & 5 & -10 \end{bmatrix}$

$\begin{matrix} \\ -\frac{1}{4}R_2 \\ \frac{1}{5}R_3 \end{matrix} \begin{bmatrix} 1 & 2 & -1 & 5 \\ 0 & 1 & -1 & 1 \\ 0 & 0 & 1 & -2 \end{bmatrix}$

43. $\begin{bmatrix} 1 & -1 & -1 & 1 \\ 5 & -4 & 1 & 8 \\ -6 & 8 & 18 & 0 \end{bmatrix}$ $\begin{matrix} \\ -5R_1 + R_2 \\ 6R_1 + R_3 \end{matrix} \begin{bmatrix} 1 & -1 & -1 & 1 \\ 0 & 1 & 6 & 3 \\ 0 & 2 & 12 & 6 \end{bmatrix}$

$\begin{matrix} \\ \\ -2R_2 + R_3 \end{matrix} \begin{bmatrix} 1 & -1 & -1 & 1 \\ 0 & 1 & 6 & 3 \\ 0 & 0 & 0 & 0 \end{bmatrix}$

45. $\begin{bmatrix} 4 & 4 & 8 \\ 1 & 2 & 2 \\ -3 & 6 & -9 \end{bmatrix}$ Interchange R_1 and R_2 $\begin{bmatrix} 1 & 2 & 2 \\ 4 & 4 & 8 \\ -3 & 6 & -9 \end{bmatrix}$

$\begin{matrix} \\ -4R_1 + R_2 \\ 3R_1 + R_3 \end{matrix} \begin{bmatrix} 1 & 2 & 2 \\ 0 & -4 & 0 \\ 0 & 12 & -3 \end{bmatrix}$

$\begin{matrix} \\ -\frac{1}{4}R_2 \\ -\frac{1}{3}R_3 \end{matrix} \begin{bmatrix} 1 & 2 & 2 \\ 0 & 1 & 0 \\ 0 & -4 & 1 \end{bmatrix}$

$\begin{matrix} -2R_2 + R_1 \\ \\ 4R_2 + R_3 \end{matrix} \begin{bmatrix} 1 & 0 & 2 \\ 0 & 1 & 0 \\ 0 & 0 & 1 \end{bmatrix}$

$\begin{matrix} -2R_3 + R_1 \\ \\ \\ \end{matrix} \begin{bmatrix} 1 & 0 & 0 \\ 0 & 1 & 0 \\ 0 & 0 & 1 \end{bmatrix}$

47. $\begin{bmatrix} 1 & 0 & 2 & 1 \\ 0 & 3 & -6 & 6 \\ 2 & 0 & 5 & -4 \\ 0 & 1 & 0 & 1 \end{bmatrix}$ $\begin{matrix} \\ \frac{1}{3}R_2 \\ -2R_1 + R_3 \\ \\ \end{matrix} \begin{bmatrix} 1 & 0 & 2 & 1 \\ 0 & 1 & -2 & 2 \\ 0 & 0 & 1 & -6 \\ 0 & 1 & 0 & 1 \end{bmatrix}$

Interchange R_2 and R_4 $\begin{bmatrix} 1 & 0 & 2 & 1 \\ 0 & 1 & 0 & 1 \\ 0 & 0 & 1 & -6 \\ 0 & 1 & -2 & 2 \end{bmatrix}$

$\begin{matrix} \\ \\ \\ -R_2 + R_4 \end{matrix} \begin{bmatrix} 1 & 0 & 2 & 1 \\ 0 & 1 & 0 & 1 \\ 0 & 0 & 1 & -6 \\ 0 & 0 & -2 & 1 \end{bmatrix}$

$\begin{matrix} \\ \\ \\ 2R_3 + R_4 \end{matrix} \begin{bmatrix} 1 & 0 & 2 & 1 \\ 0 & 1 & 0 & 1 \\ 0 & 0 & 1 & -6 \\ 0 & 0 & 0 & -11 \end{bmatrix}$

$\begin{matrix} \\ \\ \\ -\frac{1}{11}R_4 \end{matrix} \begin{bmatrix} 1 & 0 & 2 & 1 \\ 0 & 1 & 0 & 1 \\ 0 & 0 & 1 & -6 \\ 0 & 0 & 0 & 1 \end{bmatrix}$

$\begin{matrix} \\ -R_4 + R_2 \\ \\ \\ \end{matrix} \begin{bmatrix} 1 & 0 & 2 & 1 \\ 0 & 1 & 0 & 0 \\ 0 & 0 & 1 & -6 \\ 0 & 0 & 0 & 1 \end{bmatrix}$

$\begin{matrix} -2R_3 + R_1 \\ \\ \\ \\ \end{matrix} \begin{bmatrix} 1 & 0 & 0 & 13 \\ 0 & 1 & 0 & 0 \\ 0 & 0 & 1 & -6 \\ 0 & 0 & 0 & 1 \end{bmatrix}$

$\begin{matrix} -13R_4 + R_1 \\ \\ 6R_4 + R_3 \\ \\ \end{matrix} \begin{bmatrix} 1 & 0 & 0 & 0 \\ 0 & 1 & 0 & 0 \\ 0 & 0 & 1 & 0 \\ 0 & 0 & 0 & 1 \end{bmatrix}$

49. $\begin{bmatrix} 2 & 1 \\ 1 & 4 \\ -2 & -1 \end{bmatrix}$ $\begin{matrix} \\ \\ R_1 + R_3 \end{matrix} \begin{bmatrix} 2 & 1 \\ 1 & 4 \\ 0 & 0 \end{bmatrix}$

Interchange R_1 and R_2 $\begin{bmatrix} 1 & 4 \\ 2 & 1 \\ 0 & 0 \end{bmatrix}$

$\begin{matrix} \\ -2R_1 + R_2 \\ \\ \end{matrix} \begin{bmatrix} 1 & 4 \\ 0 & 1 \\ 0 & 0 \end{bmatrix}$

$\begin{matrix} -4R_2 + R_1 \\ \\ \\ \end{matrix} \begin{bmatrix} 1 & 0 \\ 0 & 1 \\ 0 & 0 \end{bmatrix}$

51. $\begin{bmatrix} 1 & 0 & \vdots & -4 \\ 0 & 1 & \vdots & 6 \end{bmatrix}$

$x = -4$

$y = 6$

Answer: $(-4, 6)$

53. $\begin{bmatrix} 1 & 0 & 0 & \vdots & -4 \\ 0 & 1 & 0 & \vdots & -8 \\ 0 & 0 & 1 & \vdots & 2 \end{bmatrix}$

$x = -4$

$y = -8$

$z = 2$

Answer: $(-4, -8, 2)$

55. $\begin{bmatrix} 1 & 0 & 2 & \vdots & -4 \\ 0 & 1 & 1 & \vdots & 6 \\ 0 & 0 & 0 & \vdots & 0 \end{bmatrix}$

$x + 2z = -4 \Rightarrow x = -4 - 2z \Rightarrow x = -4 - 2a$

$y + z = 6 \Rightarrow y = 6 - z \Rightarrow y = 6 - a$

Answer: $(-2a - 4, -a + 6, a)$

57. $x + 2y = 7$

$2x + y = 8$

$\begin{bmatrix} 1 & 2 & \vdots & 7 \\ 2 & 1 & \vdots & 8 \end{bmatrix}$ $\begin{matrix} \\ -2R_1 + R_2 \end{matrix}\begin{bmatrix} 1 & 2 & \vdots & 7 \\ 0 & -3 & \vdots & -6 \end{bmatrix}$

$\begin{matrix} \\ -\frac{1}{3}R_2 \end{matrix}\begin{bmatrix} 1 & 2 & \vdots & 7 \\ 0 & 1 & \vdots & 2 \end{bmatrix}$

$y = 2$

$x + 2(2) = 7 \Rightarrow x = 3$

Answer: $(3, 2)$

59. $-3x + 5y = -22$

$3x + 4y = 4$

$4x - 8y = 32$

$\begin{bmatrix} -3 & 5 & \vdots & -22 \\ 3 & 4 & \vdots & 4 \\ 4 & -8 & \vdots & 32 \end{bmatrix}$ $\begin{matrix} R_3 + R_1 \\ \\ \\ \end{matrix}\begin{bmatrix} 1 & -3 & \vdots & 10 \\ 3 & 4 & \vdots & 4 \\ 4 & -8 & \vdots & 32 \end{bmatrix}$

$\begin{matrix} \\ -3R_1 + R_2 \\ -4R_1 + R_3 \end{matrix}\begin{bmatrix} 1 & -3 & \vdots & 10 \\ 0 & 13 & \vdots & -26 \\ 0 & 4 & \vdots & -8 \end{bmatrix}$

$\begin{matrix} \\ \frac{1}{13}R_2 \\ -4R_2 + R_3 \end{matrix}\begin{bmatrix} 1 & -3 & \vdots & 10 \\ 0 & 1 & \vdots & -2 \\ 0 & 0 & \vdots & 0 \end{bmatrix}$

$y = -2$

$x - 3(-2) = 10 \Rightarrow x = 4$

Answer: $(4, -2)$

61. $8x - 4y = 7$

$5x + 2y = 1$

$\begin{bmatrix} 8 & -4 & \vdots & 7 \\ 5 & 2 & \vdots & 1 \end{bmatrix}$ $\begin{matrix} 3R_1 \\ 5R_2 \end{matrix}\begin{bmatrix} 24 & -12 & \vdots & 21 \\ 25 & 10 & \vdots & 5 \end{bmatrix}$

$\begin{matrix} -R_2 + R_1 \\ \\ \end{matrix}\begin{bmatrix} -1 & -22 & \vdots & 16 \\ 25 & 10 & \vdots & 5 \end{bmatrix}$

$\begin{matrix} \\ 25R_1 + R_2 \end{matrix}\begin{bmatrix} -1 & -22 & \vdots & 16 \\ 0 & -540 & \vdots & 405 \end{bmatrix}$

$\begin{matrix} -R_1 \\ -\frac{1}{540}R_2 \end{matrix}\begin{bmatrix} 1 & 22 & \vdots & -16 \\ 0 & 1 & \vdots & -\frac{3}{4} \end{bmatrix}$

$y = -\frac{3}{4}$

$x + 22\left(-\frac{3}{4}\right) = -16 \Rightarrow x = \frac{1}{2}$

Answer: $\left(\frac{1}{2}, -\frac{3}{4}\right)$

63. $-x + 2y = 1.5$

$2x - 4y = 3.0$

$\begin{bmatrix} -1 & 2 & \vdots & 1.5 \\ 2 & -4 & \vdots & 3.0 \end{bmatrix}$ $\begin{matrix} \\ 2R_1 + R_2 \end{matrix}\begin{bmatrix} -1 & 2 & \vdots & 1.5 \\ 0 & 0 & \vdots & 6.0 \end{bmatrix}$

The system is inconsistent and there is no solution.

65. $x + 2y + z = 8$

$3x + 7y + 6z = 26$

$\begin{bmatrix} 1 & 2 & 1 & \vdots & 8 \\ 3 & 7 & 6 & \vdots & 26 \end{bmatrix}$ $\begin{matrix} \\ -3R_1 + R_2 \end{matrix}\begin{bmatrix} 1 & 2 & 1 & \vdots & 8 \\ 0 & 1 & 3 & \vdots & 2 \end{bmatrix}$

$\begin{matrix} -2R_2 + R_1 \\ \\ \end{matrix}\begin{bmatrix} 1 & 0 & -5 & \vdots & 4 \\ 0 & 1 & 3 & \vdots & 2 \end{bmatrix}$

$z = a$

$y = -3a + 2$

$x = 5a + 4$

Answer: $(5a + 4, -3a + 2, a)$

67. $3x + 3y + 12z = 6$

$x + y + 4z = 2$

$2x + 5y + 20z = 10$

$-x + 2y + 8z = 4$

$$\begin{bmatrix} 3 & 3 & 12 & \vdots & 6 \\ 1 & 1 & 4 & \vdots & 2 \\ 2 & 5 & 20 & \vdots & 10 \\ -1 & 2 & 8 & \vdots & 4 \end{bmatrix} \begin{matrix} -2R_2 + R_1 \\ \\ \\ \\ \end{matrix} \begin{bmatrix} 1 & 1 & 4 & \vdots & 2 \\ 1 & 1 & 4 & \vdots & 2 \\ 2 & 5 & 20 & \vdots & 10 \\ -1 & 2 & 8 & \vdots & 4 \end{bmatrix}$$

$$\begin{matrix} \\ -R_1 + R_2 \\ -2R_1 + R_3 \\ R_1 + R_4 \end{matrix} \begin{bmatrix} 1 & 1 & 4 & \vdots & 2 \\ 0 & 0 & 0 & \vdots & 0 \\ 0 & 3 & 12 & \vdots & 6 \\ 0 & 3 & 12 & \vdots & 6 \end{bmatrix}$$

$$\begin{matrix} \\ \\ \\ -R_3 + R_4 \end{matrix} \begin{bmatrix} 1 & 1 & 4 & \vdots & 2 \\ 0 & 0 & 0 & \vdots & 0 \\ 0 & 3 & 12 & \vdots & 6 \\ 0 & 0 & 0 & \vdots & 0 \end{bmatrix}$$

$$\begin{matrix} \\ \\ \frac{1}{3}R_3 \\ \\ \end{matrix} \begin{bmatrix} 1 & 1 & 4 & \vdots & 2 \\ 0 & 0 & 0 & \vdots & 0 \\ 0 & 1 & 4 & \vdots & 2 \\ 0 & 0 & 0 & \vdots & 0 \end{bmatrix}$$

$$\begin{matrix} -R_3 + R_1 \\ \\ \\ \\ \end{matrix} \begin{bmatrix} 1 & 0 & 0 & \vdots & 0 \\ 0 & 0 & 0 & \vdots & 0 \\ 0 & 1 & 4 & \vdots & 2 \\ 0 & 0 & 0 & \vdots & 0 \end{bmatrix}$$

$z = a$

$y = -4a + 2$

$x = 0$

Answer: $(0, -4a + 2, a)$

69. $4x + 12y - 7z - 20w = 22$

$3x + 9y - 5z - 28w = 30$

$$\begin{bmatrix} 4 & 12 & -7 & -20 & \vdots & 22 \\ 3 & 9 & -5 & -28 & \vdots & 30 \end{bmatrix} \begin{matrix} -R_2 + R_1 \\ \\ \end{matrix} \begin{bmatrix} 1 & 3 & -2 & 8 & \vdots & -8 \\ 3 & 9 & -5 & -28 & \vdots & 30 \end{bmatrix}$$

$$\begin{matrix} \\ -3R_1 + R_2 \end{matrix} \begin{bmatrix} 1 & 3 & -2 & 8 & \vdots & -8 \\ 0 & 0 & 1 & -52 & \vdots & 54 \end{bmatrix}$$

$$\begin{matrix} 2R_2 + R_1 \\ \\ \end{matrix} \begin{bmatrix} 1 & 3 & 0 & -96 & \vdots & 100 \\ 0 & 0 & 1 & -52 & \vdots & 54 \end{bmatrix}$$

$w = a$

$z = 52a + 54$

$y = b$

$x = -3b + 96a + 100$

Answer: $(-3b + 96a + 100, b, 52a + 54, a)$

71. $\begin{bmatrix} 1 & 2 & \vdots & 0 \\ -1 & -1 & \vdots & 0 \end{bmatrix}$ $\begin{matrix} \\ R_1 + R_2 \end{matrix}\begin{bmatrix} 1 & 2 & \vdots & 0 \\ 0 & 1 & \vdots & 0 \end{bmatrix}$

$\begin{matrix} -2R_2 + R_1 \\ \\ \end{matrix}\begin{bmatrix} 1 & 0 & \vdots & 0 \\ 0 & 1 & \vdots & 0 \end{bmatrix}$

$x = 0$

$y = 0$

Answer: $(0, 0)$

73. $\begin{bmatrix} 1 & 1 & 1 & \vdots & 0 \\ 2 & 3 & 1 & \vdots & 0 \\ 3 & 5 & 1 & \vdots & 0 \end{bmatrix}$ $\begin{matrix} \\ -2R_1 + R_2 \\ -3R_1 + R_3 \end{matrix}\begin{bmatrix} 1 & 1 & 1 & \vdots & 0 \\ 0 & 1 & -1 & \vdots & 0 \\ 0 & 2 & -2 & \vdots & 0 \end{bmatrix}$

$\begin{matrix} -R_2 + R_1 \\ \\ -2R_2 + R_3 \end{matrix}\begin{bmatrix} 1 & 0 & 2 & \vdots & 0 \\ 0 & 1 & -1 & \vdots & 0 \\ 0 & 0 & 0 & \vdots & 0 \end{bmatrix}$

$z = a$

$y = a$

$x = -2a$

Answer: $(-2a, a, a)$

75. $\begin{cases} x \qquad\quad - 3z = -2 \\ 3x + y - 2z = 5 \\ 2x + 2y + z = 4 \end{cases} \Rightarrow \begin{bmatrix} 1 & 0 & -3 & \vdots & -2 \\ 3 & 1 & -2 & \vdots & 5 \\ 2 & 2 & 1 & \vdots & 4 \end{bmatrix}$

$-3R_1 + R_2 \rightarrow \begin{bmatrix} 1 & 0 & -3 & \vdots & -2 \\ 0 & 1 & 7 & \vdots & 11 \\ 2 & 2 & 1 & \vdots & 4 \end{bmatrix}$

$-2R_1 + R_3 \rightarrow \begin{bmatrix} 1 & 0 & -3 & \vdots & -2 \\ 0 & 1 & 7 & \vdots & 11 \\ 0 & 2 & 7 & \vdots & 8 \end{bmatrix}$

$-2R_2 + R_3 \rightarrow \begin{bmatrix} 1 & 0 & -3 & \vdots & -2 \\ 0 & 1 & 7 & \vdots & 11 \\ 0 & 0 & -7 & \vdots & -14 \end{bmatrix}$

$-\frac{1}{7}R_3 \rightarrow \begin{bmatrix} 1 & 0 & -3 & \vdots & -2 \\ 0 & 1 & 7 & \vdots & 11 \\ 0 & 0 & 1 & \vdots & 2 \end{bmatrix}$

$3R_3 + R_1 \rightarrow \begin{bmatrix} 1 & 0 & 0 & \vdots & 4 \\ 0 & 1 & 7 & \vdots & 11 \\ 0 & 0 & 1 & \vdots & 2 \end{bmatrix}$

$-7R_3 + R_2 \rightarrow \begin{bmatrix} 1 & 0 & 0 & \vdots & 4 \\ 0 & 1 & 0 & \vdots & -3 \\ 0 & 0 & 1 & \vdots & 2 \end{bmatrix}$

Solution: $(4, -3, 2)$

77. $\begin{cases} x + y - 5z = 3 \\ x \qquad - 2z = 1 \\ 2x - y - z = 0 \end{cases} \Rightarrow \begin{bmatrix} 1 & 1 & -5 & \vdots & 3 \\ 1 & 0 & -2 & \vdots & 1 \\ 2 & -1 & -1 & \vdots & 0 \end{bmatrix}$

$R_2 \leftrightarrow R_3 \rightarrow \begin{bmatrix} 1 & 1 & -5 & \vdots & 3 \\ 2 & -1 & -1 & \vdots & 0 \\ 1 & 0 & -2 & \vdots & 1 \end{bmatrix}$

$-2R_1 + R_2 \rightarrow \begin{bmatrix} 1 & 1 & -5 & \vdots & 3 \\ 0 & -3 & 9 & \vdots & -6 \\ 1 & 0 & -2 & \vdots & 1 \end{bmatrix}$

$-R_1 + R_3 \rightarrow \begin{bmatrix} 1 & 1 & -5 & \vdots & 3 \\ 0 & -3 & 9 & \vdots & -6 \\ 0 & -1 & 3 & \vdots & -2 \end{bmatrix}$

$-\frac{1}{3}R_2 \rightarrow \begin{bmatrix} 1 & 1 & -5 & \vdots & 3 \\ 0 & 1 & -3 & \vdots & 2 \\ 0 & -1 & 3 & \vdots & -2 \end{bmatrix}$

$R_2 + R_3 \rightarrow \begin{bmatrix} 1 & 1 & -5 & \vdots & 3 \\ 0 & 1 & -3 & \vdots & 2 \\ 0 & 0 & 0 & \vdots & 0 \end{bmatrix}$

$\begin{cases} x + y - 5z = 3 \\ \quad\;\; y - 3z = 2 \end{cases}$

Let $z = a$: $y = 3z + 2 \Rightarrow\ = 3a + 2$

$x = -y + 5z + 3 \Rightarrow\ = -(3a + 2) + 5a + 3 \Rightarrow\ = 2a + 1$

The solution is $(2a + 1, 3a + 2, a)$ where a is any real number.

79. $\begin{cases} -x + y - z = -14 \\ 2x - y + z = 21 \\ 3x + 2y + z = 19 \end{cases} \Rightarrow \left[\begin{array}{ccc:c} -1 & 1 & -1 & -14 \\ 2 & -1 & 1 & 21 \\ 3 & 2 & 1 & 19 \end{array}\right]$

$-R_1 \rightarrow \left[\begin{array}{ccc:c} 1 & -1 & 1 & 14 \\ 2 & -1 & 1 & 21 \\ 3 & 2 & 1 & 19 \end{array}\right]$

$-2R_1 + R_2 \rightarrow \left[\begin{array}{ccc:c} 1 & -1 & 1 & 14 \\ 0 & 1 & -1 & -7 \\ 3 & 2 & 1 & 19 \end{array}\right]$

$-3R_1 + R_3 \rightarrow \left[\begin{array}{ccc:c} 1 & -1 & 1 & 14 \\ 0 & 1 & -1 & -7 \\ 0 & 5 & -2 & -23 \end{array}\right]$

$R_2 + R_1 \rightarrow \left[\begin{array}{ccc:c} 1 & 0 & 0 & 7 \\ 0 & 1 & -1 & -7 \\ 0 & 5 & -2 & -23 \end{array}\right]$

$-5R_2 + R_3 \rightarrow \left[\begin{array}{ccc:c} 1 & 0 & 0 & 7 \\ 0 & 1 & -1 & -7 \\ 0 & 0 & 3 & 12 \end{array}\right]$

$\frac{1}{3}R_3 \rightarrow \left[\begin{array}{ccc:c} 1 & 0 & 0 & 7 \\ 0 & 1 & -1 & -7 \\ 0 & 0 & 1 & 4 \end{array}\right]$

$R_3 + R_2 \rightarrow \left[\begin{array}{ccc:c} 1 & 0 & 0 & 7 \\ 0 & 1 & 0 & -3 \\ 0 & 0 & 1 & 4 \end{array}\right]$

Solution: $(7, -3, 4)$

81. (a) $\begin{cases} x - 2y + z = -6 \\ y - 5z = 16 \\ z = -3 \end{cases}$

$y - 5(-3) = 16$

$y = 1$

$x - 2(1) + (-3) = -6$

$x = -1$

Solution: $(-1, 1, -3)$

Both systems yield the same solution, $(-1, 1, -3)$.

(b) $\begin{cases} x + y - 2z = 6 \\ y + 3z = -8 \\ z = -3 \end{cases}$

$y + 3(-3) = -8$

$y = 1$

$x + (1) - 2(-3) = 6$

$x = -1$

Solution: $(-1, 1, -3)$

83. (a) $\begin{cases} x - 4y + 5z = 27 \\ y - 7z = -54 \\ z = 8 \end{cases}$

$y - 7(8) = -54$

$y = 2$

$x - 4(2) + 5(8) = 27$

$x = -5$

Solution: $(-5, 2, 8)$

The systems do *not* yield the same solution.

(b) $\begin{cases} x - 6y + z = 15 \\ y + 5z = 42 \\ z = 8 \end{cases}$

$y + 5(8) = 42$

$y = 2$

$x - 6(2) + (8) = 15$

$x = 19$

Solution: $(19, 2, 8)$

85. Let x = amount at 8%.
Let y = amount at 9%.
Let z = amount at 12%.

$x + y + z = 2{,}000{,}000$

$0.08x + 0.09y + 0.12z = 186{,}000$

$-x + 2z = 0$

$\left[\begin{array}{ccc:c} 1 & 1 & 1 & 2{,}000{,}000 \\ 0.08 & 0.09 & 0.12 & 186{,}000 \\ -1 & 0 & 2 & 0 \end{array}\right]$

$\begin{matrix} \\ 100R_2 \\ R_1 + R_3 \end{matrix} \left[\begin{array}{ccc:c} 1 & 1 & 1 & 2{,}000{,}000 \\ 8 & 9 & 12 & 18{,}600{,}000 \\ 0 & 1 & 3 & 2{,}000{,}000 \end{array}\right]$

$-8R_1 + R_2 \left[\begin{array}{ccc:c} 1 & 1 & 1 & 2{,}000{,}000 \\ 0 & 1 & 4 & 2{,}600{,}000 \\ 0 & 1 & 3 & 2{,}000{,}000 \end{array}\right]$

Interchange R_2 and R_3 $\left[\begin{array}{ccc:c} 1 & 1 & 1 & 2{,}000{,}000 \\ 0 & 1 & 3 & 2{,}000{,}000 \\ 0 & 1 & 4 & 2{,}600{,}000 \end{array}\right]$

$-R_2 + R_3 \left[\begin{array}{ccc:c} 1 & 1 & 1 & 2{,}000{,}000 \\ 0 & 1 & 3 & 2{,}000{,}000 \\ 0 & 0 & 1 & 600{,}000 \end{array}\right]$

$-3R_3 + R_2 \left[\begin{array}{ccc:c} 1 & 1 & 1 & 2{,}000{,}000 \\ 0 & 1 & 0 & 200{,}000 \\ 0 & 0 & 1 & 600{,}000 \end{array}\right]$

$z = 600{,}000$

$y = 200{,}000$

$x + 200{,}000 + 600{,}000 = 2{,}000{,}000 \Rightarrow x = 1{,}200{,}000$

Answer: (1,200,000, 200,000, 600,000)

So, \$1,200,000 was borrowed at 8%, \$200,000 was borrowed at 9%, and \$600,000 was borrowed at 12%.

87. Both are correct. Because there are infinitely many ordered triples that are solutions to this system, a solution can be written many different ways. If $a = 3$, the ordered triple would be $(3, -3, 5)$ which is the same ordered triple if $b = 5$ is substituted in $\left(\frac{1}{2}b + \frac{1}{2}, -\frac{1}{2}b - \frac{1}{2}, b\right)$.

89. $\begin{cases} 12b + 66a = 831 \\ 66b + 506a = 5643 \end{cases} \Rightarrow \begin{bmatrix} 12 & 66 & \vdots & 831 \\ 66 & 506 & \vdots & 5643 \end{bmatrix}$

$$\frac{1}{12}R_1 \rightarrow \begin{bmatrix} 1 & 5.5 & \vdots & 69.25 \\ 66 & 506 & \vdots & 5643 \end{bmatrix}$$

$$-66R_1 + R_2 \rightarrow \begin{bmatrix} 1 & 5.5 & \vdots & 69.25 \\ 0 & 143 & \vdots & 1072.5 \end{bmatrix}$$

$$\frac{1}{143}R_2 \rightarrow \begin{bmatrix} 1 & 5.5 & \vdots & 69.25 \\ 0 & 1 & \vdots & 7.5 \end{bmatrix}$$

$$-5.5R_2 + R_1 \rightarrow \begin{bmatrix} 1 & 0 & \vdots & 28 \\ 0 & 1 & \vdots & 7.5 \end{bmatrix}$$

$y = 7.5t + 28$

For 2012, $t = 14$:

$y = 7.5(14) + 28 = 133$

There will be 133 new cases in 2012.

Because the data values increased in a linear pattern, this estimate seems reasonable.

Section 6.2 Operations with Matrices

Skills Warm Up

1. $-3\left(-\frac{5}{6}\right) + 10\left(-\frac{3}{4}\right) = \frac{5}{2} + \left(-\frac{15}{2}\right) = -5$

2. $-22\left(\frac{5}{2}\right) + 6(8) = -55 + 48 = -7$

3. No

4. No

5. $\begin{bmatrix} -5 & 10 & \vdots & 12 \\ 7 & -3 & \vdots & 0 \end{bmatrix}$

6. $\begin{bmatrix} 10 & 15 & -9 & \vdots & 42 \\ 6 & -5 & 0 & \vdots & 0 \end{bmatrix}$

7. $(0, 2)$

8. $y = 3 - a$

$x = a + 2$

$(a + 2, 3 - a, a)$

9. $(1 - 2a, a, -1)$

10. $\begin{bmatrix} 1 & -1 & 0 & \vdots & 3 \\ 0 & 1 & -2 & \vdots & 1 \\ 0 & 0 & 1 & \vdots & -1 \end{bmatrix}$

$$2R_3 + R_2 \begin{bmatrix} 1 & -1 & 0 & \vdots & 3 \\ 0 & 1 & 0 & \vdots & -1 \\ 0 & 0 & 1 & \vdots & -1 \end{bmatrix}$$

$$R_2 + R_1 \begin{bmatrix} 1 & 0 & 0 & \vdots & 2 \\ 0 & 1 & 0 & \vdots & -1 \\ 0 & 0 & 1 & \vdots & -1 \end{bmatrix}$$

$(2, -1, -1)$

1. $x = -3, y = 2$

3. $x - 2 = -4, 2y - 1 = 9$

$x = -2, \quad y = 5$

5. $A = \begin{bmatrix} 7 & -3 \end{bmatrix}, B = \begin{bmatrix} -4 & 2 \end{bmatrix}$

(a) $A + B = \begin{bmatrix} 7 & -3 \end{bmatrix} + \begin{bmatrix} -4 & 2 \end{bmatrix}$
$= \begin{bmatrix} 7 + (-4) & -3 + 2 \end{bmatrix} = \begin{bmatrix} 3 & -1 \end{bmatrix}$

(b) $A - B = \begin{bmatrix} 7 & -3 \end{bmatrix} - \begin{bmatrix} -4 & 2 \end{bmatrix}$
$= \begin{bmatrix} 7 - (-4) & -3 - 2 \end{bmatrix} = \begin{bmatrix} 11 & -5 \end{bmatrix}$

(c) $6A = 6\begin{bmatrix} 7 & -3 \end{bmatrix} = \begin{bmatrix} 42 & -18 \end{bmatrix}$

(d) $4A - 3B = 4\begin{bmatrix} 7 & -3 \end{bmatrix} - 3\begin{bmatrix} -4 & 2 \end{bmatrix}$
$= \begin{bmatrix} 28 & -12 \end{bmatrix} - \begin{bmatrix} -12 & 6 \end{bmatrix}$
$= \begin{bmatrix} 28 - (-12) & -12 - 6 \end{bmatrix}$
$= \begin{bmatrix} 40 & -18 \end{bmatrix}$

7. $A = \begin{bmatrix} 5 & -2 \\ 3 & 1 \end{bmatrix}, B = \begin{bmatrix} 3 & 1 \\ -2 & 6 \end{bmatrix}$

(a) $A + B = \begin{bmatrix} 5 + 3 & -2 + 1 \\ 3 + (-2) & 1 + 6 \end{bmatrix} = \begin{bmatrix} 8 & -1 \\ 1 & 7 \end{bmatrix}$

(b) $A - B = \begin{bmatrix} 5 - 3 & -2 - 1 \\ 3 - (-2) & 1 - 6 \end{bmatrix} = \begin{bmatrix} 2 & -3 \\ 5 & -5 \end{bmatrix}$

(c) $6A = \begin{bmatrix} 6(5) & 6(-2) \\ 6(3) & 6(1) \end{bmatrix} = \begin{bmatrix} 30 & -12 \\ 18 & 6 \end{bmatrix}$

(d) $4A - 3B = \begin{bmatrix} 4(5) & 4(-2) \\ 4(3) & 4(1) \end{bmatrix} - \begin{bmatrix} 3(3) & 3(1) \\ 3(-2) & 3(6) \end{bmatrix}$
$= \begin{bmatrix} 11 & -11 \\ 18 & -14 \end{bmatrix}$

9. $A = \begin{bmatrix} 6 & -1 \\ 2 & 4 \\ -3 & 5 \end{bmatrix}, B = \begin{bmatrix} 1 & 4 \\ -1 & 5 \\ 1 & 10 \end{bmatrix}$

(a) $A + B = \begin{bmatrix} 6 + 1 & -1 + 4 \\ 2 + (-1) & 4 + 5 \\ -3 + 1 & 1 + 10 \end{bmatrix} = \begin{bmatrix} 7 & 3 \\ 1 & 9 \\ -2 & 15 \end{bmatrix}$

(b) $A - B = \begin{bmatrix} 6 - 1 & -1 - 4 \\ 2 - (-1) & 4 - 5 \\ -3 - 1 & 5 - 10 \end{bmatrix} = \begin{bmatrix} 5 & -5 \\ 3 & -1 \\ -4 & -5 \end{bmatrix}$

(c) $6A = \begin{bmatrix} 6(6) & 6(-1) \\ 6(2) & 6(4) \\ 6(-3) & 6(5) \end{bmatrix} = \begin{bmatrix} 36 & -6 \\ 12 & 24 \\ -18 & 30 \end{bmatrix}$

(d) $4A - 3B = \begin{bmatrix} 4(6) & 4(-1) \\ 4(2) & 4(4) \\ 4(-3) & 4(5) \end{bmatrix} - \begin{bmatrix} 3(1) & 3(4) \\ 3(-1) & 3(5) \\ 3(1) & 3(10) \end{bmatrix} = \begin{bmatrix} 21 & -16 \\ 11 & 1 \\ -15 & -10 \end{bmatrix}$

11. $A = \begin{bmatrix} 2 & 2 & -1 \\ 1 & 1 & -2 \\ 1 & -1 & 3 \end{bmatrix}, B = \begin{bmatrix} 1 & 1 & -1 \\ -3 & 4 & 9 \\ 0 & -7 & 8 \end{bmatrix}$

(a) $A + B = \begin{bmatrix} 2 + 1 & 2 + 1 & -1 + (-1) \\ 1 + (-3) & 1 + 4 & -2 + 9 \\ 1 + 0 & -1 + (-7) & 3 + 8 \end{bmatrix} = \begin{bmatrix} 3 & 3 & -2 \\ -2 & 5 & 7 \\ 1 & -8 & 11 \end{bmatrix}$

(b) $A - B = \begin{bmatrix} 2 - 1 & 2 - 1 & -1 - (-1) \\ 1 - (-3) & 1 - 4 & -2 - 9 \\ 1 - 0 & -1 - (-7) & 3 - 8 \end{bmatrix} = \begin{bmatrix} 1 & 1 & 0 \\ 4 & -3 & -11 \\ 1 & 6 & -5 \end{bmatrix}$

(c) $6A = \begin{bmatrix} 6(2) & 6(2) & 6(-1) \\ 6(1) & 6(1) & 6(-2) \\ 6(1) & 6(-1) & 6(3) \end{bmatrix} = \begin{bmatrix} 12 & 12 & -6 \\ 6 & 6 & -12 \\ 6 & -6 & 18 \end{bmatrix}$

(d) $4A - 3B = \begin{bmatrix} 4(2) & 4(2) & 4(-1) \\ 4(1) & 4(1) & 4(-2) \\ 4(1) & 4(-1) & 4(3) \end{bmatrix} - \begin{bmatrix} 3(1) & 3(1) & 3(-1) \\ 3(-3) & 3(4) & 3(9) \\ 3(0) & 3(-7) & 3(8) \end{bmatrix} = \begin{bmatrix} 5 & 5 & -1 \\ 13 & -8 & -35 \\ 4 & 17 & -12 \end{bmatrix}$

13. $A = \begin{bmatrix} -13 & 24 & 20 & -11 \end{bmatrix}$, $B = \begin{bmatrix} 15 & 16 & 10 & -12 \end{bmatrix}$

(a) $A + B = \begin{bmatrix} -13 & 24 & 20 & -11 \end{bmatrix} + \begin{bmatrix} 15 & 16 & 10 & -12 \end{bmatrix}$

$= \begin{bmatrix} (-13) + 15 & 24 + 16 & 20 + 10 & (-11) + (-12) \end{bmatrix}$

$= \begin{bmatrix} 2 & 40 & 30 & -23 \end{bmatrix}$

(b) $A - B = \begin{bmatrix} -13 & 24 & 20 & -11 \end{bmatrix} - \begin{bmatrix} 15 & 16 & 10 & -12 \end{bmatrix}$

$= \begin{bmatrix} -13 - 15 & 24 - 16 & 20 - 10 & -11 - (-12) \end{bmatrix}$

$= \begin{bmatrix} -28 & 8 & 10 & 1 \end{bmatrix}$

(c) $6A = 6\begin{bmatrix} -13 & 24 & 20 & -11 \end{bmatrix} = \begin{bmatrix} -78 & 144 & 120 & -66 \end{bmatrix}$

(d) $4A - 3B = 4\begin{bmatrix} -13 & 24 & 20 & -11 \end{bmatrix} - 3\begin{bmatrix} 15 & 16 & 10 & -12 \end{bmatrix}$

$= \begin{bmatrix} -52 & 96 & 80 & -44 \end{bmatrix} - \begin{bmatrix} 45 & 48 & 30 & -36 \end{bmatrix}$

$= \begin{bmatrix} -52 - 45 & 96 - 48 & 80 - 30 & -44 - (-36) \end{bmatrix}$

$= \begin{bmatrix} -97 & 48 & 50 & -8 \end{bmatrix}$

15. $\begin{bmatrix} -5 & 0 \\ 3 & -6 \end{bmatrix} + \begin{bmatrix} 7 & 1 \\ -2 & -1 \end{bmatrix} + \begin{bmatrix} -10 & -8 \\ 14 & 6 \end{bmatrix} = \begin{bmatrix} -5 + 7 + (-10) & 0 + 1 + (-8) \\ 3 + (-2) + 14 & -6 + (-1) + 6 \end{bmatrix} = \begin{bmatrix} -8 & -7 \\ 15 & -1 \end{bmatrix}$

17. $\begin{bmatrix} 4 & -8 \end{bmatrix} + \begin{bmatrix} -3 & 1 \end{bmatrix} + \begin{bmatrix} 2 & 0 \end{bmatrix} + \begin{bmatrix} 9 & 1 \end{bmatrix} = \begin{bmatrix} 4 + (-3) + 2 + 9 & -8 + 1 + 0 + 1 \end{bmatrix} = \begin{bmatrix} 12 & -6 \end{bmatrix}$

19. $5\left(\begin{bmatrix} 1 & -3 \\ -2 & -8 \end{bmatrix} + \begin{bmatrix} 10 & -7 \\ 5 & 4 \end{bmatrix}\right) = 5\begin{bmatrix} 1 + 10 & -3 + (-7) \\ -2 + 5 & -8 + 4 \end{bmatrix} = 5\begin{bmatrix} 11 & -10 \\ 3 & -4 \end{bmatrix} = \begin{bmatrix} 55 & -50 \\ 15 & -20 \end{bmatrix}$

21. $8\left(\begin{bmatrix} -4 & 0 & 1 \\ 0 & 2 & 3 \end{bmatrix} - \begin{bmatrix} 2 & 1 & -2 \\ 3 & -6 & 0 \end{bmatrix}\right) = 8\begin{bmatrix} -4 - 2 & 0 - 1 & 1 - (-2) \\ 0 - 3 & 2 - (-6) & 3 - 0 \end{bmatrix} = 8\begin{bmatrix} -6 & -1 & 3 \\ -3 & 8 & 3 \end{bmatrix} = \begin{bmatrix} -48 & -8 & 24 \\ -24 & 64 & 24 \end{bmatrix}$

23. $-3\left(\begin{bmatrix} 0 & -3 \\ 7 & 2 \end{bmatrix} + \begin{bmatrix} -6 & 3 \\ 8 & 1 \end{bmatrix}\right) - 2\begin{bmatrix} 4 & -4 \\ 7 & -9 \end{bmatrix} = -3\begin{bmatrix} -6 & 0 \\ 15 & 3 \end{bmatrix} - \begin{bmatrix} 8 & -8 \\ 14 & -18 \end{bmatrix} = \begin{bmatrix} 18 & 0 \\ -45 & -9 \end{bmatrix} - \begin{bmatrix} 8 & -8 \\ 14 & -18 \end{bmatrix} = \begin{bmatrix} 10 & 8 \\ -59 & 9 \end{bmatrix}$

25. $\frac{3}{7}\begin{bmatrix} 2 & 5 \\ -1 & -4 \end{bmatrix} + 6\begin{bmatrix} -3 & 0 \\ 2 & 2 \end{bmatrix} \approx \begin{bmatrix} -17.143 & 2.143 \\ 11.571 & 10.286 \end{bmatrix}$

27. $-\begin{bmatrix} 3.211 & 6.829 \\ -1.004 & 4.914 \\ 0.055 & -3.889 \end{bmatrix} - \begin{bmatrix} -1.630 & -3.090 \\ 5.256 & 8.335 \\ -9.768 & 4.251 \end{bmatrix} = \begin{bmatrix} -1.581 & -3.739 \\ -4.252 & -13.249 \\ 9.713 & -0.362 \end{bmatrix}$

29. $X = 3\begin{bmatrix} -2 & -1 \\ 1 & 0 \\ 3 & -4 \end{bmatrix} - 2\begin{bmatrix} 0 & 3 \\ 2 & 0 \\ -4 & -1 \end{bmatrix} = \begin{bmatrix} -6 & -3 \\ 3 & 0 \\ 9 & -12 \end{bmatrix} - \begin{bmatrix} 0 & 6 \\ 4 & 0 \\ -8 & -2 \end{bmatrix} = \begin{bmatrix} -6 & -9 \\ -1 & 0 \\ 17 & -10 \end{bmatrix}$

31. $X = -\frac{3}{2}A + \frac{1}{2}B = -\frac{3}{2}\begin{bmatrix} -2 & -1 \\ 1 & 0 \\ 3 & -4 \end{bmatrix} + \frac{1}{2}\begin{bmatrix} 0 & 3 \\ 2 & 0 \\ -4 & -1 \end{bmatrix} = \begin{bmatrix} 3 + 0 & \frac{3}{2} + \frac{3}{2} \\ -\frac{3}{2} + 1 & 0 + 0 \\ -\frac{9}{2} + (-2) & 6 + \left(-\frac{1}{2}\right) \end{bmatrix} = \begin{bmatrix} 3 & 3 \\ -\frac{1}{2} & 0 \\ -\frac{13}{2} & \frac{11}{2} \end{bmatrix}$

33. A is 1×2, B is $2 \times 1 \Rightarrow AB$ is 1×1.

$AB = \begin{bmatrix} -3 & 5 \end{bmatrix}\begin{bmatrix} 4 \\ -4 \end{bmatrix} = \begin{bmatrix} -3(4) + 5(-4) \end{bmatrix} = \begin{bmatrix} -32 \end{bmatrix}$

35. A is 2×2, B is $2 \times 1 \Rightarrow AB$ is 2×1.

$AB = \begin{bmatrix} -1 & 2 \\ 4 & -2 \end{bmatrix}\begin{bmatrix} 3 \\ -1 \end{bmatrix} = \begin{bmatrix} -1(3) + 2(-1) \\ 4(3) + (-2)(-1) \end{bmatrix} = \begin{bmatrix} -5 \\ 14 \end{bmatrix}$

37. A is 3×2 and B is $2 \times 4 \Rightarrow AB$ is 3×4.

$$AB = \begin{bmatrix} 3 & -2 \\ 4 & 5 \\ 1 & -1 \end{bmatrix}\begin{bmatrix} -1 & 4 & -2 & 5 \\ 2 & 1 & 3 & -1 \end{bmatrix} = \begin{bmatrix} -3-4 & 12-2 & -6-6 & 15+2 \\ -4+10 & 16+5 & -8+15 & 20-5 \\ -1-2 & 4-1 & -2-3 & 5+1 \end{bmatrix} = \begin{bmatrix} -7 & 10 & -12 & 17 \\ 6 & 21 & 7 & 15 \\ -3 & 3 & -5 & 6 \end{bmatrix}$$

39. A is 3×2, B is $2 \times 2 \Rightarrow AB$ is 3×2.

$$AB = \begin{bmatrix} -1 & 3 \\ 4 & -5 \\ 0 & 2 \end{bmatrix}\begin{bmatrix} 1 & 2 \\ 0 & 7 \end{bmatrix} = \begin{bmatrix} -1 & -2+21 \\ 4 & 8-35 \\ 0 & 14 \end{bmatrix} = \begin{bmatrix} -1 & 19 \\ 4 & -27 \\ 0 & 14 \end{bmatrix}$$

41. A is 3×3, B is $3 \times 3 \Rightarrow AB$ is 3×3.

$$AB = \begin{bmatrix} 5 & 0 & 0 \\ 0 & -8 & 0 \\ 0 & 0 & 7 \end{bmatrix}\begin{bmatrix} \frac{1}{5} & 0 & 0 \\ 0 & -\frac{1}{8} & 0 \\ 0 & 0 & \frac{1}{2} \end{bmatrix} = \begin{bmatrix} 1 & 0 & 0 \\ 0 & 1 & 0 \\ 0 & 0 & \frac{7}{2} \end{bmatrix}$$

43. A is 3×3, B is $4 \times 1 \Rightarrow AB$ is not possible.

45. $$AB = \begin{bmatrix} 5 & 6 & -3 \\ -2 & 5 & 1 \\ 10 & -5 & 5 \end{bmatrix}\begin{bmatrix} 1 & -1 & 2 \\ 8 & 1 & 4 \\ 4 & -2 & 9 \end{bmatrix} = \begin{bmatrix} 41 & 7 & 7 \\ 42 & 5 & 25 \\ -10 & -25 & 45 \end{bmatrix}$$

47. A is 3×3, B is $4 \times 2 \Rightarrow AB$ is not possible.

49. (a) $$AB = \begin{bmatrix} 1 & 2 \\ 4 & 2 \end{bmatrix}\begin{bmatrix} 2 & -1 \\ -1 & 8 \end{bmatrix} = \begin{bmatrix} 2-2 & -1+16 \\ 8-2 & -4+16 \end{bmatrix} = \begin{bmatrix} 0 & 15 \\ 6 & 12 \end{bmatrix}$$

(b) $$BA = \begin{bmatrix} 2 & -1 \\ -1 & 8 \end{bmatrix}\begin{bmatrix} 1 & 2 \\ 4 & 2 \end{bmatrix} = \begin{bmatrix} 2-4 & 4-2 \\ -1+32 & -2+16 \end{bmatrix} = \begin{bmatrix} -2 & 2 \\ 31 & 14 \end{bmatrix}$$

(c) $$A^2 = \begin{bmatrix} 1 & 2 \\ 4 & 2 \end{bmatrix}\begin{bmatrix} 1 & 2 \\ 4 & 2 \end{bmatrix} = \begin{bmatrix} 1+8 & 2+4 \\ 4+8 & 8+4 \end{bmatrix} = \begin{bmatrix} 9 & 6 \\ 12 & 12 \end{bmatrix}$$

51. (a) $$AB = \begin{bmatrix} -1 & 2 & 3 \\ 4 & 1 & -1 \end{bmatrix}\begin{bmatrix} 1 & 3 \\ -1 & -2 \\ 2 & 4 \end{bmatrix} = \begin{bmatrix} -1-2+6 & -3-4+12 \\ 4-1-2 & 12-2-4 \end{bmatrix} = \begin{bmatrix} 3 & 5 \\ 1 & 6 \end{bmatrix}$$

(b) $$BA = \begin{bmatrix} 1 & 3 \\ -1 & -2 \\ 2 & 4 \end{bmatrix}\begin{bmatrix} -1 & 2 & 3 \\ 4 & 1 & -1 \end{bmatrix} = \begin{bmatrix} -1+12 & 2+3 & 3-3 \\ 1-8 & -2-2 & -3+2 \\ -2+16 & 4+4 & 6-4 \end{bmatrix} = \begin{bmatrix} 11 & 5 & 0 \\ -7 & -4 & -1 \\ 14 & 8 & 2 \end{bmatrix}$$

(c) A^2 is not possible.

53. (a) $$AB = \begin{bmatrix} -4 & 2 & 3 \end{bmatrix}\begin{bmatrix} 1 \\ 0 \\ 5 \end{bmatrix} = \begin{bmatrix} -4(1) + 2(0) + 3(5) \end{bmatrix} = \begin{bmatrix} 11 \end{bmatrix}$$

(b) $$BA = \begin{bmatrix} 1 \\ 0 \\ 5 \end{bmatrix}\begin{bmatrix} -4 & 2 & 3 \end{bmatrix} = \begin{bmatrix} 1(-4) & 1(2) & 1(3) \\ 0(-4) & 0(2) & 0(3) \\ 5(-4) & 5(2) & 5(3) \end{bmatrix} = \begin{bmatrix} -4 & 2 & 3 \\ 0 & 0 & 0 \\ -20 & 10 & 15 \end{bmatrix}$$

(c) A^2 is not possible.

55. You cannot perform the operation because to add two matrices, their order must be the same. Because A is 2×3, you can only add A to other 2×3 matrices.

57. Because the number of columns of A is not equal to the number of rows of B, you cannot perform the operation.

59. AD is not possible because the number of columns of A, 3, is not equal to the number of rows of D, 2.

61. Because the number of columns of B is equal to the number of rows of C, you can perform BC and the result will be a 2×2 matrix. Because D is a 2×2 matrix, you can subtract D from BC. So, you can perform $BC–D$ and the final result will be a 2×2 matrix.

63. Because A and B are both 2×3 matrices, you can find $(A–3B)$ and the result would be a 2×3 matrix. Because the number of columns of D is equal to the number of rows of $(A–3B)$, you can find $D(A–3B)$ and the result will be a 2×3 matrix.

65. $-x + y = 4$

$-2x + y = 0$

(a) Matrix Equation: $\begin{bmatrix} -1 & 1 \\ -2 & 1 \end{bmatrix}\begin{bmatrix} x \\ y \end{bmatrix} = \begin{bmatrix} 4 \\ 0 \end{bmatrix}$

(b) $\left[\begin{array}{cc:c} -1 & 1 & 4 \\ -2 & 1 & 0 \end{array}\right] \Rightarrow \left[\begin{array}{cc:c} 1 & -1 & -4 \\ 0 & -1 & -8 \end{array}\right] \Rightarrow \left[\begin{array}{cc:c} 1 & 0 & 4 \\ 0 & 1 & 8 \end{array}\right]$ so $\begin{bmatrix} x \\ y \end{bmatrix} = \begin{bmatrix} 4 \\ 8 \end{bmatrix}$

67. $x + 2y = 3$

$3x - y = 2$

(a) Matrix Equation: $\begin{bmatrix} 1 & 2 \\ 3 & -1 \end{bmatrix}\begin{bmatrix} x \\ y \end{bmatrix} = \begin{bmatrix} 3 \\ 2 \end{bmatrix}$

(b) $\left[\begin{array}{cc:c} 1 & 2 & 3 \\ 3 & -1 & 2 \end{array}\right] \Rightarrow \left[\begin{array}{cc:c} 1 & 2 & 3 \\ 0 & -7 & -7 \end{array}\right]\left[\begin{array}{cc:c} 1 & 0 & 1 \\ 0 & 1 & 1 \end{array}\right]$ so $\begin{bmatrix} x \\ y \end{bmatrix} = \begin{bmatrix} 1 \\ 1 \end{bmatrix}$

69. $\begin{cases} x - 4y + 5z = 3 \\ 2x + 5y + z = -24 \\ -x - y - z = 8 \end{cases}$

(a) Matrix Equation: $\begin{bmatrix} 1 & -4 & 5 \\ 2 & 5 & 1 \\ -1 & -1 & -1 \end{bmatrix}\begin{bmatrix} x \\ y \\ z \end{bmatrix} = \begin{bmatrix} 3 \\ -24 \\ 8 \end{bmatrix}$

(b) $\left[\begin{array}{ccc:c} 1 & -4 & 5 & 3 \\ 2 & 5 & 1 & -24 \\ -1 & -1 & -1 & 8 \end{array}\right] \Rightarrow \left[\begin{array}{ccc:c} 1 & -4 & 5 & 3 \\ 0 & 13 & -9 & -30 \\ 0 & -5 & 4 & 11 \end{array}\right] \Rightarrow \left[\begin{array}{ccc:c} 1 & -4 & 5 & 3 \\ 0 & 3 & -1 & -8 \\ 0 & -5 & 4 & 11 \end{array}\right] \Rightarrow \left[\begin{array}{ccc:c} 1 & -4 & 5 & 3 \\ 0 & 1 & -\frac{1}{3} & -\frac{8}{3} \\ 0 & 0 & \frac{7}{3} & -\frac{7}{3} \end{array}\right] \Rightarrow \left[\begin{array}{ccc:c} 1 & 0 & 5 & -9 \\ 0 & 1 & 0 & -3 \\ 0 & 0 & 1 & -1 \end{array}\right]$

$\Rightarrow \left[\begin{array}{ccc:c} 1 & 0 & 0 & -4 \\ 0 & 1 & 0 & -3 \\ 0 & 0 & 1 & -1 \end{array}\right]$ so $\begin{bmatrix} x \\ y \\ z \end{bmatrix} = \begin{bmatrix} -4 \\ -3 \\ -1 \end{bmatrix}$

71. $x - 2y + 3z = 9$

$-x + 3y - z = -6$

$2x - 5y + 5z = 17$

(a) Matrix Equation: $\begin{bmatrix} 1 & -2 & 3 \\ -1 & 3 & -1 \\ 2 & -5 & 5 \end{bmatrix}\begin{bmatrix} x \\ y \\ z \end{bmatrix} = \begin{bmatrix} 9 \\ -6 \\ 17 \end{bmatrix}$

(b) $\left[\begin{array}{ccc:c} 1 & -2 & 3 & 9 \\ -1 & 3 & -1 & -6 \\ 2 & -5 & 5 & 17 \end{array}\right] \Rightarrow \left[\begin{array}{ccc:c} 1 & -2 & 3 & 9 \\ 0 & 1 & 2 & 3 \\ 0 & -1 & -1 & -1 \end{array}\right] \Rightarrow \left[\begin{array}{ccc:c} 1 & 0 & 7 & 15 \\ 0 & 1 & 2 & 3 \\ 0 & 0 & 1 & 2 \end{array}\right] \Rightarrow \left[\begin{array}{ccc:c} 1 & 0 & 0 & 1 \\ 0 & 1 & 0 & -1 \\ 0 & 0 & 1 & 2 \end{array}\right]$ so $\begin{bmatrix} x \\ y \\ z \end{bmatrix} = \begin{bmatrix} 1 \\ -1 \\ 2 \end{bmatrix}$

73. $1.10\begin{bmatrix} 100 & 120 & 60 & 40 \\ 140 & 160 & 200 & 80 \end{bmatrix} = \begin{bmatrix} 110 & 132 & 66 & 44 \\ 154 & 176 & 220 & 88 \end{bmatrix}$

75. $1.12\begin{bmatrix} 615 & 670 & 740 & 990 \\ 995 & 1030 & 1180 & 1105 \end{bmatrix} = \begin{bmatrix} 688.80 & 750.40 & 828.80 & 1108.80 \\ 1114.40 & 1153.60 & 1321.60 & 1237.60 \end{bmatrix}$

	Hotel w	Hotel x	Hotel y	Hotel z
Double	688.80	750.40	828.80	1108.80
Family	1114.40	1153.60	1321.60	1237.60

77. (a) $\begin{bmatrix} 3 & 2 & 2 & 3 & 0 \end{bmatrix} \begin{bmatrix} 475 \\ 700 \\ 850 \\ 950 \\ 1200 \end{bmatrix} = \begin{bmatrix} 3(475) + 2(700) + 2(850) + 3(950) + 0(1200) \end{bmatrix} = \begin{bmatrix} 7375 \end{bmatrix}$

The total retail price of inventory at outlet 1 is $7375.

(b) $\begin{bmatrix} 4 & 2 & 1 & 3 & 2 \end{bmatrix} \begin{bmatrix} 250 \\ 400 \\ 575 \\ 650 \\ 760 \end{bmatrix} = \begin{bmatrix} 4(250) + 2(400) + 1(575) + 3(650) + 2(760) \end{bmatrix} = \begin{bmatrix} 5845 \end{bmatrix}$

The total wholesale price of inventory at outlet 3 is $5845.

(c) Wholesale Retail

$ST = \begin{bmatrix} \$4650 & \$7375 \\ \$7405 & \$11{,}350 \\ \$5845 & \$9400 \end{bmatrix} \begin{matrix} 1 \\ 2 \\ 3 \end{matrix} \Big\} \text{Outlet}$

ST represents the wholesale and retail prices of laptop computer inventories at the three outlets.

79. (a) $B = \begin{bmatrix} 2 & 0.25 & 0.5 \end{bmatrix}$

(b) $BA = \begin{bmatrix} 2 & 0.25 & 0.5 \end{bmatrix} \begin{bmatrix} 472 & 563 \\ 590 & 704 \\ 177 & 211 \end{bmatrix}$

Calories burned

130-lb person 155-lb person

$BA = \begin{bmatrix} 1180 & 1407.5 \end{bmatrix}$

BA represents the total calories burned by each person.

81. (a) $D = \begin{bmatrix} 3 & 1 & 2 \\ 0 & 2 & 3 \\ 4 & 5 & 3 \end{bmatrix}$ (columns X, Y, Z; rows Sacks, Interceptions, Key tackles)

Each entry d_{ij} represents the number of each type of defensive play made by each player.

(b) Sacks Interceptions Tackles

$B = \begin{bmatrix} \$2000 & \$1000 & \$800 \end{bmatrix}$

Each entry b_{ij} represents the bonus received for each type of play.

(c) $BD = \begin{bmatrix} \$2000 & \$1000 & \$800 \end{bmatrix} \begin{bmatrix} 3 & 1 & 2 \\ 0 & 2 & 3 \\ 4 & 5 & 3 \end{bmatrix}$

X Y Z

$= \begin{bmatrix} \$9200 & \$8000 & \$9400 \end{bmatrix}$

Each entry represents the bonus each player will receive.

(d) Player Z

83. $P^2 = \begin{bmatrix} 0.6 & 0.1 & 0.1 \\ 0.2 & 0.7 & 0.1 \\ 0.2 & 0.2 & 0.8 \end{bmatrix} \begin{bmatrix} 0.6 & 0.1 & 0.1 \\ 0.2 & 0.7 & 0.1 \\ 0.2 & 0.2 & 0.8 \end{bmatrix}$

$= \begin{bmatrix} 0.40 & 0.15 & 0.15 \\ 0.28 & 0.53 & 0.17 \\ 0.32 & 0.32 & 0.68 \end{bmatrix}$

85. $A = \begin{bmatrix} 1 & 3 \\ 1 & 2 \end{bmatrix}, B = \begin{bmatrix} -2 & 3 \\ 1 & -1 \end{bmatrix}$

$AB = \begin{bmatrix} 1 & 0 \\ 0 & 1 \end{bmatrix}$, *B* is unique.

87. $AC = \begin{bmatrix} 1 & 2 & 3 \\ 0 & 5 & 4 \\ 3 & -2 & 1 \end{bmatrix} \begin{bmatrix} 0 & 0 & 0 \\ 0 & 0 & 0 \\ 4 & -2 & 3 \end{bmatrix} = \begin{bmatrix} 12 & -6 & 9 \\ 16 & -8 & 12 \\ 4 & -2 & 3 \end{bmatrix}$

$BC = \begin{bmatrix} 4 & -6 & 3 \\ 5 & 4 & 4 \\ -1 & 0 & 1 \end{bmatrix} \begin{bmatrix} 0 & 0 & 0 \\ 0 & 0 & 0 \\ 4 & -2 & 3 \end{bmatrix} = \begin{bmatrix} 12 & -6 & 9 \\ 16 & -8 & 12 \\ 4 & -2 & 3 \end{bmatrix}$

$= AC$ and $A \neq B$

89. $A = \begin{bmatrix} 0.70 & 0.15 & 0.15 \\ 0.20 & 0.80 & 0.15 \\ 0.10 & 0.05 & 0.70 \end{bmatrix}, B = \begin{bmatrix} 25{,}000 \\ 30{,}000 \\ 45{,}000 \end{bmatrix}$

$AB = \begin{bmatrix} 28{,}750 \\ 35{,}750 \\ 35{,}500 \end{bmatrix}$

After one year:

Gold subscribers: 28,750

Galaxy subscribers: 35,750

Nonsubscribers: 35,500

For the second year matrix B is $\begin{bmatrix} 28{,}750 \\ 35{,}750 \\ 35{,}500 \end{bmatrix}$.

$AB = \begin{bmatrix} 30{,}812.5 \\ 39{,}675 \\ 29{,}512.5 \end{bmatrix}$

Gold subscribers: 30,813

Galaxy subscribers: 39,675

Nonsubscribers: 29,513

For the third year matrix B is $\begin{bmatrix} 30{,}813 \\ 39{,}675 \\ 29{,}513 \end{bmatrix}$.

$AB = \begin{bmatrix} 31{,}947.3 \\ 42{,}329.55 \\ 25{,}724.15 \end{bmatrix}$

Gold subscribers: 31,947

Galaxy subscribers: 42,330

Nonsubscribers: 25,724

The number of subscribers to each company is increasing each year. The number of nonsubscribers is decreasing each year.

Section 6.3 The Inverse of a Square Matrix

Skills Warm Up

1. $4\begin{bmatrix} 1 & 6 \\ 0 & -4 \\ 12 & 2 \end{bmatrix} = \begin{bmatrix} 4(1) & 4(6) \\ 4(0) & 4(-4) \\ 4(12) & 4(2) \end{bmatrix} = \begin{bmatrix} 4 & 24 \\ 0 & -16 \\ 48 & 8 \end{bmatrix}$

2. $\frac{1}{2}\begin{bmatrix} 11 & 10 & 48 \\ 1 & 0 & 16 \\ 0 & 2 & 8 \end{bmatrix} = \begin{bmatrix} \frac{1}{2}(11) & \frac{1}{2}(10) & \frac{1}{2}(48) \\ \frac{1}{2}(1) & \frac{1}{2}(0) & \frac{1}{2}(16) \\ \frac{1}{2}(0) & \frac{1}{2}(2) & \frac{1}{2}(8) \end{bmatrix}$

$= \begin{bmatrix} \frac{11}{2} & 5 & 24 \\ \frac{1}{2} & 0 & 8 \\ 0 & 1 & 4 \end{bmatrix}$

3. $\begin{bmatrix} 1 & -10 & 3 \\ 4 & 1 & 0 \end{bmatrix} - 2\begin{bmatrix} 3 & -4 & 8 \\ 0 & 7 & 1 \end{bmatrix}$

$= \begin{bmatrix} 1 & -10 & 3 \\ 4 & 1 & 0 \end{bmatrix} - \begin{bmatrix} 6 & -8 & 16 \\ 0 & 14 & 2 \end{bmatrix}$

$= \begin{bmatrix} 1-6 & -10-(-8) & 3-16 \\ 4-0 & 1-14 & 0-2 \end{bmatrix} = \begin{bmatrix} -5 & -2 & -13 \\ 4 & -13 & -2 \end{bmatrix}$

4. $\begin{bmatrix} 5 & 20 \\ -7 & 15 \end{bmatrix} - 3\begin{bmatrix} 6 & 3 \\ 4 & -2 \end{bmatrix} = \begin{bmatrix} 5 & 20 \\ -7 & 15 \end{bmatrix} - \begin{bmatrix} 18 & 9 \\ 12 & -6 \end{bmatrix}$

$= \begin{bmatrix} 5-18 & 20-9 \\ -7-12 & 15-(-6) \end{bmatrix}$

$= \begin{bmatrix} -13 & 11 \\ -19 & 21 \end{bmatrix}$

5. $\begin{bmatrix} 1 & -2 \\ -1 & 3 \end{bmatrix}\begin{bmatrix} 3 & 2 \\ 1 & 1 \end{bmatrix} = \begin{bmatrix} 3-2 & 2-2 \\ -3+3 & -2+3 \end{bmatrix} = \begin{bmatrix} 1 & 0 \\ 0 & 1 \end{bmatrix}$

6. $\begin{bmatrix} 1 & 0 \\ 0 & 1 \end{bmatrix}\begin{bmatrix} 6 & 5 \\ 3 & -2 \end{bmatrix} = \begin{bmatrix} 6 & 5 \\ 3 & -2 \end{bmatrix}$

7. $\begin{bmatrix} 2 & 0 & 0 \\ 0 & -1 & 0 \\ 0 & 0 & 3 \end{bmatrix}\begin{bmatrix} \frac{1}{2} & 0 & 0 \\ 0 & -1 & 0 \\ 0 & 0 & \frac{1}{3} \end{bmatrix} = \begin{bmatrix} 1 & 0 & 0 \\ 0 & 1 & 0 \\ 0 & 0 & 1 \end{bmatrix}$

Skills Warm Up —continued—

8. $\begin{bmatrix} 1 & 2 & 3 \\ 3 & -1 & -2 \\ 3 & 1 & 1 \end{bmatrix}\begin{bmatrix} 1 & 1 & -1 \\ -9 & -8 & 11 \\ 6 & 5 & -7 \end{bmatrix}$

$= \begin{bmatrix} 1-18+18 & 1-16+15 & -1+22-21 \\ 3+9-12 & 3+8-10 & -3-11+14 \\ 3-9+6 & 3-8+5 & -3+11-7 \end{bmatrix}$

$= \begin{bmatrix} 1 & 0 & 0 \\ 0 & 1 & 0 \\ 0 & 0 & 1 \end{bmatrix}$

9. $\left[\begin{array}{cc|cc} 3 & -2 & 1 & 0 \\ 4 & -3 & 0 & 1 \end{array}\right] = \left[\begin{array}{cc|cc} 1 & -\frac{2}{3} & \frac{1}{3} & 0 \\ 4 & -3 & 0 & 1 \end{array}\right]$

$= \left[\begin{array}{cc|cc} 1 & -\frac{2}{3} & \frac{1}{3} & 0 \\ 0 & -\frac{1}{3} & -\frac{4}{3} & 1 \end{array}\right]$

$= \left[\begin{array}{cc|cc} 1 & -\frac{2}{3} & \frac{1}{3} & 0 \\ 0 & 1 & 4 & -3 \end{array}\right]$

$= \left[\begin{array}{cc|cc} 1 & 0 & 3 & -2 \\ 0 & 1 & 4 & -3 \end{array}\right]$

10. $\left[\begin{array}{ccc|ccc} 1 & 1 & 2 & 1 & 0 & 0 \\ -1 & 0 & 3 & 0 & 1 & 0 \\ 1 & 2 & 8 & 0 & 0 & 1 \end{array}\right] = \left[\begin{array}{ccc|ccc} 1 & 1 & 2 & 1 & 0 & 0 \\ 0 & 1 & 5 & 1 & 1 & 0 \\ 0 & 1 & 6 & -1 & 0 & 1 \end{array}\right] = \left[\begin{array}{ccc|ccc} 1 & 1 & 2 & 1 & 0 & 0 \\ 0 & 1 & 5 & 1 & 1 & 0 \\ 0 & 0 & 1 & -2 & -1 & 1 \end{array}\right] = \left[\begin{array}{ccc|ccc} 1 & 1 & 2 & 1 & 0 & 0 \\ 0 & 1 & 0 & 11 & 6 & -5 \\ 0 & 0 & 1 & -2 & -1 & 1 \end{array}\right]$

$= \left[\begin{array}{ccc|ccc} 1 & 0 & 2 & -10 & -6 & 5 \\ 0 & 1 & 0 & 11 & 6 & -5 \\ 0 & 0 & 1 & -2 & -1 & 1 \end{array}\right] = \left[\begin{array}{ccc|ccc} 1 & 0 & 0 & -6 & -4 & 3 \\ 0 & 1 & 0 & 11 & 6 & -5 \\ 0 & 0 & 1 & -2 & -1 & 1 \end{array}\right]$

1. $AB = \begin{bmatrix} 7 & 4 \\ 5 & 3 \end{bmatrix}\begin{bmatrix} 3 & -4 \\ -5 & 7 \end{bmatrix} = \begin{bmatrix} 7(3)+4(-5) & 7(-4)+4(7) \\ 5(3)+3(-5) & 5(-4)+3(7) \end{bmatrix} = \begin{bmatrix} 1 & 0 \\ 0 & 1 \end{bmatrix}$

$BA = \begin{bmatrix} 3 & -4 \\ -5 & 7 \end{bmatrix}\begin{bmatrix} 7 & 4 \\ 5 & 3 \end{bmatrix} = \begin{bmatrix} 3(7)+(-4)(5) & 3(4)+(-4)(3) \\ -5(7)+7(5) & -5(4)+7(3) \end{bmatrix} = \begin{bmatrix} 1 & 0 \\ 0 & 1 \end{bmatrix}$

3. $AB = \begin{bmatrix} 2 & -1 \\ 5 & -4 \end{bmatrix}\frac{1}{3}\begin{bmatrix} 4 & -1 \\ 5 & -2 \end{bmatrix} = \frac{1}{3}\begin{bmatrix} 2 & -1 \\ 5 & -4 \end{bmatrix}\begin{bmatrix} 4 & -1 \\ -5 & -2 \end{bmatrix} = \frac{1}{3}\begin{bmatrix} 2(4)+(-1)5 & 2(-1)+(-1)(-2) \\ 5(4)+(-4)(5) & 5(-1)+(-4)(-2) \end{bmatrix} = \frac{1}{3}\begin{bmatrix} 3 & 0 \\ 0 & 3 \end{bmatrix} = \begin{bmatrix} 1 & 0 \\ 0 & 1 \end{bmatrix}$

$BA = \frac{1}{3}\begin{bmatrix} 4 & -1 \\ 5 & -2 \end{bmatrix}\begin{bmatrix} 2 & -1 \\ 5 & -4 \end{bmatrix} = \frac{1}{3}\begin{bmatrix} 4(2)+(-1)(5) & 4(-1)+(-1)(-4) \\ 5(2)+(-2)(5) & 5(-1)+(-2)(-4) \end{bmatrix} = \frac{1}{3}\begin{bmatrix} 3 & 0 \\ 0 & 3 \end{bmatrix} = \begin{bmatrix} 1 & 0 \\ 0 & 1 \end{bmatrix}$

5. $AB = \frac{1}{3}\begin{bmatrix} -2 & 2 & 3 \\ 1 & -1 & 0 \\ 0 & 1 & 4 \end{bmatrix}\begin{bmatrix} -4 & -5 & 3 \\ -4 & -8 & 3 \\ 1 & 2 & 0 \end{bmatrix} = \frac{1}{3}\begin{bmatrix} -8+8+3 & 10-16+6 & -6+6 \\ -4+4 & -5+8 & 3-3 \\ -4+4 & -8+8 & 3 \end{bmatrix} = \frac{1}{3}\begin{bmatrix} 3 & 0 & 0 \\ 0 & 3 & 0 \\ 0 & 0 & 3 \end{bmatrix} = \begin{bmatrix} 1 & 0 & 0 \\ 0 & 1 & 0 \\ 0 & 0 & 1 \end{bmatrix}$

$BA = \frac{1}{3}\begin{bmatrix} -4 & -5 & 3 \\ -4 & -8 & 3 \\ 1 & 2 & 0 \end{bmatrix}\begin{bmatrix} -2 & 2 & 3 \\ 1 & -1 & 0 \\ 0 & 1 & 4 \end{bmatrix} = \frac{1}{3}\begin{bmatrix} 8-5 & -8+5+3 & -12+12 \\ 8-8 & -8+8+3 & -12+12 \\ -2+2 & 2-2 & 3 \end{bmatrix} = \begin{bmatrix} 1 & 0 & 0 \\ 0 & 1 & 0 \\ 0 & 0 & 1 \end{bmatrix}$

7. $AB = \begin{bmatrix} 2 & -17 & 11 \\ -1 & 11 & -7 \\ 0 & 3 & -2 \end{bmatrix}\begin{bmatrix} 1 & 1 & 2 \\ 2 & 4 & -3 \\ 3 & 6 & -5 \end{bmatrix} = \begin{bmatrix} 2-34+33 & 2-68+66 & 4+51-55 \\ -1+22-21 & -1+44-42 & -2-33+35 \\ 6-6 & 12-12 & -9+10 \end{bmatrix} = \begin{bmatrix} 1 & 0 & 0 \\ 0 & 1 & 0 \\ 0 & 0 & 1 \end{bmatrix}$

$BA = \begin{bmatrix} 1 & 1 & 2 \\ 2 & 4 & -3 \\ 3 & 6 & -5 \end{bmatrix}\begin{bmatrix} 2 & -17 & 11 \\ -1 & 11 & -7 \\ 0 & 3 & -2 \end{bmatrix} = \begin{bmatrix} 2-1 & -17+11+6 & 11-7-4 \\ 4-4 & -34+44-9 & 22-28+6 \\ 6-6 & -51+66-15 & 33-42+10 \end{bmatrix} = \begin{bmatrix} 1 & 0 & 0 \\ 0 & 1 & 0 \\ 0 & 0 & 1 \end{bmatrix}$

9. $AB = \frac{1}{3}\begin{bmatrix} 2 & 0 & 2 & 1 \\ 3 & 0 & 0 & 1 \\ -1 & 1 & -2 & 1 \\ 3 & -1 & 1 & 0 \end{bmatrix}\begin{bmatrix} -1 & 3 & -2 & -2 \\ -2 & 9 & -7 & -10 \\ 1 & 0 & -1 & -1 \\ 3 & -6 & 6 & 6 \end{bmatrix}$

$$= \frac{1}{3}\begin{bmatrix} -2+2+3 & 6-6 & -4-2+6 & -4-2+6 \\ -3+3 & 9-6 & -6+6 & -6+6 \\ 1-2-2+3 & -3+9-6 & 2-7+2+6 & 2-10+2+6 \\ -3+2+1 & 9-9 & -6+7-1 & -6+10-1 \end{bmatrix}$$

$$= \frac{1}{3}\begin{bmatrix} 3 & 0 & 0 & 0 \\ 0 & 3 & 0 & 0 \\ 0 & 0 & 3 & 0 \\ 0 & 0 & 0 & 3 \end{bmatrix} = \begin{bmatrix} 1 & 0 & 0 & 0 \\ 0 & 1 & 0 & 0 \\ 0 & 0 & 1 & 0 \\ 0 & 0 & 0 & 1 \end{bmatrix}$$

$$BA = \frac{1}{3}\begin{bmatrix} -1 & 3 & -2 & -2 \\ -2 & 9 & -7 & -10 \\ 1 & 0 & -1 & -1 \\ 3 & -6 & 6 & 6 \end{bmatrix}\begin{bmatrix} 2 & 0 & 2 & 1 \\ 3 & 0 & 0 & 1 \\ -1 & 1 & -2 & 1 \\ 3 & -1 & 1 & 0 \end{bmatrix}$$

$$= \frac{1}{3}\begin{bmatrix} -2+9+2-6 & -2+2 & -2+4-2 & -1+3-2 \\ -4+27+7-30 & -7+10 & -4+14-10 & -2+9-7 \\ 2+1-3 & -1+1 & 2+2-1 & 1-1 \\ 6-18-6+18 & 6-6 & 6-12+6 & 3-6+6 \end{bmatrix}$$

$$= \frac{1}{3}\begin{bmatrix} 3 & 0 & 0 & 0 \\ 0 & 3 & 0 & 0 \\ 0 & 0 & 3 & 0 \\ 0 & 0 & 0 & 3 \end{bmatrix} = \begin{bmatrix} 1 & 0 & 0 & 0 \\ 0 & 1 & 0 & 0 \\ 0 & 0 & 1 & 0 \\ 0 & 0 & 0 & 1 \end{bmatrix}$$

11. $A = \begin{bmatrix} 1 & 2 \\ 3 & 7 \end{bmatrix}$

$$\begin{matrix} A & X & = & I \end{matrix}$$

$$\begin{bmatrix} 1 & 2 \\ 3 & 7 \end{bmatrix}\begin{bmatrix} x_{11} & x_{12} \\ x_{21} & x_{22} \end{bmatrix} = \begin{bmatrix} 1 & 0 \\ 0 & 1 \end{bmatrix}$$

$$\begin{bmatrix} x_{11} + 2x_{21} & x_{12} + 2x_{22} \\ 3x_{11} + 7x_{21} & 3x_{12} + 7x_{22} \end{bmatrix} = \begin{bmatrix} 1 & 0 \\ 0 & 1 \end{bmatrix}$$

$$\begin{cases} x_{11} + 2x_{21} = 1 \\ 3x_{11} + 7x_{21} = 0 \end{cases} \qquad \begin{cases} x_{12} + 2x_{22} = 0 \\ 3x_{12} + 7x_{22} = 1 \end{cases}$$

$$\Downarrow \qquad\qquad \Downarrow$$

$$\begin{aligned} x_{11} &= 1 - 2x_{21} \\ 3(1 - 2x_{21}) + 7x_{21} &= 0 \\ 3 - 6x_{21} + 7x_{21} &= 0 \\ 3 + x_{21} &= 0 \\ x_{21} &= -3 \end{aligned} \qquad \begin{aligned} x_{12} &= -2x_{22} \\ 3(-2x_{22}) + 7x_{22} &= 1 \\ -6x_{22} + 7x_{22} &= 1 \\ x_{22} &= 1 \\ x_{12} &= -2 \end{aligned}$$

$$x_{11} = 1 - 2(-3)$$

$$x_{11} = 7$$

$$x = \begin{bmatrix} 7 & -2 \\ -3 & 1 \end{bmatrix}$$

13. $A = \begin{bmatrix} 8 & 4 \\ -2 & -2 \end{bmatrix}$

$$\begin{matrix} A & X & = & I \end{matrix}$$

$$\begin{bmatrix} 8 & 4 \\ -2 & -2 \end{bmatrix}\begin{bmatrix} x_{11} & x_{12} \\ x_{21} & x_{22} \end{bmatrix} = \begin{bmatrix} 1 & 0 \\ 0 & 1 \end{bmatrix}$$

$$\begin{bmatrix} 8x_{11} + 4x_{21} & 8x_{12} + 4x_{22} \\ -2x_{11} - 2x_{21} & -2x_{12} - 2x_{22} \end{bmatrix} = \begin{bmatrix} 1 & 0 \\ 0 & 1 \end{bmatrix}$$

$$\begin{cases} 8x_{11} + 4x_{21} = 1 \\ -2x_{11} - 2x_{21} = 0 \end{cases} \qquad \begin{cases} 8x_{12} + 4x_{22} = 0 \\ -2x_{12} - 2x_{22} = 1 \end{cases}$$

$$\Downarrow \qquad\qquad \Downarrow$$

$$\begin{aligned} -2x_{11} &= 2x_{21} \\ x_{11} &= -x_{21} \\ 8(-x_{21}) + 4x_{21} &= 1 \\ -4x_{21} &= 1 \\ x_{21} &= -\tfrac{1}{4} \\ x_{11} &= \tfrac{1}{4} \end{aligned} \qquad \begin{aligned} 4x_{22} &= -8x_{12} \\ x_{22} &= -2x_{12} \\ -2x_{12} - 2(-2x_{12}) &= 1 \\ -2x_{12} + 4x_{12} &= 1 \\ 2x_{12} &= 1 \\ x_{12} &= \tfrac{1}{2} \\ x_{22} &= -1 \end{aligned}$$

$$x = \begin{bmatrix} \frac{1}{4} & \frac{1}{2} \\ -\frac{1}{4} & -1 \end{bmatrix}$$

15. $[A \;\vdots\; I] = \begin{bmatrix} 11 & 1 & \vdots & 1 & 0 \\ -1 & 0 & \vdots & 0 & 1 \end{bmatrix} \Rightarrow \begin{bmatrix} 1 & 1 & \vdots & 1 & 10 \\ -1 & 0 & \vdots & 0 & 1 \end{bmatrix} \Rightarrow \begin{bmatrix} 1 & 1 & \vdots & 1 & 10 \\ 0 & 1 & \vdots & 1 & 11 \end{bmatrix} \Rightarrow \begin{bmatrix} 1 & 0 & \vdots & 0 & -1 \\ 0 & 1 & \vdots & 1 & 11 \end{bmatrix} = [I \;\vdots\; A^{-1}]$

$A^{-1} = \begin{bmatrix} 0 & -1 \\ 1 & 11 \end{bmatrix}$

17. $[A \;\vdots\; I] = \begin{bmatrix} 0 & 4 & \vdots & 1 & 0 \\ -3 & 6 & \vdots & 0 & 1 \end{bmatrix} \Rightarrow \begin{bmatrix} -3 & 6 & \vdots & 0 & 1 \\ 0 & 4 & \vdots & 1 & 0 \end{bmatrix} \Rightarrow \begin{bmatrix} 1 & -2 & \vdots & 0 & -\frac{1}{3} \\ 0 & 1 & \vdots & \frac{1}{4} & 0 \end{bmatrix} \Rightarrow \begin{bmatrix} 1 & 0 & \vdots & \frac{1}{2} & -\frac{1}{3} \\ 0 & 1 & \vdots & \frac{1}{4} & 0 \end{bmatrix}$

$A^{-1} = \begin{bmatrix} \frac{1}{2} & -\frac{1}{3} \\ \frac{1}{4} & 0 \end{bmatrix} = \frac{1}{12}\begin{bmatrix} 6 & -4 \\ 3 & 0 \end{bmatrix}$

19. $[A \;\vdots\; I] = \begin{bmatrix} 6 & 15 & \vdots & 1 & 0 \\ 4 & 10 & \vdots & 0 & 1 \end{bmatrix} \Rightarrow \begin{bmatrix} 1 & \frac{5}{2} & \vdots & \frac{1}{6} & 0 \\ 4 & 10 & \vdots & 0 & 1 \end{bmatrix} \Rightarrow \begin{bmatrix} 1 & \frac{5}{2} & \vdots & \frac{1}{6} & 0 \\ 0 & 0 & \vdots & -\frac{2}{3} & 1 \end{bmatrix}$

Because the left side bottom row is all zeros, matrix A does not have an inverse.

21. $A = \begin{bmatrix} 2 & 7 & 1 \\ -3 & -9 & 2 \end{bmatrix}$

A has no inverse because it is not square.

23. $[A \;\vdots\; I] = \begin{bmatrix} 1 & 1 & 1 & \vdots & 1 & 0 & 0 \\ 3 & 5 & 4 & \vdots & 0 & 1 & 0 \\ 3 & 6 & 5 & \vdots & 0 & 0 & 1 \end{bmatrix} \Rightarrow \begin{bmatrix} 1 & 1 & 1 & \vdots & 1 & 0 & 0 \\ 0 & 2 & 1 & \vdots & -3 & 1 & 0 \\ 0 & 3 & 2 & \vdots & -3 & 0 & 1 \end{bmatrix} \Rightarrow \begin{bmatrix} 1 & 1 & 1 & \vdots & 1 & 0 & 0 \\ 0 & 1 & 1 & \vdots & 0 & -1 & 1 \\ 0 & 3 & 2 & \vdots & -3 & 0 & 1 \end{bmatrix}$

$\Rightarrow \begin{bmatrix} 1 & 0 & 0 & \vdots & 1 & 1 & -1 \\ 0 & 1 & 1 & \vdots & 0 & -1 & 1 \\ 0 & 0 & -1 & \vdots & -3 & 3 & -2 \end{bmatrix} \Rightarrow \begin{bmatrix} 1 & 0 & 0 & \vdots & 1 & 1 & -1 \\ 0 & 1 & 0 & \vdots & -3 & 2 & -1 \\ 0 & 0 & 1 & \vdots & 3 & -3 & 2 \end{bmatrix} = [I \;\vdots\; A^{-1}]$

$A^{-1} = \begin{bmatrix} 1 & 1 & -1 \\ -3 & 2 & -1 \\ 3 & -3 & 2 \end{bmatrix}$

25. $[A \;\vdots\; I] = \begin{bmatrix} 1 & 1 & 2 & \vdots & 1 & 0 & 0 \\ 3 & 1 & 0 & \vdots & 0 & 1 & 0 \\ -2 & 0 & 3 & \vdots & 0 & 0 & 1 \end{bmatrix} \Rightarrow \begin{bmatrix} 1 & 1 & 2 & \vdots & 1 & 0 & 0 \\ 0 & -2 & -6 & \vdots & -3 & 1 & 0 \\ 0 & 2 & 7 & \vdots & 2 & 0 & 1 \end{bmatrix} \Rightarrow \begin{bmatrix} 1 & 1 & 2 & \vdots & 1 & 0 & 0 \\ 0 & -2 & -6 & \vdots & -3 & 1 & 0 \\ 0 & 0 & 1 & \vdots & -1 & 1 & 1 \end{bmatrix}$

$\Rightarrow \begin{bmatrix} 1 & 1 & 0 & \vdots & 3 & -2 & -2 \\ 0 & -2 & 0 & \vdots & -9 & 7 & 6 \\ 0 & 0 & 1 & \vdots & -1 & 1 & 1 \end{bmatrix} \Rightarrow \begin{bmatrix} 1 & 0 & 0 & \vdots & -\frac{3}{2} & \frac{3}{2} & 1 \\ 0 & 1 & 0 & \vdots & \frac{9}{2} & -\frac{7}{2} & -3 \\ 0 & 0 & 1 & \vdots & -1 & 1 & 1 \end{bmatrix} = [I \;\vdots\; A^{-1}]$

$A^{-1} = \frac{1}{2}\begin{bmatrix} -3 & 3 & 2 \\ 9 & -7 & -6 \\ -2 & 2 & 2 \end{bmatrix}$

27. $[A \;\vdots\; I] = \begin{bmatrix} 3 & 0 & 0 & \vdots & 1 & 0 & 0 \\ 0 & -2 & 0 & \vdots & 0 & 1 & 0 \\ 0 & 0 & 4 & \vdots & 0 & 0 & 1 \end{bmatrix} \Rightarrow \begin{bmatrix} 1 & 0 & 0 & \vdots & \frac{1}{3} & 0 & 0 \\ 0 & 1 & 0 & \vdots & 0 & -\frac{1}{2} & 0 \\ 0 & 0 & 1 & \vdots & 0 & 0 & \frac{1}{4} \end{bmatrix} = [I \;\vdots\; A^{-1}]$

$A^{-1} = \begin{bmatrix} \frac{1}{3} & 0 & 0 \\ 0 & -\frac{1}{2} & 0 \\ 0 & 0 & \frac{1}{4} \end{bmatrix} = \frac{1}{12}\begin{bmatrix} 4 & 0 & 0 \\ 0 & -6 & 0 \\ 0 & 0 & 3 \end{bmatrix}$

29. $[A \;\vdots\; I] = \left[\begin{array}{ccc:ccc} 1 & 0 & 0 & 1 & 0 & 0 \\ 3 & 4 & 0 & 0 & 1 & 0 \\ 2 & 5 & 5 & 0 & 0 & 1 \end{array}\right] \Rightarrow \left[\begin{array}{ccc:ccc} 1 & 0 & 0 & 1 & 0 & 0 \\ 0 & 4 & 0 & -3 & 1 & 0 \\ 0 & 5 & 5 & -2 & 0 & 1 \end{array}\right] \Rightarrow \left[\begin{array}{ccc:ccc} 1 & 0 & 0 & 1 & 0 & 0 \\ 0 & 4 & 0 & -3 & 1 & 0 \\ 0 & 0 & 5 & \frac{7}{4} & -\frac{5}{4} & 1 \end{array}\right]$

$\Rightarrow \left[\begin{array}{ccc:ccc} 1 & 0 & 0 & 1 & 0 & 0 \\ 0 & 1 & 0 & -\frac{3}{4} & \frac{1}{4} & 0 \\ 0 & 0 & 1 & \frac{7}{20} & -\frac{1}{4} & \frac{1}{5} \end{array}\right] = [I \;\vdots\; A^{-1}]$

$A^{-1} = \begin{bmatrix} 1 & 0 & 0 \\ -\frac{3}{4} & \frac{1}{4} & 0 \\ \frac{7}{20} & -\frac{1}{4} & \frac{1}{5} \end{bmatrix}$

31. $[A \;\vdots\; I] = \left[\begin{array}{cccc:cccc} 1 & 0 & 3 & 0 & 1 & 0 & 0 & 0 \\ 0 & 2 & 0 & 4 & 0 & 1 & 0 & 0 \\ 1 & 0 & 3 & 0 & 0 & 0 & 1 & 0 \\ 0 & 2 & 0 & 4 & 0 & 0 & 0 & 1 \end{array}\right] \Rightarrow \left[\begin{array}{cccc:cccc} 1 & 0 & 3 & 0 & 1 & 0 & 0 & 0 \\ 0 & 2 & 0 & 4 & 0 & 1 & 0 & 0 \\ 0 & 0 & 0 & 0 & -1 & 0 & 1 & 0 \\ 0 & 2 & 0 & 4 & 0 & 0 & 0 & 1 \end{array}\right]$

Because the first four entries of row 3 are all zeros, the inverse of A does not exist.

33. $[A \;\vdots\; I] = \left[\begin{array}{ccc:ccc} 1 & 2 & -1 & 1 & 0 & 0 \\ 3 & 7 & -10 & 0 & 1 & 0 \\ -5 & -7 & -15 & 0 & 0 & 1 \end{array}\right] \Rightarrow \left[\begin{array}{ccc:ccc} 1 & 2 & -1 & 1 & 0 & 0 \\ 0 & 1 & -7 & -3 & 1 & 0 \\ 0 & 3 & -20 & 5 & 0 & 1 \end{array}\right]$

$\Rightarrow \left[\begin{array}{ccc:ccc} 1 & 0 & 13 & 7 & -2 & 0 \\ 0 & 1 & -7 & -3 & 1 & 0 \\ 0 & 0 & 1 & 14 & -3 & 1 \end{array}\right] \Rightarrow \left[\begin{array}{ccc:ccc} 1 & 0 & 0 & -175 & 37 & -13 \\ 0 & 1 & 0 & 95 & -20 & 7 \\ 0 & 1 & 1 & 14 & -3 & 1 \end{array}\right] = [I \;\vdots\; A^{-1}]$

$A^{-1} = \begin{bmatrix} -175 & 37 & -13 \\ 95 & -20 & 7 \\ 14 & -3 & 1 \end{bmatrix}$

35. $A^{-1} = \begin{bmatrix} 0 & -\frac{20}{11} & \frac{10}{11} \\ -10 & 5 & 5 \\ 10 & -\frac{30}{11} & -\frac{40}{11} \end{bmatrix}$

37. The inverse is $\begin{bmatrix} -12 & -5 & -9 \\ -4 & -2 & -4 \\ -8 & -4 & -6 \end{bmatrix}$.

39. Does not exist

41. $A = \begin{bmatrix} 5 & -2 \\ 2 & 3 \end{bmatrix}$; $A^{-1} = \dfrac{1}{ad - bc}\begin{bmatrix} d & -b \\ -c & a \end{bmatrix}$

$= \dfrac{1}{15 - 2(-2)}\begin{bmatrix} 3 & 2 \\ -2 & 5 \end{bmatrix}$

$= \dfrac{1}{19}\begin{bmatrix} 3 & 2 \\ -2 & 5 \end{bmatrix}$

43. $A = \begin{bmatrix} -4 & -6 \\ 2 & 3 \end{bmatrix}$; $A^{-1} = \dfrac{1}{ad - bc}\begin{bmatrix} d & -b \\ -c & a \end{bmatrix}$

$= \dfrac{1}{(-4)(3) - 2(-6)}\begin{bmatrix} 3 & 6 \\ -2 & -4 \end{bmatrix}$

$= \dfrac{1}{0}\begin{bmatrix} 3 & 6 \\ -2 & -4 \end{bmatrix}$

Because division by zero is not possible, there is no A^{-1}. Note that $R_1 = -2R_2$.

45. $A = \begin{bmatrix} \frac{7}{2} & -\frac{3}{4} \\ \frac{1}{5} & \frac{4}{5} \end{bmatrix}$; $A^{-1} = \dfrac{1}{ad - bc}\begin{bmatrix} d & -b \\ -c & a \end{bmatrix}$

$= \dfrac{1}{\left(\frac{7}{2}\right)\left(\frac{4}{5}\right) - \left(-\frac{3}{4}\right)\left(\frac{1}{5}\right)}\begin{bmatrix} \frac{4}{5} & \frac{3}{4} \\ -\frac{1}{5} & \frac{7}{2} \end{bmatrix}$

$= \frac{20}{59} \cdot \frac{1}{20}\begin{bmatrix} 16 & 15 \\ -4 & 70 \end{bmatrix} = \frac{1}{59}\begin{bmatrix} 16 & 15 \\ -4 & 70 \end{bmatrix}$

47. $\begin{bmatrix} x \\ y \end{bmatrix} = A^{-1}B = \begin{bmatrix} 7 & -2 \\ -3 & 1 \end{bmatrix}\begin{bmatrix} 0 \\ 1 \end{bmatrix} = \begin{bmatrix} -2 \\ 1 \end{bmatrix}$

Solution: $(-2, 1)$

49. $\begin{bmatrix} x \\ y \end{bmatrix} = A^{-1}B = \begin{bmatrix} 7 & -2 \\ -3 & 1 \end{bmatrix}\begin{bmatrix} 8 \\ 26 \end{bmatrix} = \begin{bmatrix} 4 \\ 2 \end{bmatrix}$

Solution: $(4, 2)$

51. $\begin{bmatrix} x \\ y \end{bmatrix} = A^{-1}B = \frac{1}{5}\begin{bmatrix} 4 & -3 \\ -1 & 2 \end{bmatrix}\begin{bmatrix} 5 \\ 10 \end{bmatrix} = \begin{bmatrix} -2 \\ 3 \end{bmatrix}$

Solution: $(-2, 3)$

53. $\begin{bmatrix} x \\ y \end{bmatrix} = A^{-1}B = \frac{1}{5}\begin{bmatrix} 4 & -3 \\ -1 & 2 \end{bmatrix}\begin{bmatrix} 4 \\ 2 \end{bmatrix} = \begin{bmatrix} 2 \\ 0 \end{bmatrix}$

Solution: $(2, 0)$

55. $\begin{bmatrix} x \\ y \\ z \end{bmatrix} = \begin{bmatrix} 1 & 1 & -1 \\ -3 & 2 & -1 \\ 3 & -3 & 2 \end{bmatrix}\begin{bmatrix} 0 \\ 5 \\ 2 \end{bmatrix} = \begin{bmatrix} 3 \\ 8 \\ -11 \end{bmatrix}$

Solution: $(3, 8, -11)$

57. $\begin{bmatrix} x_1 \\ x_2 \\ x_3 \\ x_4 \end{bmatrix} = A^{-1}B = \begin{bmatrix} -24 & 7 & 1 & -2 \\ -10 & 3 & 0 & -1 \\ -29 & 7 & 3 & -2 \\ 12 & -3 & -1 & 1 \end{bmatrix}\begin{bmatrix} 0 \\ 1 \\ -1 \\ 2 \end{bmatrix} = \begin{bmatrix} 2 \\ 1 \\ 0 \\ 0 \end{bmatrix}$

Solution: $(2, 1, 0, 0)$

59. $3x + 4y = -2$

$5x + 3y = 4$

$A = \begin{bmatrix} 3 & 4 \\ 5 & 3 \end{bmatrix}; x = \begin{bmatrix} x \\ y \end{bmatrix}; B = \begin{bmatrix} -2 \\ 4 \end{bmatrix}; A^{-1} = \frac{1}{3(3) - 5(4)}\begin{bmatrix} 3 & -4 \\ -5 & 3 \end{bmatrix} = -\frac{1}{11}\begin{bmatrix} 3 & -4 \\ -5 & 3 \end{bmatrix}$

$x = A^{-1}B = -\frac{1}{11}\begin{bmatrix} 3 & -4 \\ -5 & 3 \end{bmatrix} \cdot \begin{bmatrix} -2 \\ 4 \end{bmatrix} = \begin{bmatrix} 2 \\ -2 \end{bmatrix}$ so $x = 2$, $y = -2$

61. $-0.4x + 0.8y = 1.6$

$2x - 4y = 5$

$A = \begin{bmatrix} -0.4 & 0.8 \\ 2 & -4 \end{bmatrix}; x = \begin{bmatrix} x \\ y \end{bmatrix}; B = \begin{bmatrix} 1.6 \\ 5 \end{bmatrix}; A^{-1} = \frac{1}{(-4)(-0.4) - 2(0.8)}\begin{bmatrix} -4 & -0.8 \\ -2 & -0.4 \end{bmatrix}$

Because $ad - bc = 0$, A^{-1} does not exist. The system is inconsistent and there is no solution.

63. $-\frac{1}{4}x + \frac{3}{8}y = -2$

$\frac{3}{2}x + \frac{3}{4}y = -12$

$A = \begin{bmatrix} -\frac{1}{4} & \frac{3}{8} \\ \frac{3}{2} & \frac{3}{4} \end{bmatrix}; x = \begin{bmatrix} x \\ y \end{bmatrix}; B = \begin{bmatrix} -2 \\ -12 \end{bmatrix};$

$A^{-1} = \frac{1}{\left(-\frac{1}{4}\right)\left(\frac{3}{4}\right) - \left(\frac{3}{2}\right)\left(\frac{3}{8}\right)}\begin{bmatrix} \frac{3}{4} & -\frac{3}{8} \\ -\frac{3}{2} & -\frac{1}{4} \end{bmatrix}$

$= -\frac{4}{3} \cdot \frac{1}{8}\begin{bmatrix} 6 & -3 \\ -12 & -2 \end{bmatrix}$

$= -\frac{1}{6}\begin{bmatrix} 6 & -3 \\ -12 & -2 \end{bmatrix}$

$= \begin{bmatrix} -1 & \frac{1}{2} \\ 2 & \frac{1}{3} \end{bmatrix}$

so

$x = A^{-1}B$

$= \begin{bmatrix} -1 & \frac{1}{2} \\ 2 & \frac{1}{3} \end{bmatrix}\begin{bmatrix} -2 \\ -12 \end{bmatrix}$

$= \begin{bmatrix} -4 \\ -8 \end{bmatrix}$ so $x = -4$, $y = -8$

65. $4x - y + z = -5$

$2x + 2y + 3z = 10$

$5x - 2y + 6z = 1$

$A = \begin{bmatrix} 4 & -1 & 1 \\ 2 & 2 & 3 \\ 5 & -2 & 6 \end{bmatrix}; x = \begin{bmatrix} x \\ y \\ z \end{bmatrix};$

$B = \begin{bmatrix} -5 \\ 10 \\ 1 \end{bmatrix}; A^{-1} = \frac{1}{55}\begin{bmatrix} 18 & 4 & -5 \\ 3 & 19 & -10 \\ -14 & 3 & 10 \end{bmatrix}$

$x = A^{-1}B$

$= \frac{1}{55}\begin{bmatrix} 18 & 4 & -5 \\ 3 & 19 & -10 \\ -14 & 3 & 10 \end{bmatrix}\begin{bmatrix} -5 \\ 10 \\ 1 \end{bmatrix}$

$= \begin{bmatrix} -1 \\ 3 \\ 2 \end{bmatrix}$ so $x = -1$, $y = 3$, $z = 2$

67. $A = \begin{bmatrix} 7 & -3 & 0 & 2 \\ -2 & 1 & 0 & -1 \\ 4 & 0 & 1 & -2 \\ -1 & 1 & 0 & -1 \end{bmatrix}$

$A^{-1} = \begin{bmatrix} 0 & -1 & 0 & 1 \\ -1 & -5 & 0 & 3 \\ -2 & -4 & 1 & -2 \\ -1 & -4 & 0 & 1 \end{bmatrix}$

$$\begin{bmatrix} x \\ y \\ z \\ w \end{bmatrix} = \begin{bmatrix} 0 & -1 & 0 & 1 \\ -1 & -5 & 0 & 3 \\ -2 & -4 & 1 & -2 \\ -1 & -4 & 0 & 1 \end{bmatrix}\begin{bmatrix} 41 \\ -13 \\ 12 \\ -8 \end{bmatrix} = \begin{bmatrix} 5 \\ 0 \\ -2 \\ 3 \end{bmatrix}$$

Solution: $(5, 0, -2, 3)$

69. $2(2) + 1(-3) + 3(5) = 16 \Rightarrow 2x + y + 3z = 16$

$4(2) + 0(-3) - 2(5) = -2 \Rightarrow 4x - 2z = -2$

$0(2) + 3(-3) + 2(5) = 1 \Rightarrow 3y + 2z = 1$

$$A = \begin{bmatrix} 2 & 1 & 3 \\ 4 & 0 & -2 \\ 0 & 3 & 2 \end{bmatrix};\ A^{-1} = \begin{bmatrix} \frac{3}{20} & \frac{7}{40} & -\frac{1}{20} \\ -\frac{1}{5} & \frac{1}{10} & \frac{2}{5} \\ \frac{3}{10} & -\frac{3}{20} & -\frac{1}{10} \end{bmatrix};$$

$$A^{-1}B = \begin{bmatrix} \frac{3}{20} & \frac{7}{40} & -\frac{1}{20} \\ -\frac{1}{5} & \frac{1}{10} & \frac{2}{5} \\ \frac{3}{10} & -\frac{3}{20} & -\frac{1}{10} \end{bmatrix}\begin{bmatrix} 16 \\ -2 \\ 1 \end{bmatrix} = \begin{bmatrix} 2 \\ -3 \\ 5 \end{bmatrix} = \begin{bmatrix} x \\ y \\ z \end{bmatrix}$$

For Exercises 71 and 73:

$$[A \ \vdots \ I] = \begin{bmatrix} 1 & 1 & 1 & \vdots & 1 & 0 & 0 \\ 0.09 & 0.07 & 0.08 & \vdots & 0 & 1 & 0 \\ 0 & 2 & -1 & \vdots & 0 & 0 & 1 \end{bmatrix} \Rightarrow \begin{bmatrix} 1 & 1 & 1 & \vdots & 1 & 0 & 0 \\ 0 & -0.02 & -0.01 & \vdots & -0.09 & 1 & 0 \\ 0 & 2 & -1 & \vdots & 0 & 0 & 1 \end{bmatrix} \Rightarrow \begin{bmatrix} 1 & 0 & 0.5 & \vdots & -3.5 & 50 & 0 \\ 0 & 1 & 0.5 & \vdots & 4.5 & -50 & 0 \\ 0 & 0 & -2 & \vdots & -9 & 100 & 1 \end{bmatrix}$$

$$\Rightarrow \begin{bmatrix} 1 & 0 & 0 & \vdots & -5.75 & 75 & 0.25 \\ 0 & 1 & 0 & \vdots & 2.25 & -25 & 0.25 \\ 0 & 0 & 1 & \vdots & 4.5 & -50 & -0.50 \end{bmatrix} = [\mathbf{I} \ \vdots \ \mathbf{A}^{-1}]$$

$$\mathbf{A}^{-1} = \begin{bmatrix} -5.75 & 75 & 0.25 \\ 2.25 & -25 & 0.25 \\ 4.5 & -50 & -0.50 \end{bmatrix}$$

71. $\begin{bmatrix} x \\ y \\ z \end{bmatrix} = A^{-1}B = \begin{bmatrix} -5.75 & 75 & 0.25 \\ 2.25 & -25 & 0.25 \\ 4.5 & -50 & -0.50 \end{bmatrix}\begin{bmatrix} 35{,}000 \\ 2{,}450 \\ 0 \end{bmatrix} = \begin{bmatrix} 20{,}000 \\ 5{,}000 \\ 10{,}000 \end{bmatrix}$

Answer: AAA bonds: \$20,000
A bonds: \$5,000
B bonds: \$10,000

73. $\begin{bmatrix} x \\ y \\ z \end{bmatrix} = A^{-1}B = \begin{bmatrix} -5.75 & 75 & 0.25 \\ 2.25 & -25 & 0.25 \\ 4.5 & -50 & -0.50 \end{bmatrix}\begin{bmatrix} 36{,}000 \\ 3{,}040 \\ 0 \end{bmatrix} = \begin{bmatrix} 21{,}000 \\ 5{,}000 \\ 10{,}000 \end{bmatrix}$

Answer: AAA bonds: \$21,000
A bonds: \$5,000
B bonds: \$10,000

For Exercise 75:

$$A = \begin{bmatrix} 2 & 0 & 4 \\ 0 & 1 & 4 \\ 1 & 1 & -1 \end{bmatrix};\ A^{-1} = \frac{1}{14}\begin{bmatrix} 5 & -4 & 4 \\ -4 & 6 & 8 \\ 1 & 2 & -2 \end{bmatrix}$$

75. $\begin{bmatrix} I_1 \\ I_2 \\ I_3 \end{bmatrix} = \frac{1}{14}\begin{bmatrix} 5 & -4 & 4 \\ -4 & 6 & 8 \\ 1 & 2 & -2 \end{bmatrix}\begin{bmatrix} 28 \\ 21 \\ 0 \end{bmatrix} = \begin{bmatrix} 4 \\ 1 \\ 5 \end{bmatrix}$

Solution: $I_1 = 4$ amperes, $I_2 = 1$ ampere, $I_3 = 5$ amperes

77. $2x + y + 2z = 500$
$x + 2y + 2z = 500$
$x + y + 2z = 400$

$$A = \begin{bmatrix} 2 & 1 & 2 \\ 1 & 2 & 2 \\ 1 & 1 & 2 \end{bmatrix};\ A^{-1} = \begin{bmatrix} 1 & 0 & -1 \\ 0 & 1 & -1 \\ -0.5 & -0.5 & 1.5 \end{bmatrix};\ B = \begin{bmatrix} 500 \\ 500 \\ 400 \end{bmatrix}$$

$$\begin{bmatrix} x \\ y \\ z \end{bmatrix} = A^{-1}B = \begin{bmatrix} 1 & 0 & -1 \\ 0 & 1 & -1 \\ -0.5 & -0.5 & 1.5 \end{bmatrix}\begin{bmatrix} 500 \\ 500 \\ 400 \end{bmatrix} = \begin{bmatrix} 100 \\ 100 \\ 100 \end{bmatrix}$$

Answer: 100 bags of potting soil for seedlings
100 bags of potting soil for general potting
100 bags of potting soil for hardwood plants

79. $2x + y + 2z = 350$
$x + 2y + 2z = 445$
$x + y + 2z = 345$

$$A = \begin{bmatrix} 2 & 1 & 2 \\ 1 & 2 & 2 \\ 1 & 1 & 2 \end{bmatrix};\ A^{-1} = \begin{bmatrix} 1 & 0 & -1 \\ 0 & 1 & -1 \\ -0.5 & -0.5 & 1.5 \end{bmatrix};\ B = \begin{bmatrix} 350 \\ 445 \\ 345 \end{bmatrix}$$

$$\begin{bmatrix} x \\ y \\ z \end{bmatrix} = A^{-1}B = \begin{bmatrix} 1 & 0 & -1 \\ 0 & 1 & -1 \\ -0.5 & -0.5 & 1.5 \end{bmatrix}\begin{bmatrix} 350 \\ 445 \\ 345 \end{bmatrix}\begin{bmatrix} 5 \\ 100 \\ 120 \end{bmatrix}$$

Answer: 5 bags of potting soil for seedlings
100 bags of potting soil for general potting
120 bags of potting soil for hardwood plants

81. (a) $\begin{cases} 2.5x + 4y + 2z = 300 \\ x + y + z = 120 \\ -x + 2y + 2z = 0 \end{cases}$

(b) $\begin{bmatrix} 2.5 & 4 & 2 \\ 1 & 1 & 1 \\ -1 & 2 & 2 \end{bmatrix}\begin{bmatrix} x \\ y \\ z \end{bmatrix} = \begin{bmatrix} 300 \\ 120 \\ 0 \end{bmatrix}$

$A \quad X \quad B$

(c) $AX = B$

$X = A^{-1}B$

$$X = \begin{bmatrix} x \\ y \\ z \end{bmatrix} = A^{-1}B = \begin{bmatrix} 0 & \frac{2}{3} & -\frac{1}{3} \\ \frac{1}{2} & -\frac{7}{6} & \frac{1}{12} \\ -\frac{1}{2} & \frac{3}{2} & \frac{1}{4} \end{bmatrix}\begin{bmatrix} 300 \\ 120 \\ 0 \end{bmatrix} = \begin{bmatrix} 80 \\ 10 \\ 30 \end{bmatrix}$$

The florist will need 80 roses, 10 lilies and 30 irises.

83. (a) $AB = \begin{bmatrix} 13 & 4 \\ 1 & 8 \end{bmatrix},\ BA = \begin{bmatrix} 8 & 1 \\ 4 & 13 \end{bmatrix}$

$AB = C$ and in the product BA, the diagonal entries are the reverse of those in C.

(b) $C^{-1} = \begin{bmatrix} \frac{2}{25} & -\frac{1}{25} \\ -\frac{1}{100} & \frac{13}{100} \end{bmatrix}$

$A^{-1} \cdot B^{-1} = \begin{bmatrix} \frac{13}{100} & -\frac{1}{100} \\ -\frac{1}{25} & \frac{2}{25} \end{bmatrix}$

$B^{-1} \cdot A^{-1} = \begin{bmatrix} \frac{2}{25} & -\frac{1}{25} \\ -\frac{1}{100} & \frac{13}{100} \end{bmatrix};\ C^{-1} = B^{-1} \cdot A^{-1}$

(c) The inverse of a product AB is equal to the inverse of B times the inverse of A.

85. Let $k = 3$.

Then $A = \begin{bmatrix} 4 & 3 \\ -2 & 3 \end{bmatrix}$ and $A^{-1} = \begin{bmatrix} \frac{1}{6} & -\frac{1}{6} \\ \frac{1}{9} & \frac{2}{9} \end{bmatrix} = \frac{1}{18}\begin{bmatrix} 3 & -3 \\ 2 & 4 \end{bmatrix} \Rightarrow A$ is invertible.

To find the value of k that makes A singular:

$$[A \vdots I] = \left[\begin{array}{cc:cc} 4 & 3 & 1 & 0 \\ -2 & k & 0 & 1 \end{array}\right] \Rightarrow \left[\begin{array}{cc:cc} 1 & \frac{3}{4} & \frac{1}{4} & 0 \\ 0 & k + \frac{3}{2} & \frac{1}{2} & 1 \end{array}\right]$$

If $k + \frac{3}{2} = 0$, then $k = -\frac{3}{2}$ and A will be singular.

Note, any $k \neq \frac{3}{2}$ will mean that A will be invertible.

87. True. The inverse of I_n is I_n.

89. $$AA^{-1} = \begin{bmatrix} a & b \\ c & d \end{bmatrix}\left(\frac{1}{ad - bc}\right)\begin{bmatrix} d & -b \\ -c & a \end{bmatrix} = \frac{1}{ad - bc}\begin{bmatrix} a & b \\ c & d \end{bmatrix}\begin{bmatrix} d & -b \\ -c & a \end{bmatrix} = \frac{1}{ad - bc}\begin{bmatrix} ad - bc & 0 \\ 0 & ad - bc \end{bmatrix} = \begin{bmatrix} 1 & 0 \\ 0 & 1 \end{bmatrix}$$

$$A^{-1}A = \frac{1}{ad - bc}\begin{bmatrix} d & -b \\ -c & a \end{bmatrix}\begin{bmatrix} a & b \\ c & d \end{bmatrix} = \frac{1}{ad - bc}\begin{bmatrix} ad - bc & 0 \\ 0 & ad - bc \end{bmatrix} = \begin{bmatrix} 1 & 0 \\ 0 & 1 \end{bmatrix}$$

Chapter 6 Quiz Yourself

1. Any matrix with 4 rows and 3 columns will satisfy this requirement. Some examples are

$$\begin{bmatrix} 1 & 0 & 0 \\ 0 & 1 & 0 \\ 0 & 1 & 1 \\ 1 & 0 & 1 \end{bmatrix}, \begin{bmatrix} 1 & 2 & 3 \\ 4 & 5 & 6 \\ 7 & 8 & 9 \\ 10 & 11 & 12 \end{bmatrix}, \begin{bmatrix} -2 & 1 & 1 \\ 7 & -1 & 0 \\ 13 & 5 & 6 \\ 0 & 0 & 0 \end{bmatrix}$$

2. Any matrix with 3 rows and 1 column will satisfy this requirement. Some examples are

$$\begin{bmatrix} 1 \\ 2 \\ 3 \end{bmatrix}, \begin{bmatrix} 0 \\ -1 \\ 5 \end{bmatrix}, \begin{bmatrix} a \\ b \\ c \end{bmatrix}$$

3. $\begin{bmatrix} 3 & 2 & \vdots & -2 \\ 5 & -1 & \vdots & 19 \end{bmatrix}$

4. $\begin{bmatrix} 1 & 0 & 3 & \vdots & -5 \\ 1 & 2 & -1 & \vdots & 3 \\ 3 & 0 & 4 & \vdots & 0 \end{bmatrix}$

5. $$\begin{bmatrix} 3 & 2 & \vdots & -2 \\ 5 & -1 & \vdots & 19 \end{bmatrix} \begin{matrix} \frac{1}{3}R_1 \rightarrow \\ -5R_1 + R_2 \rightarrow \end{matrix} \begin{bmatrix} 1 & \frac{2}{3} & \vdots & -\frac{2}{3} \\ 0 & -\frac{13}{3} & \vdots & \frac{67}{3} \end{bmatrix}$$

$$-\tfrac{3}{13}R_2 \rightarrow \begin{bmatrix} 1 & \frac{2}{3} & \vdots & -\frac{2}{3} \\ 0 & 1 & \vdots & -\frac{67}{13} \end{bmatrix}$$

$y = -\frac{67}{13}$

$x + \frac{2}{3}\left(-\frac{67}{13}\right) = -\frac{2}{3} \Rightarrow x = \frac{36}{13}$

Thus, $x = \frac{36}{13} \approx 2.769$ and $y = -\frac{67}{13} \approx -5.154$.

Answer: $(2.769, -5.154)$

6. $$\begin{bmatrix} 1 & 0 & 3 & \vdots & -5 \\ 1 & 2 & -1 & \vdots & 3 \\ 3 & 0 & 4 & \vdots & 0 \end{bmatrix} \begin{matrix} \\ -R_1 + R_2 \rightarrow \\ -3R_1 + R_3 \rightarrow \end{matrix} \begin{bmatrix} 1 & 0 & 3 & \vdots & -5 \\ 0 & 2 & -4 & \vdots & 8 \\ 0 & 0 & -5 & \vdots & 15 \end{bmatrix}$$

$$\begin{matrix} \\ -\frac{1}{2}R_2 \rightarrow \\ -\frac{1}{5}R_3 \rightarrow \end{matrix} \begin{bmatrix} 1 & 0 & 3 & \vdots & -5 \\ 0 & 1 & -2 & \vdots & 4 \\ 0 & 0 & 1 & \vdots & -3 \end{bmatrix}$$

$$\begin{matrix} -3R_3 + R_1 \rightarrow \\ 2R_3 + R_2 \rightarrow \\ \end{matrix} \begin{bmatrix} 1 & 0 & 0 & \vdots & 4 \\ 0 & 1 & 0 & \vdots & -2 \\ 0 & 0 & 1 & \vdots & -3 \end{bmatrix}$$

Thus, $x = 4$, $y = -2$, and $z = -3$.

Answer: $(4, -2, -3)$

7. $$2A + 3C = 2\begin{bmatrix} -1 & 4 \\ -2 & 6 \end{bmatrix} + 3\begin{bmatrix} 0 & 1 \\ 3 & 1 \end{bmatrix} = \begin{bmatrix} -2 & 8 \\ -4 & 12 \end{bmatrix} + \begin{bmatrix} 0 & -3 \\ 9 & 3 \end{bmatrix} = \begin{bmatrix} -2 + 0 & 8 + (-3) \\ -4 + 9 & 12 + 3 \end{bmatrix} = \begin{bmatrix} -2 & 5 \\ 5 & 15 \end{bmatrix}$$

8. $$AB = \begin{bmatrix} -1 & 4 \\ -2 & 6 \end{bmatrix}\begin{bmatrix} -1 & 2 & -3 \\ 2 & 0 & 5 \end{bmatrix} = \begin{bmatrix} 1 + 8 & -2 + 0 & 3 + 20 \\ 2 + 12 & -4 + 0 & 6 + 30 \end{bmatrix} = \begin{bmatrix} 9 & -2 & 23 \\ 14 & -4 & 36 \end{bmatrix}$$

9. $$A - 3C = \begin{bmatrix} -1 & 4 \\ -2 & 6 \end{bmatrix} - 3\begin{bmatrix} 0 & -1 \\ 3 & 1 \end{bmatrix} = \begin{bmatrix} -1 & 4 \\ -2 & 6 \end{bmatrix} - \begin{bmatrix} 0 & -3 \\ 9 & 3 \end{bmatrix} = \begin{bmatrix} -1 - 0 & 4 - (-3) \\ -2 - 9 & 6 - 3 \end{bmatrix} = \begin{bmatrix} -1 & 7 \\ -11 & 3 \end{bmatrix}$$

10. $$C^2 = \begin{bmatrix} 0 & -1 \\ 3 & 1 \end{bmatrix}\begin{bmatrix} 0 & -1 \\ 3 & 1 \end{bmatrix} = \begin{bmatrix} 0 + (-3) & 0 + (-1) \\ 0 + 3 & -3 + 1 \end{bmatrix} = \begin{bmatrix} -3 & -1 \\ 3 & -2 \end{bmatrix}$$

11. $$A^{-1} = \begin{bmatrix} -1 & 4 \\ -2 & 6 \end{bmatrix}^{-1} = \frac{1}{2}\begin{bmatrix} 6 & -4 \\ 2 & -1 \end{bmatrix} = \begin{bmatrix} 3 & -2 \\ 1 & -\frac{1}{2} \end{bmatrix}$$

12. Not possible

13. $X = 3A - 2C$

$$= 3\begin{bmatrix} -1 & 4 \\ -2 & 6 \end{bmatrix} - 2\begin{bmatrix} 0 & -1 \\ 3 & 1 \end{bmatrix}$$

$$= \begin{bmatrix} -3 & 12 \\ -6 & 18 \end{bmatrix} - \begin{bmatrix} 0 & -2 \\ 6 & 2 \end{bmatrix}$$

$$= \begin{bmatrix} -3 - 0 & 12 - (-2) \\ -6 - 6 & 18 - 2 \end{bmatrix}$$

$$= \begin{bmatrix} -3 & 14 \\ -12 & 16 \end{bmatrix}$$

14. $2X + 4A = 2C$

$2X = 2C - 4A$

$X = C - 2A$

$$= \begin{bmatrix} 0 & -1 \\ 3 & 1 \end{bmatrix} - 2\begin{bmatrix} -1 & 4 \\ -2 & 6 \end{bmatrix}$$

$$= \begin{bmatrix} 0 & -1 \\ 3 & 1 \end{bmatrix} - \begin{bmatrix} -2 & 8 \\ -4 & 12 \end{bmatrix}$$

$$= \begin{bmatrix} 0 - (-2) & -1 - 8 \\ 3 - (-4) & 1 - 12 \end{bmatrix}$$

$$= \begin{bmatrix} 2 & -9 \\ 7 & -11 \end{bmatrix}$$

15. $\begin{bmatrix} 1.0 & 0.6 & 0.2 \end{bmatrix}\begin{bmatrix} 15 \\ 10 \\ 9 \end{bmatrix} = 1.0(15) + 0.6(10) + 0.2(9) = \$22.80$

16. $\begin{bmatrix} 2.4 & 1.0 & 0.2 \end{bmatrix}\begin{bmatrix} 12 \\ 11 \\ 8 \end{bmatrix} = 2.4(12) + 1.0(11) + 0.2(8) = \$41.40$

17. $\begin{bmatrix} 2.8 & 2.0 & 0.5 \end{bmatrix}\begin{bmatrix} 12 \\ 11 \\ 8 \end{bmatrix} = 2.8(12) + 2.0(11) + 0.5(8) = \$59.60$

18. $$\begin{bmatrix} 1.0 & 0.6 & 0.2 \\ 2.4 & 1.0 & 0.2 \\ 2.8 & 2.0 & 0.5 \end{bmatrix}\begin{bmatrix} 15 & 12 \\ 10 & 11 \\ 9 & 8 \end{bmatrix} = \begin{bmatrix} 1.0(15) + 0.6(10) + 0.2(9) & 1.0(12) + 0.6(11) + 0.2(8) \\ 2.4(15) + 1.0(10) + 0.2(9) & 2.4(12) + 1.0(11) + 0.2(8) \\ 2.8(15) + 2.0(10) + 0.5(9) & 2.8(12) + 2.0(11) + 0.5(8) \end{bmatrix}$$

$$= \begin{array}{c} \text{Plant 1} \quad \text{Plant 2} \\ \begin{bmatrix} \$22.80 & \$20.20 \\ \$47.80 & \$41.40 \\ \$66.50 & \$59.60 \end{bmatrix} \end{array} \begin{array}{l} \text{Model A} \\ \text{Model B} \\ \text{Model C} \end{array}$$

LW represents the total labor costs for each model at each plant.

19. $A = \begin{bmatrix} 1 & -3 \\ -2 & 1 \end{bmatrix}, X = \begin{bmatrix} x \\ y \end{bmatrix}, B = \begin{bmatrix} 10 \\ -10 \end{bmatrix}$

$$X = A^{-1}B = \begin{bmatrix} -0.2 & -0.6 \\ -0.4 & -0.2 \end{bmatrix}\begin{bmatrix} 10 \\ -10 \end{bmatrix} = \begin{bmatrix} 4 \\ -2 \end{bmatrix}$$

Answer: $x = 4$, $y = -2$

20. $A = \begin{bmatrix} 2 & -1 & 1 \\ 3 & 0 & -1 \\ 0 & 4 & 3 \end{bmatrix}, X = \begin{bmatrix} x \\ y \\ z \end{bmatrix}, B = \begin{bmatrix} 3 \\ 15 \\ -1 \end{bmatrix}$

$$X = A^{-1}B = \begin{bmatrix} \frac{4}{29} & \frac{7}{29} & \frac{1}{29} \\ -\frac{9}{29} & \frac{6}{29} & \frac{5}{29} \\ \frac{12}{29} & -\frac{8}{29} & \frac{3}{29} \end{bmatrix}\begin{bmatrix} 3 \\ 15 \\ -1 \end{bmatrix} = \begin{bmatrix} 4 \\ 2 \\ -3 \end{bmatrix}$$

So, $x = 4$, $y = 2$, and $z = -3$.

Section 6.4 The Determinant of a Square Matrix

Skills Warm Up

1. $\begin{bmatrix} 1 & -2 \\ 0 & 3 \end{bmatrix} + \begin{bmatrix} 2 & 7 \\ 4 & -3 \end{bmatrix} = \begin{bmatrix} 1+2 & -2+7 \\ 0+4 & 3+(-3) \end{bmatrix} = \begin{bmatrix} 3 & 5 \\ 4 & 0 \end{bmatrix}$

2. $\begin{bmatrix} -2 & 5 \\ 3 & -2 \end{bmatrix} - \begin{bmatrix} 0 & -3 \\ 1 & 2 \end{bmatrix} = \begin{bmatrix} -2-0 & 5-(-3) \\ 3-1 & -2-2 \end{bmatrix}$
$= \begin{bmatrix} -2 & 8 \\ 2 & -4 \end{bmatrix}$

3. $3\begin{bmatrix} 3 & -4 & 2 \\ 1 & 0 & -1 \\ 0 & 1 & -2 \end{bmatrix} = \begin{bmatrix} 3(3) & 3(-4) & 3(2) \\ 3(1) & 3(0) & 3(-1) \\ 3(0) & 3(1) & 3(-2) \end{bmatrix} = \begin{bmatrix} 9 & -12 & 6 \\ 3 & 0 & -3 \\ 0 & 3 & -6 \end{bmatrix}$

4. $4\begin{bmatrix} 0 & 2 & 3 \\ -1 & 2 & 3 \\ -2 & 1 & -2 \end{bmatrix} = \begin{bmatrix} 4(0) & 4(2) & 4(3) \\ 4(-1) & 4(2) & 4(3) \\ 4(-2) & 4(1) & 4(-2) \end{bmatrix} = \begin{bmatrix} 0 & 8 & 12 \\ -4 & 8 & 12 \\ -8 & 4 & -8 \end{bmatrix}$

5. $[(1)(3) + (-3)(2)] - [(1)(4) + (3)(5)]$
$= (3 - 6) - (4 + 15)$
$= -3 - 19 = -22$

6. $[(4)(4) + (-1)(-3)] - [(-1)(2) + (-2)(7)]$
$= (16 + 3) - (-2 - 14) = 19 - (-16) = 35$

7. $\dfrac{4(7) - 1(-2)}{(-5)(-2) - 3(4)} = \dfrac{28 + 2}{10 - 12} = \dfrac{30}{-2} = -15$

8. $\dfrac{3(6) - 2(7)}{6(-5) - 2(1)} = \dfrac{18 - 14}{-30 - 2} = \dfrac{4}{-32} = -\dfrac{1}{8}$

9. $-5(-1)^2[6(-2) - 7(-3)] = -5(-12 + 21)$
$= -5(9) = -45$

10. $4(-1)^3[3(6) - 2(7)] = -4(18 - 14) = -16$

1. -5

3. $\begin{vmatrix} 1 & 3 \\ 2 & 7 \end{vmatrix} = 7(1) - 2(3) = 7 - 6 = 1$

5. $\begin{vmatrix} 5 & 6 \\ 2 & 3 \end{vmatrix} = 5(3) - 2(6) = 15 - 12 = 3$

7. $\begin{vmatrix} 9 & 3 \\ 12 & 4 \end{vmatrix} = 9(4) - 3(12) = 36 - 36 = 0$

9. $\begin{vmatrix} 2 & 1 \\ 3 & 4 \end{vmatrix} = 4(2) - 1(3) = 5$

11. $\begin{vmatrix} \frac{2}{3} & 0 \\ -1 & 6 \end{vmatrix} = 6(\tfrac{2}{3}) - 0(-1) = 4 - 0 = 4$

13. $\begin{vmatrix} -\frac{1}{2} & \frac{1}{3} \\ -6 & \frac{1}{3} \end{vmatrix} = \tfrac{1}{3}(-\tfrac{1}{2}) - (-6)(\tfrac{1}{3}) = \tfrac{11}{6}$

15. $\begin{vmatrix} 2.5 & 0.4 \\ -3.5 & -1.6 \end{vmatrix} = 2.5(-1.6) - (-3.5)(0.4) = -2.6$

17. $\begin{vmatrix} 5 & -3 & 2 \\ 7 & 5 & -7 \\ 0 & 6 & -1 \end{vmatrix} = 248$

19. $\begin{vmatrix} 0.1 & 0.3 & 0.2 \\ -0.3 & -0.2 & 0.1 \\ 1 & 2 & 3 \end{vmatrix} = 0.14$

21. $\begin{vmatrix} 0.9 & 0.7 & 0 \\ -0.1 & 0.3 & 1.3 \\ 2.2 & 4.2 & 6.1 \end{vmatrix} = -0.838$

23. $\begin{vmatrix} 3 & 4 \\ 2 & -5 \end{vmatrix}$

(a) $M_{11} = -5$
$M_{12} = 2$
$M_{21} = 4$
$M_{22} = 3$

(b) $C_{11} = M_{11} = -5$
$C_{12} = -M_{12} = -2$
$C_{21} = -M_{21} = -4$
$C_{22} = M_{22} = 3$

25. $\begin{bmatrix} 3 & 1 \\ -2 & -4 \end{bmatrix}$

(a) $M_{11} = -4$
$M_{12} = -2$
$M_{21} = 1$
$M_{22} = 3$

(b) $C_{11} = M_{11} = -4$
$C_{12} = -M_{12} = 2$
$C_{21} = -M_{21} = -1$
$C_{22} = M_{22} = 3$

27. $\begin{bmatrix} 4 & 0 & 2 \\ -3 & 2 & 1 \\ 1 & -1 & 1 \end{bmatrix}$

(a) $M_{11} = \begin{vmatrix} 2 & 1 \\ -1 & 1 \end{vmatrix} = 2 + 1 = 3$

$M_{12} = \begin{vmatrix} -3 & 1 \\ 1 & 1 \end{vmatrix} = -3 - 1 = -4$

$M_{13} = \begin{vmatrix} -3 & 2 \\ 1 & -1 \end{vmatrix} = 3 - 2 = 1$

$M_{21} = \begin{vmatrix} 0 & 2 \\ -1 & 1 \end{vmatrix} = 0 + 2 = 2$

$M_{22} = \begin{vmatrix} 4 & 2 \\ 1 & 1 \end{vmatrix} = 4 - 2 = 2$

$M_{23} = \begin{vmatrix} 4 & 0 \\ 1 & -1 \end{vmatrix} = -4 - 0 = -4$

$M_{31} = \begin{vmatrix} 0 & 2 \\ 2 & 1 \end{vmatrix} = 0 - 4 = -4$

$M_{32} = \begin{vmatrix} 4 & 2 \\ -3 & 1 \end{vmatrix} = 4 + 6 = 10$

$M_{33} = \begin{vmatrix} 4 & 0 \\ -3 & 2 \end{vmatrix} = 8 - 0 = 8$

(b) $C_{11} = (-1)^2 M_{11} = 3$

$C_{12} = (-1)^3 M_{12} = 4$

$C_{13} = (-1)^4 M_{13} = 1$

$C_{21} = (-1)^3 M_{21} = -2$

$C_{22} = (-1)^4 M_{22} = 2$

$C_{23} = (-1)^5 M_{23} = 4$

$C_{31} = (-1)^4 M_{31} = -4$

$C_{32} = (-1)^5 M_{32} = -10$

$C_{33} = (-1)^6 M_{33} = 8$

29. $\begin{bmatrix} 3 & -2 & 8 \\ 3 & 2 & -6 \\ -1 & 3 & 6 \end{bmatrix}$

(a) $M_{11} = \begin{vmatrix} 2 & -6 \\ 3 & 6 \end{vmatrix} = 12 + 18 = 30$

$M_{12} = \begin{vmatrix} 3 & -6 \\ -1 & 6 \end{vmatrix} = 18 - 6 = 12$

$M_{13} = \begin{vmatrix} 3 & 2 \\ -1 & 3 \end{vmatrix} = 9 + 2 = 11$

$M_{21} = \begin{vmatrix} -2 & 8 \\ 3 & 6 \end{vmatrix} = -12 - 24 = -36$

$M_{22} = \begin{vmatrix} 3 & 8 \\ -1 & 6 \end{vmatrix} = 18 + 8 = 26$

$M_{23} = \begin{vmatrix} 3 & -2 \\ -1 & 3 \end{vmatrix} = 9 - 2 = 7$

$M_{31} = \begin{vmatrix} -2 & 8 \\ 2 & -6 \end{vmatrix} = 12 - 16 = -4$

$M_{32} = \begin{vmatrix} 3 & 8 \\ 3 & -6 \end{vmatrix} = -18 - 24 = -42$

$M_{33} = \begin{vmatrix} 3 & -2 \\ 3 & 2 \end{vmatrix} = 6 + 6 = 12$

(b) $C_{11} = (-1)^2 M_{11} = 30$

$C_{12} = (-1)^3 M_{12} = -12$

$C_{13} = (-1)^4 M_{13} = 11$

$C_{21} = (-1)^3 M_{21} = 36$

$C_{22} = (-1)^4 M_{22} = 26$

$C_{23} = (-1)^5 M_{23} = -7$

$C_{31} = (-1)^4 M_{31} = -4$

$C_{32} = (-1)^5 M_{32} = 42$

$C_{33} = (-1)^6 M_{33} = 12$

31. (a) $\begin{vmatrix} 4 & 1 & -3 \\ 6 & 5 & -2 \\ -1 & 3 & -4 \end{vmatrix} = -1\begin{vmatrix} 1 & -3 \\ 5 & -2 \end{vmatrix} - 3\begin{vmatrix} 4 & -3 \\ 6 & -2 \end{vmatrix} + (-4)\begin{vmatrix} 4 & 1 \\ 6 & 5 \end{vmatrix}$

$= -1(13) - 3(10) - 4(14) = -99$

(b) $\begin{vmatrix} 4 & 1 & -3 \\ 6 & 5 & -2 \\ -1 & 3 & -4 \end{vmatrix} = -1\begin{vmatrix} 6 & -2 \\ -1 & -4 \end{vmatrix} + 5\begin{vmatrix} 4 & -3 \\ -1 & -4 \end{vmatrix} - 3\begin{vmatrix} 4 & -3 \\ 6 & -2 \end{vmatrix}$

$= -1(-26) + 5(-19) - 3(10) = -99$

33. (a) $\begin{vmatrix} 7 & 0 & -4 \\ 2 & -3 & 0 \\ 5 & 8 & 1 \end{vmatrix} = 7\begin{vmatrix} -3 & 0 \\ 8 & 1 \end{vmatrix} - 0\begin{vmatrix} 2 & 0 \\ 5 & 1 \end{vmatrix} + (-4)\begin{vmatrix} 2 & -3 \\ 5 & 8 \end{vmatrix} = 7(-3) - 4(31) = -145$

(b) $\begin{vmatrix} 7 & 0 & -4 \\ 2 & -3 & 0 \\ 5 & 8 & 1 \end{vmatrix} = -4\begin{vmatrix} 2 & -3 \\ 5 & 8 \end{vmatrix} - 0\begin{vmatrix} 7 & 0 \\ 5 & 8 \end{vmatrix} + 1\begin{vmatrix} 7 & 0 \\ 2 & -3 \end{vmatrix} = -4(31) + 1(-21) = -145$

35. (a) $$\begin{vmatrix} 6 & 0 & -3 & 5 \\ 4 & 13 & 6 & -8 \\ -1 & 0 & 7 & 4 \\ 8 & 6 & 0 & 2 \end{vmatrix} = -4\begin{vmatrix} 0 & -3 & 5 \\ 0 & 7 & 4 \\ 6 & 0 & 2 \end{vmatrix} + 13\begin{vmatrix} 6 & -3 & 5 \\ -1 & 7 & 4 \\ 8 & 0 & 2 \end{vmatrix} - 6\begin{vmatrix} 6 & 0 & 5 \\ -1 & 0 & 4 \\ 8 & 6 & 2 \end{vmatrix} - 8\begin{vmatrix} 6 & 0 & -3 \\ -1 & 0 & 7 \\ 8 & 6 & 0 \end{vmatrix}$$

$$= -4(-282) + 13(-298) - 6(-174) - 8(-234) = 170$$

(b) $$\begin{vmatrix} 6 & 0 & -3 & 5 \\ 4 & 13 & 6 & -8 \\ -1 & 0 & 7 & 4 \\ 8 & 6 & 0 & 2 \end{vmatrix} = 0\begin{vmatrix} 4 & 6 & -8 \\ -1 & 7 & 4 \\ 8 & 0 & 2 \end{vmatrix} + 13\begin{vmatrix} 6 & -3 & 5 \\ -1 & 7 & 4 \\ 8 & 0 & 2 \end{vmatrix} + 0\begin{vmatrix} 6 & -3 & 5 \\ 4 & 6 & -8 \\ 8 & 0 & 2 \end{vmatrix} + 6\begin{vmatrix} 6 & -3 & 5 \\ 4 & 6 & -8 \\ -1 & 7 & 4 \end{vmatrix}$$

$$= 0 + 13(-298) + 0 + 6(674) = 170$$

37. Expand by Column 3.

$$\begin{vmatrix} 1 & 4 & -2 \\ 3 & 2 & 0 \\ -1 & 4 & 3 \end{vmatrix} = -2\begin{vmatrix} 3 & 2 \\ -1 & 4 \end{vmatrix} + 3\begin{vmatrix} 1 & 4 \\ 3 & 2 \end{vmatrix}$$

$$= -2(14) + 3(-10) = -58$$

39. $$\begin{vmatrix} 2 & 4 & 6 \\ 0 & 3 & 1 \\ 0 & 0 & -5 \end{vmatrix} = (2)(3)(-5) = -30 \text{ (Upper Triangular)}$$

41. $$\begin{vmatrix} 2 & -1 & 0 \\ 4 & 2 & 1 \\ 4 & 2 & 1 \end{vmatrix} = 2\begin{vmatrix} 2 & 1 \\ 2 & 1 \end{vmatrix} - 4\begin{vmatrix} -1 & 0 \\ 2 & 1 \end{vmatrix} + 4\begin{vmatrix} -1 & 0 \\ 2 & 1 \end{vmatrix}$$

$$= 2(0) - 4(-1) + 4(-1) = 0$$

43. $$\begin{vmatrix} 1 & 4 & -2 \\ 3 & 6 & -6 \\ -2 & 1 & 4 \end{vmatrix} = \begin{vmatrix} 6 & -6 \\ 1 & 4 \end{vmatrix} - 3\begin{vmatrix} 4 & -2 \\ 1 & 4 \end{vmatrix} - 2\begin{vmatrix} 4 & -2 \\ 6 & -6 \end{vmatrix}$$

$$= 30 - 3(18) - 2(-12) = 0$$

45. $$\begin{vmatrix} 0.3 & 0.2 & 0.2 \\ 0.2 & 0.2 & 0.2 \\ -0.4 & 0.4 & 0.3 \end{vmatrix} = 0.3\begin{vmatrix} 0.2 & 0.2 \\ 0.4 & 0.3 \end{vmatrix} - 0.2\begin{vmatrix} 0.2 & 0.2 \\ -0.4 & 0.3 \end{vmatrix} + 0.2\begin{vmatrix} 0.2 & 0.2 \\ -0.4 & 0.4 \end{vmatrix}$$

$$= 0.3(-0.02) - 0.2(0.14) + 0.2(0.16) = -0.002$$

47. $$\begin{vmatrix} 6 & 3 & -7 \\ 0 & 0 & 0 \\ 4 & -6 & 3 \end{vmatrix} = 0\begin{vmatrix} 3 & -7 \\ -6 & 3 \end{vmatrix} - 0\begin{vmatrix} 6 & -7 \\ 4 & 3 \end{vmatrix} + 0\begin{vmatrix} 6 & 3 \\ 4 & -6 \end{vmatrix} = 0$$

49. Expand by Row 2.

$$\begin{vmatrix} 3 & 6 & -5 & 4 \\ -2 & 0 & 6 & 0 \\ 1 & 1 & 2 & 2 \\ 0 & 3 & -1 & -1 \end{vmatrix} = 2\begin{vmatrix} 6 & -5 & 4 \\ 1 & 2 & 2 \\ 3 & -1 & -1 \end{vmatrix} - 6\begin{vmatrix} 3 & 6 & 4 \\ 1 & 1 & 2 \\ 0 & 3 & -1 \end{vmatrix}$$

$$= 2(-63) - 6(-3) = -108$$

51. Expand by Column 2.

$$\begin{vmatrix} 3 & 2 & 4 & -1 & 5 \\ -2 & 0 & 1 & 3 & 2 \\ 1 & 0 & 0 & 4 & 0 \\ 6 & 0 & 2 & -1 & 0 \\ 3 & 0 & 5 & 1 & 0 \end{vmatrix} = -2\begin{vmatrix} -2 & 1 & 3 & 2 \\ 1 & 0 & 4 & 0 \\ 6 & 2 & -1 & 0 \\ 3 & 5 & 1 & 0 \end{vmatrix} = (-2)(-2)\begin{vmatrix} 1 & 0 & 4 \\ 6 & 2 & -1 \\ 3 & 5 & 1 \end{vmatrix} = 4(103) = 412$$

53. $$\begin{vmatrix} 3 & 8 & -7 \\ 0 & -5 & 4 \\ 8 & 1 & 6 \end{vmatrix} = -126$$

55. $$\begin{vmatrix} 1 & -1 & 8 & 4 \\ 2 & 6 & 0 & -4 \\ 2 & 0 & 2 & 6 \\ 0 & 2 & 8 & 0 \end{vmatrix} = -336$$

57. (a) $\begin{vmatrix} -1 & 0 \\ 0 & 3 \end{vmatrix} = -3 - 0 = -3$

(b) $\begin{vmatrix} 2 & 0 \\ 0 & -1 \end{vmatrix} = -2 - 0 = -2$

(c) $\begin{bmatrix} -1 & 0 \\ 0 & 3 \end{bmatrix}\begin{bmatrix} 2 & 0 \\ 0 & -1 \end{bmatrix} = \begin{bmatrix} -2 & 0 \\ 0 & -3 \end{bmatrix}$

(d) $\begin{vmatrix} -2 & 0 \\ 0 & -3 \end{vmatrix} = 6 - 0 = 6$

59. (a) $\begin{vmatrix} 4 & 0 \\ 3 & -2 \end{vmatrix} = -8 - 0 = -8$

(b) $\begin{vmatrix} -1 & 1 \\ -2 & 2 \end{vmatrix} = -2 + 2 = 0$

(c) $\begin{bmatrix} 4 & 0 \\ 3 & -2 \end{bmatrix}\begin{bmatrix} -1 & 1 \\ -2 & 2 \end{bmatrix} = \begin{bmatrix} -4 & 4 \\ -3 + 4 & 3 - 4 \end{bmatrix} = \begin{bmatrix} -4 & 4 \\ 1 & -1 \end{bmatrix}$

(d) $\begin{bmatrix} -4 & 4 \\ 1 & -1 \end{bmatrix} = 4 - 4 = 0$

61. (a) $\begin{bmatrix} 0 & 1 & 2 & \vdots & 0 & 1 \\ -3 & -2 & 1 & \vdots & -3 & -2 \\ 0 & 4 & 1 & \vdots & 0 & 4 \end{bmatrix}$

$$\begin{aligned} |A| &= 0(-2)(1) + 1(1)(0) + 2(-3)(4) - 0(-2)(2) - 4(1)(0) - 1(-3)(1) \\ &= -24 - (-3) \\ &= -21 \end{aligned}$$

(b) $\begin{bmatrix} 3 & -2 & 0 & \vdots & 3 & -2 \\ 1 & -1 & 2 & \vdots & 1 & -1 \\ 3 & 1 & 1 & \vdots & 3 & 1 \end{bmatrix}$

$$\begin{aligned} |B| &= 3(-1)(1) + (-2)(2)(3) + 0(1)(1) - 3(-1)(0) - 1(2)(3) - 1(1)(-2) \\ &= -3 - 12 - 6 + 2 \\ &= -19 \end{aligned}$$

(c) $AB = \begin{bmatrix} 0 & 1 & 2 \\ -3 & -2 & 1 \\ 0 & 4 & 1 \end{bmatrix}\begin{bmatrix} 3 & -2 & 0 \\ 1 & -1 & 2 \\ 3 & 1 & 1 \end{bmatrix} = \begin{bmatrix} 1 + 6 & -1 + 2 & 2 + 2 \\ -9 - 2 + 3 & 6 + 2 + 1 & -4 + 1 \\ 4 + 3 & -4 + 1 & 8 + 1 \end{bmatrix} = \begin{bmatrix} 7 & 1 & 4 \\ -8 & 9 & -3 \\ 7 & -3 & 9 \end{bmatrix}$

(d) $\begin{bmatrix} 7 & 1 & 4 & \vdots & 7 & 1 \\ -8 & 9 & -3 & \vdots & -8 & 9 \\ 7 & -3 & 9 & \vdots & 7 & -3 \end{bmatrix}$

$$\begin{aligned} |AB| &= 7(9)(9) + 1(-3)(7) + 4(-8)(-3) - 7(9)(4) - (-3)(-3)(7) - 9(-8)(1) \\ &= 567 - 21 + 96 - 252 - 63 + 72 \\ &= 399 \end{aligned}$$

63. (a) $\begin{bmatrix} -1 & 2 & 1 & \vdots & -1 & 2 \\ 1 & 0 & 1 & \vdots & 1 & 0 \\ 0 & 1 & 0 & \vdots & 0 & 1 \end{bmatrix}$

$$|A| = -1(0)(0) + 2(1)(0) + 1(1)(1) - 0(0)(1) - 1(1)(-1) - 0(1)(2) = 1 + 1 = 2$$

(b) $|B| = \begin{vmatrix} -1 & 0 & 0 \\ 0 & 2 & 0 \\ 0 & 0 & 3 \end{vmatrix} = -1(2)(3) = -6$

(c) $AB = \begin{bmatrix} -1 & 2 & 1 \\ 1 & 0 & 1 \\ 0 & 1 & 0 \end{bmatrix}\begin{bmatrix} -1 & 0 & 0 \\ 0 & 2 & 0 \\ 0 & 0 & 3 \end{bmatrix} = \begin{bmatrix} 1 & 4 & 3 \\ -1 & 0 & 3 \\ 0 & 2 & 0 \end{bmatrix}$

(d) $\begin{bmatrix} 1 & 4 & 3 & \vdots & 1 & 4 \\ -1 & 0 & 3 & \vdots & -1 & 0 \\ 0 & 2 & 0 & \vdots & 0 & 2 \end{bmatrix}$

$|AB| = 1(0)(0) + 4(3)(0) + 3(-1)(2) - 0(0)(3) - 2(3)(1) - 0(-1)(4) = -6 - 6 = -12$

65. $\begin{vmatrix} 9 & 0 \\ 0 & 5 \end{vmatrix} = 45$

The determinant is the product of the matrix's diagonal entries.

67. $\begin{vmatrix} 5 & 0 & 0 \\ 0 & -5 & 0 \\ 0 & 0 & -3 \end{vmatrix} = 75$

The determinant is the product of the matrix's diagonal entries.

69. The matrix is lower triangular.

$\begin{vmatrix} 2 & 0 & 0 \\ 4 & -3 & 0 \\ 6 & 5 & 1 \end{vmatrix} = 2(-3)(1) = -6$

71. The matrix is upper triangular.

$\begin{vmatrix} 2 & 3 & -1 & -1 \\ 0 & -1 & -3 & 5 \\ 0 & 0 & -2 & 7 \\ 0 & 0 & 0 & -4 \end{vmatrix} = 2(-1)(-2)(-4) = -16$

73. The matrix is upper triangular.

$\begin{vmatrix} -6 & 7 & 2 & 0 & 5 \\ 0 & -1 & 3 & 4 & -3 \\ 0 & 0 & -7 & 0 & 4 \\ 0 & 0 & 0 & -2 & 1 \\ 0 & 0 & 0 & 0 & -2 \end{vmatrix} = (-6)(-1)(-7)(-2)(-2) = -168$

75. Answers will vary. Sample answers:

Upper triangular $\begin{bmatrix} 1 & 3 & 4 & 10 \\ 0 & -2 & 8 & -3 \\ 0 & 0 & 3 & 7 \\ 0 & 0 & 0 & -1 \end{bmatrix}$ Lower triangular $\begin{bmatrix} -3 & 0 & 0 & 0 \\ 9 & 2 & 0 & 0 \\ 2 & -4 & 1 & 0 \\ 6 & 0 & 5 & -1 \end{bmatrix}$

The determinant is the product of the entries on the main diagonal, which in this case equals 6.

77. Answers will vary.

Sample answers:

Upper triangular $\begin{bmatrix} 3 & 5 & 11 & 7 \\ 0 & -2 & -3 & 9 \\ 0 & 0 & 1 & -8 \\ 0 & 0 & 0 & 3 \end{bmatrix}$ Lower triangular $\begin{bmatrix} -1 & 0 & 0 & 0 \\ 3 & -2 & 0 & 0 \\ 14 & -17 & -3 & 0 \\ -9 & 6 & 5 & 3 \end{bmatrix}$

The determinant is the product of the entries on the main diagonal, which in this case equals −18.

79. Answers will vary.

Sample answers:

Upper triangular $\begin{bmatrix} 2 & 5 & -11 & 20 \\ 0 & -7 & 17 & 3 \\ 0 & 0 & -1 & 9 \\ 0 & 0 & 0 & 2 \end{bmatrix}$ Lower triangular $\begin{bmatrix} 7 & 0 & 0 & 0 \\ 5 & -2 & 0 & 0 \\ 12 & 4 & -2 & 0 \\ -3 & 7 & -9 & 1 \end{bmatrix}$

The determinant is the product of the entries on the main diagonal, which in this case equals 28.

81. −20, Answers will vary. Sample answer: Product of upward and downward diagonals.

$\begin{vmatrix} 2 & 1 & 3 & \vdots & 2 & 1 \\ 7 & 3 & -2 & \vdots & 7 & 3 \\ 4 & 1 & 1 & \vdots & 4 & 1 \end{vmatrix} = 2(3)(1) + 1(-2)(4) + 3(7)(1) - [4(3)(3) + 1(-2)(2) + 1(7)(1)] = -20$

83. −18, Answers will vary.

Sample answer:

$$\begin{vmatrix} 3 & 0 & 0 \\ 4 & -2 & 0 \\ 5 & 4 & 3 \end{vmatrix} = 3(-2)(3) = -18$$

(Lower Triangular) Product of the diagonal.

85. Because the matrix is diagonal, the determinant equals the product of the diagonal entries.

$$\begin{vmatrix} 2 & 0 & 0 & 0 \\ 0 & 5 & 0 & 0 \\ 0 & 0 & 8 & 0 \\ 0 & 0 & 0 & 1 \end{vmatrix} = 2(5)(8)(1) = 80$$

87. Answers will vary. Sample answer:

$$\begin{vmatrix} w & x \\ y & z \end{vmatrix} = wz - yx \qquad -\begin{vmatrix} y & z \\ w & x \end{vmatrix} = -(yx - wz)$$
$$= wz - yx$$

89. Answers will vary. Sample answer:

$$\begin{vmatrix} w & x \\ y & z \end{vmatrix} = wz - yx \qquad \begin{vmatrix} w & x + cw \\ y & z + cy \end{vmatrix}$$
$$= w(z + cy) - y(x + cw)$$
$$= wz + wcy - yx - ycw$$
$$= wz - yx$$

91. $\begin{vmatrix} x & 2 \\ 1 & x \end{vmatrix} = 2$

$$x^2 - 2 = 2$$
$$x^2 = 4$$
$$x = \pm 2$$

93. $\begin{vmatrix} 2x & -3 \\ -2 & 2x \end{vmatrix} = 3$

$$4x^2 - 6 = 3$$
$$4x^2 = 9$$
$$x^2 = \frac{9}{4}$$
$$x = \pm\frac{3}{2}$$

95. $\begin{vmatrix} x & 1 \\ 2 & x - 2 \end{vmatrix} = -1$

$$x(x - 2) - 2 = -1$$
$$x^2 - 2x - 2 = -1$$
$$x^2 - 2x - 1 = 0$$
$$x = \frac{-(-2) \pm \sqrt{(-2)^2 - 4(1)(-1)}}{2(1)}$$
$$x = 1 \pm \sqrt{2}$$

97. $\begin{vmatrix} 2x & 1 \\ -1 & x - 1 \end{vmatrix} = x$

$$2x(x - 1) - (-1) = x$$
$$2x^2 - 2x + 1 = x$$
$$2x^2 - 3x + 1 = 0$$
$$(2x - 1)(x - 1) = 0$$
$$2x - 1 = 0 \qquad x - 1 = 0$$
$$x = \frac{1}{2} \qquad x = 1$$

99. $\begin{vmatrix} 1 & 2 & x \\ -1 & 3 & 2 \\ 3 & -2 & 1 \end{vmatrix} = 0$

Expanding using column 1:

$$1\begin{vmatrix} 3 & 2 \\ -2 & 1 \end{vmatrix} - (-1)\begin{vmatrix} 2 & x \\ -2 & 1 \end{vmatrix} + 3\begin{vmatrix} 2 & x \\ 3 & 2 \end{vmatrix} = 0$$
$$(3 + 4) + (2 + 2x) + 3(4 - 3x) = 0$$
$$7 + 2 + 2x + 12 - 9x = 0$$
$$21 - 7x = 0$$
$$-7x = -21$$
$$x = 3$$

101. Answers will vary. For an $n \times n$ matrix $(n > 2)$ with consecutive integer entries, the determinant appears to be zero.

103. True. If an entire row is zeros, then each cofactor in the expansion is multiplied by zero.

Section 6.5 Applications of Matrices and Determinants

Skills Warm Up

1. $\begin{vmatrix} 4 & 3 \\ -3 & -2 \end{vmatrix} = 4(-2) - 3(-3) = 1$

2. $\begin{vmatrix} 10 & -20 \\ -1 & 2 \end{vmatrix} = 10(2) - (-20)(-1) = 0$

3. $\begin{vmatrix} 4 & 0 \\ -3 & -2 \end{vmatrix} = 4(-2) - 0(-3) = -8$

4. $\begin{vmatrix} x & x^2 \\ 1 & 2x \end{vmatrix} = x(2x) - x^2(1) = x^2$

5. $\begin{vmatrix} 4 & 0 & -2 \\ 3 & 1 & 2 \\ -8 & 0 & 6 \end{vmatrix} = -0\begin{vmatrix} 3 & 2 \\ -8 & 6 \end{vmatrix} + 1\begin{vmatrix} 4 & -2 \\ -8 & 6 \end{vmatrix} - 0\begin{vmatrix} 4 & -2 \\ 3 & 2 \end{vmatrix}$

$= 0 + 1(8) - 0 = 8$

6. $\begin{vmatrix} 3 & 2 & 5 \\ 0 & 0 & -4 \\ -6 & 1 & 1 \end{vmatrix} = -0\begin{vmatrix} 2 & 5 \\ 1 & 1 \end{vmatrix} + 0\begin{vmatrix} 3 & 5 \\ -6 & 1 \end{vmatrix} - (-4)\begin{vmatrix} 3 & 2 \\ -6 & 1 \end{vmatrix}$

$= 0 - 0 + 4(15) = 60$

7. $A = \begin{bmatrix} 1 & 3 \\ 2 & 7 \end{bmatrix}$ and $A^{-1} = \begin{bmatrix} 7 & -3 \\ -2 & 1 \end{bmatrix}$

8. $A = \begin{bmatrix} 10 & 5 & -2 \\ -4 & -2 & 1 \\ 1 & 1 & 0 \end{bmatrix}$ and $A^{-1} = \begin{bmatrix} 1 & 2 & -1 \\ -1 & -2 & 2 \\ 2 & 5 & 0 \end{bmatrix}$

9. $\begin{bmatrix} 0.1 & 0.2 \\ 0.4 & 0.3 \end{bmatrix}\begin{bmatrix} 0.4 \\ 0.5 \end{bmatrix} = \begin{bmatrix} 0.04 + 0.1 \\ 0.16 + 0.15 \end{bmatrix} = \begin{bmatrix} 0.14 \\ 0.31 \end{bmatrix}$

10. $\begin{bmatrix} 2 & 5 \end{bmatrix}\begin{bmatrix} 1 & 2 \\ 1 & 2 \end{bmatrix} = \begin{bmatrix} 2+5 & 4+10 \end{bmatrix} = \begin{bmatrix} 7 & 14 \end{bmatrix}$

1. Vertices: $(0, 0), (1, 6), (4, 2)$

$$\text{Area} = \pm\tfrac{1}{2}\begin{vmatrix} 0 & 0 & 1 \\ 1 & 6 & 1 \\ 4 & 2 & 1 \end{vmatrix} = \pm\tfrac{1}{2}\begin{bmatrix} 1 & 6 \\ 4 & 2 \end{bmatrix} = 11 \text{ square units}$$

3. Vertices: $(-3, -3), (1, 5), (4, -3)$

$$\text{Area} = \pm\tfrac{1}{2}\begin{vmatrix} -3 & -3 & 1 \\ 1 & 5 & 1 \\ 4 & -3 & 1 \end{vmatrix} = \pm\tfrac{1}{2}\left(\begin{vmatrix} 1 & 5 \\ 4 & -3 \end{vmatrix} - \begin{vmatrix} -3 & -3 \\ 4 & -3 \end{vmatrix} + \begin{vmatrix} -3 & -3 \\ 1 & 5 \end{vmatrix}\right) = 28 \text{ square units}$$

5. Vertices: $(0, \frac{1}{2}), (\frac{5}{2}, 0), (4, 3)$

$$\text{Area} = \pm\tfrac{1}{2}\begin{vmatrix} 0 & \frac{1}{2} & 1 \\ \frac{5}{2} & 0 & 1 \\ 4 & 3 & 1 \end{vmatrix} = \pm\tfrac{1}{2}\left(1\begin{vmatrix} \frac{5}{2} & 0 \\ 4 & 3 \end{vmatrix} - 1\begin{vmatrix} 0 & \frac{1}{2} \\ 4 & 3 \end{vmatrix} + 1\begin{vmatrix} 0 & \frac{1}{2} \\ \frac{5}{2} & 0 \end{vmatrix}\right) = \pm\tfrac{1}{2}\left(\tfrac{15}{2} + 2 - \tfrac{5}{4}\right) = \tfrac{33}{8} \text{ square units}$$

7. Vertices: $(-2, 4), (2, 3), (-1, 5)$

$$\text{Area} = \pm\tfrac{1}{2}\begin{vmatrix} -2 & 4 & 1 \\ 2 & 3 & 1 \\ -1 & 5 & 1 \end{vmatrix} = \pm\tfrac{1}{2}\begin{vmatrix} -2 & 4 & 1 \\ 4 & -1 & 0 \\ 1 & 1 & 0 \end{vmatrix} = \pm\tfrac{1}{2}\begin{vmatrix} 4 & -1 \\ 1 & 1 \end{vmatrix} = \tfrac{5}{2} \text{ square units}$$

9. Vertices: $(-3, 5), (2, 6), (3, -5)$

$$\text{Area} = \pm\tfrac{1}{2}\begin{vmatrix} -3 & 5 & 1 \\ 2 & 6 & 1 \\ 3 & -5 & 1 \end{vmatrix} = \pm\tfrac{1}{2}\begin{vmatrix} -3 & 5 & 1 \\ 5 & 1 & 0 \\ 6 & -10 & 0 \end{vmatrix} = \pm\tfrac{1}{2}\begin{vmatrix} 5 & 1 \\ 6 & -10 \end{vmatrix} = 28 \text{ square units}$$

11. $4 = \pm\frac{1}{2}\begin{vmatrix} -5 & 1 & 1 \\ 0 & 2 & 1 \\ -2 & y & 1 \end{vmatrix}$

$$\pm 8 = -5\begin{vmatrix} 2 & 1 \\ y & 1 \end{vmatrix} + (-2)\begin{vmatrix} 1 & 1 \\ 2 & 1 \end{vmatrix}$$
$$\pm 8 = -5(2 - y) - 2(-1)$$
$$\pm 8 = 5y - 8$$
$$y = \frac{8 \pm 8}{5}$$
$$y = \frac{16}{5} \text{ or } y = 0$$

13. $6 = \pm\frac{1}{2}\begin{vmatrix} -2 & -3 & 1 \\ 1 & -1 & 1 \\ -8 & y & 1 \end{vmatrix}$

$$\pm 12 = -2\begin{vmatrix} -1 & 1 \\ y & 1 \end{vmatrix} - \begin{vmatrix} -3 & 1 \\ y & 1 \end{vmatrix} + (-8)\begin{vmatrix} -3 & 1 \\ -1 & 1 \end{vmatrix}$$
$$\pm 12 = -2(-1 - y) - (-3 - y) - 8(-2)$$
$$\pm 12 = 3y + 21$$
$$y = -\frac{21 \pm 12}{3}$$
$$y = -11 \text{ or } y = -3$$

15. $A(0, 30)$, $B(15, 0)$, $C(33, 5)$

$$\text{Area} = \pm\frac{1}{2}\begin{vmatrix} 0 & 30 & 1 \\ 15 & 0 & 1 \\ 33 & 5 & 1 \end{vmatrix} = \pm\frac{1}{2}(615) = 307.5$$

The area of the region is 307.5 square miles.

17. Points: $(-4, -7)$, $(0, -4)$, $(4, -1)$

$$\begin{vmatrix} -4 & -7 & 1 \\ 0 & -4 & 1 \\ 4 & -1 & 1 \end{vmatrix} = -4\begin{vmatrix} -4 & 1 \\ 4 & 1 \end{vmatrix} - \begin{vmatrix} -4 & -7 \\ 4 & -1 \end{vmatrix}$$
$$= -4(-8) - 32 = 0$$

The points are collinear.

19. Points: $(-1, -7)$, $(0, -3)$, $(1, 2)$

$$\begin{vmatrix} -1 & -7 & 1 \\ 0 & -3 & 1 \\ 1 & 2 & 1 \end{vmatrix} = -3\begin{vmatrix} -1 & 1 \\ 1 & 1 \end{vmatrix} - \begin{vmatrix} -1 & -7 \\ 1 & 2 \end{vmatrix}$$
$$= -3(-2) - 5 = -11 \neq 0$$

The points are not collinear.

21. Points: $(-2, -11)$, $(4, 13)$, $(2, 5)$

$$\begin{vmatrix} -2 & -11 & 1 \\ 4 & 13 & 1 \\ 2 & 5 & 1 \end{vmatrix} = -2\begin{vmatrix} 13 & 1 \\ 5 & 1 \end{vmatrix} - (-11)\begin{vmatrix} 4 & 1 \\ 2 & 1 \end{vmatrix} + \begin{vmatrix} 4 & 13 \\ 2 & 5 \end{vmatrix}$$
$$= -2(8) + 11(2) + (-6) = 0$$

The points are collinear.

23. Points: $(-2, 3)$, $(2, -1)$, $(7, -4)$

$$\begin{vmatrix} -2 & 3 & 1 \\ 2 & -1 & 1 \\ 7 & -4 & 1 \end{vmatrix} = -2\begin{vmatrix} -1 & 1 \\ -4 & 1 \end{vmatrix} - 3\begin{vmatrix} 2 & 1 \\ 7 & 1 \end{vmatrix} + 1\begin{vmatrix} 2 & -1 \\ 7 & -4 \end{vmatrix}$$
$$= -2(3) - 3(-5) + 1(-1) = 8 \neq 0$$

The points are not collinear.

25. Points: $\left(\frac{1}{2}, 3\right)$, $\left(-\frac{3}{2}, -5\right)$, $(0, 1)$

$$\begin{vmatrix} \frac{1}{2} & 3 & 1 \\ -\frac{3}{2} & -5 & 1 \\ 0 & 1 & 1 \end{vmatrix} = \frac{1}{2}\begin{vmatrix} -5 & 1 \\ 1 & 1 \end{vmatrix} - \left(-\frac{3}{2}\right)\begin{vmatrix} 3 & 1 \\ 1 & 1 \end{vmatrix} + 0\begin{vmatrix} 3 & 1 \\ -5 & 1 \end{vmatrix}$$
$$= \frac{1}{2}(-5 - 1) + \frac{3}{2}(3 - 1)$$
$$= \frac{1}{2}(-6) + \frac{3}{2}(2) = -3 + 3 = 0$$

The points are collinear.

27. $\begin{vmatrix} 2 & -5 & 1 \\ 3 & y & 1 \\ 5 & -2 & 1 \end{vmatrix} = 0 \Rightarrow 2\begin{vmatrix} y & 1 \\ -2 & 1 \end{vmatrix} - (-5)\begin{vmatrix} 3 & 1 \\ 5 & 1 \end{vmatrix} + \begin{vmatrix} 3 & y \\ 5 & -2 \end{vmatrix} = 0$

$$2(y + 2) + 5(-2) + (-6 - 5y) = 0$$
$$-3y - 12 = 0$$
$$y = -4$$

29. $\begin{vmatrix} -4 & -1 & 1 \\ -1 & 2 & 1 \\ x & 6 & 1 \end{vmatrix} = 0 \Rightarrow -4\begin{vmatrix} 2 & 1 \\ 6 & 1 \end{vmatrix} - (-1)\begin{vmatrix} -1 & 1 \\ x & 1 \end{vmatrix} + 1\begin{vmatrix} -1 & 2 \\ x & 6 \end{vmatrix} = 0$

$$-4(-4) + (-1 - x) + (-6 - 2x) = 0$$
$$-3x + 9 = 0$$
$$x = 3$$

31. Points: $(-1, 2), (5, 3)$

$$\begin{vmatrix} x & y & 1 \\ -1 & 2 & 1 \\ 5 & 3 & 1 \end{vmatrix} = 2x + 5y - 3 - 10 + y - 3x = 0 \text{ OR } 6y - x - 13 = 0 \text{ OR } x - 6y + 13 = 0$$

33. Points: $(-4, 3), (2, 1)$

$$\begin{vmatrix} x & y & 1 \\ -4 & 3 & 1 \\ 2 & 1 & 1 \end{vmatrix} = 3x + 2y - 4 - 6 + 4y - x = 0 \text{ OR } x + 3y - 5 = 0$$

35. Points: $(0, 5), (2, 5)$

$$\begin{vmatrix} x & y & 1 \\ 0 & 5 & 1 \\ 2 & 5 & 1 \end{vmatrix} = x\begin{vmatrix} 5 & 1 \\ 5 & 1 \end{vmatrix} - 0\begin{vmatrix} y & 1 \\ 5 & 1 \end{vmatrix} + 2\begin{vmatrix} y & 1 \\ 5 & 1 \end{vmatrix} = 0$$

$$\begin{aligned} x(5 - 5) + 0 + 2(y - 5) &= 0 \\ 2y - 10 &= 0 \\ y &= 5 \end{aligned}$$

37. Points: $(-4, 5), (-4, -2)$

$$\begin{vmatrix} x & y & 1 \\ -4 & 5 & 1 \\ -4 & -2 & 1 \end{vmatrix} = 5x - 4y + 8 + 20 + 4y + 2x = 0 \text{ OR } x = -4$$

39. Points: $\left(-\frac{1}{2}, 3\right), \left(\frac{5}{2}, 1\right)$

$$\text{Equation: } \begin{vmatrix} x & y & 1 \\ -\frac{1}{2} & 3 & 1 \\ \frac{5}{2} & 1 & 1 \end{vmatrix} = 3x + \frac{5}{2}y - \frac{1}{2} - \frac{15}{2} + \frac{1}{2}y - x = 0 \text{ OR } 2x + 3y - 8 = 0$$

41. The uncoded 1×2 row matrices are the rows of the 7×2 matrix on the left.

$$\begin{matrix} \text{C} & \text{O} \\ \text{M} & \text{E} \\ & \text{H} \\ \text{O} & \text{M} \\ \text{E} & \\ \text{S} & \text{O} \\ \text{O} & \text{N} \end{matrix} \begin{bmatrix} 3 & 15 \\ 13 & 5 \\ 0 & 8 \\ 15 & 13 \\ 5 & 0 \\ 19 & 15 \\ 15 & 14 \end{bmatrix} \begin{bmatrix} 1 & 2 \\ 3 & 5 \end{bmatrix} = \begin{bmatrix} 48 & 81 \\ 28 & 51 \\ 24 & 40 \\ 54 & 95 \\ 5 & 10 \\ 64 & 113 \\ 57 & 100 \end{bmatrix}$$

Answer: 48, 81, 28, 51, 24, 40, 54, 95, 5, 10, 64, 113, 57, 100

43. $[20\ 5\ 24][20\ 0\ 13][5\ 0\ 1][20\ 0\ 23]$ $[15\ 18\ 11]$

$$\begin{bmatrix} 20 & 5 & 24 \\ 20 & 0 & 13 \\ 5 & 0 & 1 \\ 20 & 0 & 23 \\ 15 & 18 & 11 \end{bmatrix} \begin{bmatrix} 1 & -1 & 0 \\ 1 & 0 & -1 \\ -6 & 2 & 3 \end{bmatrix} = \begin{bmatrix} -119 & 28 & 67 \\ -58 & 6 & 39 \\ -1 & -3 & 3 \\ -118 & 26 & 69 \\ -33 & 7 & 15 \end{bmatrix}$$

Answer: −119, 28, 67, −58, 6, 39, −1, −3, 3, −118, 26, 69, −33, 7, 15

45. L A N D I N G _ S U C C E S S F U L

[12 1 14] [4 9 14] [7 0 19] [21 3 3] [5 19 19] [6 21 12]

$[12\ \ 1\ \ 14]A = [1\ \ -25\ \ -65]$

$[4\ \ 9\ \ 14]A = [17\ \ 15\ \ -9]$

$[7\ \ 0\ \ 19]A = [-12\ \ -62\ \ -119]$

$[21\ \ 3\ \ 3]A = [27\ \ 51\ \ 48]$

$[5\ \ 19\ \ 19]A = [43\ \ 67\ \ 48]$

$[6\ \ 21\ \ 12]A = [57\ \ 111\ \ 117]$

Cryptogram: 1, –25, –65, 17, 15, –9, –12, –62, –119, 27, 51, 48, 43, 67, 48, 57, 111, 117

47. S E E _ Y O U _ A T _ M O V I E S _

[19 5 5] [0 25 15] [21 0 1] [20 0 13] [15 22 9] [5 19 0]

$[19\ \ 5\ \ 5]A = [29\ \ 53\ \ 48]$

$[0\ \ 25\ \ 15]A = [60\ \ 115\ \ 120]$

$[21\ \ 0\ \ 1]A = [20\ \ 38\ \ 35]$

$[20\ \ 0\ \ 13]A = [7\ \ -12\ \ -51]$

$[15\ \ 22\ \ 9]A = [72\ \ 148\ \ 165]$

$[5\ \ 19\ \ 0]A = [62\ \ 143\ \ 181]$

Cryptogram: 29, 53, 48, 60, 115, 120, 20, 38, 35, 7, –12, –51, 72, 148, 165, 62, 143, 181

49. C O N G R A T U L A T I O N S

[3 15 14] [7 18 1] [20 21 12] [1 20 9] [15 14 19]

$[3\ \ 15\ \ 14]A = [34\ \ 55\ \ 43]$

$[7\ \ 18\ \ 1]A = [60\ \ 136\ \ 169]$

$[20\ \ 21\ \ 12]A = [71\ \ 139\ \ 145]$

$[1\ \ 20\ \ 9]A = [52\ \ 106\ \ 119]$

$[15\ \ 14\ \ 19]A = [38\ \ 52\ \ 23]$

Cryptogram: 34, 55, 43, 60, 136, 169, 71, 139, 145, 52, 106, 119, 38, 52, 23

51. P I C K _ M E _ U P _ A T _ N O O N

[16 9 3] [11 0 13] [5 0 21] [16 0 1] [20 0 14] [15 15 14]

$[16\ \ 9\ \ 3]A = [40\ \ 83\ \ 92]$

$[11\ \ 0\ \ 13]A = [-2\ \ -30\ \ -69]$

$[5\ \ 0\ \ 21]A = [-16\ \ -74\ \ -137]$

$[16\ \ 0\ \ 1]A = [15\ \ 28\ \ 25]$

$[20\ \ 0\ \ 14]A = [6\ \ -16\ \ -58]$

$[15\ \ 14\ \ 14]A = [46\ \ 79\ \ 67]$

Cryptogram: 40, 83, 92, –2, –30, –69, –16, –74, –137, 15, 28, 25, 6, –16, –58, 46, 79, 67

53. $A^{-1} = \begin{bmatrix}1 & 2\\ 3 & 5\end{bmatrix}^{-1} = \begin{bmatrix}-5 & 2\\ 3 & -1\end{bmatrix}$

$$\begin{bmatrix}11 & 21\\ 64 & 112\\ 25 & 50\\ 29 & 53\\ 23 & 46\\ 40 & 75\\ 55 & 92\end{bmatrix}\begin{bmatrix}-5 & 2\\ 3 & -1\end{bmatrix} = \begin{bmatrix}8 & 1\\ 16 & 16\\ 25 & 0\\ 14 & 5\\ 23 & 0\\ 25 & 5\\ 1 & 18\end{bmatrix}\begin{matrix}H & A\\ P & P\\ Y & \\ N & E\\ W & \\ Y & E\\ A & R\end{matrix}$$

Message: HAPPY NEW YEAR

55. $A^{-1} = \begin{bmatrix}4 & 2 & 1\\ -3 & -3 & -1\\ 3 & 2 & 1\end{bmatrix}^{-1} = \begin{bmatrix}1 & 0 & -1\\ 0 & -1 & -1\\ -3 & 2 & 6\end{bmatrix}$

$$\begin{bmatrix}94 & 35 & 25\\ 44 & 16 & 10\\ 4 & -10 & 1\\ 27 & 15 & 9\\ 71 & 43 & 22\end{bmatrix}\begin{bmatrix}1 & 0 & -1\\ 0 & -1 & -1\\ -3 & 2 & 6\end{bmatrix} = \begin{bmatrix}19 & 15 & 21\\ 14 & 4 & 0\\ 1 & 12 & 12\\ 0 & 3 & 12\\ 5 & 1 & 18\end{bmatrix}\begin{matrix}S & O & U\\ N & D & \\ A & L & L\\ & C & L\\ E & A & R\end{matrix}$$

Message: SOUND ALL CLEAR

57. $A^{-1} = \begin{bmatrix}1 & 2 & 2\\ 3 & 7 & 9\\ -1 & -4 & -7\end{bmatrix}^{-1} = \begin{bmatrix}-13 & 6 & 4\\ 12 & -5 & -3\\ -5 & 2 & 1\end{bmatrix}$

$$\begin{bmatrix}20 & 17 & -15\\ -9 & -44 & -83\\ 64 & 136 & 157\\ 24 & 31 & 12\\ 4 & -37 & -102\end{bmatrix}\begin{bmatrix}-13 & 6 & 4\\ 12 & -5 & -3\\ -5 & 2 & 1\end{bmatrix} = \begin{bmatrix}19 & 5 & 14\\ 4 & 0 & 13\\ 15 & 18 & 5\\ 0 & 13 & 15\\ 14 & 5 & 25\end{bmatrix}\begin{matrix}S & E & N\\ D & & M\\ O & R & E\\ & M & O\\ N & E & Y\end{matrix}$$

Message: SEND MORE MONEY

59. Let A be the 2×2 matrix needed to decode the message.

$$\begin{bmatrix}-18 & -18\\ 1 & 16\end{bmatrix}A = \begin{bmatrix}0 & 18\\ 15 & 14\end{bmatrix}\begin{matrix} & R\\ O & N\end{matrix}$$

$$A = \begin{bmatrix}-18 & -18\\ 1 & 16\end{bmatrix}^{-1}\begin{bmatrix}0 & 18\\ 15 & 14\end{bmatrix}$$

$$= \begin{bmatrix}-\frac{8}{135} & -\frac{1}{15}\\ \frac{1}{270} & \frac{1}{15}\end{bmatrix}\begin{bmatrix}0 & 18\\ 15 & 14\end{bmatrix}$$

$$= \begin{bmatrix}-1 & -2\\ 1 & 1\end{bmatrix}$$

$$\begin{bmatrix}8 & 21\\ -15 & -10\\ -13 & -13\\ 5 & 10\\ 5 & 25\\ 5 & 19\\ -1 & 6\\ 20 & 40\\ -18 & -18\\ 1 & 16\end{bmatrix}\begin{bmatrix}-1 & -2\\ 1 & 1\end{bmatrix} = \begin{bmatrix}13 & 5\\ 5 & 20\\ 0 & 13\\ 5 & 0\\ 20 & 15\\ 14 & 9\\ 7 & 8\\ 20 & 0\\ 0 & 18\\ 15 & 14\end{bmatrix}\begin{matrix}M & E\\ E & T\\ & M\\ E & \\ T & O\\ N & I\\ G & H\\ T & \\ & R\\ O & N\end{matrix}$$

Message: MEET ME TONIGHT RON

61. $\begin{bmatrix}45 & -35\end{bmatrix}\begin{bmatrix}w & x\\ y & z\end{bmatrix} = \begin{bmatrix}10 & 15\end{bmatrix}$ and

$\begin{bmatrix}38 & -30\end{bmatrix}\begin{bmatrix}w & x\\ y & z\end{bmatrix} = \begin{bmatrix}8 & 14\end{bmatrix}$

So, we have $\begin{bmatrix}45 & -35\\ 38 & -30\end{bmatrix}\begin{bmatrix}w & x\\ y & z\end{bmatrix} = \begin{bmatrix}10 & 15\\ 8 & 14\end{bmatrix}$.

So, $\begin{aligned}45w - 35y &= 10\\ 38w - 30y &= 8\end{aligned}$ and $\begin{aligned}45x - 35z &= 15\\ 38x - 30z &= 14\end{aligned}$.

Using techniques from Chapter 5 or matrix methods from the chapter, $w = 1, y = 1, x = -2$, and $z = -3$.

Then the message matrix times A^{-1} is

$$\begin{bmatrix}45 & -35\\ 38 & -30\\ 18 & -18\\ 35 & -30\\ 81 & -60\\ 42 & -28\\ 75 & -55\\ 2 & -2\\ 22 & -21\\ 15 & -10\end{bmatrix}\begin{bmatrix}1 & -2\\ 1 & -3\end{bmatrix} = \begin{bmatrix}10 & 15\\ 8 & 14\\ 0 & 18\\ 5 & 20\\ 21 & 18\\ 14 & 0\\ 20 & 15\\ 0 & 2\\ 1 & 19\\ 5 & 0\end{bmatrix}\begin{matrix}J & O\\ H & N\\ & R\\ E & T\\ U & R\\ N & \\ T & O\\ & B\\ A & S\\ E & \end{matrix}$$

Message: JOHN RETURN TO BASE

63. Area $= \pm\frac{1}{2}\begin{vmatrix} x_1 & y_1 & 1 \\ x_2 & y_2 & 1 \\ x_3 & y_3 & 1 \end{vmatrix} = \pm\frac{1}{2}\begin{vmatrix} 3 & -1 & 1 \\ 7 & -1 & 1 \\ 7 & 5 & 1 \end{vmatrix} = \frac{1}{2}(24) = 12$ square units

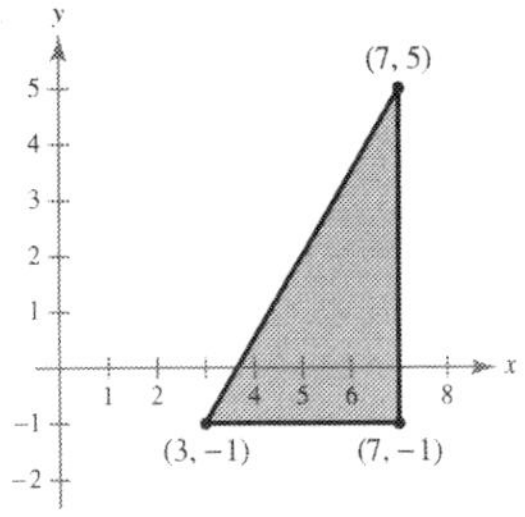

$A = \frac{1}{2}(4)(6) = 12$ square units

65. Answers will vary.

Review Exercises for Chapter 6

1. 1×2

3. 2×4

5. $\begin{bmatrix} 1 & 3 & 0 & 2 \\ 3 & 10 & 1 & 8 \\ 2 & 3 & 3 & 10 \end{bmatrix}$ $\begin{matrix} \\ -3R_1 + R_2 \\ -2R_1 + R_3 \end{matrix}\begin{bmatrix} 1 & 3 & 0 & 2 \\ 0 & 1 & 1 & 2 \\ 0 & -3 & 3 & 6 \end{bmatrix}$

$\begin{matrix} \\ \\ 3R_2 + R_3 \end{matrix}\begin{bmatrix} 1 & 3 & 0 & 2 \\ 0 & 1 & 1 & 2 \\ 0 & 0 & 6 & 12 \end{bmatrix}$

$\begin{matrix} \\ \\ \frac{1}{6}R_3 \end{matrix}\begin{bmatrix} 1 & 3 & 0 & 2 \\ 0 & 1 & 1 & 2 \\ 0 & 0 & 1 & 2 \end{bmatrix}$

7. $\begin{bmatrix} 1 & 2 & 3 \\ -2 & 0 & 2 \\ 2 & 1 & 2 \end{bmatrix}$ $\begin{matrix} \\ 2R_1 + R_2 \\ -2R_1 + R_3 \end{matrix}\begin{bmatrix} 1 & 2 & 3 \\ 0 & 4 & 8 \\ 0 & -3 & -4 \end{bmatrix}$

$\begin{matrix} -2R_1 + R_2 \\ \frac{1}{4}R_2 \\ \frac{3}{4}R_2 + R_3 \end{matrix}\begin{bmatrix} 1 & 0 & 2 \\ 0 & 1 & 2 \\ 0 & 0 & 2 \end{bmatrix}$

$\begin{matrix} R_1 - R_3 \\ R_2 - R_3 \\ \frac{1}{2}R_3 \end{matrix}\begin{bmatrix} 1 & 0 & 0 \\ 0 & 1 & 0 \\ 0 & 0 & 1 \end{bmatrix}$

9. $\begin{cases} 4x - 3y = 18 \\ x + y = 1 \end{cases}$

$\begin{matrix} R_2 \leftrightarrow R_1 \\ R_1 \leftrightarrow R_2 \end{matrix}\begin{bmatrix} 1 & 1 & \vdots & 1 \\ 4 & -3 & \vdots & 18 \end{bmatrix} \Rightarrow \begin{matrix} \\ -4R_1 + R_2 \end{matrix}\begin{bmatrix} 1 & 1 & \vdots & 1 \\ 0 & -7 & \vdots & 14 \end{bmatrix}$

$-7y = 14 \Rightarrow y = -2$

$x + (-2) = 1 \Rightarrow x = 3$

Answer: $(3, -2)$

11. $\begin{cases} 2x + 3y - z = 13 \\ 3x \qquad + z = 8 \\ x - 2y + 3z = -4 \end{cases}$

$\begin{matrix} R_3 \to R_1 \\ R_1 \to R_2 \\ R_2 \to R_3 \end{matrix}\begin{bmatrix} 1 & -2 & 3 & \vdots & -4 \\ 2 & 3 & -1 & \vdots & 13 \\ 3 & 0 & 1 & \vdots & 8 \end{bmatrix} \Rightarrow \begin{matrix} \\ -2R_1 + R_2 \\ -3R_1 + R_3 \end{matrix}\begin{bmatrix} 1 & -2 & 3 & \vdots & -4 \\ 0 & 7 & -7 & \vdots & 21 \\ 0 & 6 & -8 & \vdots & 20 \end{bmatrix} \Rightarrow \begin{matrix} \\ \frac{1}{7}R_2 \\ -\frac{6}{7}R_2 + R_3 \end{matrix}\begin{bmatrix} 1 & -2 & 3 & \vdots & -4 \\ 0 & 1 & -1 & \vdots & 3 \\ 0 & 0 & -2 & \vdots & 2 \end{bmatrix}$

$\Rightarrow \begin{matrix} 2R_2 + R_1 \\ \\ \\ \end{matrix}\begin{bmatrix} 1 & 0 & 1 & \vdots & 2 \\ 0 & 1 & -1 & \vdots & 3 \\ 0 & 0 & -2 & \vdots & 2 \end{bmatrix} \Rightarrow \begin{matrix} \frac{1}{2}R_3 + R_1 \\ -\frac{1}{2}R_3 + R_2 \\ -\frac{1}{2}R_3 \end{matrix}\begin{bmatrix} 1 & 0 & 0 & \vdots & 3 \\ 0 & 1 & 0 & \vdots & 2 \\ 0 & 0 & 1 & \vdots & -1 \end{bmatrix}$

$x = 3$

$y = 2$

$z = -1$

Answer: $(3, 2, -1)$

13. $\begin{cases} x + 2y + 2z = 10 \\ 2x + 3y + 5z = 20 \end{cases}$

$$\begin{bmatrix} 1 & 2 & 2 & \vdots & 10 \\ 2 & 3 & 5 & \vdots & 20 \end{bmatrix} \Rightarrow \begin{matrix} \\ -2R_1 + R_2 \end{matrix}\begin{bmatrix} 1 & 2 & 2 & \vdots & 10 \\ 0 & -1 & 1 & \vdots & 0 \end{bmatrix}$$

$$\Rightarrow \begin{matrix} 2R_2 + R_1 \\ -1R_2 \end{matrix}\begin{bmatrix} 1 & 0 & 4 & \vdots & 10 \\ 0 & 1 & -1 & \vdots & 0 \end{bmatrix}$$

$$x + 4z = 10$$
$$y - z = 0$$

Let $z = a$. Then $y = a$ and $x = 10 - 4a$, for all a.

Answer: $(10 - 4a,\ a,\ a)$, for all a

15. $\begin{cases} 2x + y - 3z = 4 \\ x + 2y + 2z = 10 \\ x \qquad - 2z = 12 \\ x + y + z = 6 \end{cases}$

$$\begin{matrix} R_4 \to R_1 \\ R_1 \to R_2 \\ R_2 \to R_3 \\ R_3 \to R_4 \end{matrix}\begin{bmatrix} 1 & 1 & 1 & \vdots & 6 \\ 2 & 1 & -3 & \vdots & 4 \\ 1 & 2 & 2 & \vdots & 10 \\ 1 & 0 & -2 & \vdots & 12 \end{bmatrix} \Rightarrow \begin{matrix} \\ -2R_1 + R_2 \\ R_3 - R_4 \\ \\ \end{matrix}\begin{bmatrix} 1 & 1 & 1 & \vdots & 6 \\ 0 & -1 & -5 & \vdots & -8 \\ 0 & 2 & 4 & \vdots & -2 \\ 1 & 0 & -2 & \vdots & 12 \end{bmatrix}$$

$$\Rightarrow \begin{matrix} \\ -R_2 \\ 2R_2 + R_3 \\ R_4 - R_1 \end{matrix}\begin{bmatrix} 1 & 1 & 1 & \vdots & 6 \\ 0 & 1 & 5 & \vdots & 8 \\ 0 & 0 & -6 & \vdots & -18 \\ 0 & -1 & -3 & \vdots & 6 \end{bmatrix}$$

$$\Rightarrow \begin{matrix} \\ \\ -\frac{1}{6}R_3 \\ R_4 + R_2 \end{matrix}\begin{bmatrix} 1 & 1 & 1 & \vdots & 6 \\ 0 & 1 & 5 & \vdots & 8 \\ 0 & 0 & 1 & \vdots & 3 \\ 0 & 0 & 2 & \vdots & 14 \end{bmatrix}$$

$$z = 3 \text{ and } 2z = 14$$
$$z = 7$$

Inconsistent

17. Let x = the amount borrowed at 8%, y = the amount borrowed at 10%, and z = the amount borrowed at 12%.

$$\begin{cases} x + y + z = 200{,}000 \\ 0.08x + 0.1y + 0.12z = 20{,}000 \\ -3x + y = 0 \end{cases}$$

$$\begin{bmatrix} 1 & 1 & 1 & \vdots & 200{,}000 \\ 0.08 & 0.1 & 0.12 & \vdots & 20{,}000 \\ -3 & 1 & 0 & \vdots & 0 \end{bmatrix} \Rightarrow 50R_2 - 4R_1 \begin{bmatrix} 1 & 1 & 1 & \vdots & 200{,}000 \\ 0 & 1 & 2 & \vdots & 200{,}000 \\ -3 & 1 & 0 & \vdots & 0 \end{bmatrix}$$

$$\Rightarrow \begin{matrix} \\ \\ R_3 + 3R_1 \end{matrix} \begin{bmatrix} 1 & 1 & 1 & \vdots & 200{,}000 \\ 0 & 1 & 2 & \vdots & 200{,}000 \\ 0 & 4 & 3 & \vdots & 600{,}000 \end{bmatrix}$$

$$\Rightarrow \begin{matrix} \\ \\ R_3 - 4R_2 \end{matrix} \begin{bmatrix} 1 & 1 & 1 & \vdots & 200{,}000 \\ 0 & 1 & 2 & \vdots & 200{,}000 \\ 0 & 0 & -5 & \vdots & -200{,}000 \end{bmatrix}$$

$$\Rightarrow \begin{matrix} \\ \\ -\frac{1}{5}R_3 \end{matrix} \begin{bmatrix} 1 & 1 & 1 & \vdots & 200{,}000 \\ 0 & 1 & 2 & \vdots & 200{,}000 \\ 0 & 0 & 1 & \vdots & 40{,}000 \end{bmatrix}$$

So,

$x + y + z = 200{,}000$, $y + 2z = 200{,}000$, and $z = 40{,}000$.

$y + 2z = 200{,}000 \Rightarrow y + 2(40{,}000) = 200{,}000 \Rightarrow y = 120{,}000$

$x + y + z = 200{,}000 \Rightarrow x + 120{,}000 + 40{,}000 = 200{,}000 \Rightarrow x = 40{,}000$

\$40,000 was borrowed at 8%, \$120,000 was borrowed at 10%, and \$40,000 was borrowed at 12%.

19. $\begin{bmatrix} -7 & -5 \\ 2 & -1 \end{bmatrix} = \begin{bmatrix} -7 & x + 1 \\ y & -1 \end{bmatrix}$

$y = 2$ $\qquad x + 1 = -5$

$\qquad x = -6$

21. (a) $A + B = \begin{bmatrix} -1 & 5 \\ 2 & 1 \end{bmatrix} + \begin{bmatrix} 4 & 2 \\ -6 & 3 \end{bmatrix} = \begin{bmatrix} 3 & 7 \\ -4 & 4 \end{bmatrix}$

(b) $A - B = \begin{bmatrix} -1 & 5 \\ 2 & 1 \end{bmatrix} - \begin{bmatrix} 4 & 2 \\ -6 & 3 \end{bmatrix} = \begin{bmatrix} -5 & 3 \\ 8 & -2 \end{bmatrix}$

(c) $4A = 4\begin{bmatrix} -1 & 5 \\ 2 & 1 \end{bmatrix} = \begin{bmatrix} -4 & 20 \\ 8 & 4 \end{bmatrix}$

(d) $4A - 3B = \begin{bmatrix} -4 & 20 \\ 8 & 4 \end{bmatrix} - \begin{bmatrix} 12 & 6 \\ -18 & 9 \end{bmatrix} = \begin{bmatrix} -16 & 14 \\ 26 & -5 \end{bmatrix}$

23. $A = \begin{bmatrix} 1 & -4 \\ 0 & -1 \\ 4 & 2 \end{bmatrix}$, $B = \begin{bmatrix} 6 & 5 \\ 3 & 7 \\ -1 & 4 \end{bmatrix}$

(a) $A + B = \begin{bmatrix} 1 & -4 \\ 0 & -1 \\ 4 & 2 \end{bmatrix} + \begin{bmatrix} 6 & 5 \\ 3 & 7 \\ -1 & 4 \end{bmatrix} = \begin{bmatrix} 1+6 & -4+5 \\ 0+3 & -1+7 \\ 4+(-1) & 2+4 \end{bmatrix} = \begin{bmatrix} 7 & 1 \\ 3 & 6 \\ 3 & 6 \end{bmatrix}$

(b) $A - B = \begin{bmatrix} 1 & -4 \\ 0 & -1 \\ 4 & 2 \end{bmatrix} - \begin{bmatrix} 6 & 5 \\ 3 & 7 \\ -1 & 4 \end{bmatrix} = \begin{bmatrix} 1-6 & -4-5 \\ 0-3 & -1-7 \\ 4-(-1) & 2-4 \end{bmatrix} = \begin{bmatrix} -5 & -9 \\ -3 & -8 \\ 5 & -2 \end{bmatrix}$

(c) $4A = 4\begin{bmatrix} 1 & -4 \\ 0 & -1 \\ 4 & 2 \end{bmatrix} = \begin{bmatrix} 4 & -16 \\ 0 & -4 \\ 16 & 8 \end{bmatrix}$

(d) $5A - 2B = 5\begin{bmatrix} 1 & -4 \\ 0 & -1 \\ 4 & 2 \end{bmatrix} - 2\begin{bmatrix} 6 & 5 \\ 3 & 7 \\ -1 & 4 \end{bmatrix} = \begin{bmatrix} 5 & -20 \\ 0 & -5 \\ 20 & 10 \end{bmatrix} - \begin{bmatrix} 12 & 10 \\ 6 & 14 \\ -2 & 8 \end{bmatrix} = \begin{bmatrix} 5-12 & -20-10 \\ 0-6 & -5-14 \\ 20-(-2) & 10-8 \end{bmatrix} = \begin{bmatrix} -7 & -30 \\ -6 & -19 \\ 22 & 2 \end{bmatrix}$

25. (a) $A + B = \begin{bmatrix} 1 & 3 & -2 & 6 \\ 0 & 1 & 3 & 2 \end{bmatrix} + \begin{bmatrix} 2 & 1 & 4 & -5 \\ 3 & -6 & 3 & -2 \end{bmatrix} = \begin{bmatrix} 3 & 4 & 2 & 1 \\ 3 & -5 & 6 & 0 \end{bmatrix}$

(b) $A - B = \begin{bmatrix} 1 & 3 & -2 & 6 \\ 0 & 1 & 3 & 2 \end{bmatrix} - \begin{bmatrix} 2 & 1 & 4 & -5 \\ 3 & -6 & 3 & -2 \end{bmatrix} = \begin{bmatrix} -1 & 2 & -6 & 11 \\ -3 & 7 & 0 & 4 \end{bmatrix}$

(c) $4A = 4\begin{bmatrix} 1 & 3 & -2 & 6 \\ 0 & 1 & 3 & 2 \end{bmatrix} = \begin{bmatrix} 4 & 12 & -8 & 24 \\ 0 & 4 & 12 & 8 \end{bmatrix}$

(d) $4A - 3B = \begin{bmatrix} 4 & 12 & -8 & 24 \\ 0 & 4 & 12 & 8 \end{bmatrix} - \begin{bmatrix} 6 & 3 & 12 & -15 \\ 9 & -18 & 9 & -6 \end{bmatrix} = \begin{bmatrix} -2 & 9 & -20 & 39 \\ -9 & 22 & 3 & 14 \end{bmatrix}$

27. $\begin{bmatrix} 2 & 0 \\ -5 & -4 \end{bmatrix} + \begin{bmatrix} -10 & 3 \\ 3 & -5 \end{bmatrix} + \begin{bmatrix} -4 & -6 \\ 7 & 4 \end{bmatrix} = \begin{bmatrix} 2+(-10)+(-4) & 0+3+(-6) \\ -5+3+7 & -4+(-5)+4 \end{bmatrix} = \begin{bmatrix} -12 & -3 \\ 5 & -5 \end{bmatrix}$

29. $\frac{1}{4}\left(\begin{bmatrix} 3 & -1 & -2 & 18 \end{bmatrix} + \begin{bmatrix} 5 & -3 & -10 & 2 \end{bmatrix}\right) = \frac{1}{4}\begin{bmatrix} 3+5 & -1+(-3) & -2+(-10) & 18+2 \end{bmatrix}$

$= \frac{1}{4}\begin{bmatrix} 8 & -4 & -12 & 20 \end{bmatrix}$

$= \begin{bmatrix} 2 & -1 & -3 & 5 \end{bmatrix}$

31. $X = 4A - 3B = 4\begin{bmatrix} 1 & -2 \\ 0 & 1 \\ 2 & 3 \end{bmatrix} - 3\begin{bmatrix} 0 & 1 \\ 1 & 1 \\ 3 & 5 \end{bmatrix} = \begin{bmatrix} 4 & -11 \\ -3 & 1 \\ -1 & -3 \end{bmatrix}$

33. $2X - 3A = B \Rightarrow X = \frac{1}{2}(B + 3A)$

$= \frac{1}{2}\left(\begin{bmatrix} 0 & 1 \\ 1 & 1 \\ 3 & 5 \end{bmatrix} + 3\begin{bmatrix} 1 & -2 \\ 0 & 1 \\ 2 & 3 \end{bmatrix}\right)$

$= \frac{1}{2}\begin{bmatrix} 3 & -5 \\ 1 & 4 \\ 9 & 14 \end{bmatrix}$

35. $AB = \begin{bmatrix} 1 & 4 \\ -2 & -1 \\ 3 & 2 \end{bmatrix}\begin{bmatrix} -4 \\ 3 \end{bmatrix} = \begin{bmatrix} 8 \\ 5 \\ -6 \end{bmatrix}$

37. $AB = \begin{bmatrix} 4 & 0 & 0 \\ 0 & 3 & 0 \\ 0 & 0 & -2 \end{bmatrix}\begin{bmatrix} \frac{1}{4} & 0 & 0 \\ 0 & \frac{1}{3} & 0 \\ 0 & 0 & -\frac{1}{2} \end{bmatrix} = \begin{bmatrix} 1 & 0 & 0 \\ 0 & 1 & 0 \\ 0 & 0 & 1 \end{bmatrix}$

39. (a) $AB = \begin{bmatrix} 1 & -3 & 4 \end{bmatrix}\begin{bmatrix} 2 \\ -2 \\ -1 \end{bmatrix} = \begin{bmatrix} 4 \end{bmatrix}$

(b) $BA = \begin{bmatrix} 2 \\ -2 \\ -1 \end{bmatrix}\begin{bmatrix} 1 & -3 & 4 \end{bmatrix} = \begin{bmatrix} 2 & -6 & 8 \\ -2 & 6 & -8 \\ -1 & 3 & -4 \end{bmatrix}$

(c) A^2 is not possible.

41. New production levels $= 1.2\begin{bmatrix} 80 & 120 & 20 & 40 \\ 40 & 60 & 80 & 20 \\ 140 & 60 & 100 & 80 \end{bmatrix} = \begin{bmatrix} 96 & 144 & 24 & 48 \\ 48 & 72 & 96 & 24 \\ 168 & 72 & 120 & 96 \end{bmatrix}$

43. (a) Outlet 3: Total retail price $= 5(600) + 3(705) + 2(455) + 2(1150) = \$8325$

(b) Outlet 1: Total wholesale price $= 3(350) + 2(425) + 1(300) + 4(750) = \$5200$

(c) $ST = \begin{bmatrix} 3 & 2 & 1 & 4 \\ 1 & 3 & 4 & 3 \\ 5 & 3 & 2 & 2 \end{bmatrix}\begin{bmatrix} \$350 & \$600 \\ \$425 & \$705 \\ \$300 & \$455 \\ \$750 & \$1150 \end{bmatrix} = \begin{bmatrix} \$5200 & \$8265 \\ \$5075 & \$7985 \\ \$5125 & \$8325 \end{bmatrix}\begin{matrix} 1 \\ 2 \\ 3 \end{matrix}\Bigg\}\text{Outlet}$ (columns: Wholesale, Retail)

ST represents the total wholesale and retail values of the car sound system inventory at each outlet.

45. $\begin{bmatrix} 1 & 2 & 1 \\ 3 & 6 & 4 \\ 0 & 1 & 3 \end{bmatrix}\begin{bmatrix} -14 & 5 & -2 \\ 9 & -3 & 1 \\ -3 & 1 & 0 \end{bmatrix} = \begin{bmatrix} -14 & 5 & -2 \\ 9 & -3 & 1 \\ -3 & 1 & 0 \end{bmatrix}\begin{bmatrix} 1 & 2 & 1 \\ 3 & 6 & 4 \\ 0 & 1 & 3 \end{bmatrix} = \begin{bmatrix} 1 & 0 & 0 \\ 0 & 1 & 0 \\ 0 & 0 & 1 \end{bmatrix}$

$AB = BA = I$

47. $[A \,\vdots\, I] = \left[\begin{array}{ccc|ccc} -1 & 0 & 0 & 1 & 0 & 0 \\ 0 & 2 & 0 & 0 & 1 & 0 \\ 0 & 0 & 4 & 0 & 0 & 1 \end{array}\right] \Rightarrow \begin{matrix} -R_1 \\ \frac{1}{2}R_2 \\ \frac{1}{4}R_3 \end{matrix}\left[\begin{array}{ccc|ccc} 1 & 0 & 0 & -1 & 0 & 0 \\ 0 & 1 & 0 & 0 & \frac{1}{2} & 0 \\ 0 & 0 & 1 & 0 & 0 & \frac{1}{4} \end{array}\right]; \; A^{-1} = \begin{bmatrix} -1 & 0 & 0 \\ 0 & \frac{1}{2} & 0 \\ 0 & 0 & \frac{1}{4} \end{bmatrix}$

49. $A^{-1} = \dfrac{1}{ad - bc}\begin{bmatrix} d & -b \\ -c & a \end{bmatrix} = \dfrac{1}{5 - 6}\begin{bmatrix} 5 & -3 \\ -2 & 1 \end{bmatrix}$

$= \begin{bmatrix} -5 & 3 \\ 2 & -1 \end{bmatrix}$

$A^{-1} = \begin{bmatrix} -5 & 3 \\ 2 & -1 \end{bmatrix}$

51. $\begin{bmatrix} 1 & 3 \\ 2 & 5 \end{bmatrix}\begin{bmatrix} x \\ y \end{bmatrix} = \begin{bmatrix} 15 \\ 26 \end{bmatrix}$

$\begin{bmatrix} x \\ y \end{bmatrix} = \begin{bmatrix} -5 & 3 \\ 2 & -1 \end{bmatrix}\begin{bmatrix} 15 \\ 26 \end{bmatrix} = \begin{bmatrix} 3 \\ 4 \end{bmatrix}$

Answer: $(3, 4)$

53. $\begin{bmatrix} 3 & 2 & 2 \\ 0 & 2 & 1 \\ 1 & 0 & 1 \end{bmatrix}\begin{bmatrix} x \\ y \\ z \end{bmatrix} = \begin{bmatrix} 13 \\ 4 \\ 5 \end{bmatrix}$

$\begin{bmatrix} x \\ y \\ z \end{bmatrix} = \frac{1}{4}\begin{bmatrix} 2 & -2 & -2 \\ 1 & 1 & -3 \\ -2 & 2 & 6 \end{bmatrix}\begin{bmatrix} 13 \\ 4 \\ 5 \end{bmatrix} = \begin{bmatrix} 2 \\ \frac{1}{2} \\ 3 \end{bmatrix}$

Answer: $\left(2, \frac{1}{2}, 3\right)$

55. $\begin{cases} -3x + 10y = 8 \\ 5x - 17y = -13 \end{cases}$

$A = \begin{bmatrix} -3 & 10 \\ 5 & -17 \end{bmatrix}, \; x = \begin{bmatrix} x \\ y \end{bmatrix}, \; B = \begin{bmatrix} 8 \\ -13 \end{bmatrix},$

$A^{-1} = \begin{bmatrix} -17 & -10 \\ -5 & -3 \end{bmatrix}$

$x = A^{-1}B = \begin{bmatrix} -6 \\ -1 \end{bmatrix}$

Answer: $(-6, -1)$

57. $\begin{cases} 9x + 10y + 14z = 240 \\ x + 3y + 2z = 44 \\ x + y + 2z = 28 \end{cases}$

$$A = \begin{bmatrix} 9 & 10 & 14 \\ 1 & 3 & 2 \\ 1 & 1 & 2 \end{bmatrix}, X = \begin{bmatrix} X \\ Y \\ Z \end{bmatrix}, B = \begin{bmatrix} 240 \\ 44 \\ 28 \end{bmatrix},$$

$$A^{-1} = \begin{bmatrix} \frac{1}{2} & -\frac{3}{4} & -\frac{11}{4} \\ 0 & \frac{1}{2} & -\frac{1}{2} \\ -\frac{1}{4} & \frac{1}{8} & \frac{17}{8} \end{bmatrix}$$

$$X = A^{-1}B = \begin{bmatrix} \frac{1}{2} & -\frac{3}{4} & -\frac{11}{4} \\ 0 & \frac{1}{2} & -\frac{1}{2} \\ -\frac{1}{4} & \frac{1}{8} & \frac{17}{8} \end{bmatrix}\begin{bmatrix} 240 \\ 44 \\ 28 \end{bmatrix} = \begin{bmatrix} 10 \\ 8 \\ 5 \end{bmatrix}$$

You can produce 10 units of fluid X, 8 units of fluid Y, and 5 units of fluid Z.

59. (a) $\begin{cases} 5c + 10b + 30a = 5003 \\ 10c + 30b + 100a = 10{,}856 \\ 30c + 100b + 354a = 33{,}332 \end{cases}$

$$A = \begin{bmatrix} 5 & 10 & 30 \\ 10 & 30 & 100 \\ 30 & 100 & 354 \end{bmatrix}, X = \begin{bmatrix} c \\ b \\ a \end{bmatrix}, B = \begin{bmatrix} 5003 \\ 10{,}856 \\ 33{,}332 \end{bmatrix}$$

$$X = A^{-1}B = \begin{bmatrix} \frac{31}{35} & -\frac{27}{35} & \frac{1}{7} \\ -\frac{27}{35} & \frac{87}{70} & -\frac{2}{7} \\ \frac{1}{7} & -\frac{2}{7} & \frac{1}{14} \end{bmatrix}\begin{bmatrix} 5003 \\ 10{,}856 \\ 33{,}332 \end{bmatrix} = \begin{bmatrix} 818.31 \\ 109.57 \\ -6.14 \end{bmatrix}$$

$y = -6.14t^2 + 109.6t + 818$

(b) For 2010, $t = 5$: $y = -6.14(5)^2 + 109.6(5) + 818 = 1212.5$

According to the model, in 2010 PetSmart had 1213 stores in operation.

(c) The prediction in part (b) is too high.

61. $\begin{vmatrix} 8 & 4 \\ 3 & 2 \end{vmatrix} = 16 - 12 = 4$

63. $\begin{vmatrix} 5 & 2 \\ 0 & 0 \end{vmatrix} = 0 - 0 = 0$

65. $\begin{bmatrix} 2 & -1 \\ 7 & 4 \end{bmatrix}$

(a) $M_{11} = 4, M_{12} = 7, M_{21} = -1, M_{22} = 2$

(b) $C_{11} = 4, C_{12} = -7, C_{21} = 1, C_{22} = 2$

67. $\begin{bmatrix} 3 & 2 & -1 \\ -2 & 5 & 0 \\ 1 & 8 & 6 \end{bmatrix}$

(a) $M_{11} = 30, M_{12} = -12, M_{13} = -21$
$M_{21} = 20, M_{22} = 19, M_{23} = 22$
$M_{31} = 5, M_{32} = -2, M_{33} = 19$

(b) $C_{11} = 30, C_{12} = 12, C_{13} = -21$
$C_{21} = -20, C_{22} = 19, C_{23} = -22$
$C_{31} = 5, C_{32} = 2, C_{33} = 19$

69. $\begin{vmatrix} 1 & 2 & 3 \\ 8 & 6 & 7 \\ 0 & 2 & -1 \end{vmatrix} = \begin{vmatrix} 6 & 7 \\ 2 & -1 \end{vmatrix} - 8\begin{vmatrix} 2 & 3 \\ 2 & -1 \end{vmatrix} = -20 - 8(-8) = 44$

Expansion by minors.

Answers will vary.

71. $\begin{vmatrix} 1 & 3 & 2 & 4 \\ 0 & -1 & 2 & 2 \\ 0 & 0 & 3 & 0 \\ 0 & 0 & 0 & 4 \end{vmatrix} = (1)(-1)(3)(4) = -12$

Product of the diagonal.

Answers will vary.

73. $\begin{vmatrix} -2 & 4 & 1 \\ 6 & 1 & 2 \\ 5 & 3 & 4 \end{vmatrix}\begin{matrix} -2 & 4 \\ 6 & 1 \\ 5 & 3 \end{matrix} = (-8 + 40 + 18) - (5 - 12 + 96)$

$= -39$

Upward and downward diagonals.

Answers will vary.

75. $A = \begin{bmatrix} -1 & 2 \\ -3 & -2 \end{bmatrix}, B = \begin{bmatrix} 4 & 0 \\ 1 & -3 \end{bmatrix}$

(a) $|A| = \begin{vmatrix} -1 & 2 \\ -3 & -2 \end{vmatrix} = -1(-2) - (-3)(2) = 8$

(b) $|B| = \begin{vmatrix} 4 & 0 \\ 1 & -3 \end{vmatrix} = 4(-3) - 1(0) = -12$

(c) $AB = \begin{bmatrix} -1 & 2 \\ -3 & -2 \end{bmatrix}\begin{bmatrix} 4 & 0 \\ 1 & -3 \end{bmatrix} = \begin{bmatrix} -1(4) + 2(1) & -1(0) + 2(-3) \\ -3(4) + (-2)(1) & -3(0) + (-2)(-3) \end{bmatrix} = \begin{bmatrix} -2 & -6 \\ -14 & 6 \end{bmatrix}$

(d) $|AB| = \begin{bmatrix} -2 & -6 \\ -14 & 6 \end{bmatrix} = (-2)(6) - (-14)(-6) = -96$

77. Vertices: $(3, 4), (2, -3), (-1, -4)$

$\text{Area} = \pm\frac{1}{2}\begin{bmatrix} 3 & 4 & 1 \\ 2 & -3 & 1 \\ -1 & -4 & 1 \end{bmatrix} = \pm\frac{1}{2}((-11) - (-8) + (-17))$

$= 10$ square units

79. Vertices: $(-1, 2), (3, 2), (-2, -3)$

$\text{Area} = \pm\frac{1}{2}\begin{vmatrix} -1 & 2 & 1 \\ 3 & 2 & 1 \\ -2 & -3 & 1 \end{vmatrix} = \pm\frac{1}{2}((-5) - 7 + (-8))$

$= 10$ square units

81. Points: $(0, 3), (1, 5), (2, 8)$

$\begin{vmatrix} 0 & 3 & 1 \\ 1 & 5 & 1 \\ 2 & 8 & 1 \end{vmatrix} = -3(-1) + 1(-2) = 1 \neq 0$

The points are not collinear.

83. Points: $(-4, 1), (6, 6), (0, 3)$

$\begin{vmatrix} -4 & 1 & 1 \\ 6 & 6 & 1 \\ 0 & 3 & 1 \end{vmatrix} = -3(-10) + 1(-30) = 0$

The points are collinear.

85. Points: $(-7, 3), (8, 2)$

Equation: $\begin{vmatrix} x & y & 1 \\ -7 & 3 & 1 \\ 8 & 2 & 1 \end{vmatrix} = x(1) - y(-15) + 1(-38) = 0$ OR $x + 15y - 38 = 0$

87. Points: $(2, 4), (2, -7)$

Equation: $\begin{vmatrix} x & y & 1 \\ 2 & 4 & 1 \\ 2 & -7 & 1 \end{vmatrix} = x(11) - y(0) + 1(-22) = 0$ OR $x = 2$

89. The uncoded 1×2 row matrices are the rows of the 6×2 matrix on the left.

$$\begin{matrix}\text{T} & \text{R}\\ \text{A} & \text{N}\\ \text{S} & \text{M}\\ \text{I} & \text{T}\\ & \text{N}\\ \text{O} & \text{W}\end{matrix}\begin{bmatrix}20 & 18\\ 1 & 14\\ 19 & 13\\ 9 & 20\\ 0 & 14\\ 15 & 23\end{bmatrix}\begin{bmatrix}2 & 3\\ 3 & 4\end{bmatrix} = \begin{bmatrix}94 & 132\\ 44 & 59\\ 77 & 109\\ 78 & 107\\ 42 & 56\\ 99 & 137\end{bmatrix}$$

Encoded: 94, 132, 44, 59, 77, 109, 78, 107, 42, 56, 99, 137

91. $A^{-1} = \begin{bmatrix}1 & 2\\ -1 & 3\end{bmatrix}^{-1} = \frac{1}{5}\begin{bmatrix}3 & -2\\ 1 & 1\end{bmatrix}$

$$\begin{bmatrix}14 & 53\\ -17 & 96\\ 5 & 10\\ 12 & 64\\ 5 & 10\\ 3 & 11\\ 25 & 50\end{bmatrix} \cdot \frac{1}{5}\begin{bmatrix}3 & -2\\ 1 & 1\end{bmatrix} = \begin{bmatrix}19 & 5\\ 9 & 26\\ 5 & 0\\ 20 & 8\\ 5 & 0\\ 4 & 1\\ 25 & 0\end{bmatrix}\begin{matrix}\text{S} & \text{E}\\ \text{I} & \text{Z}\\ \text{E} & \\ \text{T} & \text{H}\\ \text{E} & \\ \text{D} & \text{A}\\ \text{Y} & \end{matrix}$$

Message: SEIZE THE DAY

93. (a) Because $\begin{bmatrix}-57 & -13\end{bmatrix}\begin{bmatrix}w & x\\ y & z\end{bmatrix} = \begin{bmatrix}23 & 5\end{bmatrix}$ and $\begin{bmatrix}91 & 26\end{bmatrix}\begin{bmatrix}w & x\\ y & z\end{bmatrix} = \begin{bmatrix}0 & 13\end{bmatrix}$,

you can solve $\begin{bmatrix}-57 & -13\\ 91 & 26\end{bmatrix}\begin{bmatrix}w & x\\ y & z\end{bmatrix} = \begin{bmatrix}23 & 5\\ 0 & 13\end{bmatrix}$.

(b) So, $\begin{aligned}-57w - 13y &= 23\\ 91w + 26y &= 0\end{aligned}$ and $\begin{aligned}-57x - 13z &= 5\\ 91x + 26z &= 13\end{aligned}$.

Using techniques from Chapter 5 or matrix methods from this chapter, $w = -2$, $y = 7$, $x = -1$, and $z = 4$.

Then the message matrix times A^{-1} is

$$\begin{bmatrix}-57 & -13\\ 91 & 26\\ 97 & 29\\ -76 & -19\\ 5 & 5\\ -84 & -21\\ 55 & 16\\ -28 & -7\\ 97 & 28\\ -8 & -2\end{bmatrix}\begin{bmatrix}-2 & -1\\ 7 & 4\end{bmatrix} = \begin{bmatrix}23 & 5\\ 0 & 13\\ 9 & 19\\ 19 & 0\\ 25 & 15\\ 21 & 0\\ 2 & 9\\ 7 & 0\\ 2 & 15\\ 2 & 0\end{bmatrix}\begin{matrix}\text{W} & \text{E}\\ & \text{M}\\ \text{I} & \text{S}\\ \text{S} & \\ \text{Y} & \text{O}\\ \text{U} & \\ \text{B} & \text{I}\\ \text{G} & \\ \text{B} & \text{O}\\ \text{B} & \end{matrix}$$

Message: WE MISS YOU BIG BOB

Chapter 6 Test Yourself

1. $\begin{cases} 2x + y + 4z = 2 \\ x + 4y - z = 0 \\ -x + 3y + 3z = -1 \end{cases} \Rightarrow \begin{bmatrix} 2 & 1 & 4 & \vdots & 2 \\ 1 & 4 & -1 & \vdots & 0 \\ -1 & 3 & 3 & \vdots & -1 \end{bmatrix}$

2. $\begin{cases} 3x + 4y + 2z = 4 \\ 2x + 3y = -2 \\ 2y - 3z = -13 \end{cases} \Rightarrow \begin{bmatrix} 3 & 4 & 2 & \vdots & 4 \\ 2 & 3 & 0 & \vdots & -2 \\ 0 & 2 & -3 & \vdots & -13 \end{bmatrix}$

3. $\begin{bmatrix} 1 & 2 & 3 & \vdots & 16 \\ 5 & 4 & -1 & \vdots & 22 \end{bmatrix}$

$$-5R_1 + R_2 \begin{bmatrix} 1 & 2 & 3 & \vdots & 16 \\ 0 & -6 & -16 & \vdots & -58 \end{bmatrix}$$

$$-\tfrac{1}{6}R_2 \begin{bmatrix} 1 & 2 & 3 & \vdots & 16 \\ 0 & 1 & \frac{8}{3} & \vdots & \frac{29}{3} \end{bmatrix}$$

$$-2R_2 + R_1 \begin{bmatrix} 1 & 0 & -\frac{7}{3} & \vdots & -\frac{10}{3} \\ 0 & 1 & \frac{8}{3} & \vdots & \frac{29}{3} \end{bmatrix}$$

Dependent System

Solution: $\left(\frac{7}{3}a - \frac{10}{3}, -\frac{8}{3}a + \frac{29}{3}, a\right)$ a is any real number.

Answers will vary.

4. $\begin{bmatrix} 1 & -2 & 1 & \vdots & 14 \\ 0 & 1 & -3 & \vdots & 2 \\ 0 & 0 & 1 & \vdots & -6 \end{bmatrix}$

$$2R_2 + R_1 \begin{bmatrix} 1 & 0 & -5 & \vdots & 18 \\ 0 & 1 & -3 & \vdots & 2 \\ 0 & 0 & 1 & \vdots & -6 \end{bmatrix}$$

$$\begin{matrix} 5R_3 + R_1 \\ 3R_3 + R_2 \\ \end{matrix} \begin{bmatrix} 1 & 0 & 0 & \vdots & -12 \\ 0 & 1 & 0 & \vdots & -16 \\ 0 & 0 & 1 & \vdots & -6 \end{bmatrix}$$

$x = -12,\ y = -16,\ z = -6$

Answer: $(-12, -16, -6)$

5. $\begin{bmatrix} 2 & -3 & 1 & \vdots & 14 \\ 1 & 2 & 0 & \vdots & -4 \\ 0 & 1 & -1 & \vdots & -4 \end{bmatrix}$

$$-R_2 + R_1 \begin{bmatrix} 1 & -5 & 1 & \vdots & 18 \\ 1 & 2 & 0 & \vdots & -4 \\ 0 & 1 & -1 & \vdots & -4 \end{bmatrix}$$

$$\begin{matrix} \\ -R_1 + R_2 \\ \end{matrix} \begin{bmatrix} 1 & -5 & 1 & \vdots & 18 \\ 0 & 7 & -1 & \vdots & -22 \\ 0 & 1 & -1 & \vdots & -4 \end{bmatrix}$$

$$\begin{matrix} \\ R_3 \leftrightarrow R_2 \\ R_2 \leftrightarrow R_3 \end{matrix} \begin{bmatrix} 1 & -5 & 1 & \vdots & 18 \\ 0 & 1 & -1 & \vdots & -4 \\ 0 & 7 & -1 & \vdots & -22 \end{bmatrix}$$

$$\begin{matrix} 5R_2 + R_1 \\ \\ -7R_2 + R_3 \end{matrix} \begin{bmatrix} 1 & 0 & -4 & \vdots & -2 \\ 0 & 1 & -1 & \vdots & -4 \\ 0 & 0 & 6 & \vdots & 6 \end{bmatrix}$$

$$\begin{matrix} \frac{2}{3}R_3 + R_1 \\ \frac{1}{6}R_3 + R_2 \\ \frac{1}{6}R_3 \end{matrix} \begin{bmatrix} 1 & 0 & 0 & \vdots & 2 \\ 0 & 1 & 0 & \vdots & -3 \\ 0 & 0 & 1 & \vdots & 1 \end{bmatrix}$$

$x = 2,\ y = -3,\ z = 1$

Answer: $(2, -3, 1)$

6. $2A + C = 2\begin{bmatrix} 1 & 3 \\ 2 & 4 \end{bmatrix} + \begin{bmatrix} 0 & -2 \\ 3 & 5 \end{bmatrix}$

$$= \begin{bmatrix} 2 & 6 \\ 4 & 8 \end{bmatrix} + \begin{bmatrix} 0 & -2 \\ 3 & 5 \end{bmatrix}$$

$$= \begin{bmatrix} 2 & 4 \\ 7 & 13 \end{bmatrix}$$

7. $CA = \begin{bmatrix} 0 & -2 \\ 3 & 5 \end{bmatrix}\begin{bmatrix} 1 & 3 \\ 2 & 4 \end{bmatrix} = \begin{bmatrix} -4 & -8 \\ 13 & 29 \end{bmatrix}$

8. $BD = \begin{bmatrix} 2 & -1 & 3 \\ 4 & 0 & 1 \end{bmatrix}\begin{bmatrix} 3 \\ 2 \\ -1 \end{bmatrix} = \begin{bmatrix} 1 \\ 11 \end{bmatrix}$

9. $C^2 = \begin{bmatrix} 0 & -2 \\ 3 & 5 \end{bmatrix}\begin{bmatrix} 0 & -2 \\ 3 & 5 \end{bmatrix} = \begin{bmatrix} -6 & -10 \\ 15 & 19 \end{bmatrix}$

10.
$$\begin{aligned} A + C + E &= \begin{bmatrix} 1 & 3 \\ 2 & 4 \end{bmatrix} + \begin{bmatrix} 0 & -2 \\ 3 & 5 \end{bmatrix} + \begin{bmatrix} 7 & -8 \\ 6 & 10 \end{bmatrix} \\ &= \begin{bmatrix} 8 & -7 \\ 11 & 19 \end{bmatrix} \end{aligned}$$

11.
$$\begin{aligned} -4(A + E) &= -4\left(\begin{bmatrix} 1 & 3 \\ 2 & 4 \end{bmatrix} + \begin{bmatrix} 7 & -8 \\ 6 & 10 \end{bmatrix}\right) \\ &= -4\begin{bmatrix} 8 & -5 \\ 8 & 14 \end{bmatrix} \\ &= \begin{bmatrix} -32 & 20 \\ -32 & -56 \end{bmatrix} \end{aligned}$$

12.
$$\begin{aligned} 5(C - A) &= 5\left(\begin{bmatrix} 0 & -2 \\ 3 & 5 \end{bmatrix} - \begin{bmatrix} 1 & 3 \\ 2 & 4 \end{bmatrix}\right) \\ &= 5\begin{bmatrix} -1 & -5 \\ 1 & 1 \end{bmatrix} \\ &= \begin{bmatrix} -5 & -25 \\ 5 & 5 \end{bmatrix} \end{aligned}$$

13.
$$\begin{aligned} 3C + 2(A - E) &= 3\begin{bmatrix} 0 & -2 \\ 3 & 5 \end{bmatrix} + 2\left(\begin{bmatrix} 1 & 3 \\ 2 & 4 \end{bmatrix} - \begin{bmatrix} 7 & -8 \\ 6 & 10 \end{bmatrix}\right) \\ &= \begin{bmatrix} 0 & -6 \\ 9 & 15 \end{bmatrix} + 2\begin{bmatrix} -6 & 11 \\ -4 & -6 \end{bmatrix} \\ &= \begin{bmatrix} 0 & -6 \\ 9 & 15 \end{bmatrix} + \begin{bmatrix} -12 & 22 \\ -8 & -12 \end{bmatrix} \\ &= \begin{bmatrix} -12 & 16 \\ 1 & 3 \end{bmatrix} \end{aligned}$$

14. $A^{-1} = \dfrac{1}{2(4) - (-1)(-3)}\begin{bmatrix} 4 & -(-1) \\ -(-3) & 2 \end{bmatrix} = \dfrac{1}{5}\begin{bmatrix} 4 & 1 \\ 3 & 2 \end{bmatrix}$

15. Because $A = I$, $A^{-1} = A = \begin{bmatrix} 1 & 0 \\ 0 & 1 \end{bmatrix}$.

16.
$$[A \ \vdots \ I] = \begin{bmatrix} 3 & 4 & 2 & \vdots & 1 & 0 & 0 \\ 2 & 3 & 0 & \vdots & 0 & 1 & 0 \\ 0 & 2 & -3 & \vdots & 0 & 0 & 1 \end{bmatrix} \Rightarrow \begin{matrix} R_1 - R_2 \\ \\ \\ \end{matrix}\begin{bmatrix} 1 & 1 & 2 & \vdots & 1 & -1 & 0 \\ 2 & 3 & 0 & \vdots & 0 & 1 & 0 \\ 0 & 2 & -3 & \vdots & 0 & 0 & 1 \end{bmatrix} \Rightarrow \begin{matrix} \\ -2R_1 + R_2 \\ \\ \end{matrix}\begin{bmatrix} 1 & 1 & 2 & \vdots & 1 & -1 & 0 \\ 0 & 1 & -4 & \vdots & -2 & 3 & 0 \\ 0 & 2 & -3 & \vdots & 0 & 0 & 1 \end{bmatrix}$$

$$\Rightarrow \begin{matrix} \\ \\ -2R_2 + R_3 \end{matrix}\begin{bmatrix} 1 & 1 & 2 & \vdots & 1 & -1 & 0 \\ 0 & 1 & -4 & \vdots & -2 & 3 & 0 \\ 0 & 0 & 5 & \vdots & 4 & -6 & 1 \end{bmatrix} \Rightarrow \begin{matrix} R_1 - R_2 \\ \frac{4}{5}R_3 + R_2 \\ \frac{1}{5}R_3 \end{matrix}\begin{bmatrix} 1 & 0 & 6 & \vdots & 3 & -4 & 0 \\ 0 & 1 & 0 & \vdots & \frac{6}{5} & -\frac{9}{5} & \frac{4}{5} \\ 0 & 0 & 1 & \vdots & \frac{4}{5} & -\frac{6}{5} & \frac{1}{5} \end{bmatrix}$$

$$\Rightarrow \begin{matrix} -6R_3 + R_1 \\ \\ \\ \end{matrix}\begin{bmatrix} 1 & 0 & 0 & \vdots & -\frac{9}{5} & \frac{16}{5} & -\frac{6}{5} \\ 0 & 1 & 0 & \vdots & \frac{6}{5} & -\frac{9}{5} & \frac{4}{5} \\ 0 & 0 & 1 & \vdots & \frac{4}{5} & -\frac{6}{5} & \frac{1}{5} \end{bmatrix}$$

$$A^{-1}\begin{bmatrix} -\frac{9}{5} & \frac{16}{5} & -\frac{6}{5} \\ \frac{6}{5} & -\frac{9}{5} & \frac{4}{5} \\ \frac{4}{5} & -\frac{6}{5} & \frac{1}{5} \end{bmatrix} = \frac{1}{5}\begin{bmatrix} -9 & 16 & -6 \\ 6 & -9 & 4 \\ 4 & -6 & 1 \end{bmatrix}$$

17. $\begin{vmatrix} -5 & 2 \\ 1 & 3 \end{vmatrix} = -5(3) - 2(1) = -17$

18.
$$\begin{aligned} \begin{vmatrix} 3 & 2 & -1 \\ 1 & 0 & 2 \\ 4 & 5 & 2 \end{vmatrix} &= -1\begin{vmatrix} 2 & -1 \\ 5 & 2 \end{vmatrix} - 2\begin{vmatrix} 3 & 2 \\ 4 & 5 \end{vmatrix} \\ &= -1(9) - 2(7) \\ &= -23 \end{aligned}$$

19. $\begin{vmatrix} 2 & 0 & 0 \\ 0 & 5 & 0 \\ 0 & 0 & -2 \end{vmatrix} = 2(5)(-2) = -20$

20. $\begin{bmatrix} 3 & 4 & 2 \\ 2 & 3 & 0 \\ 0 & 2 & -3 \end{bmatrix}\begin{bmatrix} x \\ y \\ z \end{bmatrix} = \begin{bmatrix} 4 \\ -2 \\ -13 \end{bmatrix}$

$$\begin{bmatrix} x \\ y \\ z \end{bmatrix} = \frac{1}{5}\begin{bmatrix} -9 & 16 & -6 \\ 6 & -9 & 4 \\ 4 & -6 & 1 \end{bmatrix}\begin{bmatrix} 4 \\ -2 \\ -13 \end{bmatrix} = \begin{bmatrix} 2 \\ -2 \\ 3 \end{bmatrix}$$

Answer: $(2, -2, 3)$

21. Answers will vary. One possible solution is

$A = \begin{bmatrix} 2 & -2 \\ -2 & 2 \end{bmatrix}$ and $B = \begin{bmatrix} 3 & 3 \\ 3 & 3 \end{bmatrix}$.

22. Vertices: $(-3, 1), (0, 4), (5, 2)$

$$A = \pm\frac{1}{2}\begin{bmatrix} -3 & 1 & 1 \\ 0 & 4 & 1 \\ 5 & 2 & 1 \end{bmatrix} = \pm\frac{1}{2}\left(4(-8) - (-11)\right)$$

$= \frac{21}{2}$ square units

23. Points: $(2, 1), (-3, -14), (4, 7)$

$$\begin{vmatrix} 2 & 1 & 1 \\ -3 & -14 & 1 \\ 4 & 7 & 1 \end{vmatrix} = 2(-21) - 1(-7) + 1(35) = 0$$

The points are collinear.

24. Points: $(1, -2), (5, 2)$

Equation: $\begin{vmatrix} x & y & 1 \\ 1 & -2 & 1 \\ 5 & 2 & 1 \end{vmatrix} = x(-4) - y(-4) + 1(12) = 0$ OR $-x + y + 3 = 0$

25. $BA = \begin{bmatrix} \$55 & \$40 & \$33 \end{bmatrix}\begin{bmatrix} 1500 & 4000 \\ 3000 & 4500 \\ 5500 & 7000 \end{bmatrix} = \begin{bmatrix} \$384{,}000 & \$631{,}000 \end{bmatrix}$

BA represents the total value of each product at each warehouse.

CHAPTER 7
Sequences, Series, and Probability

CHAPTER 7
Sequences, Series, and Probability

Section 7.1 Sequences and Summation Notation

Skills Warm Up

1. $f(2) = \frac{2(2)}{2^2 + 1} = \frac{4}{5}$

2. $f(3) = \frac{4}{3(3+1)} = \frac{4}{12} = \frac{1}{3}$

3. $9(x - 6)$ when $x = 2: 9(2 - 6) = 9(-4) = -36$

4. $4(6 - 2x)$ when $x = 2: 4[6 - 2(2)] = 4(2) = 8$

5. $y^3(x^3 - 3)$ when $x = 2$ and $y = -1$:

$$(-1)^3(2^3 - 3) = (-1)(8 - 3)$$
$$= -5$$

6. $xy^5(7 - 3x^2)$ when $x = 2$ and $y = -1$:

$$2(-1)^5\left[7 - 3(2)^2\right] = 2(-1)[7 - 3(4)]$$
$$= -2(7 - 12)$$
$$= -2(-5) = 10$$

7. $\left(\frac{2}{3}\right)\left(\frac{3}{4}\right)\left(\frac{4}{5}\right)\left(\frac{5}{6}\right) = \frac{120}{360} = \frac{1}{3}$

8. $\frac{2 \cdot 4 \cdot 6 \cdot 8}{2^4} = \frac{384}{16} = 24$

9. $\frac{1}{2 \cdot 2} + \frac{1}{2 \cdot 3} + \frac{1}{2 \cdot 4} = \frac{1}{4} + \frac{1}{6} + \frac{1}{8} = \frac{13}{24}$

10. $\frac{1}{1 \cdot 2} + \frac{1}{2 \cdot 3} + \frac{1}{3 \cdot 4} = \frac{1}{2} + \frac{1}{6} + \frac{1}{12} = \frac{3}{4}$

1. $a_n = 2n - 7$

$a_1 = 2(1) - 7 = -5$

$a_2 = 2(2) - 7 = -3$

$a_3 = 2(3) - 7 = -1$

$a_4 = 2(4) - 7 = 1$

$a_5 = 2(5) - 7 = 3$

3. $a_n = 2^n$

$a_1 = 2^1 = 2$

$a_2 = 2^2 = 4$

$a_3 = 2^3 = 8$

$a_4 = 2^4 = 16$

$a_5 = 2^5 = 32$

5. $a_n = \left(-\frac{1}{2}\right)^n$

$a_1 = \left(-\frac{1}{2}\right)^1 = -\frac{1}{2}$

$a_2 = \left(-\frac{1}{2}\right)^2 = \frac{1}{4}$

$a_3 = \left(-\frac{1}{2}\right)^3 = -\frac{1}{8}$

$a_4 = \left(-\frac{1}{2}\right)^4 = \frac{1}{16}$

$a_5 = \left(-\frac{1}{2}\right)^5 = -\frac{1}{32}$

7. $a_n = \frac{3}{n + 1}$

$a_1 = \frac{3}{1 + 1} = \frac{3}{2}$

$a_2 = \frac{3}{2 + 1} = \frac{3}{3} = 1$

$a_3 = \frac{3}{3 + 1} = \frac{3}{4}$

$a_4 = \frac{3}{4 + 1} = \frac{3}{5}$

$a_5 = \frac{3}{5 + 1} = \frac{3}{6} = \frac{1}{2}$

9. $a_n = \frac{(-1)^n}{n}$

$a_1 = \frac{(-1)^1}{1} = -1$

$a_2 = \frac{(-1)^2}{2} = \frac{1}{2}$

$a_3 = \frac{(-1)^3}{3} = -\frac{1}{3}$

$a_4 = \frac{(-1)^4}{4} = \frac{1}{4}$

$a_5 = \frac{(-1)^5}{5} = -\frac{1}{5}$

11. $a_n = 2 - \frac{1}{3^n}$

$a_1 = 2 - \frac{1}{3^1} = 2 - \frac{1}{3} = \frac{5}{3}$

$a_2 = 2 - \frac{1}{3^2} = 2 - \frac{1}{9} = \frac{17}{9}$

$a_3 = 2 - \frac{1}{3^3} = 2 - \frac{1}{27} = \frac{53}{27}$

$a_4 = 2 - \frac{1}{3^4} = 2 - \frac{1}{81} = \frac{161}{81}$

$a_5 = 2 - \frac{1}{3^5} = 2 - \frac{1}{243} = \frac{485}{243}$

13. $a_n = \frac{n^2 - 1}{n + 1} = n - 1$

$a_1 = 0$

$a_2 = 1$

$a_3 = 2$

$a_4 = 3$

$a_5 = 4$

15. $a_n = \frac{(-1)^n}{n^2}$

$a_1 = \frac{-1}{1} = -1$

$a_2 = \frac{1}{4}$

$a_3 = -\frac{1}{9}$

$a_4 = \frac{1}{16}$

$a_5 = -\frac{1}{25}$

17. $a_n = \frac{5n^2 - n + 1}{n^2}$

$a_1 = \frac{5(1)^2 - 1 + 1}{1^2} = 5$

$a_2 = \frac{5(2)^2 - 2 + 1}{2^2} = \frac{19}{4}$

$a_3 = \frac{5(3)^2 - 3 + 1}{3^2} = \frac{43}{9}$

$a_4 = \frac{5(4)^2 - 4 + 1}{4^2} = \frac{77}{16}$

$a_5 = \frac{5(5)^2 - 5 + 1}{5^2} = \frac{121}{25}$

19. $1, 4, 7, 10, 13, \ldots$

$a_n = 3n - 2$

21. $2, -4, 6, -8, 10, \ldots$

$a_n = (-1)^{n+1}(2n)$

23. $2, \frac{1}{2}, \frac{2}{9}, \frac{1}{8}, \frac{2}{25}, \ldots$

$a_n = \frac{2}{n^2}$

25. $\frac{2}{3}, \frac{3}{4}, \frac{4}{5}, \frac{5}{6}, \frac{6}{7}, \ldots$

$a_n = \frac{n + 1}{n + 2}$

27. $-\frac{1}{3}, \frac{2}{9}, -\frac{4}{27}, \frac{8}{81}, -\frac{16}{243}, \ldots$

$a_n = \frac{(-1)^n 2^{n-1}}{3^n}$

29. $1 + \frac{1}{1}, 1 + \frac{1}{2}, 1 + \frac{1}{3}, 1 + \frac{1}{4}, 1 + \frac{1}{5}, \ldots$

$a_n = 1 + \frac{1}{n}$

31. $a_n = \frac{3^n}{n!}$

$a_1 = \frac{3^1}{1!} = 3$

$a_2 = \frac{3^2}{2!} = \frac{9}{2}$

$a_3 = \frac{3^3}{3!} = \frac{27}{6} = \frac{9}{2}$

$a_4 = \frac{3^4}{4!} = \frac{81}{24} = \frac{27}{8}$

$a_5 = \frac{3^5}{5!} = \frac{243}{120} = \frac{81}{40}$

33. $a_n = \frac{3n!}{(n - 1)!}$

$a_1 = \frac{3(1!)}{(1 - 1)!} = \frac{3(1)}{0!} = \frac{3}{1} = 3$

$a_2 = \frac{3(2!)}{(2 - 1)!} = \frac{3 \cdot 2 \cdot 1}{1!} = \frac{6}{1} = 6$

$a_3 = \frac{3(3!)}{(3 - 1)!} = \frac{3 \cdot 3 \cdot 2 \cdot 1}{2!} = \frac{3 \cdot 3 \cdot 2 \cdot 1}{2 \cdot 1} = 9$

$a_4 = \frac{3(4!)}{(4 - 1)!} = \frac{3 \cdot 4 \cdot 3!}{3!} = 12$

$a_5 = \frac{3(5!)}{(5 - 1)!} = \frac{3 \cdot 5 \cdot 4!}{4!} = 15$

35. $a_n = (-1)^n(3n - 2)$

$a_{25} = (-1)^{25}[3(25) - 2] = -73$

37. $a_n = \dfrac{4n}{2n^2 - 3}$

$a_{11} = \dfrac{4(11)}{2(11)^2 - 3} = \dfrac{44}{239}$

39. $a_n = \dfrac{2^n}{n!}$

$a_{10} = \dfrac{2^{10}}{10!} = \dfrac{1024}{3{,}628{,}800} = \dfrac{4}{14{,}175}$

41. $a_n = \dfrac{6}{n + 2}$

$a_1 = \dfrac{6}{1 + 2} = \dfrac{6}{3} = 2 \Rightarrow (1, 2)$

$a_2 = \dfrac{6}{2 + 2} = \dfrac{6}{4} = \dfrac{3}{2} \Rightarrow \left(2, \dfrac{3}{2}\right)$

$a_3 = \dfrac{6}{3 + 2} = \dfrac{6}{5} \Rightarrow \left(3, \dfrac{6}{5}\right)$

$a_4 = \dfrac{6}{4 + 2} = \dfrac{6}{6} = 1 \Rightarrow (4, 1)$

$a_n = \dfrac{6}{n + 2}$ corresponds to graph (a).

43. $a_n = \dfrac{4^n}{n!}$

$a_1 = \dfrac{4^1}{1!} = \dfrac{4}{1} = 4 \Rightarrow (1, 4)$

$a_2 = \dfrac{4^2}{2!} = \dfrac{16}{2} = 8 \Rightarrow (2, 8)$

$a_3 = \dfrac{4^3}{3!} = \dfrac{64}{6} = \dfrac{32}{3} \Rightarrow \left(3, \dfrac{32}{3}\right)$

$a_4 = \dfrac{4^4}{4!} = \dfrac{256}{24} = \dfrac{32}{3} \Rightarrow \left(4, \dfrac{32}{3}\right)$

$a_n = \dfrac{4^n}{n!}$ corresponds to graph (c).

45. $\dfrac{8!}{7!} = \dfrac{8 \cdot 7!}{7!} = 8$

47. $\dfrac{28!}{25!} = \dfrac{28 \cdot 27 \cdot 26 \cdot 25!}{25!} = 19{,}656$

49. $\dfrac{(n + 2)!}{n!} = \dfrac{(n + 2) \cdot (n + 1) \cdot n!}{n!} = (n + 2)(n + 1)$

51. $\dfrac{(n + 1)!}{n!} = \dfrac{(n + 1)n!}{n!} = n + 1$

$a_0 = 1,\ a_1 = 2$

53. $\dfrac{2n!}{(2n + 1)!} = \dfrac{2n!}{(2n + 1) \cdot 2n!} = \dfrac{1}{2n + 1}$

$a_0 = 1,\ a_1 = \dfrac{1}{3}$

55. 1, 2, 6, 24, 120, ...

$a_n = n!$

57. $1, 2, \dfrac{2^2}{2}, \dfrac{2^3}{6}, \dfrac{2^4}{24}, \dfrac{2^5}{120}, \ldots$

$a_n = \dfrac{2^{n-1}}{(n - 1)!}$

59. $\displaystyle\sum_{k=1}^{4} 12 = 12 + 12 + 12 + 12 = 4(12) = 48$

61. $\displaystyle\sum_{i=1}^{7} -3i = -3\sum_{i=1}^{7} i$

$= -3[1 + 2 + 3 + 4 + 5 + 6 + 7]$

$= -3(28)$

$= -84$

63. $\displaystyle\sum_{i=1}^{4} (6i + 3) = 3\sum_{i=1}^{4} (2i + 1) = 3[(2 + 1) + (4 + 1) + (6 + 1) + (8 + 1)] = 3(24) = 72$

65. $\displaystyle\sum_{i=0}^{4} i^2 = 0^2 + 1^2 + 2^2 + 3^2 + 4^2 = 30$

67. $\displaystyle\sum_{k=2}^{6} \frac{1}{2k} = \frac{1}{2}\sum_{k=2}^{6} \frac{1}{k} = \frac{1}{2}\left(\frac{1}{2} + \frac{1}{3} + \frac{1}{4} + \frac{1}{5} + \frac{1}{6}\right) = \frac{1}{2}\left(\frac{29}{20}\right) = \frac{29}{40}$

69. $\displaystyle\sum_{i=1}^{4} (i - 1)^2 = (1 - 1)^2 + (2 - 1)^2 + (3 - 1)^2 + (4 - 1)^2 = 14$

71. $\displaystyle\sum_{i=1}^{4} (-1)^i (2i + 4) = -(2 + 4) + (4 + 4) - (6 + 4) + (8 + 4) = 4$

73. $\displaystyle\sum_{k=0}^{4}\frac{1}{k!} = \frac{1}{0!}+\frac{1}{1!}+\frac{1}{2!}+\frac{1}{3!}+\frac{1}{4!} = 1+1+\frac{1}{2}+\frac{1}{6}+\frac{1}{24} = \frac{65}{24}$

75.
$$\sum_{j=0}^{6}\frac{(2j+2)!}{(2j)!} = \frac{[2(0)+2]!}{[2(0)]!}+\frac{[2(1)+2]!}{[2(1)]!}+\frac{[2(2)+2]!}{[2(2)]!}+\frac{[2(3)+2]!}{[2(3)]!}+\frac{[2(4)+2]!}{[2(4)]!}+\frac{[2(5)+2]!}{[2(5)]!}+\frac{[2(6)+2]!}{[2(6)]!}$$
$$= \frac{2!}{0!}+\frac{4!}{2!}+\frac{6!}{4!}+\frac{8!}{6!}+\frac{10!}{8!}+\frac{12!}{10!}+\frac{14!}{12!}$$
$$= 2+\frac{4\cdot 3\cdot \cancel{2}!}{\cancel{2}!}+\frac{6\cdot 5\cdot \cancel{4}!}{\cancel{4}!}+\frac{8\cdot 7\cdot \cancel{6}!}{\cancel{6}!}+\frac{10\cdot 9\cdot \cancel{8}!}{\cancel{8}!}+\frac{12\cdot 11\cdot \cancel{10}!}{\cancel{10}!}+\frac{14\cdot 13\cdot \cancel{12}!}{\cancel{12}!}$$
$$= 2+12+30+56+90+132+182$$
$$= 504$$

77. $\displaystyle\frac{1}{3(1)}+\frac{1}{3(2)}+\frac{1}{3(3)}+\cdots+\frac{1}{3(9)} = \sum_{i=1}^{9}\frac{1}{3i}$

79. $\displaystyle\left[2\left(\frac{1}{8}\right)+3\right]+\left[2\left(\frac{2}{8}\right)+3\right]+\left[2\left(\frac{3}{8}\right)+3\right]+\cdots+\left[2\left(\frac{8}{8}\right)+3\right] = \sum_{i=1}^{8}\left[2\left(\frac{i}{8}\right)+3\right]$

81. $\displaystyle 3-9+27-81+243-729 = \sum_{i=1}^{6}(-1)^{i+1}3^i$

83. $\displaystyle\frac{1}{1^2}-\frac{1}{2^2}+\frac{1}{3^2}-\frac{1}{4^2}+\cdots-\frac{1}{20^2} = \sum_{i=1}^{20}\frac{(-1)^{i+1}}{i^2}$

85. $\displaystyle\frac{1}{4}+\frac{3}{8}+\frac{7}{16}+\frac{15}{32}+\frac{31}{64} = \sum_{i=1}^{5}\frac{2^i-1}{2^{i+1}}$

87. $\displaystyle A_n = 25{,}000\left(1+\frac{0.07}{12}\right)^n,\ n = 1, 2, 3, \ldots$

(a) $\displaystyle A_1 = 25{,}000\left(1+\frac{0.07}{12}\right)^1 \approx \$25{,}145.83$

$\displaystyle A_2 = 25{,}000\left(1+\frac{0.07}{12}\right)^2 \approx \$25{,}292.52$

$\displaystyle A_3 = 25{,}000\left(1+\frac{0.07}{12}\right)^3 \approx \$25{,}440.06$

$\displaystyle A_4 = 25{,}000\left(1+\frac{0.07}{12}\right)^4 \approx \$25{,}588.46$

$\displaystyle A_5 = 25{,}000\left(1+\frac{0.07}{12}\right)^5 \approx \$25{,}737.72$

$\displaystyle A_6 = 25{,}000\left(1+\frac{0.07}{12}\right)^6 \approx \$25{,}887.86$

(b) $\displaystyle A_{60} = 25{,}000\left(1+\frac{0.07}{12}\right)^{60} \approx \$35{,}440.63$

(c) In 20 years $(n = 120)$: $\displaystyle A_{120} = 25{,}000\left(1+\frac{0.07}{12}\right)^{120}$

$\approx \$50{,}241.53$

Because $50{,}241.53 \neq 2(35{,}440.63) = 70{,}881.26$, the balance after 10 years is not twice the balance after 5 years.

89. (a) $a_1 = 1.0156$

$a_2 = 0.9884$

$a_3 = 0.9684$

$a_4 = 0.9556$

$a_5 = 0.95$

$a_6 = 0.9516$

$a_7 = 0.9604$

$a_8 = 0.9764$

Each term represents the number of males divided by the number of females in the United States for the corresponding year.

(b)

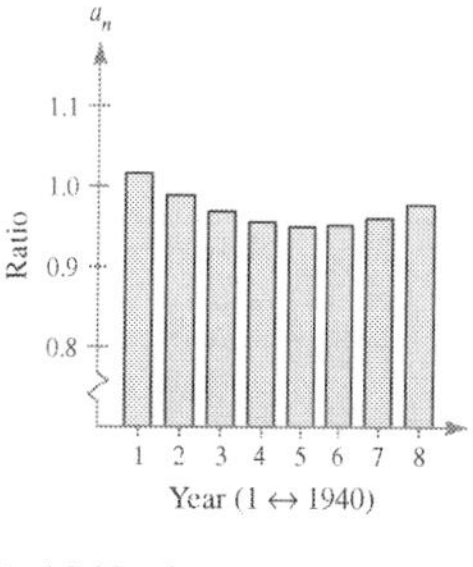

In 1940, there were more males than females. The ratio decreased until 1980 and increased thereafter.

(c) Let f represent the number of females and m represent the number of males.

In 2000 $(n = 7)$: $\displaystyle\frac{m}{f} = 0.9604$

$\displaystyle\frac{281-f}{f} = 0.9604$

$f \approx 143.3$

$m \approx 281 - 143.3 = 137.7$

In 2000, there were about 143.3 million females and 137.7 million males.

91. $a_0 = 22.9 + 3.63(0) + 2.657(0)^2 = 22.9$

$a_1 = 22.9 + 3.63(1) + 2.657(1)^2 = 29.187$

$a_2 = 22.9 + 3.63(2) + 2.657(2)^2 = 40.788$

$a_3 = 22.9 + 3.63(3) + 2.657(3)^2 = 57.703$

$a_4 = 22.9 + 3.63(4) + 2.657(4)^2 = 79.932$

$$\sum_{n=0}^{4}\left(22.9 + 3.63n + 2.657n^2\right) = 22.9 + 29.187 + 40.788 + 57.703 + 79.932 = 230.51$$

The total sales for China-Biotics from 2005 through 2009 were about \$230.51 million.

93. True; $\sum_{i=1}^{4}\left(i^2 + 2i\right) = \sum_{i=1}^{4} i^2 + 2\sum_{i=1}^{4} i$ by the Properties of Sums.

95. $a_n = \dfrac{x^n}{n!}$

$a_1 = \dfrac{x^1}{1!} = x$

$a_2 = \dfrac{x^2}{2!} = \dfrac{x^2}{2}$

$a_3 = \dfrac{x^3}{3!} = \dfrac{x^3}{6}$

$a_4 = \dfrac{x^4}{4!} = \dfrac{x^4}{24}$

$a_5 = \dfrac{x^5}{5!} = \dfrac{x^5}{120}$

97. $a_n = \dfrac{(-1)^n x^{2n}}{(2n)!}$

$a_1 = \dfrac{-x^2}{2!} = -\dfrac{x^2}{2}$

$a_2 = \dfrac{x^4}{4!} = \dfrac{x^4}{24}$

$a_3 = \dfrac{-x^6}{6!} = -\dfrac{x^6}{720}$

$a_4 = \dfrac{x^8}{8!} = \dfrac{x^8}{40{,}320}$

$a_5 = \dfrac{-x^{10}}{10!} = -\dfrac{x^{10}}{3{,}628{,}800}$

Section 7.2 Arithmetic Sequences and Partial Sums

Skills Warm Up

1. $\sum_{i=1}^{6}(2i - 1) = (2(1) - 1) + (2(2) - 1) + (2(3) - 1) + (2(4) - 1) + (2(5) - 1) + (2(6) - 1) = 1 + 3 + 5 + 7 + 9 + 11 = 36$

2. $\sum_{i=1}^{10}(4i + 2) = (4(1) + 2) + (4(2) + 2) + (4(3) + 2) + (4(4) + 2) + (4(5) + 2) + (4(6) + 2)$
$+ (4(7) + 2) + (4(8) + 2) + (4(9) + 2) + (4(10) + 2)$
$= 6 + 10 + 14 + 18 + 22 + 26 + 30 + 34 + 38 + 42 = 240$

3. $8 - \frac{5}{2} = \frac{16}{2} - \frac{5}{2} = \frac{11}{2}$

4. $\frac{14}{3} - \frac{4}{3} = \frac{10}{3}$

5. $f(3) = 10 + (3 - 1)4 = 10 + 2(4) = 18$

6. $f(10) = 1 + (10 - 1)\frac{1}{3} = 1 + (9)\frac{1}{3} = 4$

7. $\frac{11}{2}(1 + 25) = \frac{11}{2}(26) = 143$

8. $\frac{16}{2}(4 + 16) = 8(20) = 160$

9. $\frac{20}{2}[2(5) + (12 - 1)3] = 10[10 + (11)3] = 10(43) = 430$

10. $\frac{8}{2}[2(-3) + (15 - 1)5] = 4[-6 + (14)5] = 4(64) = 256$

1. 3, 7, 11, 15, 19, …

Arithmetic sequence, $d = 4$

3. $-12, -8, -4, 0, 4, 8, \dots$

Arithmetic sequence, $d = 4$

5. $\frac{1}{3}, \frac{2}{3}, \frac{4}{3}, \frac{8}{3}, \frac{16}{3}, \frac{32}{3}, \dots$

Not an arithmetic sequence

7. $3, \frac{5}{2}, 2, \frac{3}{2}, 1, \frac{1}{2}, \dots$

Arithmetic sequence, $d = -\frac{1}{2}$

9. $\log_{10} 10, \log_{10} 100, \log_{10} 1000, \log_{10} 10{,}000, \dots$
$= 1, 2, 3, 4 \dots$

Arithmetic sequence, $d = 1$

11. 5.3, 5.7, 6.2, 6.8, 7.5,...

Not an arithmetic sequence

13. $a_n = 5 + 3n$

8, 11, 14, 17, 20

Arithmetic sequence, $d = 3$

15. $a_n = (2^n)n$

2, 8, 24, 64, 160

Not an arithmetic sequence

17. $a_n = (2 + n) - (1 + n) = 1$

1, 1, 1, 1, 1

Arithmetic sequence, $d = 0$

19. $a_n = \dfrac{2}{n - 1}$

Undefined, 2, 1, $\frac{2}{3}, \frac{1}{2}$

Not an arithmetic sequence

21. $a_n = (-1)^n$

$-1, 1, -1, 1, -1$

Not an arithmetic sequence

23. $a_n = 3n - 5$

$a_1 = -2,\ a_2 = 1,\ a_3 = 4,\ a_4 = 7 \Rightarrow (1, 2),$
$(2, 1), (3, 4), (4, 7)$

$a_n = 3n - 5$ corresponds to graph (d).

25. $a_n = -\frac{3}{4}n + 8$

$a_1 = \frac{29}{4},\ a_2 = \frac{13}{2},\ a_3 = \frac{23}{4},\ a_4 = 5$

$\Rightarrow \left(1, \frac{29}{4}\right), \left(2, \frac{13}{2}\right), \left(3, \frac{23}{4}\right), (4, 5)$

$a_n = -\frac{3}{4}n + 8$ corresponds to graph (b).

27. $a_n = 15 - \frac{3}{2}n$

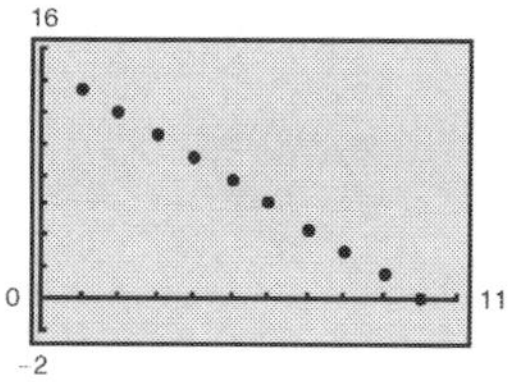

29. $a_n = 0.4n - 2$

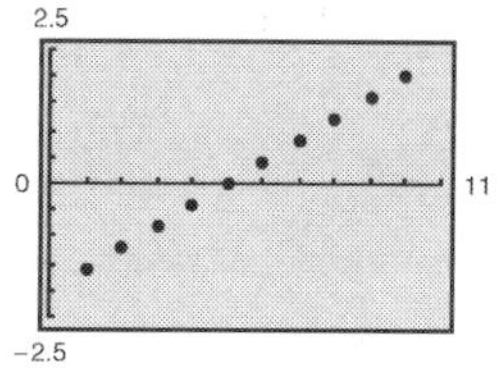

31. $a_1 = 4,\ d = 5$

$a_n = a_1 + (n - 1)d = 4 + (n - 1)5 = 5n - 1$

33. $a_1 = 50,\ d = -12$

$a_n = a_1 + (n - 1)d$
$= 50 + (n - 1)(-12)$
$= -12n + 62$

35. $a_1 = 9, a_2 = 17$

$d = a_2 - a_1 = 17 - 9 = 8$

$a_n = a_1 + (n - 1)d = 9 + (n - 1)8 = 8n + 1$

37. $a_1 = -7, a_2 = -11$

$d = a_2 - a_1 = -11 - (-7) = -4$

$a_n = a_1 + (n - 1)d = -7 + (n - 1)(-4) = -4n - 3$

39. 2, 8, 14, 20, 26, …

$a_1 = 2,\ d = 6$

$a_n = a_1 + (n - 1)d = 2 + (n - 1)(6) = 6n - 4$

41. $2, \frac{11}{4}, \frac{7}{2}, \frac{17}{4}, 5, \dots$

$a_1 = 2,\ d = \frac{3}{4}$

$a_n = a_1 + (n - 1)d = 2 + (n - 1)\frac{3}{4} = \frac{3}{4}n + \frac{5}{4}$

43. $a_7 = 29, d = 4$

(a) $a_8 = 29 + 4 = 33$

(b) $a_n = a_1 + (n - 1)d$

$a_7 = a_1 + (7 - 1)(4)$

$29 = a_1 + 6(4)$

$29 = a_1 + 24$

$5 = a_1$

45. $a_{12} = -13, d = -3$

(a) $a_{13} = -13 + (-3) = -16$

(b) $a_n = a_1 + (n - 1)d$

$a_{12} = a_1 + (12 - 1)(-3)$

$-13 = a_1 + (11)(-3)$

$-13 = a_1 - 33$

$20 = a_1$

47. $a_{14} = 15\frac{1}{3}, d = \frac{1}{3}$

(a) $a_{15} = 15\frac{1}{3} + \frac{1}{3} = 15\frac{2}{3}$

(b) $a_n = a_1 + (n - 1)d$

$a_{14} = a_1 + (14 - 1)\left(\frac{1}{3}\right)$

$15\frac{1}{3} = a_1 + 13\left(\frac{1}{3}\right)$

$15\frac{1}{3} = a_1 + \frac{13}{3}$

$11 = a_1$

49. $a_1 = 2, a_{12} = 46$

$a_{12} = a_1 + 11d$

$46 = 2 + 11d$

$44 = 11d$

$4 = d$

So, the first five terms of the sequence are 2, 6, 10, 14, and 18.

51. $a_8 = 26, a_{12} = 10$

$a_{12} = a_8 + 4d$

$10 = 26 + 4d$

$-16 = 4d$

$-4 = d$

$a_n = a_1 + (n - 1)d$

$a_8 = a_1 + (8 - 1)(-4)$

$26 = a_1 + 7(-4)$

$26 = a_1 - 28$

$54 = a_1$

So, the first 5 terms of the sequence are 54, 50, 46, 42, and 38.

53. $a_7 = 3.7, a_{13} = 7.3$

$a_{13} = a_7 + 6d$

$7.3 = 3.7 + 6d$

$3.6 = 6d$

$0.6 = d$

$a_n = a_1 + (n - 1)d$

$a_7 = a_1 + (7 - 1)(0.6)$

$3.7 = a_1 + 6(0.6)$

$3.7 = a_1 + 3.6$

$0.1 = a_1$

So, the first five terms of the sequence are 0.1, 0.7, 1.3, 1.9, and 2.5.

55. $2 + 4 + 6 + 8 + 10 + 12 + 14 + 16 + 18 + 20$

$n = 10, a_1 = 2, a_{10} = 20$

$S_n = \frac{n}{2}(a_1 + a_n)$

$S_{10} = \frac{10}{2}(2 + 20) = 5(22) = 110$

57. $-1 + (-3) + (-5) + (-7) + (-9)$

$n = 5, a_1 = -1, a_5 = -9$

$S_n = \frac{n}{2}(a_1 + a_n)$

$S_5 = \frac{5}{2}(-1 + (-9)) = \frac{5}{2}(-10) = -25$

59. The sum of the integers from 1 to 100

$n = 100, a_1 = 1, a_{100} = 100$

$S_n = \frac{n}{2}(a_1 + a_n)$

$S_{100} = \frac{100}{2}(1 + 100) = 50(101) = 5050$

61. The sum of the integers from -20 to 80

$n = 101, a_1 = -20, a_{101} = 80$

$S_n = \frac{n}{2}(a_1 + a_n)$

$S_{101} = \frac{101}{2}(-20 + 80) = \frac{101}{2}(60) = 3030$

63. $6, 12, 18, 24, \ldots, n = 8$

$a_1 = 6, d = 6, n = 8$

$a_8 = 6 + (8 - 1)6 = 48$

$S_8 = \frac{8}{2}(6 + 48) = 216$

65. $-10, -8, -6, -4, \ldots, n = 12$

$a_1 = -10, d = 2, n = 12$

$a_{12} = -10 + 2(11) = 12$

$S_{12} = \frac{12}{2}(-10 + 12) = 12$

67. $0.5, 0.9, 1.3, 1.7, \ldots$

$a_1 = 0.5, d = 0.4, n = 10$

$a_{10} = 0.5 + 9(0.4) = 4.1$

$S_{10} = \frac{10}{2}(0.5 + 4.1) = 23$

69. $a_1 = 100, a_{25} = 220, n = 25$

$S_n = \frac{n}{2}[a_1 + a_n]$

$S_{25} = \frac{25}{2}(100 + 220) = 4000$

71. $a_1 = 1, d = 1, a_{50} = 50$

$\sum_{n=1}^{50} n = \frac{50}{2}(1 + 50) = 1275$

73. $a_{32} = 160, d = 5, a_{100} = 500$

$\sum_{n=32}^{100} 5n = \frac{69}{2}(160 + 500) = 22,770$

75. $a_0 = 1000, d = -5, a_{50} = 750$

$\sum_{n=0}^{50} (100 - 5n) = \frac{51}{2}(1000 + 750) = 44,625$

77. $a_1 = 5, d = 3, a_{20} = 62$

$\sum_{n=1}^{20} (3n + 2) = \frac{20}{2}(5 + 62) = 670$

79. $a_1 = \frac{5}{2}, d = \frac{1}{2}, a_{100} = \frac{104}{2}$

$\sum_{n=1}^{100} \frac{n + 4}{2} = \frac{100}{2}\left(\frac{5}{2} + \frac{104}{2}\right) = 2725$

81. Answers will vary. Sample answer:

$3 + 6 + 9 + 12 + 15 + \ldots + 300$

$a_1 = 3, d = 3, a_n = 300$

$a_n = a_1 + (n - 1)d = 3 + (n - 1)3 = 3n$

$3n = 300 \Rightarrow n = 100$

$\sum_{n=1}^{100} 3n$

83. $\frac{1}{2} + \frac{3}{4} + 1 + \frac{5}{4} + \frac{3}{2} + \ldots + \frac{11}{4}$

Answers will vary. Sample answer:

$a_1 = \frac{1}{2}, d = \frac{1}{4}, a_n = \frac{11}{4}$

$a_n = a_1 + (n - 1)d = \frac{1}{2} + (n - 1)\frac{1}{4} = \frac{1}{4}n + \frac{1}{4}$

$\frac{1}{4}n + \frac{1}{4} = \frac{11}{4} \Rightarrow n = 10$

$\sum_{n=1}^{10} \frac{n + 1}{4}$

85. $A_n = 35,000 + (n - 1)1750$

(a) $A_6 = 35,000 + (6 - 1)1750 = \$43,750$

(b) $S_n = \frac{n}{2}(a_1 + a_n)$

$\Rightarrow S_6 = \frac{6}{2}(35,000 + 43,750) = \$236,250$

87. $n = 20, a_1 = 20, a_{20} = 58$

$S_n = \frac{n}{2}(a_1 + a_n)$

$S_{20} = \frac{20}{2}(20 + 58) = 10(78) = 780$

There are 780 seats in the auditorium.

89. $a_1 = 14, a_{18} = 31$

$S_{18} = \frac{18}{2}(14 + 31) = 405$ bricks

91. (a)

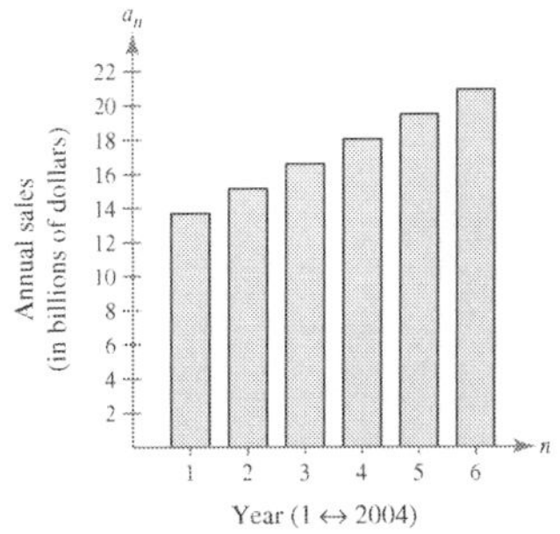

(b) $a_1 = 12.229 + 1.4471(1) = 13.6761$

$a_6 = 12.229 + 1.4471(6) = 20.9116$

$S_n = \frac{n}{2}(a_1 + a_n)$

$S_6 = \frac{6}{2}(13.6761 + 20.9116) = 3(34.5877) = 103.7631$

The total sales for Nike were about \$103.8 billion.

93. The total profit would be $50,000 + 60,000 + 70,000 + 80,000 + 90,000 = \$350,000$.

Answers will vary. Sample answer: Some economic factors that could prevent the company from meeting the profit goal include a weak economy, a recession, loss of market share to competitors, or changes in supply (raw materials).

95. $a_1 = 4.9,\ d = 9.8,\ n = 10$

$a_{10} = 4.9 + 9(9.8) = 93.1$

$S_{10} = \frac{10}{2}(4.9 + 93.1) = 490$

The object will fall 490 meters in 10 seconds.

Section 7.3 Geometric Sequences and Series

Skills Warm Up

1. $\left(\frac{4}{5}\right)^3 = \frac{4}{5}\cdot\frac{4}{5}\cdot\frac{4}{5} = \frac{64}{125}$
2. $\left(\frac{3}{4}\right)^2 = \frac{3}{4}\cdot\frac{3}{4} = \frac{9}{16}$
3. $2^{-4} = \frac{1}{2^4} = \frac{1}{2\cdot 2\cdot 2\cdot 2} = \frac{1}{16}$
4. $4(3^4) = 4(3\cdot 3\cdot 3\cdot 3) = 4(81) = 324$
5. $(2n)(3n^2) = 2\cdot 3\cdot n\cdot n^2 = 6n^{1+2} = 6n^3$
6. $n(3n)^3 = n\cdot 3^3\cdot n^3 = 3\cdot 3\cdot 3\cdot n^{1+3} = 27n^4$
7. $\frac{4n^5}{n^2} = 4n^{5-2} = 4n^3$
8. $\frac{(2n)^3}{8n} = \frac{2^3\cdot n^3}{8n} = \frac{2\cdot 2\cdot 2\cdot n^{3-1}}{8} = n^2$
9. $2 + 2(3) + 2(3^2) + 2(3^3) = \sum_{n=1}^{4} 2(3^{n-1})$
10. $3 + 3(2) + 3(2^2) + 3(2^3) = \sum_{n=1}^{4} 3(2^{n-1})$

1. $3, 15, 75, 375, 1875, \ldots$

$\frac{15}{3} = 5$

3. Because $\frac{-20}{10} = \frac{40}{-20} = \frac{-80}{40} = \frac{160}{-80} = -2$, the common ratio is $r = -2$.

5. $216, 36, 6, 1\ \frac{1}{6}, \ldots$

$\frac{36}{216} = \frac{1}{6}$

7. Because $\frac{1.8}{0.6} = \frac{5.4}{1.8} = \frac{16.2}{5.4} = \frac{48.6}{16.2} = 3$, the common ratio is $r = 3$.

9. $1, 4, 16, 64, \ldots$

Geometric sequence, $r = 4$

$a_n = (4)^{n-1}$

11. $4, -12, 36, -108, \ldots$

Geometric sequence, $r = -3$

$a_n = 4(-3)^{n-1}$

13. $1, -\frac{1}{2}, \frac{1}{4}, -\frac{1}{8}, \ldots$

Geometric sequence, $r = -\frac{1}{2}$

$a_n = \left(-\frac{1}{2}\right)^{n-1}$

15. $\frac{1}{2}, \frac{2}{3}, \frac{3}{4}, \frac{4}{5}, \ldots$

Not a geometric sequence

17. $1, \frac{1}{2}, \frac{1}{3}, \frac{1}{4}, \ldots$

Not a geometric sequence

19. $\frac{6}{4} = \frac{3}{2}, \frac{9}{6} = \frac{3}{2}, \frac{13.5}{9} = \frac{3}{2}$

Geometric sequence, $r = \frac{3}{2}$

$a_n = 4\left(\frac{3}{2}\right)^{n-1}$

21. $a_1 = 4,\ r = 2$

$a_1 = 4$

$a_2 = 4(2) = 8$

$a_3 = 8(2) = 16$

$a_4 = 16(2) = 32$

$a_5 = 32(2) = 64$

23. $a_1 = 1,\ r = \frac{1}{3}$

$a_1 = 1$

$a_2 = 1\left(\frac{1}{3}\right) = \frac{1}{3}$

$a_3 = \frac{1}{3}\left(\frac{1}{3}\right) = \frac{1}{9}$

$a_4 = \frac{1}{9}\left(\frac{1}{3}\right) = \frac{1}{27}$

$a_5 = \frac{1}{27}\left(\frac{1}{3}\right) = \frac{1}{81}$

25. $a_1 = \frac{1}{2}, r = -\frac{1}{3}$

$a_1 = \frac{1}{2}$

$a_2 = \frac{1}{2}\left(-\frac{1}{3}\right) = -\frac{1}{6}$

$a_3 = -\frac{1}{6}\left(-\frac{1}{3}\right) = \frac{1}{18}$

$a_4 = \frac{1}{18}\left(-\frac{1}{3}\right) = -\frac{1}{54}$

$a_5 = -\frac{1}{54}\left(-\frac{1}{3}\right) = \frac{1}{162}$

27. $a_1 = 200, r = 0.5$

$a_1 = 200$

$a_2 = 200(0.5)^1 = 100$

$a_3 = 200(0.5)^2 = 50$

$a_4 = 200(0.5)^3 = 25$

$a_5 = 200(0.5)^4 = 12.5$

29. $a_1 = 1, r = e$

$a_1 = 1$

$a_2 = 1(e) = e$

$a_3 = (e)(e) = e^2$

$a_4 = (e^2)(e) = e^3$

$a_5 = (e^3)(e) = e^4$

31. $a_1 = 16, r = \frac{1}{4}, n = 5$

$a_n = a_1 r^{n-1}$

$a_5 = 16\left(\frac{1}{4}\right)^4 = \left(\frac{1}{4}\right)^2 = \frac{1}{16}$

33. $a_1 = 6, r = -\dfrac{1}{3}, n = 12$

$a_n = a_1 r^{n-1}$

$a_{12} = 6\left(-\dfrac{1}{3}\right)^{11} = -\dfrac{2}{3^{10}}$

35. $a_1 = 100, r = e, n = 9$

$a_n = a_1 r^{n-1}$

$a_9 = 100(e)^8 = 100e^8$

37. $a_1 = 500, r = 1.02, n = 40$

$a_n = a_1 r^{n-1}$

$a_{40} = 500(1.02)^{39} \approx 1082.37$

39. $7, 14, 28, 56, \ldots$

The common ratio is $r = 2$, and $a_1 = 7$, so

$a_n = 7(2)^{n-1}$.

$a_7 = 7(2)^{7-1} = 7(2)^6 = 448$

41. $4, 12, 36, 108, \ldots$

The common ratio is $r = 3$, and $a_1 = 4$, so

$a_n = 4(3)^{n-1}$.

$a_{15} = 4(3)^{15-1} = 4(3)^{14} = 19{,}131{,}876$

43. $6, 30, 150, 750, \ldots$

The common ratio is $r = 5$, and $a_1 = 6$, so

$a_n = 6(5)^{n-1}$.

$a_{20} = 6(5)^{20-1} = 6(5)^{19} \approx 1.14 \times 10^{14}$

45. $1, \frac{1}{2}, \frac{1}{4}, \frac{1}{8}, \ldots$

The common ratio is $r = \frac{1}{2}$, and $a_1 = 1$, so

$a_n = 1\left(\frac{1}{2}\right)^{n-1}$.

$a_8 = 1\left(\frac{1}{2}\right)^{8-1} = \left(\frac{1}{2}\right)^7 = \frac{1}{2^7} = \frac{1}{128}$

47. $a_2 = -40, a_4 = -10, n = 3$

$a_3 = a_2 r = -40$

$a_4 = a_3 r = (a_2 r)r = (-40r)r = -10 \Rightarrow r = \pm\frac{1}{2}$

$a_3 = -40\left(\frac{1}{2}\right) = -20 \text{ or } = -40\left(-\frac{1}{2}\right) = 20$

49. $a_2 = -18, a_5 = \frac{2}{3}, n = 6$

$a_6 = a_5 r = \frac{2}{3}r$

$a_5 = a_2 r^3 = (a_2 r)r^2 = (-18r)r^2 = \frac{2}{3} \Rightarrow r = -\frac{1}{3}$

$a_6 = \frac{2}{3}\left(-\frac{1}{3}\right) = -\frac{2}{9}$

51. $a_n = 18\left(\frac{2}{3}\right)^{n-1}$

$a_1 = 18\left(\frac{2}{3}\right)^{1-1} = 18$

$a_2 = 18\left(\frac{2}{3}\right)^{2-1} = 12$

$a_3 = 18\left(\frac{2}{3}\right)^{3-1} = 8$

$a_4 = 18\left(\frac{2}{3}\right)^{4-1} = \frac{16}{3}$

$a_n = 18\left(\frac{2}{3}\right)^{n-1}$ corresponds to graph (a).

53. $a_n = 18\left(\frac{3}{2}\right)^{n-1}$

$a_1 = 18\left(\frac{3}{2}\right)^{1-1} = 18$

$a_2 = 18\left(\frac{3}{2}\right)^{2-1} = 27$

$a_3 = 18\left(\frac{3}{2}\right)^{3-1} = \frac{81}{2}$

$a_4 = 18\left(\frac{3}{2}\right)^{4-1} = \frac{243}{4}$

$a_n = 18\left(\frac{3}{2}\right)^{n-1}$ corresponds to graph (b).

55. $a_n = 12(-0.75)^{n-1}$

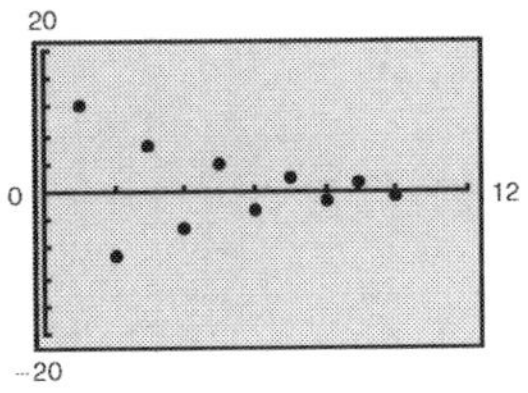

57. $a_n = 12(-0.4)^{n-1}$

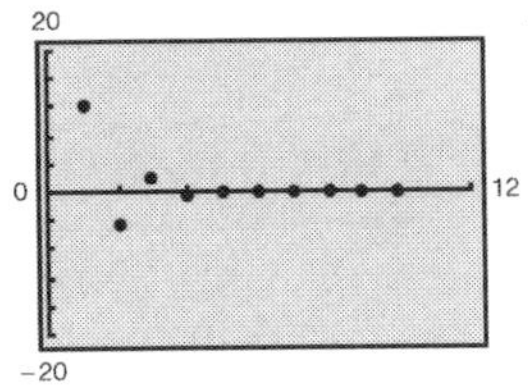

59. $a_n = 2(1.3)^{n-1}$

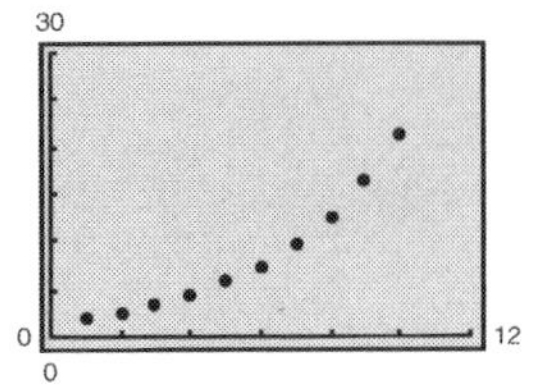

61. $\sum_{n=1}^{10} 8(2^n) = a_1\left(\frac{1-r^n}{1-r}\right) = 8(2)\left[\frac{1-2^{10}}{1-2}\right] = 16,368$

63. $\sum_{n=0}^{8} 2^n = \sum_{n=1}^{9} 2^{n-1} \Rightarrow a_1 = 1,\ r = 2$

$S_9 = \frac{1(1-2^9)}{1-2} = 511$

65. $\sum_{i=1}^{10} 8\left(\frac{1}{4}\right)^{i-1} = a_1\left(\frac{1-r^n}{1-r}\right) = 8\left[\frac{1-(1/4)^{10}}{1-(1/4)}\right] = \frac{349,525}{32,768} \approx 10.67$

67. $\sum_{i=1}^{7} 64\left(-\frac{1}{2}\right)^{i-1} = 64 + 64\left(-\frac{1}{2}\right)^1 + 64\left(-\frac{1}{2}\right)^2 + \cdots + 64\left(-\frac{1}{2}\right)^6 \Rightarrow a_1 = 64,\ r = -\frac{1}{2}$

$S_7 = 64\left[\frac{1-\left(-\frac{1}{2}\right)^7}{1-\left(-\frac{1}{2}\right)}\right] = \frac{128}{3}\left[1-\left(-\frac{1}{2}\right)^7\right] = 43$

69. $\sum_{n=0}^{40} 2\left(-\frac{1}{4}\right)^n = 2 + 2\left(-\frac{1}{4}\right) + 2\left(-\frac{1}{4}\right)^2 + \cdots + 2\left(-\frac{1}{4}\right)^{40} \Rightarrow a_1 = 2,\ r = -\frac{1}{4},\ n = 41$

$S_{41} = 2\left[\frac{1-\left(-\frac{1}{4}\right)^{41}}{1-\left(-\frac{1}{4}\right)}\right] = \frac{8}{5}\left[1-\left(-\frac{1}{4}\right)^{41}\right] \approx 1.6 = \frac{8}{5}$

71. $\sum_{n=0}^{20} 3\left(\frac{3}{2}\right)^n = \sum_{n=1}^{21} 3\left(\frac{3}{2}\right)^{n-1} \Rightarrow a_1 = 3,\ r = \frac{3}{2}$

$S_{21} = a_1\left(\frac{1-r^n}{1-r}\right) = 3\left[\frac{1-(3/2)^{21}}{1-3/2}\right] = -6\left[1-\left(\frac{3}{2}\right)^{21}\right] \approx 29,921.31$

73. $\sum_{i=1}^{40} 4\left(-\frac{5}{2}\right)^{i-1} \Rightarrow a_1 = 4,\ r = -\frac{5}{2}$

$S_{40} = 4\left[\frac{1-\left(-\frac{5}{2}\right)^{40}}{1-\left(-\frac{5}{2}\right)}\right] = \frac{8}{7}\left[1-\left(-\frac{5}{2}\right)^{40}\right] \approx -9.45 \times 10^{15}$

75. $5 + 15 + 45 + \cdots + 3645$

$r = 3$ and $3645 = 5(3)^{n-1}$

$729 = 3^{n-1} \Rightarrow 6 = n - 1 \Rightarrow n = 7$

So, the sum can be written as $\sum_{n=1}^{7} 5(3)^{n-1}$.

77. $2 - \frac{1}{2} + \frac{1}{8} - \cdots + \frac{1}{2048}$

$r = -\frac{1}{4}$ and $\frac{1}{2048} = 2\left(-\frac{1}{4}\right)^{n-1}$

By trial and error, you find that $n = 7$.

So, the sum can be written as $\sum_{n=1}^{7} 2\left(-\frac{1}{4}\right)^{n-1}$.

79. $0.1 + 0.4 + 1.6 + \cdots + 102.4$

$r = 4$ and $102.4 = 0.1(4)^{n-1}$

$1024 = 4^{n-1} \Rightarrow 5 = n - 1 \Rightarrow n = 6$

So, the sum can be written as $\sum_{n=1}^{6} 0.1(4)^{n-1}$.

81. $a_1 = 1, r = \frac{1}{2}$

$$\sum_{n=0}^{\infty}\left(\frac{1}{2}\right)^n = \frac{a_1}{1-r} = \frac{1}{1-(1/2)} = 2$$

83. $a_1 = 1, r = -\frac{1}{2}$

$$\sum_{n=1}^{\infty}\left(-\frac{1}{2}\right)^{n-1} = \frac{a_1}{1-r} = \frac{1}{1-(-1/2)} = \frac{2}{3}$$

85. $\sum_{n=0}^{\infty}\left(\frac{3}{2}\right)^n = 1 + \frac{3}{2} + \frac{9}{4} + \frac{27}{8} + \cdots$

This sum does not exist because $|r| \geq 1$.

87. $\sum_{n=0}^{\infty} 3\left(\frac{1}{10}\right)^n = 3 + 0.3 + 0.03 + 0.003 + \cdots = \frac{3}{1-1/10} = \frac{10}{3}$

89. $0.\overline{36} = \sum_{n=0}^{\infty} 0.36(0.01)^n = \frac{0.36}{1-0.01} = \frac{0.36}{0.99} = \frac{36}{99} = \frac{4}{11}$

91. $0.3\overline{18} = 0.3 + \sum_{n=0}^{\infty} 0.018(0.01)^n = \frac{3}{10} + \frac{0.018}{1-0.01} = \frac{3}{10} + \frac{0.018}{0.99} = \frac{3}{10} + \frac{18}{990} = \frac{3}{10} + \frac{2}{110} = \frac{35}{110} = \frac{7}{22}$

93. $A = 300\left(1 + \frac{0.06}{12}\right)^1 + \cdots + 300\left(1 + \frac{0.06}{12}\right)^{48}$

$$\sum_{n=1}^{48} 300\left(\frac{12.06}{12}\right)^n$$

(a) $n = 4 \cdot 12 = 48,\ a_1 = 301.5,\ r = \frac{12.06}{12}$

$$S_{48} = 301.5\left[\frac{1-\left(\frac{12.06}{12}\right)^{48}}{1-\left(\frac{12.06}{12}\right)}\right] = 301.5\left[\frac{1-\left(\frac{12.06}{12}\right)^{48}}{-0.005}\right] \approx \$16{,}310.50$$

No, after 4 years there will be approximately \$16,310.50, which is less than \$17,000.

(b) $\sum_{n=1}^{48} 300\left(\frac{12.09}{12}\right)^n$

$n = 48,\ a_1 = 302.25,\ r = \frac{12.09}{12}$

$$S_{48} = 302.25\left[\frac{1-\left(\frac{12.09}{12}\right)^{48}}{1-\left(\frac{12.09}{12}\right)}\right] = 302.25\left[\frac{1-\left(\frac{12.09}{12}\right)^{48}}{-0.0075}\right] \approx \$17{,}385.63$$

Yes, after 4 years there will be approximately \$17,385.63, which is more than \$17,000.

95. Compounded monthly:

$$A = \sum_{n=1}^{120} 100\left(1 + \frac{0.08}{12}\right)^n \approx \$18{,}416.57$$

Compounded continuously:

$$A = \sum_{n=1}^{120} 100e^{(0.08n)/12} \approx \$18{,}444.46$$

97. $a_n = 10.46e^{0.064n}, n = 0, 1, 2, ..., 5$

$$\sum_{n=0}^{5} 10.46\left(e^{0.064}\right)^n = \sum_{n=1}^{6} 10.46\left(e^{0.064}\right)^{n-1}$$

$a_1 = 10.46, r = e^{0.064}, n = 6$

$$S_n = a_1\left(\frac{1 - r^n}{1 - r}\right)$$

$$S_6 = 10.46\left(\frac{1 - e^{0.064(6)}}{1 - e^{0.064}}\right) \approx 74.09$$

The total net profit for Wal-Mart stores from 2004 through 2009 was about \$74.1 billion.

99. (a)

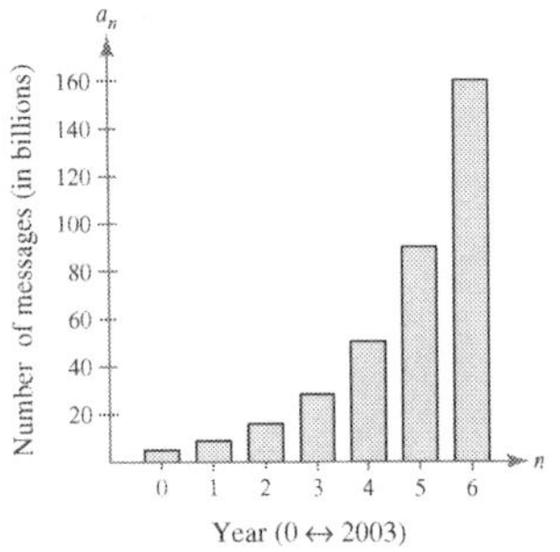

(b) $a_n = 5.09e^{0.575n}, n = 0, 1, ..., 6$

$$\sum_{n=0}^{6} 5.09\left(e^{0.575}\right)^n = \sum_{n=1}^{7} 5.09\left(e^{0.575}\right)^{n-1}$$

$a_1 = 5.09, r = e^{0.575}, n = 7$

$$S_n = a_1\left(\frac{1 - r^n}{1 - r}\right)$$

$$S_7 = 5.09\left(\frac{1 - e^{0.575(7)}}{1 - e^{0.575}}\right) \approx 360.1$$

The total number of text messages from 2003 through 2009 was about 360.1 billion.

101. $S_n = \sum_{i=1}^{n} 0.01(2)^{i-1} = \sum_{i=1}^{n} 0.005(2)^i$

$$S_{29} = 0.01\left[\frac{1 - 2^{29}}{1 - 2}\right] = \$5,368,709.11$$

$$S_{30} = 0.01\left[\frac{1 - 2^{30}}{1 - 2}\right] = \$10,737,418.23$$

$$S_{31} = 0.01\left[\frac{1 - 2^{31}}{1 - 2}\right] = \$21,474,836.47$$

The job whose wage doubles daily pays more.

Chapter 7 Quiz Yourself

1. $a_n = 2n - 3$

$a_1 = 2(1) - 3 = -1$

$a_2 = 2(2) - 3 = 1$

$a_3 = 2(3) - 3 = 3$

$a_4 = 2(4) - 3 = 5$

$a_5 = 2(5) - 3 = 7$

2. $a_n = \dfrac{n}{(n + 1)^2}$

$a_1 = \dfrac{1}{(1 + 1)^2} = \dfrac{1}{4}$

$a_2 = \dfrac{2}{(2 + 1)^2} = \dfrac{2}{9}$

$a_3 = \dfrac{3}{(3 + 1)^2} = \dfrac{3}{16}$

$a_4 = \dfrac{4}{(4 + 1)^2} = \dfrac{4}{25}$

$a_5 = \dfrac{5}{(5 + 1)^2} = \dfrac{5}{36}$

3. $\frac{5!}{3!} = \frac{5 \cdot 4 \cdot 3!}{3!} = 5 \cdot 4 = 20$

4. $\frac{30!}{27!} = \frac{30 \cdot 29 \cdot 28 \cdot 27!}{27!} = 30 \cdot 29 \cdot 28 = 24{,}360$

5. $3, \frac{3}{4}, \frac{1}{3}, \frac{3}{16}, \frac{3}{25}, \ldots$

$\frac{3}{1^2}, \frac{3}{2^2}, \frac{3}{3^2}, \frac{3}{4^2}, \frac{3}{5^2}$

$a_n = \frac{3}{n^2}$

6. $\frac{1}{2}, \frac{1}{3}, \frac{1}{4}, \frac{1}{5}, \frac{1}{6}, \ldots$

$a_n = \frac{1}{n+1}$

7. $a_n = a_1 + (n-1)d = 2 + (n-1)(5) = 5n - 3$

8. $a_n = a_1 + (n-1)d = 7 + (n-1)(3) = 3n + 4$

9. $a_1 = 10, r = 3$

$a_n = a_1 r^{n-1}$

$a_n = 10(3)^{n-1}$

10. $a_1 = 4, r = -\frac{1}{2}$

$a_n = a_1 r^{n-1}$

$a_n = 4\left(-\frac{1}{2}\right)^{n-1}$

11. $a_1 = 3, a_2 = 11, d = 11 - 3 = 8$

$a_n = a_1 + (n-1)d$

$a_n = 3 + (n-1)8$

$a_n = 3 + 8n - 8$

$a_n = 8n - 5$

When $n = 13$: $a_{13} = 8(13) - 5 = 99$

12. $a_7 = 175, d = 28$

$a_8 = 175 + 28 = 203$

$a_n = a_1 + (n-1)d$

$a_7 = a_1 + (7-1)(28)$

$175 = a_1 + 6(28)$

$175 = a_1 + 168$

$7 = a_1$

13. $a_1 = 16, r = 1.04$

$a_n = 16(1.04)^{n-1}$

$a_{20} = 16(1.04)^{20-1} = 16(1.04)^{19} \approx 33.71$

14. $a_5 = 100, a_{10} = \frac{100}{243}$

$r^5 = \frac{a_{10}}{a_5}$

$r^5 = \frac{\frac{100}{243}}{100}$

$r^5 = \frac{1}{243}$

$r = \sqrt[5]{\frac{1}{243}} = \frac{1}{3}$

$a_{12} = a_{10}r^2 = \frac{100}{243}\left(\frac{1}{3}\right)^2 = \frac{100}{2187}$

15. The first five terms of the sequence are
2, 5, 8, 11, 14, ….

The sequence is arithmetic. The common difference is 3.

16. 1, 1, 2, 3, 5, 8, 13, …

Neither; no common ratio, no common difference.

17. $1, \frac{1}{2}, \frac{1}{4}, \frac{1}{8}, \ldots$

The sequence is geometric. The common ratio is $\frac{1}{2}$.

18. The first five terms of the sequence are
6, 12, 24, 48, 96, ….

The sequence is geometric. The common ratio is 2.

19. $\sum_{i=1}^{5} 9 = 9 + 9 + 9 + 9 + 9 = 45$

20. $\sum_{i=1}^{25} 2i^2 = 2 + 8 + 18 + 32 = 60$

21. $\sum_{n=1}^{25} (4n - 3)$

$a_1 = 1, a_{25} = 97$

$S_{25} = \frac{25}{2}(1 + 97) = 1225$

22. $\sum_{n=1}^{6} 2^n = 2 + 4 + 8 + 16 + 32 + 64 = 126$

23. $\sum_{n=0}^{\infty} \left(-\frac{1}{3}\right)^n = 1 - \frac{1}{3} + \frac{1}{9} - \frac{1}{27} + \frac{1}{81} + \cdots$

$= \frac{1}{1 - (-1/3)} = \frac{3}{4}$

This is the sum of a geometric sequence with $a_1 = 1$ and $r = -\frac{1}{3}$.

24. $\sum_{n=0}^{\infty} 5\left(\frac{1}{5}\right)^n = 5 + 1 + \frac{1}{5} + \frac{1}{25} + \frac{1}{125} + \cdots = \frac{5}{1 - (1/5)} = \frac{25}{4}$

This is the sum of a geometric sequence with $a_1 = 5$ and $r = \frac{1}{5}$.

25. (a) During the sixth year of employment the person will have received five raises of $1940.

$a_6 = a_1 + (n - 1)d = 48{,}500 + (6 - 1)(1940) = 58{,}200$

The employee's salary in the sixth year will be $58,200.

(b) $n = 6,\ a_1 = 48{,}500,\ a_6 = 58{,}200$

$S_n = \frac{n}{2}(a_1 + a_n)$

$S_6 = \frac{6}{2}(48{,}500 + 58{,}200) = 3(106{,}700) = 320{,}100$

The total compensation will be $320,100.

26. $A = 200\left(1 + \frac{0.04}{12}\right) + \cdots + 200\left(1 + \frac{0.04}{12}\right)^{240}$

$\sum_{n=1}^{240} 200\left(\frac{301}{300}\right)^n,\ A_1 = 200\left(\frac{301}{300}\right) = \frac{602}{3},\ r = \frac{301}{300}$

$$S_{240} = \frac{602}{3}\left(\frac{1 - \left(\frac{301}{300}\right)^{240}}{1 - \frac{301}{300}}\right) \approx \$73{,}599.44$$

The total amount in the account will be $73,599.44.

Section 7.4 The Binomial Theorem

Skills Warm Up

1. $5x^2(x^3 + 3) = 5x^2(x^3) + 5x^2(3) = 5x^{2+3} + 5 \cdot 3 \cdot x^2 = 5x^5 + 15x^2$

2. $(x + 5)(x^2 - 3) = x(x^2) + x(-3) + 5(x^2) + 5(-3) = x^3 - 3x + 5x^2 - 15 = x^3 + 5x^2 - 3x - 15$

3. $(x + 4)^2 = (x + 4)(x + 4) = x(x) + x(4) + 4(x) + 4(4) = x^2 + 4x + 4x + 16 = x^2 + 8x + 16$

4. $(2x - 3)^2 = (2x - 3)(2x - 3) = (2x)(2x) + 2x(-3) + (-3)(2x) + (-3)(-3) = 4x^2 - 6x - 6x + 9 = 4x^2 - 12x + 9$

5. $x^2y(3xy^{-2}) = 3x^{2+1}y^{1-2} = 3x^3y^{-1} = \frac{3x^3}{y}$

6. $(-2z)^5 = (-2)^5 \cdot z^5$
$= (-2)(-2)(-2)(-2)(-2)z^5$
$= -32z^5$

7. $5! = 5 \cdot 4 \cdot 3 \cdot 2 \cdot 1 = 120$

8. $\frac{8!}{5!} = \frac{8 \cdot 7 \cdot 6 \cdot \cancel{5 \cdot 4 \cdot 3 \cdot 2 \cdot 1}}{\cancel{5 \cdot 4 \cdot 3 \cdot 2 \cdot 1}} = 8 \cdot 7 \cdot 6 = 336$

9. $\frac{10!}{7!} = \frac{10 \cdot 9 \cdot 8 \cdot \cancel{7 \cdot 6 \cdot 5 \cdot 4 \cdot 3 \cdot 2 \cdot 1}}{\cancel{7 \cdot 6 \cdot 5 \cdot 4 \cdot 3 \cdot 2 \cdot 1}}$
$= 10 \cdot 9 \cdot 8$
$= 720$

10. $\frac{6!}{3!3!} = \frac{\cancel{6} \cdot 5 \cdot 4 \cdot \cancel{3 \cdot 2 \cdot 1}}{\cancel{3} \cdot \cancel{2} \cdot 1 \cdot \cancel{3 \cdot 2 \cdot 1}} = \frac{5 \cdot 4}{1} = 20$

1. $_6C_2 = \dfrac{6!}{2!4!} = \dfrac{6 \cdot 5 \cdot \cancel{4!}}{2 \cdot 1 \cdot \cancel{4!}} = 15$

3. $_{10}C_4 = \dfrac{10!}{4!6!} = \dfrac{10 \cdot 9 \cdot 8 \cdot 7 \cdot \cancel{6!}}{4 \cdot 3 \cdot 2 \cdot 1 \cdot \cancel{6!}} = 210$

5. $_{12}C_0 = \dfrac{12!}{(12-0)!0!} = \dfrac{\cancel{12!}}{\cancel{12!} \cdot 1} = 1$

7. $_{20}C_{15} = \dfrac{20!}{15!5!} = \dfrac{20 \cdot 19 \cdot 18 \cdot 17 \cdot 16 \cdot \cancel{15!}}{\cancel{15!} \cdot 5 \cdot 4 \cdot 3 \cdot 2 \cdot 1} = 15{,}504$

9. $\binom{7}{7} = {_7C_7} = \dfrac{\cancel{7!}}{\cancel{7!}0!} = 1$

11. $\binom{100}{98} = {_{100}C_{98}} = \dfrac{100!}{98!2!} = \dfrac{100 \cdot 99 \cdot \cancel{98!}}{\cancel{98!} \cdot 2 \cdot 1} = 4950$

13. $_4C_2 = 6$

15. $_7C_4 = 35$

17. $\binom{8}{5} = {_8C_5} = 56$

19. $\binom{10}{8} = {_{10}C_8} = 45$

21. $(x+1)^4 = {_4C_0}x^4 - {_4C_1}x^3(1) + {_4C_2}x^2(1)^2 + {_4C_3}x(1)^3 + {_4C_4}(1)^4 = x^4 + 4x^3 + 6x^2 + 4x + 1$

23. $(x+6)^4 = {_4C_0}x^4 + {_4C_1}x^3(6) + {_4C_2}x^2(6)^2 + {_4C_3}x(6)^3 + {_4C_4}(6)^4$

$= x^4 + 24x^3 + 216x^2 + 864x + 1296$

25. $(x-2)^4 = {_4C_0}x^4 - {_4C_1}x^3(2) + {_4C_2}x^2(2)^2 - {_4C_3}x(2)^3 + {_4C_4}(2)^4$

$= x^4 - 4x^3(2) + 6x^2(4) - 4x(8) + 16$

$= x^4 - 8x^3 + 24x^2 - 32x + 16$

27. $(x+y)^5 = {_5C_0}x^5 + {_5C_1}x^4y + {_5C_2}x^3y^2 + {_5C_3}x^2y^3 + {_5C_4}xy^4 + {_5C_5}y^5$

$= x^5 + 5x^4y + 10x^3y^2 + 10x^2y^3 + 5xy^4 + y^5$

29. $(3s+2)^4 = {_4C_0}(3s)^4 + {_4C_1}(3s)^3(2) + {_4C_2}(35)^2(2)^2 + {_4C_3}(3s)(2)^3 + {_4C_4}(2)^4$

$= 81s^4 + 4(2)(27s^3) + 6(4)(9s^2) + 4(8)(3s) + 16$

$= 81s^4 + 216s^3 + 216s^2 + 96s + 16$

31. $(3-2s)^6 = {_6C_0}(3)^6 + {_6C_1}(3)^5(-2s) + {_6C_2}(3)^4(-2s)^2 + {_6C_3}(3)^3(-2s)^3 + {_6C_4}(3)^3(-2s)^4 + {_6C_5}(3)(-2s)^5 + {_6C_6}(-2s)^6$

$= 729 - 6(243)(2s) + 15(81)(4s^2) - 20(27)(8s^3) + 15(9)(16s^4) - 6(3)(32s^5) + 64s^6$

$= 729 - 2916s + 4860s^2 - 4320s^3 + 2160s^4 - 576s^5 + 64s^6$

or

$= 64s^6 - 576s^5 + 2160s^4 - 4320s^3 + 4860s^2 - 2916s + 729$

33. $(x-y)^5 = {_5C_0}x^5 - {_5C_1}x^4y + {_5C_2}x^3y^2 - {_5C_3}x^2y^3 + {_5C_4}xy^4 - {_5C_5}y^5$

$= x^5 - 5x^4y + 10x^3y^2 - 10x^2y^3 + 5xy^4 - y^5$

35. $(1-2x)^3 = {_3C_0}1^3 - {_3C_1}1^2(2x) + {_3C_2}1(2x)^2 - {_3C_3}(2x)^3 = 1 - 3(2x) + 3(2x)^2 - (2x)^3 = 1 - 6x + 12x^2 - 8x^3$

37. $(x^2-y^2)^5 = {_5C_0}(x^2)^5 - {_5C_1}(x^2)^4(y^2) + {_5C_2}(x^2)^3(y^2)^2 - {_5C_3}(x^2)^2(y^2)^3 + {_5C_4}(x^2)(y^2)^4 - {_5C_5}(y^2)^5$

$= x^{10} - 5x^8y^2 + 10x^6y^4 - 10x^4y^6 + 5x^2y^8 - y^{10}$

39. $\left(\dfrac{1}{x}+y\right)^5 = {_5C_0}\left(\dfrac{1}{x}\right)^5 + {_5C_1}\left(\dfrac{1}{x}\right)^4 + {_5C_2}\left(\dfrac{1}{x}\right)^3y^2 + {_5C_3}\left(\dfrac{1}{x}\right)^2y^3 + {_5C_4}\left(\dfrac{1}{x}\right)y^4 + {_5C_5}y^5$

$= \dfrac{1}{x^5} + \dfrac{5y}{x^4} + \dfrac{10y^2}{x^3} + \dfrac{10y^3}{x^2} + \dfrac{5y^4}{x} + y^5$

41. $(1+i)^4 = {}_4C_0 1^4 + {}_4C_1(1)^3 i + {}_4C_2(1)^2 i^2 + {}_4C_3 1 \cdot i^3 + {}_4C_4 i^4 = 1 + 4i - 6 - 4i + 1 = -4$

43.
$$\begin{aligned}\left(3-\sqrt{-4}\right)^6 &= (3-2i)^6\\ &= {}_6C_0 3^6 - {}_6C_1 3^5(2i) + {}_6C_2 3^4(2i)^2 - {}_6C_3 3^3(2i)^3 + {}_6C_4 3^2(2i)^4 - {}_6C_5 3(2i)^5 + {}_6C_6(2i)^6\\ &= 729 - 2916i - 4860 + 4320i + 2160 - 576i - 64\\ &= -2035 + 828i\end{aligned}$$

45.
$$\begin{aligned}\left(-\frac{1}{2}+\frac{\sqrt{3}}{2}i\right)^3 &= \frac{1}{8}\left(-1+\sqrt{3}i\right)^3\\ &= \frac{1}{8}\left[(-1)^3 + 3(-1)^2\left(\sqrt{3}i\right) + 3(-1)\left(\sqrt{3}i\right)^2 + \left(\sqrt{3}i\right)^3\right]\\ &= \frac{1}{8}\left[-1 + 3\sqrt{3}i + 9 - 3\sqrt{3}i\right]\\ &= 1\end{aligned}$$

47. 5th Row of Pascal's Triangle: 1 5 10 10 5 1

$$\begin{aligned}(2t-1)^5 &= 1(2t)^5 + 5(2t)^4(-1) + 10(2t)^3(-1)^2 + 10(2t)^2(-1)^3 + 5(2t)(-1)^4 + 1(-2)^5\\ &= 32t^5 - 80t^4 + 80t^3 - 40t^2 + 10t - 1\end{aligned}$$

49. 4th Row of Pascal's Triangle: 1 4 6 4 1

$$\begin{aligned}(3+2z)^4 &= 3^4 + 4(3)^3(2z) + 6(3)^2(2z)^2 + 4(3)(2z)^3 + (2z)^4\\ &= 81 + 216z + 216z^2 + 96z^3 + 16z^4\end{aligned}$$

51. $(x+y)^{10}$, $n = 4$

${}_{10}C_3 x^7 y^3 = 120x^7y^3$

53. $(x-6y)^5$, $n = 3$

${}_5C_2 x^3(-6y)^2 = 360x^3y^2$

55. $(4x+3y)^9$, $n = 8$

${}_9C_7(4x)^2(3y)^7 = 1{,}259{,}712x^2y^7$

57. $(10x-3y)^{12}$, $n = 10$

${}_{12}C_9(10x)^3(-3y)^9 = -4{,}330{,}260{,}000x^3y^9$

59. The coefficient of x^5 in the expansion $(x+2)^{10}$ is

$$a = {}_{10}C_5 2^5 = \frac{10!}{5!5!} \cdot 2^5 = 8064.$$

The term is $8064x^5$.

61. The coefficient of x^8y^2 in the expansion of $(x-2y)^{10}$ is

$$a = {}_{10}C_2(-2)^2 = \frac{10!}{2!8!} \cdot 4 = 180.$$

The term is $180x^8y^2$.

63. The coefficient of x^5y^4 in the expansion of $(3x-2y)^9$ is

$$a = {}_9C_4(3)^5(-2)^4 = \frac{9!}{4!5!} \cdot 3^5 \cdot 2^4 = 489{,}888.$$

The term is $489{,}888x^5y^4$.

65. The coefficient of $x^6 = \left(x^2\right)^3$ in the expansion of $\left(x^2-1\right)^8$ is

$$a = {}_8C_3(-1)^3 = -\frac{8!}{3!5!} = -56.$$

The term is $-56x^6$.

67.
$$\begin{aligned}\left(\tfrac{1}{2}+\tfrac{1}{2}\right)^7 &= \left(\tfrac{1}{2}\right)^7 + 7\left(\tfrac{1}{2}\right)^6\left(\tfrac{1}{2}\right) + 21\left(\tfrac{1}{2}\right)^5\left(\tfrac{1}{2}\right)^2 + 35\left(\tfrac{1}{2}\right)^4\left(\tfrac{1}{2}\right)^3 + 35\left(\tfrac{1}{2}\right)^3\left(\tfrac{1}{2}\right)^4 + 21\left(\tfrac{1}{2}\right)^2\left(\tfrac{1}{2}\right)^5 + 7\left(\tfrac{1}{2}\right)\left(\tfrac{1}{2}\right)^6 + \left(\tfrac{1}{2}\right)^7\\ &= \tfrac{1}{128} + \tfrac{7}{128} + \tfrac{21}{128} + \tfrac{35}{128} + \tfrac{35}{128} + \tfrac{21}{128} + \tfrac{7}{128} + \tfrac{1}{128}\\ &= \tfrac{128}{128}\\ &= 1\end{aligned}$$

69. $(0.6 + 0.4)^5 = (0.6)^5 + 5(0.6)^4(0.4) + 10(0.6)^3(0.4)^2 + 10(0.6)^2(0.4)^3 + 5(0.6)(0.4)^4 + (0.4)^5$

$= 0.07776 + 0.2592 + 0.3456 + 0.2304 + 0.0768 + 0.01024$

$= 1$

71.

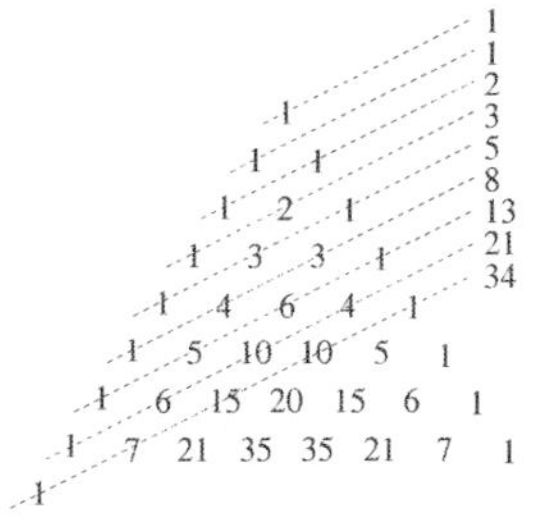

The first nine terms of the sequence are 1, 1, 2, 3, 5, 8, 13, 21, 34, ….

After the first two terms, the next terms are formed by adding the previous two terms.

$a_1 = 1,\ a_2 = 1$

$a_3 = a_1 + a_2 = 1 + 1 = 2$

$a_4 = a_2 + a_3 = 1 + 2 = 3$

$a_5 = a_3 + a_4 = 2 + 3 = 5$

$a_6 = a_4 + a_5 = 3 + 5 = 8$

$a_7 = a_5 + a_6 = 5 + 8 = 13$

This is called the Fibonacci sequence.

73. $a_1 = \begin{vmatrix} 1 & 1 \\ 1 & 2 \end{vmatrix} = 1$

$a_2 = \begin{vmatrix} 2 & 1 \\ 3 & 3 \end{vmatrix} = 3$

$a_3 = \begin{vmatrix} 3 & 1 \\ 6 & 4 \end{vmatrix} = 6$

$a_4 = \begin{vmatrix} 4 & 1 \\ 10 & 5 \end{vmatrix} = 10$

$a_5 = \begin{vmatrix} 5 & 1 \\ 15 & 6 \end{vmatrix} = 15$

In general, $a_n = a_{n-1} + n$. This is also one of the diagonals of Pascal's Triangle.

75. Answers will vary.

Section 7.5 Counting Principles

Skills Warm Up

1. $13 \cdot 8^2 \cdot 2^3 = 13 \cdot 64 \cdot 8 = 6656$

2. $10^2 \cdot 9^3 \cdot 4 = 100 \cdot 729 \cdot 4 = 291{,}600$

3. $\dfrac{12!}{2!(7!)(3!)} = \dfrac{12 \cdot 11 \cdot 10 \cdot 9 \cdot 8 \cdot 7!}{2 \cdot 1 \cdot 7! \cdot 3 \cdot 2 \cdot 1} = 11 \cdot 10 \cdot 9 \cdot 8 = 7920$

4. $\dfrac{25!}{22!} = \dfrac{25 \cdot 24 \cdot 23 \cdot 22!}{22!} = 25 \cdot 24 \cdot 23 = 13{,}800$

5. ${}_{12}C_7 = \dfrac{12!}{5! \cdot 7!} = \dfrac{12 \cdot 11 \cdot 10 \cdot 9 \cdot 8 \cdot 7!}{5 \cdot 4 \cdot 3 \cdot 2 \cdot 1 \cdot 7!} = 11 \cdot 9 \cdot 8 = 792$

6. ${}_{25}C_{22} = \dfrac{25!}{3! \cdot 22!} = \dfrac{25 \cdot 24 \cdot 23 \cdot 22!}{3 \cdot 2 \cdot 1 \cdot 22!} = 25 \cdot 4 \cdot 23 = 2300$

7. $\dfrac{n!}{(n-4)!} = \dfrac{n \cdot (n-1) \cdot (n-2) \cdot (n-3) \cdot (n-4)!}{(n-4)!} = n(n-1)(n-2)(n-3)$

8. $\dfrac{(2n)!}{4(2n-3)!} = \dfrac{2n(2n-1)(2n-2)(2n-3)!}{4(2n-3)!} = \dfrac{2n(2n-1)(2n-2)}{4} = n(n-1)(2n-1)$

Skills Warm Up —*continued*—

9. $\dfrac{2 \cdot 4 \cdot 6 \cdot 8 \cdots (2n)}{2^n} = \dfrac{2(1) \cdot 2(2) \cdot 2(3) \cdot 2(4) \cdots 2(n)}{\underbrace{2 \cdot 2 \cdot 2 \cdot 2 \cdots 2}_{n \text{ times}}} = 1 \cdot 2 \cdot 3 \cdot 4 \cdots n = n!$

10. $\dfrac{3 \cdot 6 \cdot 9 \cdot 12 \cdots (3n)}{3^n} = \dfrac{1(3) \cdot 2(3) \cdot 3(3) \cdot 4(3) \cdots 3(n)}{\underbrace{3 \cdot 3 \cdot 3 \cdot 3 \cdots 3}_{n \text{ times}}} = 1 \cdot 2 \cdot 3 \cdot 4 \cdots n = n!$

For Exercises 1–9 refer to 1, 2, 3, 4, 5, 6, 7, 8, 9, 10, 11, 12, 13, 14, 15 where 1 is considered an odd number.

1. The computer can pick 1, 3, 5, 7, 9, 11, 13, or 15; so 8 ways.

3. The computer can pick 2, 3, 5, 7, 11, or 13, so 6 ways.

5. The computer can pick 4, 8, or 12; so 3 ways.

7. The computer can pick $5 + 15$, $6 + 14$, $7 + 13$, $8 + 12$, $9 + 11$, $10 + 10$, $11 + 9$, $12 + 8$, $13 + 7$, $14 + 6$, or $15 + 5$; so 11 ways.

9. The computer can pick $5 + 15$, $6 + 14$, $7 + 13$, $8 + 12$, $9 + 11$, $11 + 9$, $12 + 8$, $13 + 7$, $14 + 6$, or $15 + 5$; so 10 ways.

11. Chemist: 5 choices

Statistician: 6 choices

Total: $5 \cdot 6 = 30$ ways

13. 1st Position: 2 choices

2nd Position: 5 choices

3rd Position: 4 choices

4th Position: 3 choices

5th Position: 2 choices

6th Position: 1 choice

Total: $2 \cdot 5 \cdot 4 \cdot 3 \cdot 2 \cdot 1 = 240$ ways

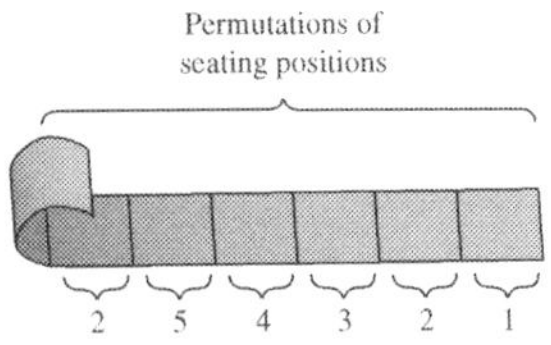

15. $26 \cdot 26 \cdot 10 \cdot 10 \cdot 10 \cdot 10 = 6{,}760{,}000$

17. $2^{10} = 1024$

19. (a) $9 \cdot 10 \cdot 10 = 900$

(b) $9 \cdot 9 \cdot 8 = 648$

(c) $9 \cdot 10 \cdot 2 = 180$

21. $40^3 = 64{,}000$

23. $5 \cdot 4 \cdot 3 \cdot 2 \cdot 1 = 120$

25. (a) $8 \cdot 7 \cdot 6 \cdot 5 \cdot 4 \cdot 3 \cdot 2 \cdot 1 = 40{,}320$

(b) $8 \cdot 1 \cdot 6 \cdot 1 \cdot 4 \cdot 1 \cdot 2 \cdot 1 = 384$

27. ${}_nP_r = \dfrac{n!}{(n-r)!}$

${}_4P_3 = \dfrac{4!}{1!} = 24$

29. ${}_4P_4 = \dfrac{4!}{(4-4)!} = \dfrac{4!}{0!} = 4! = 24$

31. ${}_6P_2 = \dfrac{6!}{(6-2)!} = \dfrac{6!}{4!} = \dfrac{6 \cdot 5 \cdot 4!}{4!} = 30$

33. ${}_8P_0 = \dfrac{8!}{(8-0)!} = \dfrac{8!}{8!} = 1$

35. ${}_{12}P_4 = \dfrac{12!}{(12-4)!} = \dfrac{12!}{8!} = \dfrac{12 \cdot 11 \cdot 10 \cdot 9 \cdot 8!}{8!} = 11{,}880$

37. ${}_{20}P_6 = 27{,}907{,}200$

39. ${}_{120}P_4 = 197{,}149{,}680$

41. B, R, O, O, M

$\dfrac{5!}{2!} = \dfrac{5 \cdot 4 \cdot 3 \cdot \cancel{2}!}{\cancel{2}!} = 60$

43. A, S, S, E, T, S

$\dfrac{6!}{3!} = 6 \cdot 5 \cdot 4 = 120$

45. S, P, E, C, I, E, S

$\dfrac{7!}{2!2!} = \dfrac{7 \cdot 6 \cdot 5 \cdot 4 \cdot 3 \cdot 2 \cdot 1}{2 \cdot 1 \cdot 2 \cdot 1} = 1260$

47. A, L, G, E, B, R, A

$\dfrac{7!}{2!} = 7 \cdot 6 \cdot 5 \cdot 4 \cdot 3 = 2520$

49. $_nC_r = \dfrac{n!}{(n-r)!r!}$

$_4C_3 = \dfrac{4!}{1!3!} = 4$

51. $_4C_4 = \dfrac{4!}{(4-4)!\,4!} = \dfrac{\cancel{4}!}{0!\cdot\cancel{4}!} = \dfrac{1}{1} = 1$

53. $_6C_2 = \dfrac{6!}{(6-2)!\cdot 2!} = \dfrac{6!}{4!\cdot 2!} = \dfrac{\overset{3}{\cancel{6}}\cdot 5\cdot\cancel{4}!}{\cancel{4}!\cdot\cancel{2}\cdot 1} = 15$

55. $_8C_0 = \dfrac{8!}{(8-0)!\,0!} = \dfrac{\cancel{8}!}{\cancel{8}!\cdot 1} = \dfrac{1}{1} = 1$

57. $_{12}C_4 = \dfrac{12!}{8!4!} = 495$

59. $_{20}C_5 = 15{,}504$

61. $_{42}C_5 = 850{,}668$

63. $_nP_3 = \dfrac{n!}{(n-3)!} = \dfrac{n(n-1)(n-2)(n-3)!}{(n-3)!}$

$= n(n-1)(n-2)$

65. $_nP_{n-1} = \dfrac{n!}{[n-(n-1)]!} = n!$

67. Because order matters, permutations should be used.

69. Because order does not matter, combinations should be used.

71. $_{15}P_4 = \dfrac{15!}{11!} = 32{,}760$

73. $_{30}C_5 = \dfrac{30!}{25!5!} = \dfrac{30\cdot 29\cdot 28\cdot 27\cdot 26}{5\cdot 4\cdot 3\cdot 2\cdot 1} = 142{,}506$

75. Because order is not important and the numbers must be distinct, use combinations.

$_{39}C_5 = 575{,}757$

There are 575,757 ways.

77. $_{100}C_4 = \dfrac{100!}{96!4!} = \dfrac{100\cdot 99\cdot 98\cdot 97}{4\cdot 3\cdot 2\cdot 1} = 3{,}921{,}225$

79. $_8C_4 = \dfrac{8!}{4!4!} = \dfrac{8\cdot 7\cdot 6\cdot 5\cdot 4\cdot 3\cdot 2\cdot 1}{4\cdot 3\cdot 2\cdot 1\cdot 4\cdot 3\cdot 2\cdot 1} = 70$

81. Select type of card for three-of-a-kind: $_{13}C_1$

Select 3 of 4 cards for three-of-a-kind: $_4C_3$

Select type of card for pair: $_{12}C_1$

Select 2 of 4 cards for pair: $_4C_2$

Total: $_{13}C_1\ {}_4C_3\ {}_{12}C_1\ {}_4C_2 = 13\cdot 4\cdot 12\cdot 6 = 3744$

83. $_7C_1\cdot{}_{12}C_3\cdot{}_{20}C_2 = \dfrac{7!}{(7-1)!1!}\cdot\dfrac{12!}{(12-3)!3!}\cdot\dfrac{20!}{(20-2)!2!}$

$= 292{,}600$

85. $_{100}P_{80} = \dfrac{100!}{(100-80)!} = \dfrac{100!}{20!}$

For some calculators, the answer is too large.

87. Answers will vary.

Section 7.6 Probability

Skills Warm Up

1. $\frac{1}{4} + \frac{5}{8} - \frac{5}{16} = \frac{4}{16} + \frac{10}{16} - \frac{5}{16} = \frac{9}{16}$

2. $\frac{4}{15} + \frac{3}{5} - \frac{1}{3} = \frac{4}{15} + \frac{9}{15} - \frac{5}{15} = \frac{8}{15}$

3. $\dfrac{5\cdot 4}{5!} = \dfrac{5\cdot 4}{5\cdot 4\cdot 3\cdot 2\cdot 1} = \dfrac{1}{6}$

4. $\dfrac{5!22!}{27!} = \dfrac{5\cdot 4\cdot 3\cdot 2\cdot 1\cdot 22!}{27\cdot 26\cdot 25\cdot 24\cdot 23\cdot 22!} = \dfrac{1}{80{,}730}$

5. $\dfrac{4!}{8!12!} = \dfrac{4!}{8\cdot 7\cdot 6\cdot 5\cdot 4!\cdot 12!} = \dfrac{1}{804{,}722{,}688{,}000}$

6. $\dfrac{9\cdot 8\cdot 7\cdot 6\cdot 5}{9\cdot 8\cdot 7\cdot 6\cdot 5\cdot 4!} = \dfrac{1}{24}$

7. $\dfrac{_5C_3}{_{10}C_3} = \dfrac{\frac{5!}{2!3!}}{\frac{10!}{7!3!}} = \dfrac{5!}{2!3!}\cdot\dfrac{7!3!}{10!} = \dfrac{7\cdot 6\cdot 5\cdot 4\cdot 3}{10\cdot 9\cdot 8\cdot 7\cdot 6} = \dfrac{1}{12}$

Skills Warm Up —*continued*—

8. $\dfrac{{}_{10}C_2 \cdot {}_{10}C_2}{{}_{20}C_4} = \dfrac{\dfrac{10!}{8!2!} \cdot \dfrac{10!}{8!2!}}{\dfrac{20!}{16!4!}} = \dfrac{10! \cdot 10!}{8!2!8!2!} \cdot \dfrac{16!4!}{20!} = \dfrac{2025}{1} \cdot \dfrac{4 \cdot 3 \cdot 2 \cdot 1}{20 \cdot 19 \cdot 18 \cdot 17} = \dfrac{135}{323}$

9. $\left(\dfrac{99}{100}\right)^{100} = \left(0.99^{10}\right)^{10} = \left(0.904382075\right)^{10} = 0.366$

10. $1 - \left(\dfrac{89}{100}\right)^{50} = 1 - 0.0029478157 = 0.997$

1. $S = \{H1, T1, H2, T2, H3, T3, H4, T4, H5, T5, H6, T6\}$

3. $S = \{ABC, ACB, BAC, BCA, CAB, CBA\}$

5. $S = \{AB, AC, AD, AE, BC, BD, BE, CD, CE, DE\}$

7.

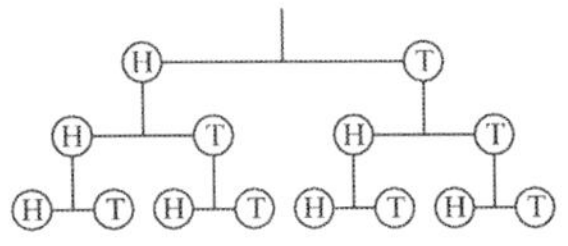

$\{HHH, HHT, HTH, HTT, THH, THT, TTH, TTT\}$

9.

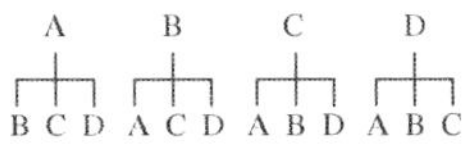

{AB, AC, AD, BA, BC, BD, CA, CB, CD, DA, DB, DC}

11. $P(E) = \dfrac{n(E)}{n(S)} = \dfrac{25}{50} = \dfrac{1}{2}$

13. $P(E) = \dfrac{n(E)}{n(S)} = \dfrac{15}{50} = \dfrac{3}{10}$

In Exercises 15–17, $S = \{HHH, HHT, HTH, HTT, THH, THT, TTH, TTT\}$ and $n(S) = 8$.

15. $E = \{TTH, THT, HTT\}$

$P(E) = \dfrac{n(E)}{n(S)} = \dfrac{3}{8}$

17. $E = \{TTT, TTH, THT, THH, HTT, HTH, HHT\}$

$P(E) = \dfrac{n(E)}{n(S)} = \dfrac{7}{8}$

In Exercises 19–23, $n(S) = 6^2 = 36$.

19. $E = (1, 4), (2, 3), (3, 2), (4, 1)$

$P(E) = \dfrac{n(E)}{n(S)} = \dfrac{4}{36} = \dfrac{1}{9}$

21. $E_7 = \{(1, 6), (2, 5), (3, 4), (4, 3), (5, 2), (6, 1)\}, n(E_7) = 6$

$E_8 = \{(2, 6), (3, 5), (4, 4), (5, 3), (6, 2)\}, n(E_8) = 5$

$E_9 = \{(3, 6), (4, 5), (5, 4), (6, 3)\}, n(E_9) = 4$

$E_{10} = \{(4, 6), (5, 5), (6, 4)\}, n(E_{10}) = 3$

$E_{11} = \{(5, 6), (6, 5)\}, n(E_{11}) = 2$

$E_{12} = \{(6, 6)\}, n(E_{12}) = 1$

$E = E_7 \cup E_8 \cup E_9 \cup E_{10} \cup E_{11} \cup E_{12}$

$n(E) = 6 + 5 + 4 + 3 + 2 + 1 = 21$

$P(E) = \dfrac{n(E)}{n(S)} = \dfrac{21}{36} = \dfrac{7}{12}$

23. $E_3 = \{(1, 2), (2, 1)\}, n(E_3) = 2$

$E_5 = \{(1, 4), (2, 3), (3, 2), (4, 1)\}, n(E_5) = 4$

$E_7 = \{(1, 6), (2, 5), (3, 4), (4, 3), (5, 2), (6, 1)\}, n(E_7) = 6$

$E = E_3 \cup E_5 \cup E_7$

$n(E) = 2 + 4 + 6 = 12$

$P(E) = \dfrac{n(E)}{n(S)} = \dfrac{12}{36} = \dfrac{1}{3}$

In Exercises 25–29, $n(s) = 52$.

25. $E = \{A, A, 1, 1, \ldots, K, K\}$

$P(E) = \dfrac{n(E)}{n(S)} = \dfrac{26}{52} = \dfrac{1}{2}$

27. $P(E) = \dfrac{n(E)}{n(S)} = \dfrac{52 - 12}{n(S)} = \dfrac{40}{52} = \dfrac{10}{13}$

29. $E = \{A, A, A, A, 2, 2, 2, 2, \ldots, 6, 6, 6, 6\}$

$P(E) = \dfrac{n(E)}{n(S)} = \dfrac{24}{52} = \dfrac{6}{13}$

In Exercises 31–35, $n(S) = {}_9C_2 = 36$.

31. $n(E) = {}_4C_1 \cdot {}_5C_1 = 20$

$P(E) = \dfrac{n(E)}{n(S)} = \dfrac{20}{36} = \dfrac{5}{9}$

33. $n(E) = {}_2C_2 = 1$

$P(E) = \dfrac{n(E)}{n(S)} = \dfrac{1}{36}$

35. $n(E) = {}_6C_2 = 15$

$P(E) = \dfrac{15}{36} = \dfrac{5}{12}$

37. $\dfrac{\text{18 or younger}}{\text{total}} = \dfrac{66}{2200} = \dfrac{3}{100}$

39. $\dfrac{\text{19–39 years old}}{\text{total}} = \dfrac{264 + 528}{2200} = \dfrac{792}{2200} = \dfrac{9}{25}$

41. The card is a club or a king.

Event A is drawing a club.

Event B is drawing a king.

$P(A \cup B) = P(A) + P(B) - P(A \cap B)$

$= \frac{13}{52} + \frac{4}{52} - \frac{1}{52} = \frac{16}{52} = \frac{4}{13}$

43. The card is a face card or a 2.

Event A is drawing a face card.

Event B is drawing a 2.

Events A and B are mutually exclusive.

$P(A \cup B) = P(A) + P(B)$

$= \frac{12}{52} + \frac{4}{52} = \frac{4}{13}$

45. $P(\text{20 or 21 years old}) = \dfrac{\text{\# of students 20 or 21}}{\text{total}}$

$= \dfrac{18}{32}$

$= \dfrac{9}{16}$

47. $P(\text{older than 21}) = \dfrac{\text{\# of students older than 21}}{\text{total}} = \dfrac{3}{32}$

49. Event A = All three numbers are even

$P(\text{even}) = \frac{5}{10} = \frac{1}{2}$

$P(A) = P(\text{even}) \cdot P(\text{even}) \cdot P(\text{even})$

$= \frac{1}{2} \cdot \frac{1}{2} \cdot \frac{1}{2} = \frac{1}{8}$

51. Event A = Two numbers are less than 5 and the other number is 10

$P(n < 5) = \frac{2}{5}, P(n = 10) = \frac{1}{10}$

$P(A) = P(n < 5) \cdot P(n < 5) \cdot P(n = 10) = \frac{2}{5} \cdot \frac{2}{5} \cdot \frac{1}{10} = \frac{4}{250} = \frac{2}{125}$

53. $P(E') = 1 - P(E) = 1 - 0.75 = 0.25$

55. $P(E') = 1 - P(E) = 1 - \frac{2}{3} = \frac{1}{3}$

57. $P(E) = 1 - P(E') = 1 - 0.12 = 0.88$

59. $P(E) = 1 - P(E') = 1 - \frac{13}{20} = \frac{7}{20}$

61. $1 - (0.46 + 0.32) = 0.22$

63. (a) $\dfrac{37 + 102}{198} = \dfrac{139}{198}$

(b) $1 - \dfrac{139}{198} = \dfrac{59}{198}$

(c) $\dfrac{43 - 37}{198} = \dfrac{6}{198} = \dfrac{1}{33}$

65. (a) $\left(\frac{1}{2}\right)\left(\frac{1}{2}\right) = \frac{1}{4}$

(b) $1 - p(\text{both even}) - p(\text{both odd}) = 1 - \frac{1}{4} - \frac{1}{4} = \frac{1}{2}$

(c) $\left(\frac{9}{30}\right)\left(\frac{9}{30}\right) = \frac{9}{100}$

(d) $(1)\left(\frac{1}{30}\right) = \frac{1}{30}$

67. (a) $\frac{13}{52} \cdot \frac{13}{52} = \frac{1}{16}$

(b) $\frac{13}{52} \cdot \frac{12}{1} = \frac{1}{17}$

69. (a) $\dfrac{1}{{}_6P_6} = \dfrac{1}{720}$

(b) $\dfrac{1}{{}_5P_5} = \dfrac{1}{120}$

(c) $\dfrac{1}{{}_4P_4} = \dfrac{1}{24}$

71.

Combination	*Number correct*
1 2 3	3
1 3 2	1
2 1 3	1
2 3 1	0
3 1 2	0
3 2 1	1

(a) $p(\text{exactly } 1) = \frac{3}{6} = \frac{1}{2}$

(b) $p(\text{at least } 1) = p(1) + p(3) = \frac{3}{6} + \frac{1}{6} = \frac{4}{6} = \frac{2}{3}$

73. $\dfrac{40}{{}_{52}C_5} = \dfrac{40}{2{,}598{,}960} = \dfrac{1}{64{,}974}$

75. (a) $\dfrac{{}_9C_4}{{}_{12}C_4} = \dfrac{126}{495} = \dfrac{14}{55}$

(b) $\dfrac{{}_9C_2\ {}_3C_2}{{}_{12}C_4} = \dfrac{108}{495} = \dfrac{12}{55}$

(c) $\dfrac{{}_9C_2\ {}_3C_2 + {}_9C_3\ {}_3C_1 + {}_9C_4}{{}_{12}C_4} = \dfrac{108 + 252 + 126}{495}$

$= \dfrac{486}{495}$

$= \dfrac{54}{55}$

77. (a) $\frac{1}{3} \cdot \frac{1}{3} \cdot \frac{1}{3} \cdot \frac{1}{3} = \frac{1}{81}$

(b) $\left(\frac{2}{3}\right)^4 = \frac{16}{81}$

(c) $1 - \frac{16}{81} = \frac{65}{81}$

79. ${}_8C_5\left(\dfrac{1}{2}\right)^5\left(\dfrac{1}{2}\right)^3 = \dfrac{8!}{5!3!}\left(\dfrac{1}{2^5}\right)\left(\dfrac{1}{2^3}\right)$

$= 8 \cdot 7 \cdot \dfrac{1}{2^8}$

$= \dfrac{2^3 \cdot 7}{2^8}$

$= \dfrac{7}{32}$

81. ${}_{10}C_4\left(\dfrac{1}{4}\right)^4\left(\dfrac{3}{4}\right)^6 = \dfrac{10!}{6!4!}\left(\dfrac{1}{4}\right)^4\left(\dfrac{3}{4}\right)^6$

$= \dfrac{76{,}545}{524{,}288}$

≈ 0.1459980011

83. Answers will vary. Sample answer: Mutually exclusive events have no outcomes in common. Two events are independent when the occurrence of one event has no effect on the occurrence of the other event.

85. (a) The probability of at least a few times per week is

$0.12 + 0.06 + 0.15 = 0.33.$

The probability that both people sent or received work-related texts at least a few times a week is $(0.33)(0.33) = 0.1089.$

(b) $P(\text{neither}) = (0.51)(0.51) = 0.2601$

Review Exercises for Chapter 7

1. $a_n = 4n + 1$

$a_1 = 4(1) + 1 = 5$

$a_2 = 4(2) + 1 = 9$

$a_3 = 4(3) + 1 = 13$

$a_4 = 4(4) + 1 = 17$

$a_5 = 4(5) + 1 = 21$

3. $a_n = \dfrac{n}{n+1}$

$a_1 = \dfrac{(1)}{(1)+1} = \dfrac{1}{2}$

$a_2 = \dfrac{(2)}{(2)+1} = \dfrac{2}{3}$

$a_3 = \dfrac{(3)}{(3)+1} = \dfrac{3}{4}$

$a_4 = \dfrac{(4)}{(4)+1} = \dfrac{4}{5}$

$a_5 = \dfrac{(5)}{(5)+1} = \dfrac{5}{6}$

5. $a_n = (-1)^{n+1}\dfrac{2^n}{n+1}$

$a_1 = (-1)^{1+1} \cdot \dfrac{2^1}{1+1} = 1$

$a_2 = (-1)^{2+1} \cdot \dfrac{2^2}{2+1} = -\dfrac{4}{3}$

$a_3 = (-1)^{3+1} \cdot \dfrac{2^3}{3+1} = 2$

$a_4 = (-1)^{4+1} \cdot \dfrac{2^4}{4+1} = -\dfrac{16}{5}$

$a_5 = (-1)^{5+1} \cdot \dfrac{2^5}{5+1} = \dfrac{16}{3}$

7. $1, 2, 4, 8, \ldots$

$a_n = 2^{n-1}$

9. $1, \dfrac{1}{2}, \dfrac{1}{3}, \dfrac{1}{4}, \dfrac{1}{5}, \ldots$

$a_n = \dfrac{1}{n}$

11. $3, -9, 27, -81, 243, \ldots$

$a_n = (-1)^{n+1}3^n$

13. $\dfrac{16!}{12!} = \dfrac{16 \cdot 15 \cdot 14 \cdot 13 \cdot \cancel{12!}}{\cancel{12!}}$

$= 16 \cdot 15 \cdot 14 \cdot 13$

$= 43{,}680$

15. $\dfrac{(n+5)!}{(n+4)!} = \dfrac{(n+5)\cancel{(n+4)!}}{\cancel{(n+4)!}} = n + 5$

17. $\displaystyle\sum_{i=1}^{6}(3i - 2) = [3(1) - 2] + [3(2) - 2] + [3(3) - 2] + [3(4) - 2] + [3(5) - 2] + [3(6) - 2]$

$= 1 + 4 + 7 + 10 + 13 + 16 = 51$

19. $\displaystyle\sum_{k=1}^{6}(k - 1)(k + 2)$

$= (1 - 1)(1 + 2) + (2 - 1)(2 + 2) + (3 - 1)(3 + 2) + (4 - 1)(4 + 2) + (5 - 1)(5 + 2) + (6 - 1)(6 + 2)$

$= 0 + 4 + 10 + 18 + 28 + 40 = 100$

21. $\dfrac{3}{1+1} + \dfrac{3}{1+2} + \dfrac{3}{1+3} + \dfrac{3}{1+4} + \cdots + \dfrac{3}{1+12} = \displaystyle\sum_{k=1}^{12}\frac{3}{1+k}$

23. $2 - 4 + 8 - 16 + 32 - 64 + \cdots + 8192 = \displaystyle\sum_{k=1}^{13}(-1)^{k-1}2^k$

25. $A_n = 4000\left(1 + \frac{0.06}{4}\right)^n$

(a) $A_1 = \$4060.00$

$A_2 = \$4120.90$

$A_3 \approx \$4182.71$

$A_4 \approx \$4245.45$

$A_5 \approx \$4309.14$

$A_6 \approx \$4373.77$

$A_7 \approx \$4439.38$

$A_8 \approx \$4505.97$

(b) $A_{40} \approx \$7256.07$

(c) $A_{120} = \$23,877.29$

27. 3, 8, 13, 18, …

Arithmetic, $d = 5$

29. $4, \frac{7}{2}, 3, \frac{5}{2}, 2, \ldots$

Arithmetic, $d = -\frac{1}{2}$

31. $a_1 = 9, d = 5$

$a_n = a_1 + (n-1)d = 9 + (n-1)5 = 5n + 4$

$a_1 = 5(1) + 4 = 9$

$a_2 = 5(2) + 4 = 14$

$a_3 = 5(3) + 4 = 19$

$a_4 = 5(4) + 4 = 24$

$a_5 = 5(5) + 4 = 29$

33. $a_5 = 14, a_8 = 29, d = \frac{29 - 14}{8 - 5} = 5$

$a_n = a_1 + (n-1)d$

$a_5: 14 = a_1 + (5-1)5 \Rightarrow a_1 = -6$

$a_n = -6 + (n-1)5 = 5n - 11$

$a_1 = 5(1) - 11 = -6$

$a_2 = 5(2) - 11 = -1$

$a_3 = 5(3) - 11 = 4$

$a_4 = 5(4) - 11 = 9$

$a_5 = 5(5) - 11 = 14$

35. $a_7 = 12, a_{10} = 18, d = \frac{18 - 12}{10 - 7} = 2$

$a_n = a_1 + (n-1)d$

$a_7 = a_1 + (7-1)(2)$

$12 = a_1 + 12$

$0 = a_1$

So, $a_n = 0 + 2(n-1) = 2n - 2$

$a_1 = 2(1) - 2 = 0$

$a_2 = 2(2) - 2 = 2$

$a_3 = 2(3) - 2 = 4$

$a_4 = 2(4) - 2 = 6$

$a_5 = 2(5) - 2 = 8$

37. $a_1 = 2, d = 3$

$a_n = a_1 + (n-1)d = 2 + (n-1)3 = 3n - 1$

39. 2, 7, 12, 17, 22, …

$a_1 = 2, d = 7 - 2 = 5$

$a_n = a_1 + (n-1)d = 2 + (n-1)5 = 5n - 3$

41. $a_1 = 7, a_5 = 27$

$a_5 = a_1 + 4d$

$27 = 7 + 4d$

$20 = 4d$

$5 = d$

$a_n = a_1 + (n-1)d = 7 + (n-1)5 = 5n + 2$

43. 3, 9, 15, 21, 27, …, $n = 8$

$a_n = 6n - 3$

$a_1 = 3, a_8 = 45$

$S_8 = \frac{8}{2}(a_1 + a_8) = 4(3 + 45) = 192$

45. $a_1 = 9, a_{15} = 93, n = 15$

$S_{15} = \frac{15}{2}(9 + 93) = 765$

47. $A_n = 32,500 + (n-1)1600 = 1600n + 30,900$

$B_n = 28,000 + (n-1)2500 = 2500n + 25,500$

(a) $A_6 = \$40,500, B_6 = \$40,500$

(b) $\sum_{i=1}^{6} 1600n + 30,900 = \$219,000$

(c) $\sum_{i=1}^{6} 2500n + 25,500 = \$205,500$

49. 1, −3, 9, −27, …

Geometric, $r = -3$

$a_n = (-3)^{n-1}$

51. $16, 8, 4, 2, 1, \frac{1}{2}, \frac{1}{4}, \frac{1}{8}, \ldots$

Geometric, $r = \frac{1}{2}$

$$a_n = 16\left(\frac{1}{2}\right)^{n-1}$$

53. $a_1 = 3,\ r = 2$

$$a_1 = 3$$
$$a_2 = 6$$
$$a_3 = 12$$
$$a_4 = 24$$
$$a_5 = 48$$

55. $a_1 = 10,\ r = -\frac{1}{5}$

$$a_1 = 10$$
$$a_2 = -2$$
$$a_3 = \frac{2}{5}$$
$$a_4 = -\frac{2}{25}$$
$$a_5 = \frac{2}{125}$$

57. $a_1 = 8,\ r = \dfrac{1}{2}$

$$a_{40} = a_1 r^{39} = 8\left(\frac{1}{2}\right)^{39} = \left(\frac{1}{2^{36}}\right)$$

59. $a_2 = -100,\ a_5 = 12.5$

$$a_5 = a_2(r)^3$$
$$12.5 = -100r^3 \Rightarrow r = -\tfrac{1}{2}$$
$$a_2 = a_1\left(-\tfrac{1}{2}\right)^1$$
$$-100 = a_1\left(-\tfrac{1}{2}\right) \Rightarrow a_1 = 200$$

So, $a_n = 200\left(-\frac{1}{2}\right)^{n-1}$ then $a_{10} = 200\left(-\frac{1}{2}\right)^{10-1} = -\frac{25}{64}$.

61. $$\sum_{i=1}^{10} 4(2^n) = a_1\left(\frac{1 - r^n}{1 - r}\right) = 8\left(\frac{1 - 2^{10}}{1 - 2}\right) = 8184$$

63. $$\sum_{i=1}^{15} 20(0.1)^{i-1} = \sum_{i=1}^{15} 200(0.1)^i$$
$$= a_1\left(\frac{1 - r^n}{1 - r}\right)$$
$$= 20\left(\frac{1 - (0.1)^{15}}{1 - 0.1}\right)$$
$$\approx 22.22$$

65. $$\sum_{n=0}^{\infty}\left(\frac{1}{3}\right)^n = \frac{a}{1 - r} = \frac{1}{1 - (1/3)} = \frac{3}{2} = 1.5$$

67. $\displaystyle\sum_{n=0}^{\infty}\left(\tfrac{5}{4}\right)^n$; the sum cannot be found because $\left|\tfrac{5}{4}\right| > 1$.

69. (a) $$A = \sum_{i=1}^{120} 100\left(1 + \frac{0.06}{12}\right)^i \approx \$16{,}469.87$$

(b) No, it is more than twice the amount because the interest is compounded.

71. (a)

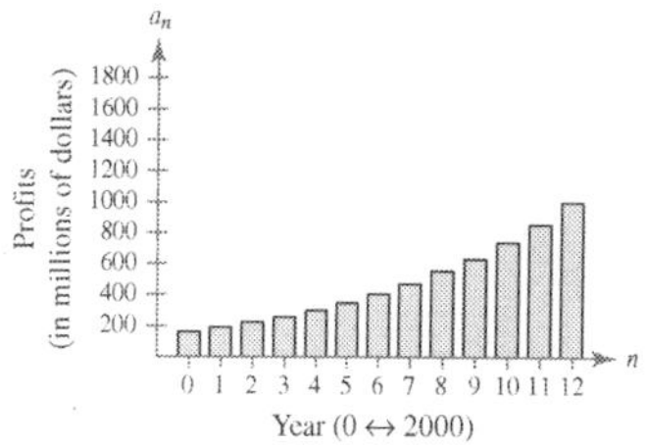

(b) $$\sum_{n=0}^{12} 165.3\left(e^{0.15}\right)^n = \sum_{n=1}^{13} 165.3\left(e^{0.15}\right)^{n-1}$$

$$n = 13,\ a_1 = 165.3,\ r = e^{0.15}$$

$$S_{13} = 165.3\left(\frac{1 - \left(e^{0.15}\right)^{13}}{1 - e^{0.15}}\right) \approx 6157.79$$

The total profit was about \$6157.79 million.

73. $${}_7C_4 = \frac{7!}{4!3!} = \frac{7 \cdot 6 \cdot 5 \cdot 4!}{4! \cdot 3 \cdot 2 \cdot 1} = 35$$

75. $$\binom{20}{20} = {}_{20}C_{20} = \frac{20!}{20!0!} = 1$$

77. ${}_{12}C_5 = 792$

79. $$(x + 1)^6 = {}_6C_0x^6 + {}_6C_1x^5(1) + {}_6C_2x^4(1)^2 + {}_6C_3x^3(1)^3 + {}_6C_4x^2(1)^4 + {}_6C_5x(1)^5 + {}_6C_6(1)^6$$
$$= x^6 + 6x^5 + 15x^4 + 20x^3 + 15x^2 + 6x + 1$$

81. $$(3x + y)^5 = {}_cC_0(3x)^5 = {}_5C_1(3x)^4y + {}_5C_2(3x)^3y^2 + {}_5C_3(3x)^2y^3 + {}_5C_4(3x)y^4 + {}_5C_5y^5$$
$$= 243x^5 + 405x^4y + 270x^3y^2 + 90x^2y^3 + 15y^4 + y^5$$

83. $(x - 2y)^8 = {}_8C_0x^8 - {}_8C_1x^7(2y) + {}_8C_2x^6(2y)^2 - {}_8C_3x^5(2y)^3 + {}_8C_4x^4(2y)^4 - {}_8C_5x^3(2y)^5$

$+ {}_8C_6x^2(2y)^6 - {}_8C_7x(2y)^7 + {}_8C_8(2y)^8$

$= x^8 - 8x^7(2y) + 28x^6(4y^2) - 56x^5(8y^3) + 70x^4(16y^4) - 56x^3(32y^5) + 28x^2(64y^6) - 8x(128y^7) + 256y^8$

$= x^8 - 16x^7y + 112x^6y^2 - 448x^5y^3 + 1120x^4y^4 - 1792x^3y^5 + 1792x^2y^6 - 1024xy^7 + 256y^8$

85. $(a - 3b)^5$, $n = 3$; The third term of $(a - 3b)^5$ would be ${}_5C_2(a)^3(-3b)^2 = 90a^3b^2$.

87. The term involving x^7 in the expansion of $(x + 2)^{12}$ is

${}_{12}C_5x^7(2)^5 = 792x^7(32) = 25{,}344x^7$. The coefficient is 25,344.

89. $6 \cdot 5 \cdot 7 = 210$ different systems

91. (a) $7 \cdot 10 \cdot 10 \cdot 10 = 7000$ numbers

(b) $7 \cdot 9 \cdot 8 \cdot 7 = 3528$ numbers

(c) $9 \cdot 10 \cdot 10 \cdot 5 = 4500$ numbers

93. $5 \cdot 4 \cdot 3 \cdot 2 \cdot 1 = 120$ orders

95. ${}_9P_3 = \dfrac{9!}{6!} = \dfrac{9 \cdot 8 \cdot 7 \cdot 6!}{6!} = 504$

97. ${}_8P_8 = \dfrac{8!}{0!} = \dfrac{8 \cdot 7 \cdot 6 \cdot 5 \cdot 4 \cdot 3 \cdot 2 \cdot 1}{1} = 40{,}320$

99. $\dfrac{12!}{4! \cdot 3! \cdot 2!} = \dfrac{12 \cdot 11 \cdot 10 \cdot 9 \cdot 8 \cdot 7 \cdot 6 \cdot 5 \cdot 4!}{4! \cdot 3 \cdot 2 \cdot 1 \cdot 2 \cdot 1} = 1{,}663{,}200$

101. (a) ${}_{24}C_5 = \dfrac{24!}{5!19!} = \dfrac{24 \cdot 23 \cdot 22 \cdot 21 \cdot 20 \cdot 19!}{5 \cdot 4 \cdot 3 \cdot 2 \cdot 1 \cdot 19!} = 42{,}504$

(b) ${}_{24}C_{10} = \dfrac{24!}{10!14!} = \dfrac{24 \cdot 23 \cdot 22 \cdot 21 \cdot 20 \cdot 19 \cdot 18 \cdot 17 \cdot 16 \cdot 15 \cdot 14!}{10 \cdot 9 \cdot 8 \cdot 7 \cdot 6 \cdot 5 \cdot 4 \cdot 3 \cdot 2 \cdot 1 \cdot 14!} = 1{,}961{,}256$

(c) ${}_{24}C_{18} = \dfrac{24!}{18!6!} = \dfrac{24 \cdot 23 \cdot 22 \cdot 21 \cdot 20 \cdot 19 \cdot 18!}{18! \cdot 6 \cdot 5 \cdot 4 \cdot 3 \cdot 2 \cdot 1} = 134{,}596$

103. $S = \{0, 1, 2, 3, 4, 5, 6, 7, 8, 9\}$

105. $\frac{12}{52} = \frac{3}{13}$

107. $P(280 \text{ to } 319) = \frac{8}{85}$

109. $P(\text{greater than } 199) = \dfrac{31 + 34 + 8}{85} = \dfrac{73}{85}$

111. Event A = selecting a red card

Event B = selecting an ace

$P(A) = \frac{26}{52} = \frac{1}{2}$, $P(B) = \frac{4}{52} = \frac{1}{13}$

$P(A \cap B) = \frac{2}{52} = \frac{1}{26}$

$P(A \cup B) = P(A) + P(B) - P(A \cap B)$

$= \frac{1}{2} + \frac{1}{13} - \frac{1}{26} = \frac{7}{13}$

113. $1 - 0.57 = 0.43$

115. $P(\text{odd}) = \frac{10}{20} = \frac{1}{2}$

$P(\text{all three are odd}) = P(\text{odd}) \cdot P(\text{odd}) \cdot P(\text{odd})$

$= \frac{1}{2} \cdot \frac{1}{2} \cdot \frac{1}{2}$

$= \frac{1}{8}$

117. (a) $\frac{30}{72} = \frac{5}{12}$

(b) $\frac{42}{72} = \frac{7}{12}$

(c) $\frac{10}{72} = \frac{5}{36}$

119. ${}_{13}C_2$ = ways to choose the two different face values for the two pairs

${}_4C_2$ = ways to choose two suits for one pair

${}_4C_2$ = ways to choose two suits for the second pair

${}_{11}C_1$ = ways to choose the face value for the 5th card

${}_4C_1$ = ways to choose the suit for the 5th card

$\dfrac{{}_{13}C_2 \cdot {}_4C_2 \cdot {}_4C_2 \cdot {}_{11}C_1 \cdot {}_4C_1}{{}_{52}C_5} = \dfrac{123{,}552}{2{,}598{,}960} = \dfrac{198}{4165}$

121.

Combination	*Number correct*	*Combination*	*Number correct*
1 2 3 4	4	3 2 1 4	2
1 2 4 3	2	3 2 4 1	1
1 3 2 4	2	3 4 1 2	0
1 3 4 2	1	3 4 2 1	0
1 4 2 3	1	3 1 4 2	0
1 4 3 2	2	3 1 2 4	1
2 1 3 4	2	4 2 1 3	0
2 1 4 3	0	4 1 3 2	1
2 4 1 3	0	4 2 3 1	2
2 4 3 1	1	4 2 1 3	1
2 3 4 1	0	4 3 1 2	0
2 3 1 4	1	4 3 2 1	0

(a) $p(\text{exactly } 1) = \frac{8}{24} = \frac{1}{3}$

(b) $p(\text{at least } 1) = p(1) + p(2) + p(4) = \frac{8}{24} + \frac{6}{24} + \frac{1}{24} = \frac{15}{24} = \frac{5}{8}$

123. (a) $\left(\frac{1}{2}\right)^5 = \frac{1}{32}$ (b) $\frac{1}{32} + \frac{1}{32} = \frac{1}{16}$ (c) $1 - \frac{1}{32} = \frac{31}{32}$

Chapter 7 Test Yourself

1. $a_n = 2n + 1$

$a_1 = 2(1) + 1 = 3$

$a_2 = 2(2) + 1 = 5$

$a_3 = 2(3) + 1 = 7$

$a_4 = 2(4) + 1 = 9$

$a_5 = 2(5) + 1 = 11$

2. $a_1 = (-1)^1(1)^2 = -1$

$a_2 = (-1)^2(2)^2 = 4$

$a_3 = (-1)^3(3)^2 = -9$

$a_4 = (-1)^4(4)^2 = 16$

$a_5 = (-1)^5(5)^2 = -25$

3. $a_n = (n + 1)!$

$a_1 = 2! = 2$

$a_2 = 3! = 6$

$a_3 = 4! = 24$

$a_4 = 5! = 120$

$a_5 = 6! = 720$

4. $a_1 = \left(\frac{1}{2}\right)^1 = \frac{1}{2}$

$a_2 = \left(\frac{1}{2}\right)^2 = \frac{1}{4}$

$a_3 = \left(\frac{1}{2}\right)^3 = \frac{1}{8}$

$a_4 = \left(\frac{1}{2}\right)^4 = \frac{1}{16}$

$a_5 = \left(\frac{1}{2}\right)^5 = \frac{1}{32}$

5. 3, 8, 13, 18, … is arithmetic. The common difference is 5.

6. 4, 6, 10, 18, 34, … is neither arithmetic nor geometric.

7. 10, 20, 40, 80, 160, … is geometric with a common ratio of 2.

8. 1, 10, 25, 46, 73, … is neither arithmetic nor geometric.

9. 5, 13, 21, ...

$a_1 = 5$ and $d = 8$

$a_n = a_1 + (n - 1)d$

$a_n = 5 + (n - 1)8$

$a_n = 5 + 8n - 8$

$a_n = 8n - 3$

$a_{20} = 8(20) - 3 = 157$

10. $a_{11} = 360$ and $d = 35$

$$a_{11} = a_1 + 10d$$
$$360 = a_1 + 10(35)$$
$$360 = a_1 + 350$$
$$10 = a_1$$

11. 4, 28, 196, ... so $r = 7$

$$a_n = a_1(r)^{n-1}$$
$$a_n = 4(7)^{n-1}$$
$$a_8 = 4(7)^{8-1} = 4(7)^7 = 3{,}294{,}172$$

12. $a_1 = 4$ and $r = 1.1$

$$a_n = 4(1.1)^{n-1}$$
$$a_{18} = 4(1.1)^{18-1} = 4(1.1)^{17} \approx 20.22$$

13. Arithmetic

$$a_1 = 3,\ a_{50} = 101$$
$$\sum_{n=1}^{50}(2n+1) = \tfrac{50}{2}(3+101) = 2600$$

14. Geometric

$$a_1 = \frac{9}{2},\ r = \frac{3}{2}$$
$$\sum_{n=1}^{20} 3\left(\frac{3}{2}\right)^n = \frac{\frac{9}{2}\left[1-\left(\frac{3}{2}\right)^{20}\right]}{1-\frac{3}{2}} \approx 29{,}918.311$$

15. Geometric

$$a_1 = \frac{1}{3},\ r = \frac{1}{3}$$
$$\sum_{n=1}^{\infty}\left(\frac{1}{3}\right)^n = \frac{\frac{1}{3}}{1-\frac{1}{3}} = \frac{1}{2}$$

16. 10 years = 120 months

$$A_{120} = 8000\left(1+\frac{0.05}{12}\right)^{120} \approx \$13{,}176.08$$

17. $(2x-3)^5 = {}_5C_0(2x)^5 + {}_5C_1(2x)^4(-3) + {}_5C_2(2x)^3(-3)^2 + {}_5C_3(2x)^2(-3)^3 + {}_5C_4(2x)(-3)^4 + {}_5C_5(-3)^5$
$= 32x^5 - 240x^4 + 720x^3 - 1080x^2 + 810x - 243$

18. The coefficient of x^7 is ${}_{12}C_5(-3)^5 = -192{,}456$.

19. (a) $${}_9P_2 = \frac{9!}{7!} = \frac{9\cdot 8\cdot 7!}{7!} = 72$$

(b) $${}_{70}P_3 = \frac{70!}{67!} = \frac{70\cdot 69\cdot 68\cdot 67!}{67!} = 328{,}440$$

20. (a) $${}_{11}C_4 = \frac{11!}{4!7!} = \frac{11\cdot 10\cdot 9\cdot 8\cdot 7!}{4\cdot 3\cdot 2\cdot 1\cdot 7!} = 330$$

(b) $${}_{66}C_5 = \frac{66!}{5!61!} = \frac{66\cdot 65\cdot 64\cdot 63\cdot 62\cdot 61!}{5\cdot 4\cdot 3\cdot 2\cdot 1\cdot 61!} = 8{,}936{,}928$$

21. P, I, Z, Z, A

$n = 5$ and Z is repeated

$$\frac{5!}{2!} = \frac{5\cdot 4\cdot 3\cdot \cancel{2}!}{\cancel{2}!} = 60$$

22. I, N, T, E, R, N, E, T

$n = 8$ and N, T, and E are repeated

$$\frac{8!}{2!\,2!\,2!} = \frac{8\cdot 7\cdot 6\cdot 5\cdot \cancel{4}\cdot 3\cdot \cancel{2}!}{\cancel{2}\cdot 1\cdot \cancel{2}\cdot 1\cdot \cancel{2}!} = 5040$$

23. $8\cdot 4\cdot 2\cdot 10 = 640$

24. $4\cdot 4\cdot 4\cdot 4\cdot 4\cdot 4\cdot 4\cdot 4\cdot 4\cdot 4\cdot 4\cdot 4 = 4^{12} = 16{,}777{,}216$ ways

25. $9\cdot 10\cdot 10\cdot 10\cdot 5 = 45{,}000$ different numbers

26. $8\cdot 7\cdot 6\cdot 5\cdot 4\cdot 3\cdot 2\cdot 1 = 40{,}320$ ways

27. $${}_{12}C_3 = \frac{12!}{3!9!} = \frac{12\cdot 11\cdot 10\cdot 9!}{3\cdot 2\cdot 1\cdot 9!} = 220 \text{ ways}$$

28. $$\left(\frac{1}{6}\right)^3 = \frac{1^3}{6^3} = \frac{1}{6\cdot 6\cdot 6} = \frac{1}{216}$$

APPENDIX B
Conic Sections

APPENDIX B
Conic Sections

Section B.1 Conic Sections

Skills Warm Up

1. $\dfrac{x^2}{16} + \dfrac{y^2}{9} = 1$

$144\left(\dfrac{x^2}{16} + \dfrac{y^2}{9}\right) = 144(1)$

$9x^2 + 16y^2 = 144$

2. $\dfrac{x^2}{32} + \dfrac{4y^2}{32} = \dfrac{32}{32}$

$32\left(\dfrac{x^2}{32} + \dfrac{4y^2}{32}\right) = 32\left(\dfrac{32}{32}\right)$

$x^2 + 4y^2 = 32$

3. $\dfrac{x^2}{1/4} - \dfrac{y^2}{4} = 1$

$4\left(\dfrac{x^2}{1/4} - \dfrac{y^2}{4}\right) = 4(1)$

$16x^2 - y^2 = 4$

4. $\dfrac{3x^2}{1/9} + \dfrac{4y^2}{9} = 1$

$9\left(\dfrac{3x^2}{1/9} + \dfrac{4y^2}{9}\right) = 9(1)$

$243x^2 + 4y^2 = 9$

5. $c^2 = 3^2 - 1^2$

$c = \sqrt{3^2 - 1^2} = \sqrt{9 - 1} = \sqrt{8} = 2\sqrt{2}$

6. $c^2 = 2^2 + 3^2$

$c = \sqrt{2^2 + 3^2} = \sqrt{4 + 9} = \sqrt{13}$

7. $c^2 + 2^2 = 4^2$

$c^2 = 4^2 - 2^2$

$c = \sqrt{4^2 - 2^2} = \sqrt{16 - 4} = \sqrt{12} = 2\sqrt{3}$

8. $c^2 - 1^2 = 2^2$

$c^2 = 2^2 + 1^2$

$c = \sqrt{2^2 + 1^2} = \sqrt{4 + 1} = \sqrt{5}$

9. $d = \sqrt{(x_2 - x_1)^2 + (y_2 - y_1)^2}$

$= \sqrt{(0 - 0)^2 + (-4 - 0)^2}$

$= \sqrt{0^2 + 4^2}$

$= \sqrt{16}$

$= 4$

10. $d = \sqrt{(x_2 - x_1)^2 + (y_2 - y_1)^2}$

$= \sqrt{(-2 - 0)^2 + (0 - 0)^2}$

$= \sqrt{2^2 + 0^2}$

$= \sqrt{4}$

$= 2$

---

1. $x^2 = 4y$

Parabola opening upward Matches (a).

3. $y^2 = 4x$

Parabola opening to the right Matches (d).

5. $\dfrac{x^2}{1} + \dfrac{y^2}{4} = 1$

Ellipse with vertical major axis Matches (g).

7. $\dfrac{x^2}{1} - \dfrac{y^2}{4} = 1$

Hyperbola with horizontal transverse axis Matches (e).

9. $y = 4x^2$

$x^2 = \frac{1}{4}y$

$x^2 = 4\left(\frac{1}{16}\right)y;$

Vertex: $(0, 0)$ $b = \frac{1}{16}$

Focus: $\left(0, \frac{1}{16}\right)$

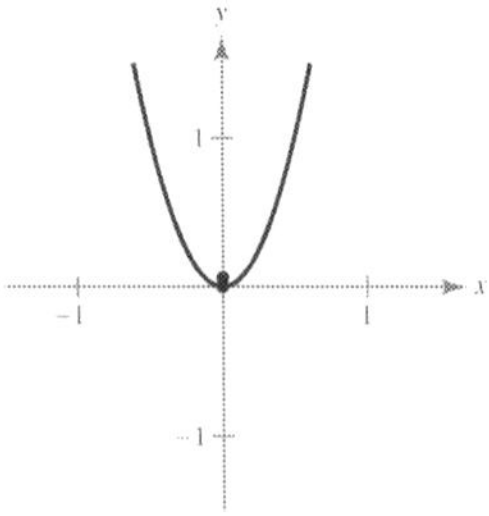

11. $y^2 = -6x$

$y^2 = 4\left(-\frac{3}{2}\right)x;\ p = -\frac{3}{2}$

Vertex: $(0, 0)$

Focus: $\left(-\frac{3}{2}, 0\right)$

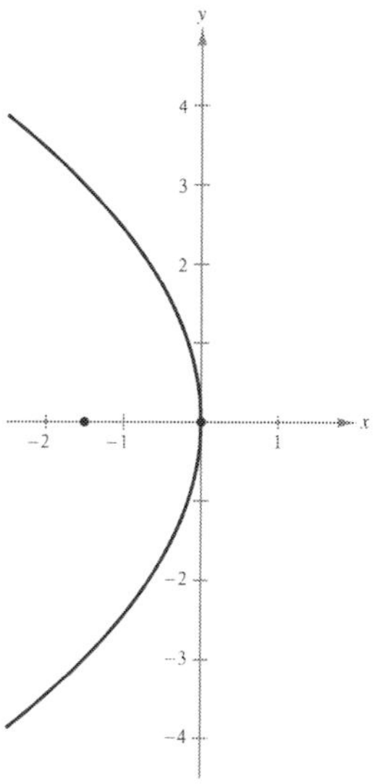

13. $x^2 + 8y = 0$

$x^2 = 4(-2)y;\ p = -2$

Vertex: $(0, 0)$

Focus: $(0, -2)$

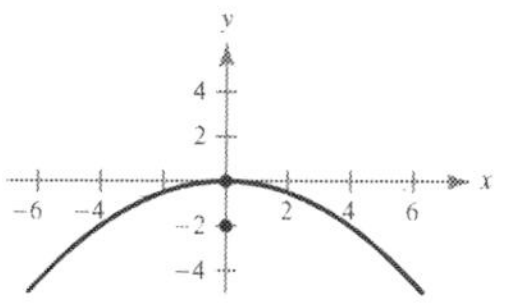

15. $x + y^2 = 0$

$y^2 = 4\left(-\frac{1}{4}\right)x;\ p = -\frac{1}{4}$

Vertex: $(0, 0)$

Focus: $\left(-\frac{1}{4}, 0\right)$

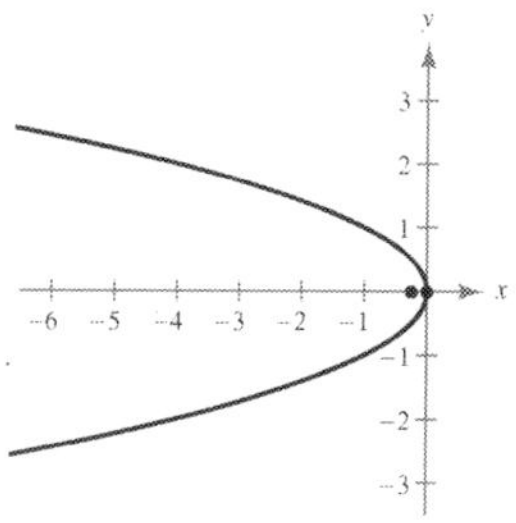

17. Focus: $\left(0, -\frac{3}{2}\right)$

$x^2 = 4\left(-\frac{3}{2}\right)y$

$x^2 = -6y$

19. Focus: $(-2, 0)$

$y^2 = 4(-2)x$

$y^2 = -8x$

21. Directrix: $y = -1$

$x^2 = 4(1)y$

$x^2 = 4y$

23. Directrix: $x = 3$

$y^2 = 4(-3)x$

$y^2 = -12x$

25. $y^2 = 4px$

$6^2 = 4p(4)$

$36 = 16p$

$p = \frac{9}{4}$

$y^2 = 4\left(\frac{9}{4}\right)x$

$y^2 = 9x$

27. Vertices: $(0, \pm 2) \Rightarrow a = 2$

Minor axis of length 2 $\Rightarrow b = 1$

Vertical major axis

$$\frac{x^2}{b^2} + \frac{y^2}{a^2} = 1$$

$$\frac{x^2}{1} + \frac{y^2}{4} = 1$$

29. Vertices: $(\pm 5, 0) \Rightarrow a = 5$

Foci: $(\pm 2, 0) \Rightarrow b = \sqrt{5^2 - 2^2} = \sqrt{21}$

Horizontal major axis

$$\frac{x^2}{a^2} + \frac{y^2}{b^2} = 1$$

$$\frac{x^2}{25} + \frac{y^2}{21} = 1$$

31. Foci: $(\pm 5, 0) \Rightarrow c = 5$

Major axis of length 12 $\Rightarrow a = 6$

$b = \sqrt{6^2 - 5^2} = \sqrt{11}$

Horizontal major axis

$$\frac{x^2}{a^2} + \frac{y^2}{b^2} = 1$$

$$\frac{x^2}{36} + \frac{y^2}{11} = 1$$

33. Vertices: $(0, \pm 5) \Rightarrow a = 5$

Vertical major axis

$$\frac{x^2}{b^2} + \frac{y^2}{25} = 1$$

$$\frac{(4)^2}{b^2} + \frac{(2)^2}{25} = 1$$

$$\frac{400}{21} = b^2$$

$$\frac{21x^2}{400} + \frac{y^2}{25} = 1 \text{ or } \frac{x^2}{400/21} + \frac{y^2}{25} = 1$$

35. $\dfrac{x^2}{25} + \dfrac{y^2}{16} = 1$

Horizontal major axis

$a = 5, b = 4$

Center: (0, 0)

Vertices: $(\pm 5, 0)$

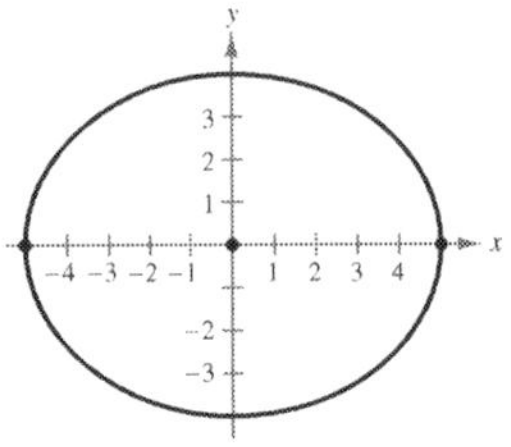

37. $\dfrac{x^2}{25/9} + \dfrac{y^2}{16/9} = 1$

Horizontal major axis

$a = \dfrac{5}{3}, b = \dfrac{4}{3}$

Center: (0, 0)

Vertices: $\left(\pm\dfrac{5}{3}, 0\right)$

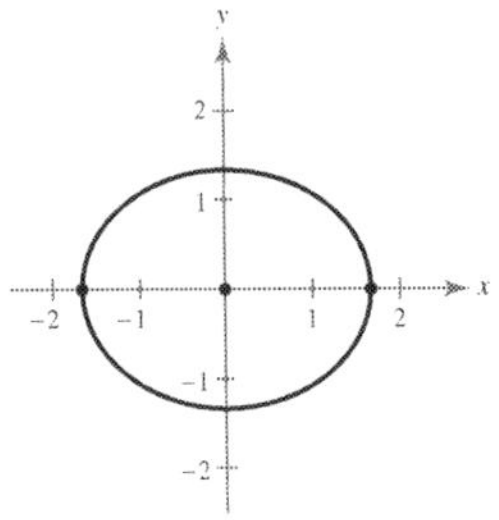

39. $\dfrac{x^2}{9} + \dfrac{y^2}{5} = 1$

Horizontal major axis

$a = 3, b = \sqrt{5}$

Center: (0, 0)

Vertices: $(\pm 3, 0)$

41. $5x^2 + 3y^2 = 15$

$\dfrac{x^2}{3} + \dfrac{y^2}{5} = 1$

$a = \sqrt{5}, b = \sqrt{3}$

Center: (0, 0)

Vertices: $\left(0, \pm\sqrt{5}\right)$

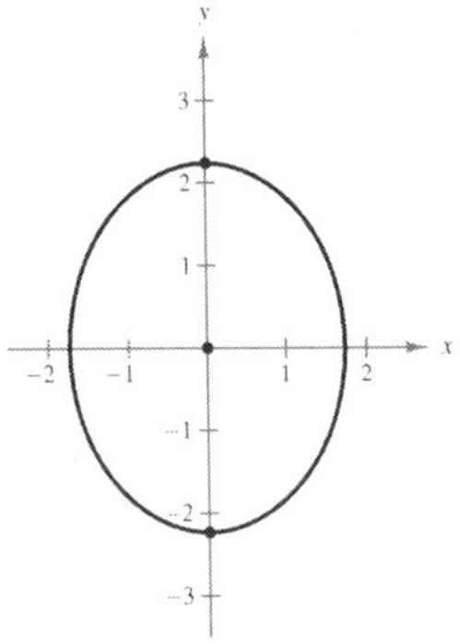

43. Vertices: $(0, \pm 2) \Rightarrow a = 2$

Foci: $(0, \pm 4) \Rightarrow c = 4$

$b^2 = c^2 - a^2 = 12$

Vertical transverse axis

$\dfrac{y^2}{a^2} - \dfrac{x^2}{b^2} = 1$

$\dfrac{y^2}{4} - \dfrac{x^2}{12} = 1$

45. Vertices: $(\pm 6, 0) \Rightarrow a = 6$

Foci: $(\pm 9, 0) \Rightarrow c = 9$

$b^2 = c^2 - a^2 = 45$

Horizontal transverse axis

$\dfrac{x^2}{a^2} - \dfrac{y^2}{b^2} = 1$

$\dfrac{x^2}{36} - \dfrac{y^2}{45} = 1$

47. $x^2 - y^2 = 1$

$a = 1, b = 1$

Center: (0, 0)

Vertices: $(\pm 1, 0)$

Asymptotes: $y = \pm x$

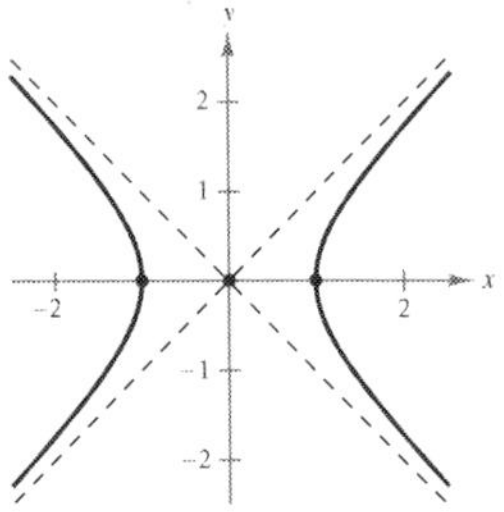

49. $\dfrac{y^2}{1} - \dfrac{x^2}{4} = 1$

$a = 1, b = 2$

Center: (0, 0)

Vertices: $(0, \pm 1)$

Asymptotes: $y = \pm\dfrac{1}{2}x$

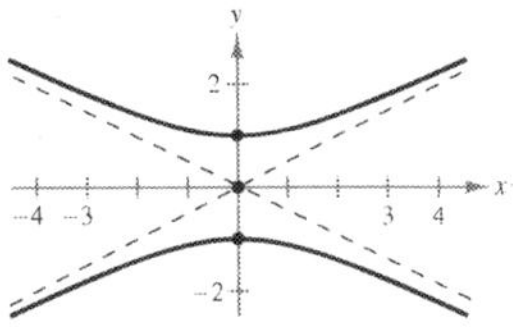

51. $\dfrac{y^2}{25} - \dfrac{x^2}{144} = 1$

$a = 5, b = 12$

Center: (0, 0)

Vertices: $(0, \pm 5)$

Asymptotes: $y = \pm\dfrac{5}{12}x$

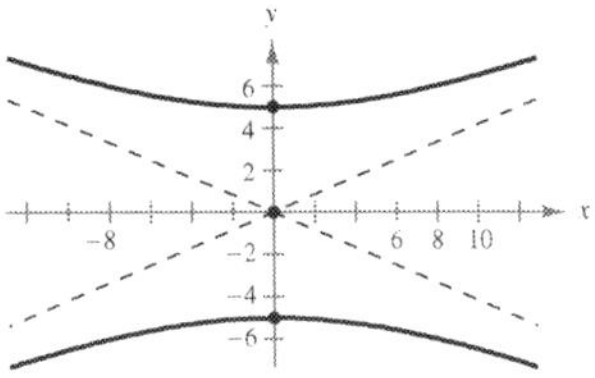

53. $2x^2 - 3y^2 = 6$

$\dfrac{x^2}{3} - \dfrac{y^2}{2} = 1$

$a = \sqrt{3}, b = \sqrt{2}$

Center: (0, 0)

Vertices: $\left(\pm\sqrt{3}, 0\right)$

Asymptotes: $y = \pm\sqrt{\dfrac{2}{3}}x$

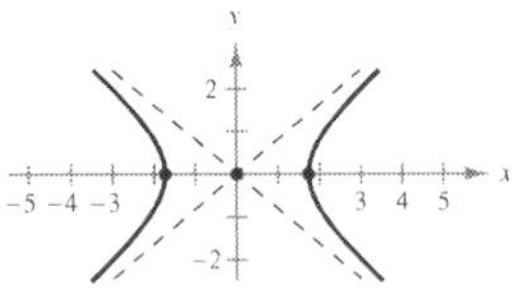

55. Vertices: $(\pm1, 0) \Rightarrow a = 1$

Asymptotes: $y = \pm3x$

Horizontal transverse axis

$$3 = \frac{b}{a} = \frac{b}{1} \Rightarrow b = 3$$

$$\frac{x^2}{a^2} - \frac{y^2}{b^2} = 1$$

$$\frac{x^2}{1} - \frac{y^2}{9} = 1$$

57. Foci: $(0, \pm4) \Rightarrow c = 4$

Asymptotes: $y = \pm\frac{1}{2}x$

Vertical transverse axis

$$\frac{1}{2} = \frac{a}{b} \Rightarrow b = 2a$$

$$a^2 + 4a^2 = 4^2$$

$$5a^2 = 16$$

$$a^2 = \frac{16}{5} \Rightarrow a = \frac{4\sqrt{5}}{5},\ b = \frac{8\sqrt{5}}{5}$$

$$\frac{y^2}{16/5} - \frac{x^2}{64/5} = 1 \text{ or } \frac{5y^2}{16} - \frac{5x^2}{64} = 1$$

59. Vertices: $(0, \pm3) \Rightarrow a = 3$

Vertical transverse axis

$$\frac{y^2}{9} - \frac{x^2}{b^2} = 1$$

$$\frac{5^2}{9} - \frac{(-2)^2}{b^2} = 1$$

$$b^2 = \frac{9}{4}$$

$$\frac{y^2}{9} - \frac{x^2}{\frac{9}{4}} = 1 \text{ or } \frac{y^2}{9} - \frac{4x^2}{9} = 1$$

61. Focus: $(0, 3) \Rightarrow p = 3$

Vertex: $(0, 0)$

$$x^2 = 4py$$

$$x^2 = 4(3)y$$

$$x^2 = 12y$$

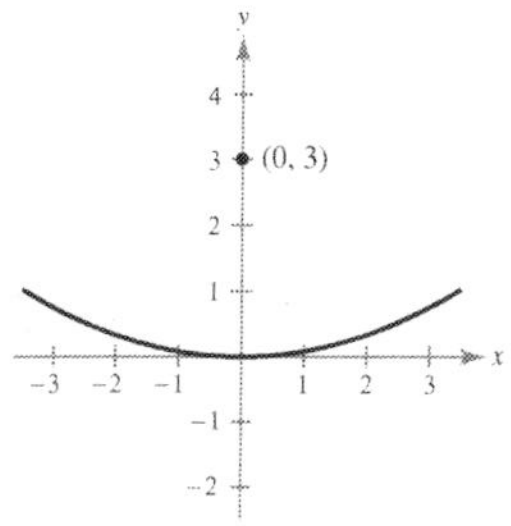

63. $2a = 5 \quad 2b = 4$

$$a = \tfrac{5}{2} \quad b = 2$$

$$c^2 = a^2 - b^2$$

$$c^2 = \tfrac{25}{4} - 2^2$$

$$c^2 = \tfrac{9}{4}$$

$$c = \pm\tfrac{3}{2}$$

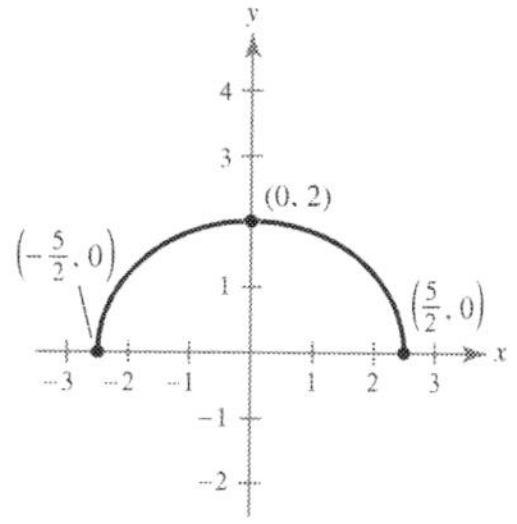

The length of the string should be 5 feet. The tacks should be located 1.5 feet from the center on either side of the fireplace.

65.

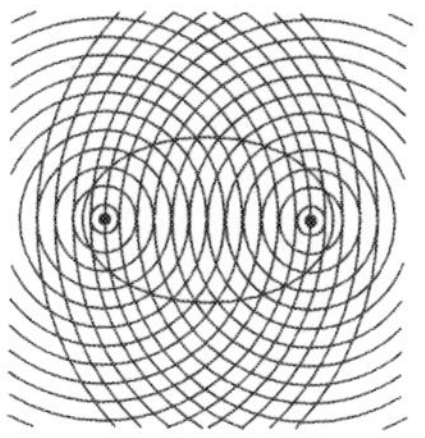

67. $\frac{x^2}{4} + \frac{y^2}{1} = 1$

$a = 2, b = 1, c = \sqrt{3}$

Points on the ellipse: $(\pm2, 0), (0, \pm1)$

Length of each latus rectum: $\frac{2b^2}{a} = 1$

Additional points: $\left(\sqrt{3}, \pm\frac{1}{2}\right), \left(-\sqrt{3}, \pm\frac{1}{2}\right)$

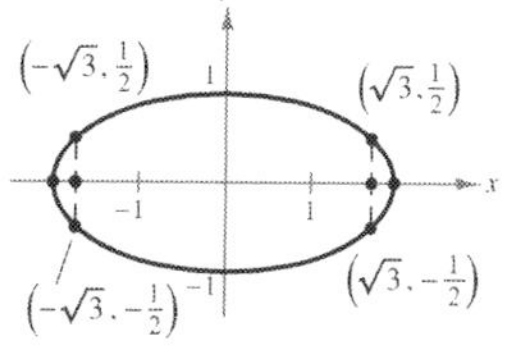

69. $9x^2 + 4y^2 = 36$

$\dfrac{x^2}{4} + \dfrac{y^2}{9} = 1$

$a = 3, b = 2, c = \sqrt{5}$

Points on the ellipse: $(\pm 2, 0), (0, \pm 3)$

Length of latus recta: $\dfrac{2b^2}{a} = \dfrac{8}{3}$

Additional points: $\left(\pm\dfrac{4}{3}, \pm\sqrt{5}\right)$

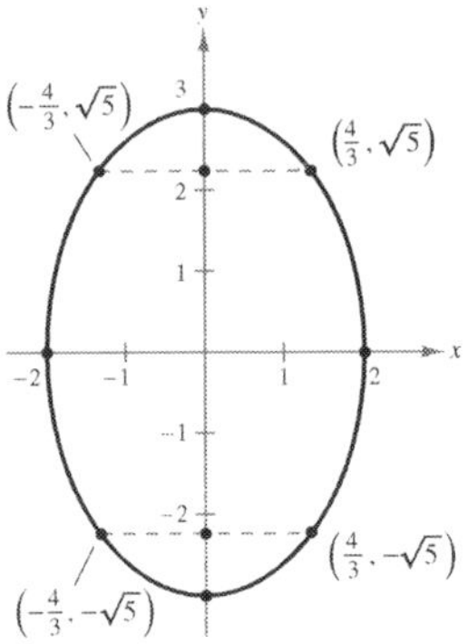

71. Answers will vary. Sample answer: $\dfrac{x^2}{328} + \dfrac{y^2}{327} = 1$

$c^2 = a^2 - b^2 = 328 - 327 = 1$

The ellipse is nearly circular because the foci are closer to the center than to the vertices.

73. Center: (0, 0)

Focus: (12, 0)

$b^2 = c^2 - a^2 = 12^2 - a^2 = 144 - a^2$

$$\frac{x^2}{a^2} + \frac{y^2}{144 - a^2} = 1$$

$$\frac{12^2}{a^2} - \frac{12^2}{144 - a^2} = 1$$

$$\frac{144}{a^2} - \frac{144}{144 - a^2} = 1$$

$$144(144 - a^2) - 144a^2 = a^2(144 - a^2)$$

$a^4 - 432a^2 + 20{,}736 = 0$

$a \approx 19.42$ or $a \approx 7.42$

Because $a < c$ and $c = 12$, you choose $a \approx 7.42$. The vertex is (7.42, 0).

[**Note:** By the Quadratic Formula, the exact value of a is $\sqrt{216 - 72\sqrt{5}}$.]

Section B.2 Conic Sections and Translations

Skills Warm Up

1. $\dfrac{x^2}{4} - \dfrac{y^2}{4} = 1$

Hyperbola

2. $\dfrac{x^2}{9} + \dfrac{y^2}{1} = 1$

Ellipse

3. $2x + y^2 = 0$

Parabola

4. $\dfrac{x^2}{4} + \dfrac{y^2}{16} = 1$

Ellipse

5. $4x^2 + 4y^2 = 25$

Circle

6. $\dfrac{y^2}{4} - \dfrac{x^2}{2} = 1$

Hyperbola

Skills Warm Up —*continued*—

7. $x^2 - 6y = 0$

   Parabola

8. $3x - y^2 = 0$

   Parabola

9. $\dfrac{9x^2}{16} + \dfrac{y^2}{4} = 1$

   Ellipse

10. $3y^2 - 3x^2 = 48$

    Hyperbola

1. The graph of $(x + 2)^2 + (y - 1)^2 = 4$ is a circle whose center is at the point $(-2, 1)$ and whose radius is 2. The graph is that of $x^2 + y^2 = 4$ shifted two units left and up one unit from standard position.

3. The graph of $(y + 3)^2/4 - (x - 1)^2 = 1$ is a hyperbola whose center is $(1, -3)$. The transverse axis is vertical with a length of $2(2) = 4$. The conjugate axis is horizontal with length $2(1) = 2$. The graph of the hyperbola has been shifted one unit to the right and downward three units.

5. The graph of $(x - 1)^2/9 + (y + 2)^2/16 = 1$ is an ellipse whose center is the point $(1, -2)$. The major axis of the ellipse is vertical with length $2(4) = 8$. The minor axis is horizontal with length $2(3) = 6$. The graph of the ellipse has been shifted one unit to the right and downward two units.

7. $(x - 1)^2 + 8(y + 2) = 0$

   $(x - 1)^2 = 4(-2)(y + 2); p = -2$

   Vertex: $(1, -2)$

   Focus: $(1, -4)$

   Directrix: $y = 0$

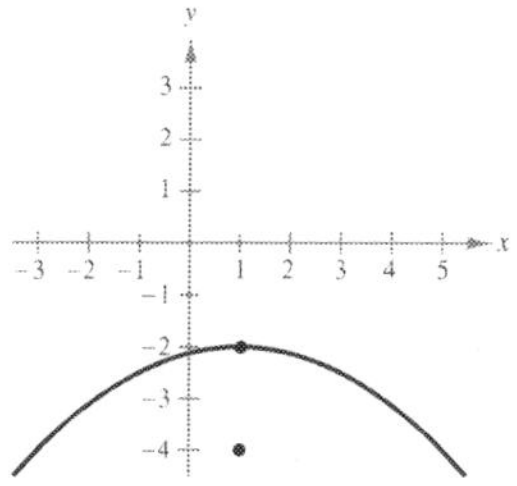

9. $\left(y + \frac{1}{2}\right)^2 = 2(x - 5)$

   $\left(y + \frac{1}{2}\right)^2 = 4\left(\frac{1}{2}\right)(x - 5);\ p = \frac{1}{2}$

   Vertex: $\left(5, -\frac{1}{2}\right)$

   Focus: $\left(\frac{11}{2}, -\frac{1}{2}\right)$

   Directrix: $x = \frac{9}{2}$

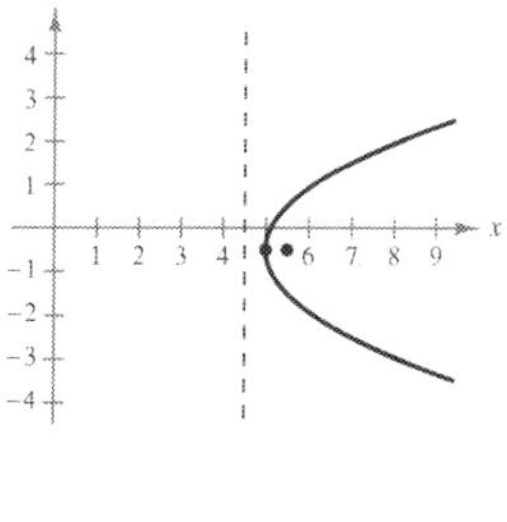

11. $y = \frac{1}{4}\left(x^2 - 2x + 5\right)$

    $y = \frac{1}{4}(x - 1)^2 + 1$

    $4(y - 1) = (x - 1)^2;\ p = 1$

    Vertex: $(1, 1)$

    Focus: $(1, 2)$

    Directrix: $y = 0$

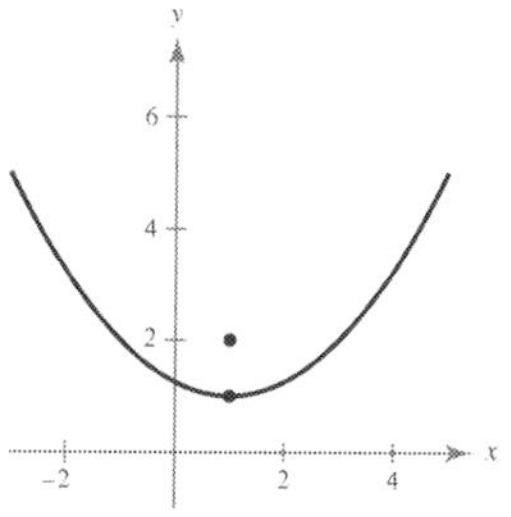

13. $4x - y^2 - 2y - 33 = 0$

    $4(x - 8) = (y + 1)^2;\ p = 1$

    Vertex: $(8, -1)$

    Focus: $(9, -1)$

    Directrix: $x = 7$

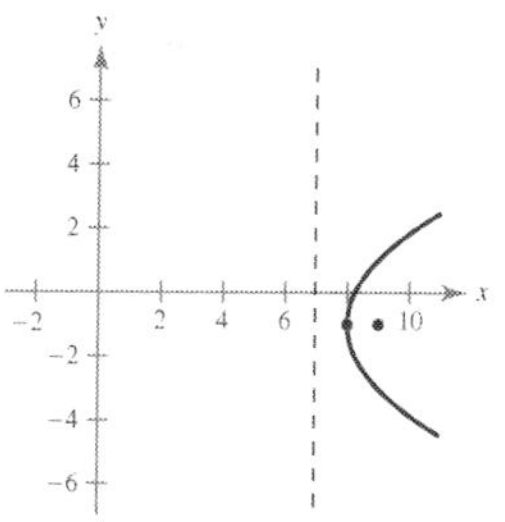

15. $y^2 + 6y + 8x + 25 = 0$

    $(y + 3)^2 = 4(-2)(x + 2);\ p = -2$

    Vertex: $(-2, -3)$

    Focus: $(-4, -3)$

    Directrix: $x = 0$

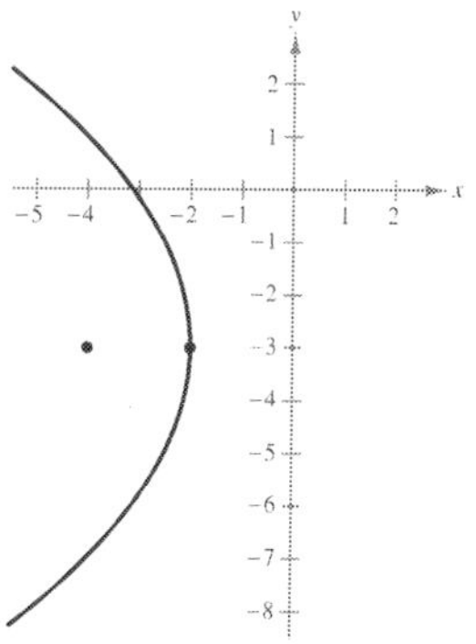

17. Vertex: $(3, 2)$

    Focus: $(1, 2)$

    Horizontal axis

    $p = 1 - 3 = -2$

    $(y - 2)^2 = 4(-2)(x - 3)$

    $(y - 2)^2 = -8(x - 3)$

19. Vertex: $(0, 4)$

Directrix: $y = 2$

Vertical axis

$$p = 4 - 2 = 2$$

$$(x - 0)^2 = 4(2)(y - 4)$$

$$x^2 = 8(y - 4)$$

21. Focus: $(2, 2)$

Directrix: $x = -2$

Horizontal axis

Vertex: $(0, 2)$

$$p = 2 - 0 = 2$$

$$(y - 2)^2 = 4(2)(x - 0)$$

$$(y - 2)^2 = 8x$$

23. $(x - h)^2 = 4p(y - k)$

$$(x - 0)^2 = 4p(y - 4)$$

Using $(2, 0)$

$$(2 - 0)^2 = 4p(0 - 4) \Rightarrow p = -\tfrac{1}{4}$$

or

$$(x - 0)^2 = -(y - 4)$$

$$x^2 = -(y - 4)$$

$$y = 4 - x^2$$

25. $\dfrac{(x - 1)^2}{9} + \dfrac{(y - 5)^2}{25} = 1$

$$a = 5, b = 3, c = \sqrt{a^2 - b^2} = 4$$

Center: $(1, 5)$

Foci: $(1, 1), (1, 9)$

Vertices: $(1, 0), (1, 10)$

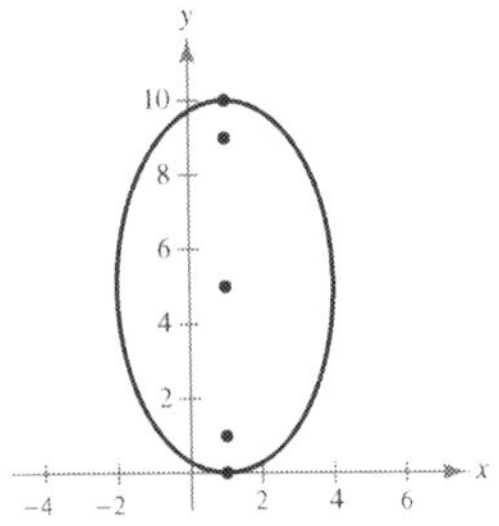

27. $9x^2 + 4y^2 + 36x - 24y + 36 = 0$

$$9(x^2 + 4x + 4) + 4(y^2 - 6y + 9) = 36$$

$$\frac{(x + 2)^2}{4} + \frac{(y - 3)^2}{9} = 1$$

$$a = 3, b = 2, c = \sqrt{a^2 - b^2} = \sqrt{5}$$

Vertical major axis

Center: $(-2, 3)$

Foci: $\left(-2, 3 - \sqrt{5}\right), \left(-2, 3 + \sqrt{5}\right)$

Vertices: $(-2, 0), (-2, 6)$

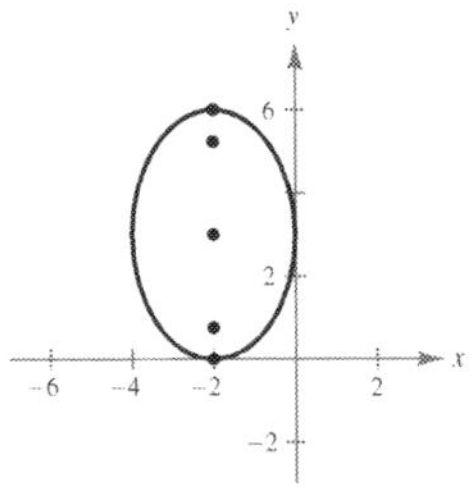

29. $16x^2 + 25y^2 - 32x + 50y + 16 = 0$

$$16(x^2 - 2x + 1) + 25(y^2 + 2y + 1) = 25$$

$$\frac{(x + 1)^2}{25/16} + (y + 1)^2 = 1$$

$$a = \frac{5}{4}, b = 1, c = \sqrt{a^2 - b^2} = \frac{3}{4}$$

Horizontal major axis

Center: $(1, -1)$

Foci: $\left(\frac{1}{4}, -1\right), \left(\frac{7}{4}, -1\right)$

Vertices: $\left(-\frac{1}{4}, -1\right), \left(\frac{9}{4}, -1\right)$

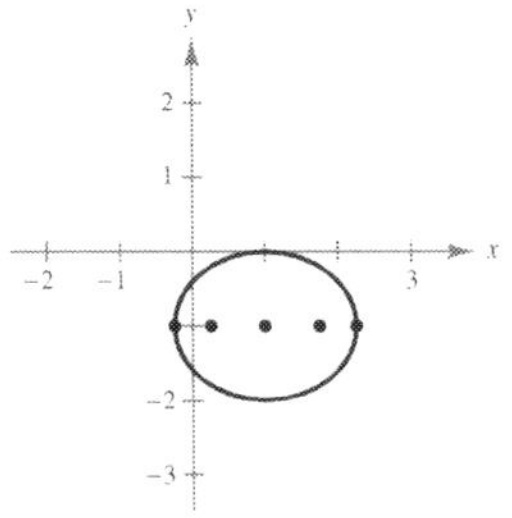

31. $12x^2 + 20y^2 - 12x + 40y - 37 = 0$

$$12\left(x^2 - x + \frac{1}{4}\right) + 20\left(y^2 + 2y + 1\right) = 60$$

$$\frac{\left(x - \frac{1}{2}\right)^2}{5} + \frac{(y+1)^2}{3} = 1$$

$a = \sqrt{5}, b = \sqrt{3}, c = \sqrt{a^2 - b^2} = \sqrt{2}$

Horizontal major axis

Center: $\left(\frac{1}{2}, -1\right)$

Foci: $\left(\frac{1}{2} - \sqrt{2}, -1\right)$, $\left(\frac{1}{2} + \sqrt{2}, -1\right)$

Vertices: $\left(\frac{1}{2} - \sqrt{5}, -1\right)$, $\left(\frac{1}{2} + \sqrt{5}, -1\right)$

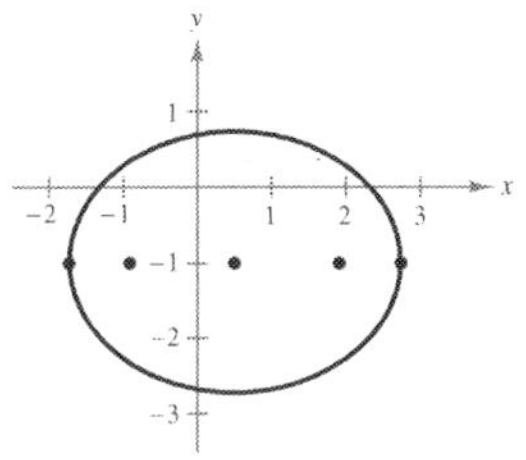

33. $\frac{(x-1)^2}{4} - \frac{(y+2)^2}{1} = 1$

$a = 2, b = 1, c = \sqrt{a^2 + b^2} = \sqrt{5}$

Center: $(1, -2)$

Horizontal transverse axis

Vertices: $(-1, -2), (3, -2)$

Foci: $\left(1 \pm \sqrt{5}, -2\right)$

Asymptotes: $y = \pm\frac{1}{2}(x - 1) - 2$

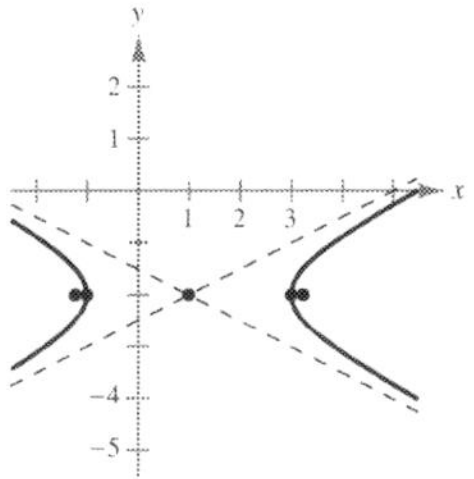

35. $(y + 6)^2 - (x - 2)^2 = 1$

$a = 1, b = 1,$

$c = \sqrt{a^2 + b^2} = \sqrt{2}$

Center: $(2, -6)$

Vertical transverse axis

Vertices: $(2, -7), (2, -5)$

Foci: $\left(2, -6 \pm \sqrt{2}\right)$

Asymptotes:

$y = \pm(x - 2) - 6$

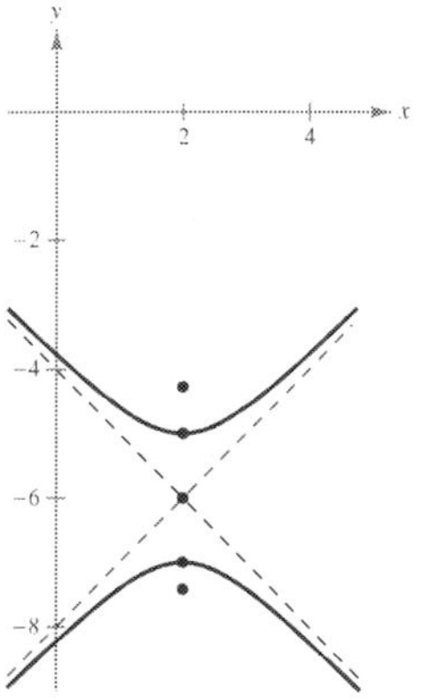

37. $9x^2 - y^2 - 36x - 6y + 18 = 0$

$$9\left(x^2 - 4x + 4\right) - \left(y^2 + 6y + 9\right) = 9$$

$$(x-2)^2 - \frac{(y+3)^2}{9} = 1$$

$a = 1, b = 3, c = \sqrt{a^2 + b^2} = \sqrt{10}$

Center: $(2, -3)$

Horizontal transverse axis

Vertices: $(1, -3)$, $(3, -3)$

Foci: $\left(2 \pm \sqrt{10}, -3\right)$

Asymptotes:

$y = \pm 3(x - 2) - 3$

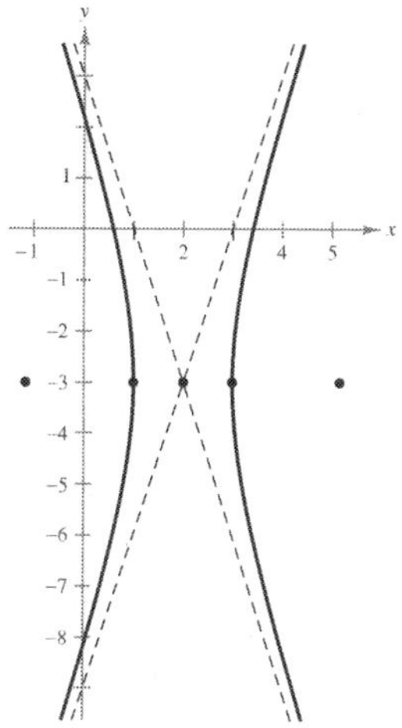

39. $9y^2 - x^2 + 2x + 54y + 62 = 0$

$$9\left(y^2 + 6y + 9\right) - \left(x^2 - 2x + 1\right) = 18$$

$$\frac{(y+3)^2}{2} - \frac{(x-1)^2}{18} = 1$$

$a = \sqrt{2}, b = 3\sqrt{2}, c = \sqrt{a^2 + b^2} = 2\sqrt{5}$

Center: $(1, -3)$

Vertical transverse axis

Vertices: $\left(1, -3 \pm \sqrt{2}\right)$

Foci: $\left(1, -3 \pm 2\sqrt{5}\right)$

Asymptotes:

$y = \pm\frac{1}{3}(x - 1) - 3$

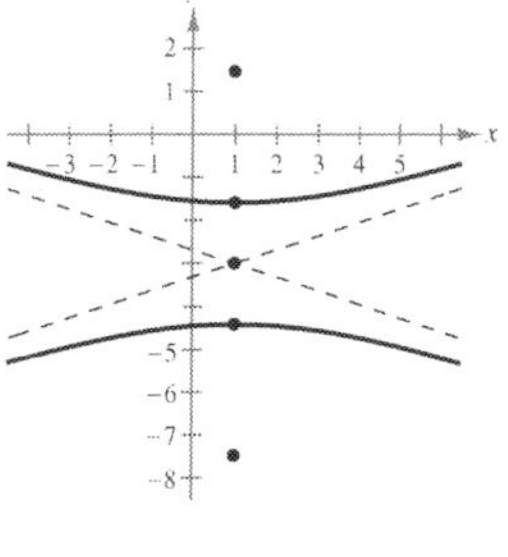

41. $x^2 - 9y^2 + 2x - 54y - 107 = 0$

$$\left(x^2 + 2x + 1\right) - 9\left(y^2 + 6y + 9\right) = 27$$

$$\frac{(x + 1)^2}{27} - \frac{(y + 3)^2}{3} = 1$$

$a = 3\sqrt{3},\ b = \sqrt{3},\ c = \sqrt{a^2 + b^2} = \sqrt{30}$

Center: $(-1, -3)$

Vertices: $\left(-1 - 3\sqrt{3}, -3\right), \left(-1 + 3\sqrt{3}, -3\right)$

Foci: $\left(-1 - \sqrt{30}, -3\right), \left(-1 + \sqrt{30}, -3\right)$

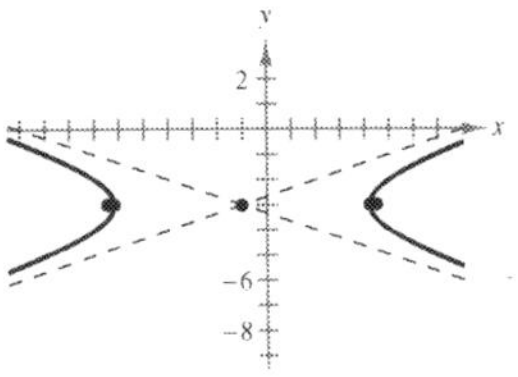

43. Vertices: $(0, 2), (4, 2)$

Minor axis of length 2

$a = \dfrac{4 - 0}{2} = 2,\ b = \dfrac{2}{2} = 1$

Center: $(2, 2)$

Horizontal major axis

$$\frac{(x - 2)^2}{4} + \frac{(y - 2)^2}{1} = 1$$

45. Foci: $(0, 0), (4, 0)$

Major axis of length 8

$a = 4,\ c = 2,\ b^2 = 12$

Center: $(2, 0)$

Horizontal major axis

$$\frac{(x - 2)^2}{16} + \frac{y^2}{12} = 1$$

47. Center: $(2, -1)$

Vertex: $\left(2, \dfrac{1}{2}\right)$

Minor axis of length 2

$a = \dfrac{3}{2},\ b = 1$

Vertical major axis

$$(x - 2)^2 + \frac{4(y + 1)^2}{9} = 1$$

49. Center: $(3, 2)$

$a = 3c$

Foci: $(1, 2), (5, 2)$

$c = 2,\ a = 6,\ b^2 = 32$

Horizontal major axis

$$\frac{(x - 3)^2}{36} + \frac{(y - 2)^2}{32} = 1$$

51. Vertices: $(5, 0), (5, 12)$

Endpoints of the minor axis: $(0, 6), (10, 6)$

$a = 6,\ b = 5$

Center: $(5, 6)$

Vertical major axis

$$\frac{(x - 5)^2}{25} + \frac{(y - 6)^2}{36} = 1$$

53. Vertices: $(2, 0), (6, 0)$

Foci: $(0, 0), (8, 0)$

Center: $(4, 0)$

Horizontal transverse axis

$a = 2,\ c = 4,\ b^2 = c^2 - a^2 = 12$

$$\frac{(x - 4)^2}{4} - \frac{y^2}{12} = 1$$

55. Vertices: $(4, 1), (4, 9)$

Foci: $(4, 0), (4, 10)$

Center: $(4, 5)$

Vertical transverse axis

$a = 4,\ c = 5,\ b^2 = c^2 - a^2 = 9$

$$\frac{(y - 5)^2}{16} - \frac{(x - 4)^2}{9} = 1$$

57. Vertices: $(2, 3), (2, -3)$

Passes through the point $(0, 5)$

Center: $(2, 0)$

Vertical transverse axis

$a = 3$

$$\frac{y^2}{9} - \frac{(x-2)^2}{b^2} = 1$$

$$\frac{5^2}{9} - \frac{(0-2)^2}{b^2} = 1$$

$$b^2 = \frac{9}{4}$$

$$\frac{y^2}{9} - \frac{4(x-2)^2}{9} = 1$$

59. Vertices: $(0, 2), (6, 2)$

Asymptotes: $y = \frac{2}{3}x,\ y = 4 - \frac{2}{3}x$

Center: $(3, 2)$

Horizontal transverse axis

$a = 3,\ b = 2$

$$\frac{(x-3)^2}{9} - \frac{(y-2)^2}{4} = 1$$

61. $x^2 + y^2 - 6x + 4y + 9 = 0$

$$(x-3)^2 + (y+2)^2 = 4$$

Circle

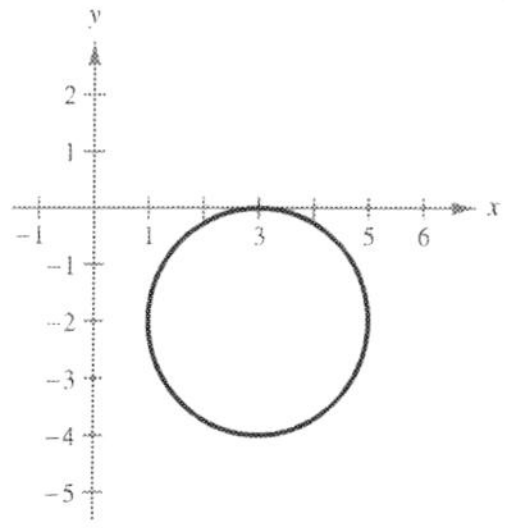

63. $4x^2 - y^2 - 4x - 3 = 0$

$$4\left(x - \frac{1}{2}\right)^2 - y^2 = 4$$

$$\left(x - \frac{1}{2}\right)^2 - \frac{y^2}{4} = 1$$

Hyperbola

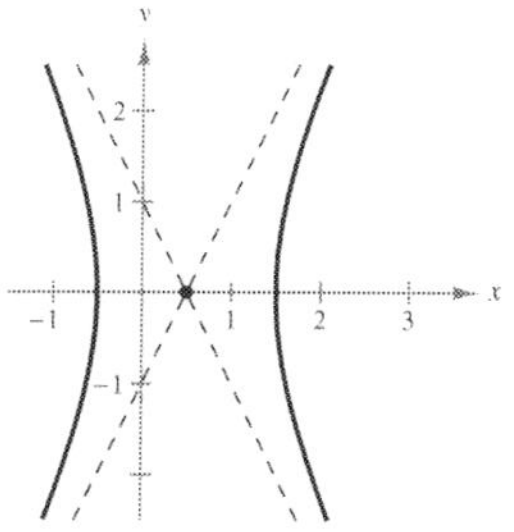

65. $4x^2 + 3y^2 + 8x - 24y + 51 = 0$

$$4(x+1)^2 + 3(y-4)^2 = 1$$

$$\frac{(x+1)^2}{1/4} + \frac{(y-4)^2}{1/3} = 1$$

Ellipse

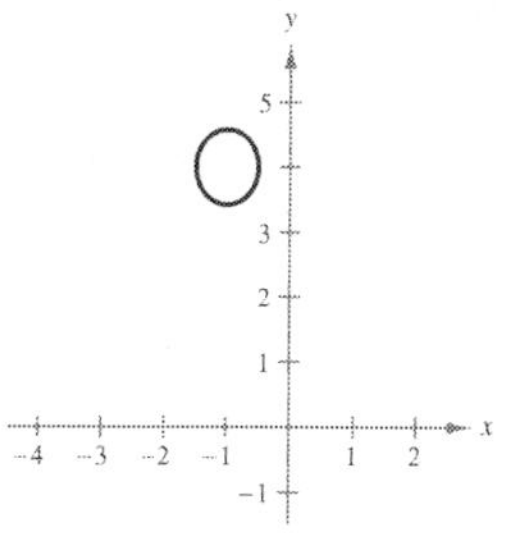

67. $25x^2 - 10x - 200y - 119 = 0$

$$25\left(x - \tfrac{1}{5}\right)^2 = 200\left(y + \tfrac{3}{5}\right)$$

$$\left(x - \tfrac{1}{5}\right)^2 = 8\left(y + \tfrac{3}{5}\right)$$

Parabola

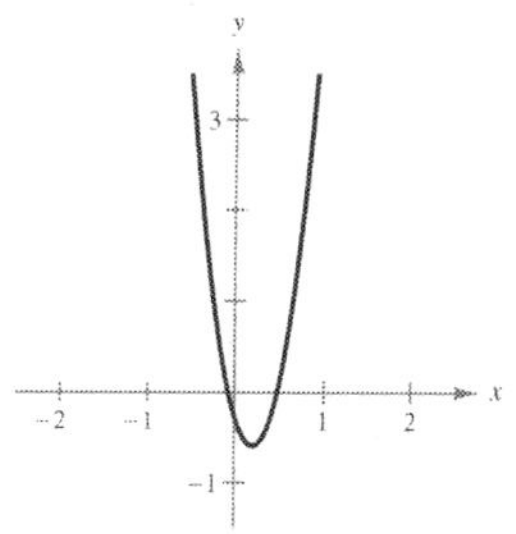

69. (a) $V = 17{,}500\sqrt{2}\text{ mi/h} \approx 24{,}750\text{ mi/h}$

(b) $p = -4100, (h, k) = (0, 4100)$

$$(x-0)^2 = 4(-4100)(y - 4100)$$

$$x^2 = -16{,}400(y - 4100)$$

71. $e = \frac{c}{a}$

Vertices: $(\pm 5, 0) \Rightarrow a = 5$

$$\frac{3}{5} = \frac{c}{5} \Rightarrow c = 3$$

$b^2 = a^2 - c^2 = 16$

$b = 4$

so the ellipse is

$$\frac{x^2}{25} + \frac{y^2}{16} = 1$$

73. $a = 92.956 \times 10^6$

$e = \dfrac{c}{a} = 0.017$

$c = 1.58 \times 10^6$

Perehelion: $a - c \approx 9.138 \times 10^7$ miles

Aphelion: $a + c \approx 9.454 \times 10^7$ miles

75. Least distance: $a - c = 1.3495 \times 10^9$

Greatest distance: $a + c = 1.5040 \times 10^9$

$a = 1.3495 \times 10^9 + c$

$\left(1.3495 \times 10^9 + c\right) + c = 1.5040 \times 10^9$

$2c = 1.545 \times 10^8$

$c = 7.725 \times 10^7$

$a = 1.3495 \times 10^9 + 7.725 \times 10^7$

$a = 1.427 \times 10^9$

$e = \dfrac{c}{a} = \dfrac{7.725 \times 10^7}{1.427 \times 10^9} \approx 5.414 \times 10^{-2}$

77.

$$\frac{(x-h)^2}{a^2} + \frac{(y-k)^2}{b^2} = 1$$

$$\frac{(x-h)^2}{a^2} + \frac{(y-k)^2}{a^2 - c^2} = 1$$

$$\frac{(x-h)^2}{a^2} + \frac{(y-k)^2}{\left(a^2/a^2\right)\left(a^2 - c^2\right)} = 1$$

$$\frac{(x-h)^2}{a^2} + \frac{(y-k)^2}{a^2\left(1 - c^2/a^2\right)} = 1$$

$$\frac{(x-h)^2}{a^2} + \frac{(y-k)^2}{a^2\left[1 - (c/a)^2\right]} = 1$$

Because $e = \dfrac{c}{a}$,

$$\frac{(x-h)^2}{a^2} + \frac{(y-k)^2}{a^2\left(1 - e^2\right)} = 1.$$

79. Center: $(0, 0)$

$2a = 0.34 + 4.08$

$a = 2.21 \Rightarrow a^2 \approx 4.88$

$c = 2.21 - 0.34 = 1.87$

$c^2 = a^2 - b^2 \Rightarrow b^2 = a^2 - c^2 \approx 1.39$

So, the equation of the ellipse is approximately

$$\frac{x^2}{4.88} + \frac{y^2}{1.39} = 1.$$

Made in United States
Orlando, FL
04 November 2026

90207777R00179